ANIMAL CELL TECHNOLOGY: PRODUCTS FROM CELLS, CELLS AS PRODUCTS

Animal Cell Technology: Products from Cells, Cells as Products

Proceedings of the 16th ESACT Meeting
April 25–29, 1999, Lugano, Switzerland

Edited by

A. Bernard
B. Griffiths
W. Noé
F. Wurm

KLUWER ACADEMIC PUBLISHERS
DORDRECHT / BOSTON / LONDON

A C.I.P. Catalogue record for this book is available from the Library of Congress.

ISBN 978-90-481-5363-3 e-ISBN 978-0-306-46875-9

Published by Kluwer Academic Publishers,
P.O. Box 17, 3300 AA Dordrecht, The Netherlands.

Sold and distributed in North, Central and South America
by Kluwer Academic Publishers,
101 Philip Drive, Norwell, MA 02061, U.S.A.

In all other countries, sold and distributed
by Kluwer Academic Publishers,
P.O. Box 322, 3300 AH Dordrecht, The Netherlands.

Printed on acid-free paper

CONTENTS

CHAPTER II. Metabolic and Process Engineering

16th ESACT Meeting Organising Committee

Florian Wurm (Chairman)	EPFL, Lausanne
Alain Bernard (Co-Chairman)	Ares-Serono, Geneva
Maria-Grazia Calì	Serono Symposia, Rome
Hans Eppenberger	ETH-Zurich, Zurich
Ruth Freitag	EPFL, Lausanne
Hansjörg Hauser	GBF, Braunschweig
Caroline MacDonald	University of Paisley, Paisley
Ferruccio Messi	Cell Culture Technologies,
Zurich	
Maurizio Morandi	Chiron, Siena
Wolfgang Noé (Trade Exhibition)	BI Pharma, Biberach
Thomas Ryll	Genentech, San Francisco
Uwe Schlokat	Baxter, Vienna
Georg Schmid (Sponsorship)	Hoffmann-La Roche, Basel
Giuseppe Viscomi	Alfa Wassermann, Bologna

ESACT EXECUTIVE COMMITTEE

Manuel CARRONDO, Chairman	IBET, Portugal
Bryan GRIFFITHS, Treasurer	Porton Down, UK
Florian WURM, Meeting Chairman	EPFL, Switzerland
Alain BERNARD	Ares-Serono, Switzerland
Francesc GODIA	Universitad Autonoma de Barcelona, Spain
Elisabeth LINDNER-OLSSON	Pharmacia & Upjohn, Sweden
Caroline MACDONALD	University of Paisley, UK
Otto-Wilhelm MERTEN	AFM-Généthon, France

SPONSORS

ESACT and the Organising Committee wish to thank the following companies for their generous support

Akzo Nobel Pharma
Amersham Pharmacia Biotech
Ares Serono SA
B. Braun Biotech International GmbH
Bayer Corporation
Bioengineering AG
BioInvent Production AB
BioWhittaker Inc.
Boehringer Ingelheim Pharma KG
Canberra Packard S.A.
Cansera International Inc.
Cantone Ticino
Crossair-Swissair
E. Merck
Genentech Inc.
Genetics Institute
Genzyme Transgenics Corp.
Hoffmann-La Roche AG
Hyclone Europe S.A.
Immuno AG (Division of Baxter Inc.)
Immunex Corp.
Institut de Recherche Pierre Fabre
Intervet International BV
Inveresk Research International Ltd.
JRH-Biosciences
Life Technologies Ltd.
Lonza Biologics Plc
MA Bioservices
Merck Research Laboratories
MicroSafe BV
Novartis Pharma AG
Novo Nordisk A/S
Nunc A/S
Pall Europe Ltd.
Pasteur Merieux Connaught
PCS Process Control Systems AG
Pharmacia and Upjohn AB
Q-One Biotech Ltd.
Sanofi Recherche
Schärfe System GmbH
Schering AG
Swiss Red Cross - ZLB
SmithKline Beecham Biologicals
Union Bank of Switzerland - UBS

COMPANIES PARTICIPATING IN THE TRADE EXHIBITION

Aber Instruments Ltd.
Amersham Pharmacia Biotech
Applikon BV
Aquasant-Messtechnik AG
Asahi
B. Braun Biotech International GmbH
Bibby Sterilin Ltd.
BioInvent Production AB
BioWhittaker
BioReliance
Boehringer Ingelheim Pharma KG
Paul Bucher Analytik und Biotechnologie
Cansera International Inc.
Cantone Ticino
Cellon Sarl
Cellex Biosciences
Chemunex
Clontech
Connectors Verbindungstechnik AG
Covance Laboratories Ltd.
Digitana SA
Dr. F. Messi Cell Culture Technologies
ECACC CAMR
Genespan Corporation
Genetic Engineering News
Genzyme Transgenics Corp.
Greiner Labortechnik GmbH
Hyclone Europe N.V.
Infors AG
Innovatis GmbH
Integra Biosciences
Inveresk Research
JRH Europe Ltd.
Kendro L.P. GmbH (Sorvall GmbH)
Life Technologies
Lonza Biologics Plc
MAVAG Verfahrenstechnik AG
Microsafe BV
Nature Magazine
New Brunswick Scientific BV
Nunc A/S
Pall Europe Ltd.
Q-One Biotech Ltd.
Quest International
Rutten Engineering
Sarstedt AG
Schärfe System GmbH
Selborne Biologicals
Serologicals Proteins
Sigma-Aldrich Ltd.
Stedim S.A.
Summit Biotechnology
TC Tech CorporationTerracell
The Automation Partnership
Wave Biotech AG
Winiger AG

LIST OF PARTICIPANTS

Mr. Sushil Abraham
Lonza Biologics
Bath Road 228
Slough, Berkshire SL1 4DY
UK

Prof. Spiros N. Agathos
University of Louvain
Place Croix du Sud 2/19
Louvain-la-Neuve B-3148
Belgium

Mr. François Aguilon
Sanofi Recherches
Labege Innopole Voie n°1 -BP 137
Labege Cedex 31676
France

Prof. Masuo Aizawa
Tokyo Institute of Technology
Dept. Bio Engineering
Nagatsuta , Midori-ku
Yokohama 226-8501
Japan

Dr. Bert Al
CLB
Dept. Biotecnology
Plesman Laan 125
Amsterdam 1066 CX
The Netherlands

Dr. Daniel Allison
ICOS Corporation
22021 - 20th Avenue SE
Bothell, WA 98021-4406
USA

Mrs. Claudia Altamirano Gómez
Universitad Autónoma de Barcelona
Dept. Enginyeria Quimica
Edifici C
Bellaterra (Barcelona) 08193
Spain

Dr. Massimo Amadori
Istituto Zooprofilattico Sperimentale
Via A. Bianchi, 7
Brescia 25124
Italy

Dr. Hanspeter Amstutz
ZLB Zentrallaboratorium
Blutspendedienst SRK
Wankdorfstrasse 10
Bern 22 3000
Switzerland

Dr. Hideharu Anazawa
Kyowa Hakko co. Ltd.
Asahi-Cho, 3-6-6
Machida, Tokyo 194-8533
Japan

Dr. Dana Andersen
Genentech, Inc.
1 DNA Way
South San Francisco, CA 94080
USA

Mr. Hans Juul Andersen
Statens Serum Institut
5, Artillerivej
Copenhagen 2300
Denmark

Dr. Carlo Andretta
Biospectra AG
Zuercherstrasse 137
Zurich - Schlieren CH-8952
Switzerland

Dr. Clarisse Antoni
Chemunex S.A.
3, Allè de la Seine
Ivry sur Seine Cedex 94854
France

Mrs. Eva Charlotte Appelgren
Meridiano
Via Mentana 2b
Rome 00185
Italy

Dr. Achille Arini
Cerbios-Pharma S.A.
Mol.Biol.
Via Pian Scairolo 6
Barbengo 6917
Switzerland

Dr. Edward Baetge
Modex Therapeutiques
Rue du Bugnon 27
Lausanne 1005
Switzerland

Dr. Gianni Baffelli
Nagual Anstalt
Via Carà 9a
Manno 6928
Switzerland

Mr. Charles Bailey
CRC for Biopharmaceutical Research, UNSW
Dept. Biotechnology, UNSW
Sydney 2052
Australia

Prof. James E. Bailey
Institute of Biotechnology, ETH Hönggerberg
ETH Hönggerberg
Zurich 8093
Switzerland

Dr. Alex Baker
Selborne Biological Services Ltd.
Goleigh Farm
Selborne, Alton Hampshire GU34 3SE
UK

Dr. Kim Baker
School of Biosciences, University of Kent
Canterbury, Kent CT2 7NJ
UK

Prof. Matthias Bally
Bioengineering AG
Sagenrainstrasse 7
Wald 8636
Switzerland

Mr. Denis Barral
Merial
254 Rue M. Merieux
Lyon 69007
France

Dr. Michael Bavand
Siegfried CMS, Ltd.
Zofingen 4800
Switzerland

Dr. Ulrich Behrendt
Roche Diagnostics GmbH
Nonnenwald 2
Penzberg 82377
Germany

Dr. Roland Beliard
LFB
59 Rue de Trevise
Lille 59011
France

Dr. Peter Belt
ID/DLO
Post Box 65
Lelystad 8200 AB
The Netherlands

Dr. Claudia Benati
Molmed S.p.a.
Via Olgettina 58
Milano 20132
Italy

Dr. Walter Beyeler
PCS Process Control System AG
Werkstrasse 8
Wetzikon, ZH CH-8623
Switzerland

Dr. Guy Berg
Covance Laboratories Ltd.
Otley Road
Harrogate, N. Yorkshire HG3 2X4
UK

Dr. Bisson
Lausanne 1011
Switzerland

Dr. Klaus Bergemann
Boehringer Ingelheim Pharma KG
Birkendorferstr 65
Biberach/Riss 88397
Germany

Prof. Henri Blachere
New Brunswick Scientific
Kerkenbos 11-01
Nijmegen 6546 BC
The Netherlands

Dr. Alain Bernard
Serono Pharmaceutical Research
Institute S.A.
14 Chemin des Aulx
Plan-les-Ouates/Geneva 1228
Switzerland

Dr. David Black
Excell Biotech
Pentland Science Park
Penicuik, Edinburgh EH26OPZ
UK

Mrs. Laure Berruex
EPFL
Lab. of Cellular Biotechnology
Lausanne 1015
Switzerland

Dr. Horst Blasey
Serono Pharmaceutical Research
Institute S.A.
14 Chemin de Aulx
Plan-Les-Ouates/Geneva 1228
Switzerland

Dr. Eric Berry
Osmonics
135, Flanders Road
Westborough, MA 01581-6046
USA

Mr. Carel F. Bode
Bodinco B.V.
Otterkoog 7
Alkmaar 1822 BW
Holland

Dr. Wolfgang Berthold
Hoffmann-La Roche Inc.
340 Kingsland St.
Nutley, NJ 07110
USA

Dr. Berthold Boedecker
Bayer AG
Abteilung PH-TO/ELB, Biotechnolige
Gebaude 46
Friedrich Erbert Strasse 217
Wuppertal 42096
Germany

Dr. Jan Boesen
IntroGene BV
P.O. Box 2048
Leiden 2301CA
The Netherlands

Prof. Claudio Bordignon
Istituto Scientifico H.S. Raffaele
Via Olgettina 60
Milano 20132
Italy

Mr. Maximilian Boldt
Innovatis GmbH
Hauptstrasse 72
Berlin 12159
Germany

Dr. Octaaf Bos
ID-DLO
P.O. Box 65
Lelystad 8200 AB
The Netherlands

Dr. Bryan Bolton
ECACC
Porton Down
Salisbury, Wiltshire SP4 0JD
UK

Mrs. Michaela Bourgeios
EPFL
Lab. of Cellular Biotechnology
Lausanne 1015
Switzerland

Dr. Michele Bomio
SAM AG
Albulastrasse 57
Zurich 8048
Switzerland

Mr. Leo Bowski
Hoffmann-La Roche, Inc.
340, Kingsland Street
Nutley, NJ 07110
USA

Dr. Hendrik Bonarius
Novo Nordisk A/S
Hagedornsvej 1
Gentofte 2820
Denmark

Mrs. Johanna Brändli
Hochschule Wädenswil
Grüntal
Wädenswil 8820
Switzerland

Mr Marco Boorsma
Cytos Biotechnology AG
Einsteinstrasse 1-5 Postfach 150
Zurich CH-8093
Switzerland

Dr. Ruud Brands
Solvay Pharmaceuticals
P.O. Box 900
Weesp 1380
The Netherlands

Miss Maryline Bordes
Virbac
Bio 4 ZI Carros
Carros Cedex 06511
France

Mr. Malcolm Brattle
Q-One Biotech Ltd
West of Scotland Science Park
Glasgow G20 0XA
UK

Dr. Henner Brett-Schneider
Bio West
Rue de la Caille
Nusille 49340
France

Dr. Herve Broly
Soregio
Bordeaux Technopolis
Martillac 33650
France

Dr. Peter Brown
Biotechnology Solutions
20 Woodcrest Drive
Orinda, CA 94563
USA

Dr. Paul Bucher
Paul Bucher Comp
Schutzengraben 7
Basel 4051
Switzerland

Mr. William Bucher
Lampire Biological Laboratories, Inc.
P.O. Box 270
Pipersville 18947
USA

Dr. Greg Buckley
TC Tech Corporation
7600 W. 27th st. Unit 202
Minneapolis, MN 55426
USA

Prof. Heino Büntemeyer
Lehrstuhl Zellkulturtechnik, University Bielefeld
PO BOX 100131
Bielefeld 33501
Germany

Dr. Christa Burger
Merck KGaA
Frankfurter Str.250
Darmstadt 64271
Germany

Mr. Gerry Burgers
New Brunswick Scientific
Kerkenbos 11-01
Nijmegen 6546
The Netherlands

Prof. Michael Butler
University of Manitoba
Dept. Microbiology
118 Buller Bldg.
Winnipeg, Manitoba R3T2N2
Canada

Mrs Marie Ange Buyse
Innogenetics N.V.
Industriepark
Ghent 9052
Belgium

Dr. Paolo Caccia
Pharmacia Up John
Viale Pasteur 10
Nerniano 20014
Italy

Dr. Cinzia Cagnoli
ETH
Universitätstr. 16
Zurich 8092
Switzerland

Mrs. Maria Grazia Calì
Serono Symposia
Via Casilina 125
Rome 00176
Italy

Mr. Eric Calvosa
Merial
254 Rue M. Merieux
Lyon 69007
France

Mr. Gian Luca Casella
Economic Promotion Ticino
Viale St.Franscini 17
Bellinzona 6500
Switzerland

Dr. Joseph Camire
Hyclone Europe NV - Belgium
Friedenstrasse 34
Asiar 35614
Germany

Mr. Adolfo José Castillo Vitlloch
Centre of Molecular Immunology
Calle 216 y 15, Atabey, Playa
Havana 11600
Cuba

Dr. Manuel Carrondo
IBET
Apartado 12
Oeiras 2780
Portugal

Dr. Aziz Cayli
Roche Diagnostics GmbH
Nonnenwald 2
Penzberg 82377
Germany

Mr. Gerald Carson
Basf Bioresearch Corporation
100 Research Drive
Worcester, MA 01605
USA

Ms. Anna Cellesi
Chiron Vaccines
Via Fiorentina, 1
Siena 53100
Italy

Miss Ana Verónica Carvalhal
IBET / ITQB
Apartado 12, 4° piso
Oeiras 2780
Portugal

Ms. Roberta Cenci
Serono Symposia
Convention Bureau
Via Casilina 125 - 00176
Rome 00100
Italy

Dr. John Carvell
ABER Instruments Ltd.
5 Science Park
Aberystwyth SY23 3AH
UK

Dr. Helen E. Chadd
Abgenix Inc.
7601 Dumbarton Circle
Fremont, CA 94555
USA

Mr. Antoni Casablancas
Universitad Autónoma de Barcelona
Dept. Enginyeria Quimica
Edifici C
Bellaterra (Barcelona) 08193
Spain

Mr. Michael Chaffee
Lonza Biologics
225, Bath Road
Slough, Berkshire SL1 4DY
UK

Mr. Patrick Chang
BioWhittaker Inc.
8830 Biggs Ford Road
Walkerville, MD 21793
USA

Mr. Ernesto Chico
Center of Molecular Immunology
(CIM)
P.O.Box 16040
Havana 11600
Cuba

Miss Susan Chapple
Oxford Brookes University
BMS Gipsy Lane Campus
Oxford OX3OBP
UK

Dr. Myung-Sam Cho
Bayer Corporation
800 Dwight Way
Berkeley, CA 94701-1986
USA

Miss Nathalie Chatzisavido
Pharmacia and Upjohn
Linhagensgatan 133
Stockholm 11287
Sweden

Dr. Veronique Chotteau
Pharmacia Upjohn
Strandbergsgatan 47
Stockholm 11287
Sweden

Dr. Jean-Francoise Chaubard
RhÔne -Poulenc
13, Quai Jules Guesde BP 14
Vitry sur Seine Cedex 94403
France

Dr. Yi-Ding Chu
Development Center for Biotechnology
81, Chang-Hsing st.
Taipei 105
Taiwan, R.O.C.

Dr. Li-How Chen
Genzyme Transgenics Corporation
5 Mountain Road
Framingham, MA 01701-9322
USA

Prof. Klaus Cichutek
Paul-Ehrlich-Institute
Medical Biotechnology
Paul Erlich Str. 51
Langen 63225
Germany

Mr. Stéphane Chenu
CNRS Nancy
2, Av. Forèt de Haye
Vandoeuvre-Les-Nancy 54505
France

Mr. Martin Clarkson
Vericore Ltd
4, Warner Drive, Springwood Ind.
Estate
Braintree Essex CM7 24W
UK

Dr. Laurent Chevalet
Centre d'Immunologie Pierre Fabre
5 Avenue Napoleon III
Saint-Julien en Genevois Cedex 74164
France

Dr. Timothy Clayton
Glaxo Wellcome R & D
Biotechnology Development
Laboratories
South Eden Park Road
Beckenham, Kent B233B5
UK

Mr. Efi Cohen-Arazi
Ares Serono
Av. Perrausaz, 131
La Tour de Peirz 1814
Switzerland

Prof. Mark Cunningham
Ares Advanced Technology Inc.
280 Pond Street
Randolph, MA 02368
USA

Mr. David Connolly
National Diagnostic Centre
N.U.I. Galway
Galway
Ireland

Mr. Marcello Cusinato
B. Braun Biotech International
Schwarzenberger Weg 73-79
Melsungen 34212
Germany

Prof. Harald S. Conradt
GBF - Braunschweig
Dept. Protein Glycosylation
Mascheroder Weg 1
Braunschweig D-38124
Germany

Mr. Samuel Cymbalista
Rue Dr. Yersin, 9
Morges 1110
Switzerland

Dr. Coppolecchia
Cerbios-Pharma S.A.
Mol. Biol.
Via Pian Scairolo, 6
Barbengo 6917
Switzerland

Mr. Martin Daescher
Zürcher Hochschule Winterthur ZHW
Postfach 805/ Chemie
Winterthur 8401
Switzerland

Dr. Angel Cruz
Schering Plough Research Institute
1011 Morris Ave (u-1-2 2200)
Union, NJ 07083
USA

Dr. Jo Dalle
Pharming N.V.
Cipalstraat 3
Geel 2440
Belgium

Mr. Helder Cruz
IBET
Apartado 12
Oeiras P-2780
Portugal

Mr. Daniel Dätwyler
Institut für Zellbiologie ETH Zurich
HPM F27 ETH Hönggerberg
Zurich 8093
Switzerland

Dr. Pedro Cruz
IBET
Apartado 12
Oeiras P-2780
Portugal

Dr. Bruno De Bortoli
BioWhittaker Europe
Parc Industriel de Petit Rechain
Verviers 4800
Belgium

Dr. Maria De Jesus
EPFL
Lab. of Cellular Biotechnology
Lausanne 1015
Switzerland

Dr. Jonathan Dempsey
Cambridge Antibody Technology
The Science Park
Melbourn, Cambs. SG8 6SS
UK

Dr. Michele De Luca
I.D.I. Ist. Dermopatico dell'Immacolata
Via dei Castelli Romani 83/85
Rome 00040
Italy

Mr Mohamed Desai
Medeva Pharma
Gaskil Road
Speke, Liverpool L24 9GR
UK

Dr. Robert De Paulis
BIOPTIM
12, Rue de la Salle
Saint Germain en Laye 78100
France

Mohamed Desay
Medeva Pharma
Gaskil Road - Speke
Liverpool L24 9GR
UK

Mr. Frank Deer
Ares Advanced Technology
27, Pacella Park Drive
Randolph, MA 02368
USA

Dr. Emmanuel Desmèziéres
Institut Pasteur
25-28 rue du Dr. Roux
Paris Cedex 15 75724
France

Dr. Nicole Deglon
Division of Surg. Research Gene
Therapy Center
Pavillon 4
Lausanne 1011
Switzerland

Dr. Kathleen Devos
N.V. Innogenetics
Industriepark 7 Box 4
Zwijnaarde (Ghent) B-9052
Belgium

Mr. Yves Dehon
Smithkline Beecham
89 Rue de l'Institute
Rixensart B- 1330
Belgium

Miss Janique Dewelle
Laboratoire Biochimie et Biologie
Cell., FUNDP
Rue de Bruxelles 61
Namur 5000
Belgium

Dr. Jean Delobel
Merial
254 rue Marcel Merieux
Lyon
France

Mr. Jean Didelez
Smithkline Beecham Biologicals
Rue de l'Institut 89
Rixensart 1330
Belgium

Mr. Fritz Diener
Rütten Engineering
Industriestrasse 9
Stäfa 8712
Switzerland

Mr. Paul Ducommon
EPFL
Lausanne 1015
Switzerland

Dr. Paul Dierickx
Institute of Public Health
Wytsmanstraat 14
Brussels B-1050
Belgium

Mr. Ulrich Dudel
Astra Biotech Laboratory
Byggnad 329
Södertälje S-151 85
Sweden

Mr. Othnar J. Dill
BLC
Biology
Ernst-Handschuch-Str.3A
Worms 67549
Germany

Mrs. Petra Eberhardt
Boehringer Ingelheim Pharma KG
Birkendorfer Str.65
Biberach/ Riss 88397
Germany

Dr. André Dinter
University of Zurich, Institute of Physiology
Winterthurerstr.190
Zurich 8057
Switzerland

Mrs. Joke Ederveen
MicroSafe B.V.
Niels Bohrweg 11-13
Leiden 2333 CA
The Netherlands

Mr. Uwe Ditzen
Kendro Laboratory GmbH
Heraeusstrasse 12-14
Hanau 63450
Germany

Ms Jacqueline Edwards
The Automation Partnership Ltd.
Melbourn Science Park
Royston, Herts SG8 6HB
UK

Mrs. Jana Dolnikova
Biogen Inc.
14, Cambridge Center
Cambridge, MA 02142
USA

Prof. Regine Eibl
Hochschule Wädenswil
Grüntal
Wädenswil 8820
Switzerland

Dr. Charles Dowding
Systemix
3155 Porter Drive
Palo Alto, CA 94304
USA

Mr. Edgar Elsner
Innovatis GmbH
Hauptstrasse 72
Berlin 12159
Germany

Mrs. Ewa Engström
SKD Konferens Service AB
P.O. Box 1252
Solna S-17124
Sweden

Prof. Hans M. Eppenberger
ETH Zürich, Cell Biology
Zurich 8093
Switzerland

Mr. Ulrich Essig
Roche Diagnostic GmbH
Nonnenwald 2
Penzberg 82372
Germany

Mrs. Katrin Esslinger
Digitana In Vitro Systems & Services GmbH
Am Kalkberg
Osterode 37520
Germany

Dr. Tina Etcheverry
Genentech MS 32
1 DNA Way
South San Francisco, CA 94080
USA

Mr. Jean-Pierre Faessler
Winiger AG
Angelikerstrasse 20
Wohlen 5610
Switzerland

Dr. Alain Fairbank
BioWhittaker Inc.
8830 Biggs Ford Road
Walkersville MD 21793
USA

Dr. Edgar Falkner
Bender & co. GmbH
Dr. Boehringergasse 5-11
Vienna 1121
Austria

Dr. Franca Fassio
RBM Istituto di Ricerche Biomediche
Via Ribes 1
Colleretto Giacosa (TO) 10010
Italy

Mr. Dieter Fassnacht
TU Hamburg - Harburg
Denickestr. 15
Hamburg 21071
Germany

Dr. Thérése Faure
Transgene
11, Rue de Molsheim
Strasbourg Cedex 67082
France

Mr. Giuseppe Ferrari
Snam Progetti Biotecnologici
Processo
Viale De Gasperi,16
San Donato (MI)
Italy

Dr. Rich Feston
JRH Biosciences
13804 W. 107th street
Lenexa KS 66215
USA

Dr. Ray Field
Cambridge Antibody Technology
The Science Park
Melbourn, Cambs SG8 6SS
UK

Mr. David Fiorentini
Biological Industries Ltd
Kibbutz Beit Haemek 25115
Israel

Mr. David Freedman
New Brunswick Scientific co. Inc.
44, Talmadge Road
Edison, NJ 08818
USA

Mr. René W. Fischer
Swiss Federal Institute of Technology
Dept. of Biochemistry
Universitätsstr. 16
Zurich 8092
Switzerland

Prof. Ruth Freitag
EPFL
IGC-DC-EPFL
Ecublens 1015
Switzerland

Mr. Tom Fletcher
Irvine Scientific
2511 Daimler St.
Santa Ana, CA 92705
USA

Dr. John Frenz
Genentech
1 DNA Way
South San Francisco, CA 94080
USA

Dr. Sean Forestell
Systemix
3155 Porter Drive
Palo Alto, CA 94304
USA

Mr. Jan-Gerd Frerichs
Institute Fuer Techn. Chemie
Callinstr. 3
Hannover 30167
Germany

Dr. Don Francis
Vax Gen. Inc c/o Genentech Inc.
1000 Marina Blvd
Brisbane, CA 94005
USA

Dr. Peter Frey
CHUV
Lausanne 1011
Switzerland

Dr. Reinhard Franze
Roche Diagnostics GmbH
Nonnenwald 2
Penzberg 82372
Germany

Dr. Bert Frohlich
Genzyme Corporation
500 Soldier's Field Road
Allston, MA 02134
USA

Dr Elisabeth Fraune
B. Braun Biotech International
Schwarzenberger Weg 73-79
Melsungen 34212
Germany

Dr. Steve Froud
Algroup Lonza
228 Bath Road
Slough, Berkshire SL1 4DY
UK

Mr. Richard Fry
Cellon SA
204 Route d'Arlon
Strassen L- 8010
Luxembourg

Dr. Gilad E. Gallili
Abic Vet Biological Lab. Teva
P.O. Box 27047
Jerusalem 97800
Israel

Mr. Victor Fung
Immunex Corporation
51 University Street
Seattle, WA 98105
USA

Dr. Christine Gandor
Eugenex Biotechnologies
Lauchefeld 31
Matzingen 9548
Switzerland

Dr. Martin Fussenegger
Institute of Biotechnology, ETH Zürich
ETH Zürich, ETH Hönggerberg
Zurich 8093
Switzerland

Dr. Herman Gaub
Lehrstuhl für Angewandte Physik,
Biophysik
Amalienstrasse 54
München 80799
Germany

Mrs. Carme Gabernet
Universitad Autónoma de Barcelona
Dept. Enginyeria Quimica
Edifici C
Bellaterra (Barcelona) 08193
Spain

Dr. Martin Gawlitzek
Genentech Inc.
1 DNA Way
South San Francisco, CA 94080
USA

Dr. Zbigniew Gadek
Talweg 14
Schmallenberg 57392
Germany

Dr. Heinz Gebbing
Novartis Pharma GmbH
Roonstrasse 25
Nuremberg D-90429
Germany

Mr. Luciano Gaggetta
Economic Promotion Ticino
Viale St. Franscini 17
Bellinzona 6500
Switzerland

Dr. Sabine Geisse
Novartis Pharma Inc.
CTA/BMP
Building S-506.304
Basel 4002
Switzerland

Dr. Daniel Galbraith
Q-One Biotech Ltd.
West of Scotland Science Park
Glasgow G20 0XA
UK

Mrs. Heidi Gerber
Institute of Virology and
Immunoprophylaxis
Sensemattstrasse
Mittelhäusern 3147
Switzerland

Dr. Catherine Gerdil
Pasteur Merieux Connaught
1541, Avenue Marcel Merieux
Marcy L'Etoile 69280
France

Mrs. Simona Germoni
Meridiano
Via Mentana 2b
Rome 00185
Italy

Mr. Motti Geron
Ares-Serono
Corsier sur Vevey 1804
Switzerland

Mr. Christoph Geserick
Novo Nordisk
HAB 108.1, Hagedornsvej 1
Gentofte 2820
Denmark

Mr. Scott Geyer
Protein Design Labs
34801 Campus Drive
Fremont, CA 94555
USA

Mr. Yves Ghislain
SmithKline Beecham Biologicals
Rue de l'Institut 89
Rixensart 1330
Belgium

Dr. Steve Gibson
Q-One Biotech Ltd
West of Scotland Science Park
Glasgow G20 OXA
UK

Mr. Roberto Giovannini
EPFL
Lausanne 1015
Switzerland

Mr. Philippe Girard
EPFL
Lab. Cell. Biotechnology
Lausanne 1015
Switzerland

Dr. Shah Girish
Glaxo Wellcome R+D
Gunnels Wood Road
Stevenage SG1 2NY
UK

Dr. Arnaud Glacet
LFB
59 Rue de Trevise
Lille 59011
France

Prof. Francesc Godia
Universitad Autónoma de Barcelona
Dept. Enginyeria Quimica
Edifici C
Bellaterra (Barcelona) 08193
Spain

Dr. Jean Louis Goergen
CNRS - Nancy
LSGC
2, Av. Forèt de Haye
Vandoeuvre-Les-Nancy 54500
France

Dr. Randal Goffe
Genespan Corporation
19310 North Creek Pkwy
Bothell, WA 98011-8006
USA

Dr. Sigrid Gonski
Hoechst Marion Roussel
Biotechnology, H825
Frankfurt am Main D-65926
Germany

Prof. Ursula Graf-Hausner
Zürcher Hochschule Winterthur
Dept. Chimie+Biotechnologie
Postfach 805
Winterthur CH-8401
Switzerland

Dr. Marie-Monique Gonze
SmithKline Beecham Biologicals
Rue de l'Institut 89
Rixensart 1330
Belgium

Dr. Stefanos Grammatikos
Boehringer Ingelheim Pharma KG
Birkendorfer str.65
Biberach / Riss 88397
Germany

Dr. Charles Goochee
Merck & Co.
P.O Box 4, WP75-300
West Point, PA 19486-0004
USA

Mr. Andrew Grant
IDEC Pharmaceuticals Corporation
11011 Torreyana Road
San Diego , CA 92121
USA

Mr. Roel Gordÿn
BioWhittaker Europe
Parc Industriel de Petit Rechain
Verviers B 4800
Belgium

Dr. Doug Gray
JRH Biosciences
13804 W. 107th Street
Lenexa, KS 66215
USA

Mr. Gerard Gourdon
Applikon BV
Brauwweg 13
Schiedam 3125
The Netherlands

Prof. Peter Gray
University of New South Wales
Dept. of Biotechnology
UNSW
Sydney 2052
Australia

Dr. Eckart Grabenhorst
Gesellschaft für Biotechnologische
Forschung mbH
Dept. of Protein Glycosylation
Mascheroder Weg 1
Braunschweig D-38124
Germany

Dr. Bryan Griffiths
5 Bourne Gardens
Porton Salisbury SP4 0NU
UK

Dr. Hermann Graf
Schering AG SBU Therapeutics
Berlin D-13342
Germany

Mr. Leopold Grillberger
Immuno AG
Uferstrasse 15
Orth/Donau 2304
Austria

Dr. Daniel Grob
Ingenieur Schule Wadenswil
Postfach 335
Wädenswil 8820
Switzerland

Mrs. Maria Guarguaglini
Meridiano
Via Mentana 2b
Rome 00185
Italy

Dr. Frank Gudermann
Innovatis GmbH
Hauptstrasse 72
Berlin 12159
Germany

Mr. Raphael Gugerli
EPFL
Ecublens 1015
Switzerland

Dr. Jürgen Haas
Boehringer Ingelheim Pharma KG
Birkendorfer Strasse 65
Biberach/ Riss 88397
Germany

Mr. Paul Haffenden
PO. BOX 250
Nobleton - Ontario LOG/NO
Canada

Dr. Haffliger
Lausanne 1011
Switzerland

Mr. Erik Hamann
Institute for Animal Science and Health
ID-DLO
P.O. Box 65
Lelystad 8200 AB
The Netherlands

Miss. Maureen Hamilton
Genetics Institute
One Burrt Road
Andover 01810
USA

Dr. Julian Hanak
Cobra Therapeutics LTD
The Science Park
Keele, Staffs, ST5 5SP
UK

Dr. Frank Hanakam
Micromet GmbH
Am Klopferspitz 19
Martinsried (Munich) 82152
Germany

Dr. Louane Hann
Genetics Institute
One Burtt Road
Andover, MA 01810
USA

Mrs. Yasmine Hannaby
Bayer Diagnostics (France)
Abt. Pentek, Weisensee 101
München 81539
Germany

Dr. Karen Hansen
Novo Nordisk A/S
Hagedornsvej 1
Gentofte 2820
Denmark

Mrs. Wiebke Hansen
GBF
Regulation and differentation
Mascheroder Weg 1
Braunschweig 38124
Germany

Mrs. Andrea Hawerkamp
Institute of Cell Culture Technologies
Bielefeld D-33501
Germany

Prof. Colin Harbour
Sydney University
Dept. Infectious Diseases
Sydney 2006
Australia

Prof. Leonard Hayflick
University of California, San Francisco
P.O. Box No. 89
The Sea Ranch, CA 95497
USA

Mrs. Ruth Harrison
Covance Laboratories Ltd.
Otley Road
Harrogate , N.Yorkshire HG3 2X4
UK

Miss Michèle Heaton
University of Kent
School of Biosciences
Canterbury, Kent CT2 7NJ
UK

Mrs. Jane Harvey
Selborne Biological Services Ltd.
Goleigh Farm
Selborne, Alton Hampshire GU34 3SE
UK

Mr. Pierre Heimendinger
Pasteur Mérieux Connaught
1541, avenue Marcel Mérieux
Marcy l'Etoile 1541
France

Dr. Diane Hatton
Cambridge Antibody Technology
The Science Park
Melbourn, Cambs. SG8 6SS
UK

Dr. Holger Heine
Serono Pharmaceutical Research
Institute S.A.
14, Chemin des Aulx
Plan-les-Ouates/Geneva 1228
Switzerland

Dr. Jacques Hatzfeld
CNRS Institute for Research on Cancer
7, Rue Guy Mocquet
Villejuif Cedex 94801
France

Dr. Horst Hellwig
Sigma-Aldrich
Grünwalder Weg. 30
Deisenhofen 82041
Germany

Prof. Hans Jörgen Hauser
GBF
Mascheroder Weg 1
Braunschweig 38124
Germany

Mr. Vincianne Hendrick
Universite Libre de Bruxelles
Laboratory of Animal Cell
Biotechnology
Faculty of Sciences - CP 160/17 Av.
F.D. Roosvelt
Bruxelles 1050

Mrs. Iris Hermanns
Winiger AG
Angelikerstrasse 20
Wohlen 5610
Switzerland

Mr. Bernard E. Horwath
Cellex Biosciences Inc.
8500 Evergreen Boulevard
Minneapolis, MN 55433
USA

Dr. Friedemann Hesse
Gesellschaft für Biotechnologische
Forschung mbH
Mascheroder Weg 1
Braunschweig D-38124
Germany

Dr. Alain Houllemare
Instrumenten Gesellschaft AG
23 Route de Jeunes
Carough- Geneve 1227
Switzerland

Miss Anna Hills
University of Kent
Dept. of Biosciences
University of Kent at Canterbury
Canterbury, Kent CT2 7NJ
UK

Dr. Martin Howald
Bioconcept
Geweberstr. 14
Allschwitz
Switzerland

Prof Helmut Hoffmann
Boerhinger Ingelheim Pharma KG
Birkendorfer Strasse 65
Biberach/ Riss 88397
Germany

Dr. Cynthia Hoy
Genentech, Inc.
1 DNA Way
South San Francisco, CA 94080
USA

Mrs. Els Hogervorst
MicroSafe B.V.
Niels Bohrweg 11-13
Leiden 2333 CA
The Netherlands

Mrs. Lisa Hunt
EPFL
DC-IGCIV-LBTC EPFL
Lausanne 1015
Switzerland

Dr. Otmar Hohenwarter
University of Agricultural Sciences
Muthgasse 18
Vienna A-1190
Austria

Dr. Laertis Ikonomou
Catholic University of Louvain
Place Croix du Sud 2/19
Louvain-La-Neuve B-1348
Belgium

Dr. Paul Holmes
School of Chemical Engineering, Univ.
Birmingham
Birmingham B15 2TT
UK

Mr. Markus Inglin
Zürcher Hochschule Winterthur
Postfach 805/ Chemie
Winterthur CH-8401
Switzerland

Dr. Yuichi Inoue
Kagoshima University
Faculty of Agriculture
1-21-24 Korimoto, Kagoshima
Kagoshima 890-0065
Japan

Dr. David Jayme
Life Technologies Inc.
3175 Staley Road
Grand Island, NY 14072
USA

Miss Noushin Irani
GBF Braunschweig
Mascheroder Weg 1
Braunschweig 38124
Germany

Mr. Joël Jean-Mairet
Swiss Federal Institute of Technology
Institute of Biotechnology
Einsteinstrasse
Zurich 8093
Switzerland

Dr. Thomas Irish
JRH Biosciences
13804 W. 107th Street
Lenexa , KS 66215
USA

Miss Nanni Jelinek
Institute of Biotechnology 2
Forschungszentrum Jülich
Jülich 52425
Germany

Dr. Barbara Jacko
BioWhittaker Inc.
8830 Biggs Ford Road
Walkersville, MD 21793
USA

Dr. Rose Marie Jönsson
Pharmacia & Upjohn Diagnostics AB
Rapsgatan 7
Uppsala SE-75182
Sweden

Dr. Volker Jaeger
Gesellschaft für Biotech. Forschung mbH
Mascheroder Weg 1
Braunschweig 38124
Germany

Dr. Martin Jordan
Ecole Polytechnique Féderale de Lausanne
DC - LBTC
Lausanne 1015
Switzerland

Mr. Jean-Marc Jalby
Pasteur Merieux Connaught
1541 Avenue marcel Merieux
Marcy l'Etoile 69280
France

Mr. Chris Julien
New Brunswick Scientific
Kerkenbos 11-01
Nijmegen 6546 BC
The Netherlands

Mrs. Vittoria Javicoli
Meridiano
Via Mentana 2b
Rome 00185
Italy

Dr. Avinoam Kadouri
Laboratoires Serono Corsier sur Vevey
Zt En Fonil
Corsier sur Vevey 1804
Switzerland

Mr. Erik Kakes
Applikon BV
De Brauwweg 13
Schiedam 3125
The Netherlands

Dr. Hela Kallel
Institut Pasteur
13, Place Pasteur B.P. 74
Tunis 1002
Tunis

Dr. Robert Kallmeier
Lonza Biologics plc
224 Bath Road
Slough, Berkshire SLI 4DT
UK

Dr. Frank Kasteliz
Bender & Co. - Gesellenschaft m.b. H.
Dr. Boehringer Gasse 5-11
Wien 1120
Austria

Prof. Yoshinori Katakura
Grad. School of Genetic Resources Technology
Kyushu University
6/10/1 Hakozaki, Higashi-Ku
Fukuoka 812-8581
Japan

Prof. Hermann Katinger
Institute of Applied Microbiology
Muthgasse 18
Vienna A-1190
Austria

Prof. Keiichi Kato
Ehime University, Faculty of Engineering
Dept. of Applied Chemistry
3, Bunnkyou cyou
Matsuyama, Ehime 790-8577
Japan

Dr. Denis Kelsch
Merial
254 Rue Marcel Merieux
Lyon 69007
France

Dr. Richard B. Kemp
Institute Biological Sciences,
University of Wales
Penglais
Aberystwyth SY2 333DA
UK

Dr. Ralph Kempken
Boheringer Ingelheim Pharma KG
Birkendorfer str. 65
Biberach / Riss 88397
Germany

Mrs. Ina Kerkloh
DASGIP mbH
Karl-Heinz-Beckurts-Strasse 13
Jülich 52428
Germany

Dr. Nicole Kessler
BioWhittaker Europe
Parc Industriel de Petit Rechain
Verviers B 4800
Belgium

Mr. Thomas Kessler
Bayer Diagnostics (UK)
Abt. Pentek, Weisensee 101
München 81539
Germany

Dr. Peter Ketelaar
DSM Biologics
P.O. Box 454
Groningen 9700 AL
The Netherlands

Dr. Ken Ketley
JRH Biosciences
13804 W. 107th Street
Lenexa, KS 66215
USA

Dr. Miomir Knezevic
Educell d.o.o.
Teslova 30
Ljubljana 1000
Slovenia

Mrs. Caroline Kewney
Q-One Biotech Ltd.
West of Scotland Science Park
Glasgow G20 0XA
UK

Mrs Susann Koch
Pro Bio Gen GmbH
Rudower Chaussee 5
Berlin 12489
Germany

Mr. Makoto Kitano
Nichirei Corporation
6-19-20 Tsukiji, Chuo-ku
Tokyo 104-8402
Japan

Dr. Marieke Koedood-Zhao
Genetics Institute
One Burtt Rd.
Andover, MA 01810
USA

Mr. Gerhard Klement
Laboratoires Serono Corsier sur Vevey
ZI En Fonil
Corsier sur Vevey 1804
Switzerland

Dr. Florian Koelle
Paul Bucher Company
Schutzengraben 7
Basel 4002
Switzerland

Dr. Beate Kleuser
Novartis Pharma Inc.
Research CTA-BMP
S-506.304
Basel 4002
Switzerland

Mr. Daniel Koller
Cytos Biotechnology AG
Einsteinstrasse 1-5 Postfach 150
Zurich CH-8093
Switzerland

Miss Claudia Kloth
University of Leinzig
Dept. Medical Biotechnology
Delitzscher str.141
Leipzig D-4129
Germany

Mrs. Tamara Kolokoltsova
State Research Centre of Virology and Biotechnolog
Novosibirisk Region
Koltsovo 639159
Russia

Dr. Franz Knauseder
Biochemie GmbH
Biochemiestrasse 10
Kundl 6250
Austria

Prof. Dhinakar Kompala
University of Colorado
Dept. of Chemical Engineering
Boulder, Colorado 80309-0424
USA

Dr. Leif Kongerslev
Novo Nordisk A/S
Hagedornsvej 1
Gentofte 2820
Denmark

Dr. Lynne Krummen
Genentech FWC
1 DNA Way
South San Francisco, CA 94080
USA

Dr. Kurt Konopitzky
Boehringer Ingelheim GmbH
Binger Strasse
Ingelheim 55216
Germany

Dr. Renate Kunert
Institute of Applied Microbiology
Muthgasse 18
Vienna A-1190
Austria

Dr. Konstantin Konstantinov
Bayer Corporation
800 Dwight Way
Berkeley. CA 94710
USA

Ms. Giovanna Labemano
Serono Pharmaceutical Research
Institute S.A.
14, Chemin des Aulx
Plan-les-Ouates/Geneva 1228
Switzerland

Dr. Thomas Kost
Glaxo Wellcome
5 Moore Drive
Research Triangle Park, NC 27709
USA

Mr. Renè Lardenoye
IntroGene
Zernikedreef 6
Leiden 2333CL
The Netherlands

Ms. Judy Kramer
The Automation Partnership Ltd.
Melbourn Science Park
Royston, Herts SG8 6HB
UK

Dr. Arye Lazar
Israel Institute for Biological Research
P.O. Box 19
Ness-Ziona 74100
Israel

Mrs. Daniella Kranjac
Wave Biotech AG
Ringstrasse 24
Reigelsangen 8317
Switzerland

Dr. Christian Leist
Novartis Pharma Ltd.
K-681.1.45
Basel 4002
Switzerland

Dr. Gerlinde Kretzmer
Institut für Technische Chemie
Callinstr. 3
Hannover 30167
Germany

Mrs. Christine Lettenbauer
Hochschule Wädenswil
Grüntal
Wädenswil 8820
Switzerland

Dr. Roger Lias
Covance Laboratory
Otley Road
Harrogate, N. Yorkshire HG3 2XH
UK

Ms. Elisabeth Lindner - Olsson
Pharmacia & Upjohn
Stockholm SE-11287
Sweden

Miss Ena K. Linnau
University of Agricultural Sciences Vienna
Muenzwadeingasse 11/7
Vienna A-1060
Austria

Miss Monika Linz
Gesellschaft für Biotechnologische Forschung mbH
Mascheroder Weg 1
Braunschweig D-38124
Germany

Dr. Chao-Min Liu
Hoffmann-La Roche
340, Kingsland St.
Nutley, NJ 07110
USA

Dr. Jan Ljunggren
Karobio AB, Cell Culture and Fermentation Tech.
Novum
Huddinge 14157
Sweden

Mrs. Eva Ljungkvist
SKD Konferens Service AB
P.O. Box 1252
Solna S-17124
Sweden

Prof David Lloyd
School of Chemical Eng., University of Birmingham
Edgbaston, Birmingham
Birmingham B15 2TT
UK

Mr. Alexander Loa
BioWhittaker Europe
Parc Industriel de Petit Rechain
Verviers 4800
Belgium

Mr. René Lohser
Bio Pro International Inc.
265 Conklin street
Farmingdale, NY 11735
USA

Dr. Denis Looby
CAMR
Porton Down
Salisbury SP4 0SG
UK

Mrs. Monika Loperiol
EPFL
Lab. de Biotechnologie Cellulaire
Lausanne 1015
Switzerland

Dr. Thomas Lorenz
Roche Diagnostics GmbH
Nonnenwald 2
Penzberg 82372
Germany

Mr. Christophe Losberger
SPRI
14, Chemin des Aulx
Plan-les-Ouates/Geneva 1228
Switzerland

Mr. Hans-Jürgen Lotz
Kendro Laboratory Products GmbH
Harseustrasse 12-14
Hanau 63450
Germany

Prof. Caroline MacDonald
University of Paisley Research Dept.
High Street
Paisley PA1 2BE
Scotland

Dr. Archie Lovatt
Q-One Biotech Ltd.
West of Scotland Science Park
Glasgow G20 0XA
UK

Dr. Malcolm Macnaughton
Inveresk Research
Tranent, East Lothian EH33 2NE
UK

Dr. Holger Lübben
Chiron-Behring
Emel. V. Behring Str. 76
Marburg 35041
Germany

Mrs. Carine Maggetto
SmithKline Beecham Biologicals
Rue de l'Institut 89
Rixensart 1330
Belgium

Dr. Anthony Lubiniecki
Smithkline Beecham Pharm.
709 Swedeland Rd. - P.O. Box 1539
King of Prussia, PA 19406
USA

Dr. Josef P. Magyar
Swiss Federal Institute of Technology
Institute of Cell Biology
ETH - Hönggerberg HPM F27
Zurich CH-8093
Switzerland

Dr. Dirk Luetkemeyer
Institute of Cell Culture Technologies
Faculty of Technology
P.O. Box 100131
Bielefeld 33501
Germany

Dr. Fabio Malavasi
University of Torino
Lab. Biologia Cellulare
Via Santena 19
Torino 10126
Italy

Dr. Elke Lüllau
Astra Biotech Laboratory
Byggnad 329
Södertälje S-15185
Sweden

Dr. Pawan Malhotra
Serono Pharmaceutical Research
Institute S.A.
14, Chemin des Aulx
Plan-les-Ouates/Geneva 1228
Switzerland

Dr. Jan Lupker
Sanofi Recherches
Labege Innopole Voie n°1 - BP 137
Labege Cedex 31676
France

Dr. Chris Mannix
SmithKline Beecham
H31, Nesp-N, Third Ave
Harlow CM19 5AW
UK

Dr. Annie Marc
CNRS - Nancy
LSGC
2, Av. Forèt de Haye
Vandoeuvre-Les-Nancy 54500
France

Dr. Dino Marcus
Israel Institute Biological Research
P.O. Box 19
Ness-Ziona 74100
Israel

Dr. Dirk E. Martens
University Wageningen
Bomenweg 2
Wageningen 6703 HD
The Netherlands

Dr. Carl Martin
Cancelled
Otley Road
Harrogate, N.Yorkshire HG3 2XH
UK

Dr. Uwe Marx
University of Leinzig
Institute Clinical Immunology &
Transfusion Med.
Delitzscher Str.141
Leipzig D-4129
Germany

Mr. Rainer Marzahl
Integra Biosciences
Industriestrasse 44
Wadisellen 8304
Switzerland

Dr. Bernard Massie
Institut de Recherche en Biotechnologie
CNRC)
6100, Royalmount Ave.
Montreal H48 2R2
Canada

Mr. Ricaredo Matanguihan
Bayer Corporation
800 Dwight Way
Berkeley, CA 94701
USA

Dr. Eric Mathieu
SmithKline Beecham
Rue de l'Institut 89
Rixensart 1330
Belgium

Dr. Hiroshi Matsuoka
Teikyo University of Science &
Technology
2525 Vanohara
Yamanashi 409-0193
Japan

Dr. Paolo Mattana
Alfa Wassermann S.p.A.
Via Ragazzi del '99, 5
Bologna 40133
Italy

Mrs. Rosalind McAllister
SmithKline Beecham
Third Avenue
Harlow, Essex CM19 5AW
UK

Dr. Craig McDonald
JRH Biosciences
13804 W. 107th Street
Lenexa , KS 66215
USA

Mr. Peter McGrady
Cellex Biosciences Inc.
8500 Evergreen Boulevard
Minneapolis, MN 55433
USA

Mr. Heiko Meents
Boehringer Ingelheim Pharma KG
Birkendorfer Str. 65
Biberach / Riss 88397
Germany

Dr. Otto - Wilhelm Merten
Genethon III
1, Rue de l'Internationale, BP 60
Evry - Cedex 9100
France

Dr. Hans Peter Meier
Mavag
Zürcherstr. 94
Altendorf 8852
Switzerland

Dr. Ferruccio Messi
Cell Culture Technologies
Buhnrain 14
Zurich 8052
Switzerland

Dr. Johann Meinhart
Lainz Hospital
Wolkersbergenstr. 1
Vienna 1130
Austria

Dr. Hilary Metcalfe
Lonza Biologics plc
224 Bath Road
Slough, Berkshire SLI 4DY
UK

Dr. Petra Meissner
EPFL , Centre of Biotechnology
UNIL - EPFL
Lausanne 1015
Switzerland

Mr. Masayasu Mie
Tokyo Institute of Technology
Dept. Bio-Engineering
Nagatsuta, Midori-Ku
Yokohama 226-8501
Japan

Dr. Angelica Meneses
Istituto de Biotecnologia Unam.
Av. Universidad 2001, Col. Chamilpa.
Cuoernavaca, Morelos 62210
Mexico

Dr. Douglas Miller
Immunex Corporation
51 University St.
Seattle, WA 98101
USA

Dr. Lee Mermelstein
Scios Inc.
2450 Bayshore Parkway
Mountain View, CA 94043
USA

Mr. Takumi Miura
Grad. School of Genetic Resources
Technology
Kyushu University
6-10-1 Hakozaki Higashi-Ku
Fukuoka 8128581
Japan

Prof. Nicolas Mermod
Uni Lausanne CBUE
CBUE
DC - IGC - EPFL
Lausanne 1015
Switzerland

Mr. Dieter Moebest
Uniklinik
Has steter Str.
Freiburg 79102
Germany

Dr. Lucia Monaco
DIBIT - San Raffaele
Via Olgettina, 58
Milano 20132
Italy

Dr. Diana Morgan
Bioreliance
Innovation Park Hillfoots Road
Stirling FK9 4NF
Scotland

Mrs. Miriam Monge
Cancelled
Aubagne Cedex 13781
France

Dr. Sandro Mori
Molmed S.p.a.
Via Olgettina 58
Milano 20132
Italy

Mr. Jean-Claude Monnier
Aquasant-Messtechnik AG
Hauptstrasse 20
Bubendorf 4416
Switzerland

Dr. Ana Maria Moro
Instituto Butantan
Centro de Biotecnologia
Av. Vital Brasil 1500
Sao Paolo, SP 05503-900
Brasil

Dr. Bryan Monroe
Zymogenetics
1201 Eastlake Ave
Seattle, WA 98102
USA

Mr. Jon Martin Mowles
Valbiotech
57, Boulevard de la villette
Paris 75010
France

Dr. Roberto Montini
Biowhittaker
Via G. Galilei 6
Bergamo
Italy

Dr. Peter P. Mueller
National Research Centre for
Biotechnology
Mascheroder Weg 1
Braunschweig D-38124
Germany

Dr. Enda Moran
Glaxo Wellcome R & D
South Eden Park Road
Beckenham Kent BR33BS
UK

Dr. Beate Mueller-Tiemann
Schering AG/Proteinchemistry
Berlin D-13342
Germany

Dr. Maurizio Morandi
Chiron S.p.a.
Via Fiorentina
Siena 53100
Italy

Mr. Dethardt Müller
Institute of Applied Microbiology
Muthgasse 18
Vienna A-1190
Austria

Dr. Amy Murnane
SmithKline Beecham Pharmaceuticals
709 Swedeland Road
King of Prussia, PA 19406-0939
USA

Dr. David Naveh
Bayer Corporation
800, Dwight Way
Berkeley, CA 94710
USA

Mr. N. N.
Bayer Diagnostics (USA)
Abt. Pentek, Weisensee 101
München 81539
Germany

Mrs. Elena Nechaeva
State Research Centre of Virology & Biotechnology
Novosibirisk Region
Koltsovo 633159
Russia

Prof. Kazuo Nagai
Tokyo Institute of Technology
Nagatsuta Midori-Ku
Yokohama 226-8501
Japan

Dr. Lars Nieba
Cytos Biotechnology AG
Einsteinstrasse 1-5 Postfach 150
Zurich CH-8093
Switzerland

Mr. Tsutomu Nagira
Grad. School of Genetic Resources Technology
Kyushu University
6-10-1 Hakozaki Higashi-Ku
Fukuoka 8128581
Japan

Mrs. Annelie Niemi
Astra Biotech Laboratory
Byggnad 329
Södertälje S-151 85
Sweden

Mr. Stefan Nahrgang
EPFL
Lab. De Genie Clinique & Biologie
Lausanne 1015
Switzerland

Dr. Carmen Nievergelt
Ist. ETH Zurich
Teknische Chemie
Universitätstr. 6
Zurich 8092
Switzerland

Dr. P. Kumar Namdev
Schering Plough Research Institute
Mail Stop 4-14-2, 1011 Morris Ave.
Union, NJ 07083
USA

Dr. Inge Nilsson
Bio Invent Production AB
Lund 22370
Sweden

Dr. Tatiana Natashenko
Wyeth Ayerst Labs.
P.O. Box 304
Marietta, PA 17547
USA

Dr. Kjell Nilsson
Percell Biolytica
Ji-Te Gatan 9
Åstorp S-26538
Sweden

Dr. Josè-Jorge Nobre
Valbiotech
57, Boulevard de la Villette
Paris 75010
France

Mrs. Yvonne Ögren
BioInvent Production AB
Lund 22370
Sweden

Dr. Wolfgang Noé
BI Pharma
Birkendorfer Str. 65
Biberach/Riss 88397
Germany

Dr. Melvin Oka
Bristol-Myers Squibb
P.O. Box 5400
Princeton, NJ 08543-5400
USA

Dr. Thomas Noll
Institute of Biotechnology
Forschungszentraum Jülich
Jülich 52425
Germany

Dr. Bertram Opalka
Innere Klinik (Tumorforschung)
Hufelandstr. 55
Essen D-45122
Germany

Dr. Maria Luisa Nolli
Biotec Consultant, Biosearch Italia
Via R. Lepetit 34
Gerenzano (VA) 21040
Italy

Dr. Catherine Ovdot
Hyclone Europe NV - Belgium
Friedenstrasse 34
Asiar 35614
Germany

Mrs. Lottie Norrsén
Active Biotech Research AB
Scheelevägen 22
Lund 22363
Sweden

Mrs. Laurie Overton
Glaxo Wellcome
5 Moore Drive
Research Triangle Park, NC 27709
USA

Dr. Anette Ocklind
Amersham Pharmacia Biotech.
Björkgatan 30
Uppsala 75184
Sweden

Dr. Meran Owen
Trend In Biotechnology
68, Hills Road
Cambride CB2 1LA
UK

Mr. Philip Offin
The Automation Partnership Ltd.
Melbourn Science Park
Royston, Herts SG8 6HB
UK

Dr. Sadettin S. Ozturk
Bayer Corporation
800 Dwight Way
Berkeley, CA 94701
USA

Dr. Ute Pägelow
GBF/RDIF
Mascheroder weg 1
Braunschweig 38124
Germany

Dr. Pierre Perrin
Institut Pasteur
Labo. Lyssavirus
25 rue du Dr. Roux
Paris Cedex 15 75724
France

Dr. Stefan Papadileris
Fa. PAA Laboratories GmbH
Wienerstr. 151
Linz 4020
Austria

Mrs. Rosanna Pescini-Gobert
SPRI- Ares Serono International S.A.
14, Chemin des Auxl
Plan-les-Ouates/Geneva 1228
Switzerland

Dr. Manfred Papaspyrou
Papaspyrou Biotechnologie GMBH
Karl-Heinz Beckhurts Str. 13
Jülich D-52428
Germany

Mr. Jörn-Meidahl Petersen
Novo Nordisk AB
Novo Alle, 3BM1
Bagsvaerd DK-2880
Denmark

Mr. Gabriele Passador
Istituto Zooprofilattico Sperimentale
Via A. Bianchi 7
Brescia 25124
Italy

Dr. Eckhardt Petri
Greiner Labortechnik
Goethestrasse 6-8
Hirschberg 69493
Germany

Dr. Alastair Paton
Bioreliance
Innovation Park, Hillfoots Road
Stirling FK9 4NF
Scotland

Dr. Gary Pettman
SmithKline Beecham
Third Avenue
Harlow, Essex CM 19 5AW
UK

Dr. Jean-Charles Pelanchon
Stedim
Z.I. Des Paluds, BP 1051
Aubagne Cedex 13781
France

Mrs. Louise Phillips
Selborne Biological Services Ltd.
Goleigh Farm
Selborne, Alton Hampshire GU34 3SE
UK

Dr. Angelo Perani
Lonza Biologics
228 Bath Road
Slough, Berkshire SL1 4DY
UK

Mr. Laurent Pierard
SmithKline Beecham Biologicals SA
Rue de l'Institut 85
Rixensart 1330
Belgium

Mrs. Laurence Pineau
Schärfe System GmbH
Krämerstrasse 22
Reutlingen 72764
Germany

Mrs Lourdes Porquet Garanto
Laboratorios Hipra, S.A.
Avda. La Selva 135
Amer (Girona) 17170
Spain

Prof. James Piret
University of British Columbia,
Biotechnology Lab
6174 University Blvd.
Vancouver, British Columbia V6T
1Z3
Canada

Mr. Raymond Portenier
INFORS AG
Rittergasse 27
Bottmingen 4103
Switzerland

Dr. Sabine Pirotton
Laboratoire Biochimie Cellulaire ,
FUNDP
61, rue de Bruxelles
Namur 5000
Belgium

Dr. Gerhard Poßeckert
Rentschler Biotechnologie GmbH
Etwin-Rentschler-Str. 21
Laupheim 88471
Germany

Mr. Gian Polastri
Genentech Inc.
1 DNA Way
South San Francisco, CA 94080-4990
USA

Mrs. Marlise Potelle
Genzyme Transgenics Corporation
5 Mountain Road
Framingham , MA 01701
USA

Dr. Stephen Pollitt
Scios Inc.
820 West Maude Ave
Sunny Valley, Ca 94086
USA

Mr. Martin Potgeter
Clontech AG
Tullastr. 4
Heidelberg 69126
Germany

Miss Lyudmila Polonchuk
Institute for Cell Biology ETH-Zurich
ETH-Hönggerberg
Zurich 8093
Switzerland

Mr. Alain Pralong
Novartis Pharma AG
Gebande K-681.1.07
Basel 4002
Switzerland

Mr. Neville Pope
Selborne Biological Services Ltd.
Goleigh Farm
Alton, Hampshire GU34 3SE
UK

Dr. Jim Prendergast
Unisyn Technologies
25 South Street
Nopkinton, MA 01748
USA

Dr. Holly Prentice
Biogen Inc.
14, Cambridge Center
Cambridge, MA 02142
USA

Prof. Octavio Ramirez
Instituto de Biotecnologia, UNAM
Av. Universidad 2001, Col. Chamilpa
Cuernavaca, Morelos 62210
Mexico

Dr. Ingrid Rapp
Labor Dr. Koch - Dr. Merk
Schloss Strasse, 9
Ochsenhausen 88416
Germany

Dr. Frans Reek
ID-DLO
Post Box 65
Lelystad 8200 AB
The Netherlands

Mr. Bernd Rehberger
Medigene AG
Lochhamerstr. 11
Martinsried 82152
Germany

Mr Hubert Rehm
LaborJournal
Rathausgasse 20
Freiburg 79098
Germany

Dr. Jon Reid
Hyclone Europe NV - Belgium
Friedenstrasse 34
Asiar 35614
Germany

Mr. Georg Renemann
Institut für Technische Chemie
Callinstr. 3
Hannover 30167
Germany

Dr. Wolfang Renner
Cytos Biotechnology AG
Einsteinstrasse 1-5 Postfach 150
Zurich CH-8093
Switzerland

Dr. Shaul Reuveny
Israel Institute for Biological Research
P.O. Box 19
Ness-Ziona 74100
Israel

Mr. Per Rexen
Novo Nordisk A/S
Hallas Alle Bld. Edi 05
Kalundborg 4400
Denmark

Dr. Martin Rhiel
LGCB, EPFL
Lausanne 1015
Switzerland

Dr. Andreas Richter
Newlab Diagnostic Systems GmbH
Max-Planok-Str 15a
Erkrath 40699
Germany

Dr. Thomas Rigenstrup
Medi-cult A/S
Møllehaven 12
Jyllinge 4040
Denmark

Mr. Nigel Rimmer
YSI Limited
Lynchford
Farnborough GU14 6LT
UK

Mr. Marcel Roell
Wave Biotech AG
Ringstr. 24
Reigelsangen
Switzerland

Dr. Chantal Robadey
Digitana AG
Av. des Boveresses 44
Lausanne 1010
Switzerland

Mrs. Anja Romeijnders
MicroSafe B.V.
Niels Bohrweg 11-13
Leiden 2333 CA
The Netherlands

Dr. Graham Roberts
British Biotech.
Watlington Road, Cowley
Oxford OX4 5LY
UK

Prof. Marty Rosenberg
Smith Kline Beecham Pharmaceutical
709, Swedeland Road P.O. Box Box
1539
King of Prussia, PA 19406
USA

Mr. John Robertson
Selborne Biological Services Ltd.
Goleigh Farm
Selborne, Alton Hampshire GU34 3SE
UK

Dr. Morris Rosenberg
Eli Lilly and Company
Lilly Corp. Center
Indianapolis, IN 46285 D.C.3322
USA

Dr. Olivier Rocher
Chemunex
3, Alleè de la Seine
Ivry sur Seine Cedex 94854
France

Dr. Simon Rothen
Swiss Serum & Vaccine Institute Berne
Virology Bioprocess Engineering
P.O. Box
Berna 3001
Switzerland

Dr. Teresita Rodriguez
Centro de Immunologia Molecular
Habana
Cuba

Mr. Kurt Russ
Rentschler Biotechnologie GmbH
Etwin-Rentschler-Str. 21
Laupheim 88471
Germany

Dr. Susanna Roe
Genetics Institute
One Burtt Road
Andover , MA 01810
USA

Mrs. Annette Russell
Bayer PLC.
Stoke Court, Stoke Poges
Slough, Berkshire
UK

Mr. Kurt Rütten
Rütten Engineering
Industriestrasse 9
Stäfa 8712
Switzerland

Mr. Paul W. Sauer
Protein Design Labs
34801 Campus Drive
Fremont, CA 94555
USA

Dr. Thomas Ryll
Genentech Inc.
1 DNA Way
South San Francisco, CA 94080
USA

Mr. Peter Savas
Unisyn Technologies
25 South Steet
Hopkinton, MA 01748
USA

Mr. Andrea Sampieri
Chiron Vaccines
Via Fiorentina , 1
Siena 53100
Italy

Dr. Simon Saxby
Unisyn Technologies
25 South Street
Nopkinton, MA 01748
USA

Prof. Ryuzo Sasaki
Kyoto University
Graduate School of Agricolture
Kyoto 606-8502
Japan

Mrs. Melinda Scanlen
Onderstepoort Veterinary Institute ,
South Africa
Private Bag XS , Onderstepoort
Pretoria 0110
South Africa

Prof. Takeshi Sasaki
College of Agriculture
Dept. Biotechnology. Ehime Univ.
3-5-7 Tarumi
Matsuyama 790-8566
Japan

Mr. Eugene Schaefer
Schering Plough Research Institute
1011 Morris Avenue
Union, NJ 07083
USA

Mr. Tetsuo Sato
Asahi Chemical Ind.
Planova Division
9-1 Kanda Mitoshirocho Chiyoda-Ku
Tokyo 101-8481
Japan

Dr. Klaus Scharfenberg
Hoechst Marion Roussel
Emil Von Behringstrasse 76
Marburg 35041
Germany

Mrs. Yuko Sato
cancelled
9-1 Kanda Mitoshirocho, Chiyoda-ku
Tokyo 101-8481
Japan

Dr. Bart V. Schie
Quest International
P.O. Box 2
Bussum 1400
The Netherlands

Dr. Ernst-J Schlaeger
F. Hoffmann - La Roche AG
Basel 4070
Switzerland

Dr. Jacky Schmitt
Pharmacia & UpJohn
Munzinger Str 7
Freiburg 79111
Germany

Dr. H. Rudolf Schläfli
Paul Bucher
Schutzen Str.
Basel 4051
Switzerland

Mrs. Evelyn Schmucker
Boerhinger Ingelheim Pharma KG
Birkendorfer Strasse 65
Biberach /Riss 88397
Germany

Dr. Uwe Schlokat
Hyland-Immuno
Div. of Baxter Inc.
Uferstr.15
Orth/Donau 2304
Austria

Dr. Wolfgang Schneider
Hyclone Europe NV - Belgium
Friedenstrasse 34
Asiar 35614
Germany

Mrs. Annette Schmid
University of Zürich
Institute of Anatomy
Winterthurerstr.190
Zurich 8057
Switzerland

Prof. Yves-Jacques Schneider
Universitè Catolique de Louvain
Biochimie Cellulaire
L. Pasteur 1
Louvain-La-Neuve B-1348
Belgium

Dr. Georg Schmid
F. Hoffmann - La Roche AG
Grenzacherstr. 124
Basel 4070
Switzerland

Dr. Richard Schoenfeld
Genzyme
P.O. Box 9322
Framingham, MA 01701-9322
USA

Mr. Jörg Schmidt
Novartis - Pharmacia
WKL - 681.101
Basel 4002
Switzerland

Prof. Bernd Schröder
Maingen Biotechnologie GmbH
Weismüller Strasse 45
Frankfurt am Main 60314
Germany

Mr. Sebastian Schmidt
Institute of Biotechnology 2
Jülich 52425
Germany

Mr. Christof Schulz
GBF
Cell Culture Technology Dept.
Mascheroder Weg 1
Braunschweig D-38124
Germany

Mr. Norbert Schulze
GBF
Cell Culture Technology Dept.
Mascheroder Weg 1
Braunschweig D-38124
Germany

Mr. Jan-Oliver Schwabe
TU Hamburg - Harburg
Bioprozess und Bioverfahrenstechnik
(2-09)
Denickestr. 15
Hamburg D-21071
Germany

Dr. Zivia Schwarzbard
Laboratoires Serono SA
Aubonne 1170
Switzerland

Prof. Brian Seed
Harvard Medical School Molecular Biology
Massachusetts General Hospital
Wellman 911
Boston, MA 02114
USA

Dr. Thomas Seewoester
Basf Bioresearch Corporation
100 Research Drive
Worcester, MA 01605
USA

Mr. Ian Sellick
Pall Corporation
50 Bearfoot Road
Northborough, MA MAO 1532
USA

Dr. Cecilia Sendresen
Science Park Raf
Via Olgettina 58
Milano 20132
Italy

Miss. Alisa Shepherd
Inveresk Research
Tranent EH33 2NE
UK

Dr. Joseph Shiloach
N.I.H.
Bldg.6 BI-33
Bethesda, MD 20892-2715
USA

Prof. Sanetaka Shirahata
Grad. School of Genetic Resources Technology
Kyushu University
6/10/1 Hakozaki, Higashi-Ku
Fukuoka 812-8581
Japan

Mr. Wilhelm Siebertz
Greiner Labortechnik
Goethestrasse 6-8
Hirschberg 69493
Germany

Dr. Martin Sinacore
Genetics Institute
One Burtt Road
Andover, MA 01810
USA

Dr. Vijay Singh
Schering Plough Research Institute
1011 Morris Avenue
Union, NJ 07083
USA

Dr. Annelie Sjöberg
Active Biotech Research AB
Scheelevägen 22
Lund 222363
Sweden

Miss Randi N. Skovgaard
Novo Nordisk AB
Novo Alle, 3BM1
Bagsvaerd DK-2880
Denmark

Prof. Marco Soria
DIBIT - San Raffaele
Via Olgettina , 58
Milano 20132
Italy

Dr. Ole Skyggebjerg
Novo Nordisk A/S
Hagedornsvej 1
Gentofte 2820
Denmark

Prof. Ray Spier
University of Surrey
Stag Hill
Guildford GU2 5X17
UK

Mr. Diavor Sladic
Pliva Research Institute
Priliaz 6 Filipovica 25
Zagreb 10000
Croatia

Dr. Benjamin Spindler
Digitana AG
Tödistrasse 50
Horgen 8810
Switzerland

Mr. Dick Smit
ID-DLO
Post box 65
Lelystad 8200 AB
The Netherlands

Mr. Reto Spinnler
Zürcher Hochschule Winterthur
Postfach 805/ Chemie
Winterthur CH-8401
Switzerland

Dr. Lorraine M. Smith
Inveresk Research
Tranent EH 33 2NE
UK

Dr. Jan Sta
Quest International
P.O. Box 2
Bussum 1400
The Netherlands

Dr. Rodney Smith
Cantab Pharmaceuticals
Bio Cambridge Science Park
Cambridge CB4 1LM
UK

Dr. Andreas Stärk
Hoechst Marion Roussel Deutschland
Gmb.H.
P.O. Box 1140
Marburg 35001
Germany

Miss Heidi Sörensen
Novo Nordisk A/S
Hallas Alle Bld. Edi 05
Kalundborg 4400
Denmark

Mr. Harald Steeb
Aquasant Messtechnik AG
Hauptstrasse 20
Bubendorf 4416
Switzerland

Dr. Ulrich Steiner
Bayer Corporation
800 Dwight Way
Berkeley, CA 94598
USA

Dr. Noelle-Ann Sunstrom
University of New South Wales
Dept. of Biotechnology
UNSW
Sydney 2052
Australia

Mr. John Sterling
Genetic Engineering News
2 Madison Avenue
Larchmont, NY 10538
USA

Dr. Ivan Svendsen
Novo Nordisk A/S
Novo Alle
Bagsvaerd 2880
Denmark

Miss Gabriele Stiegler
Institute of Applied Microbiology
Muthgasse 18
Vienna A-1190
Austria

Dr. Berthold Szperalski
Roche Diagnostics GmbH
Nonnenwald 2
Penzberg 82372
Germany

Dr. Roland Stötzel
Novatech
Blochmonterstr. 8
Basel 4054
Switzerland

Mr. Shinya Takuma
Chugai Pharmaceutical co. Ltd.
5-1 5-chome Ukima, Kitaku
Tokyo 115-0051
Japan

Mr. Thomas Struckmeyer
Socochim
Marketing
Ch. Du Trabaudan 28
Lausanne CH-1006
Switzerland

Dr. Ronald Taticek
Genentech, Inc.
1 DNA Way, ms *32
South San Francisco, CA 44080
USA

Mrs. Désirée Studer
Cytos Biotechnology AG
Einsteinstrasse
Zurich 8093
Switzerland

Dr. Patricia Tavernier
University Wageningen
Bomenweg 2
Wageningen 6703 HD
The Netherlands

Dr. Takuya Sugahara
College of Agriculture, Ehime University
3-5-7 Tarumi, Matsuyama
Ehime 790-8566
Japan

Mr. Ian Taylor
Zeneca Pharmaceuticals
Room 13S28 Mereside
Macclesfield (Alderley Park) Cheshire
SK10 4TG
UK

Miss Tuija Teerinen
VTT Biotechnology and Food Research
FIN-02044 VTT
Espoo
Finland

Dr. Jean- Marc Teissier
Stedim
Z.I. Des Paluds,BP 1051
Aubagne Cedex 13781
France

Dr. William Tente
Chimeric Therapies Inc.
Elmwood court I, 409 Elmwood
Avenue
Sharon Hill, PA 19079
USA

Dr. Satoshi Terada
Fukui University
Dept. of Applied Chemistry &
Biotechnology
3-9-1, Bunkyo
Fukui 910-8507
Japan

Dr. Kiichiro Teruya
Kyushu
Graduate School of Genetic Resources
Technology
6-10-1 Hakozaki, Higashi-ku
Fukuoka 812-8581
Japan

Mr. Beng Ti Tey
The University of Birmingham
Edgbaston
Birmingham B15 299
UK

Mr. Klaus Theiss
Integra Biosciences AG
Industriestrasse 44
Wallisellen 8304
Switzerland

Mr. John Thrift
Bayer Corporation
4th Parker Street
Berkeley, CA 94701-5451
USA

Mr. Pierre Thysman
Smithkline Beecham Biologicals S.A.
89, Rue de l'Institut
Rixensart 1330
Belgium

Mr. Tim Tiemann
Cancelled - Bayer Corporation (USA)
Abt. Pentek, Weisensee 101
München 81539
Germany

Dr. Jerry Tong
Lonza Biologics
228 Bath Road
Slough, Berkshire SLI 4DY
UK

Dr. Helmut Trautmann
Biospectra AG
Zuerchestrasse 137
Zurich - Schlieren CH-8952
Switzerland

Dr. Julia Tree
Centre for Applied Microbiology and
Research
Porton Down
Salisbury, Wiltshire SP4 0JG
UK

Dr. Philip Tsai
Biogen Inc.
14, Cambridge Center
Cambridge, MA 02142
USA

Dr. Mary Tsao
BioWhittaker Inc.
8830 Biggs Ford Road
Walkersville, MD 21793
USA

Dr. Leo Van Der Pol
DSM - Biologics
P.O. Box 454
Groningen 9700 AL
The Netherlands

Mrs. Hilary Turnbull
Genetic Engineering News
2, Madison Ave
Larchmont, NY 10538
USA

Mrs. Tiny Van Der Velden - de Groot
RIVM
P.O. Box 1
Bilthoven 3720 BA
The Netherlands

Mr. Rodolfo Valdes
Swiss Federal Institute of Technology
Dept. of Biochemistry
Universitatsstr. 16
Zurich 8092
Switzerland

Dr. Miranda Van Iersel
Quest International
28 Huizerstraatweg
Naarden 1411 GP
The Netherlands

Dr. Johanna H.M. Van Adrichem
EPFL - CBUE
DC - IGC - LBTC
Lausanne 1015
Switzerland

Dr. Kirk Van Ness
Immunex Corporation
51, University Street
Seattle, WA 98101
USA

Dr. André Van Beekhuizen
New Brunswick Scientific
Kerkennbos 11-01
Nijmegen 6546 BC
The Netherlands

Mr. Mark Van Trotter
TC Tech Corporation
7600 West 27th Street 202
Minneapolis MN 55426
USA

Dr. R.J. Van de Griend
Biocult
Niels Bohrweg 11-13 K2
Leiden
The Netherlands

Dr. Jos Van Weperen
Waterloolaan 27
Groningen 9725
The Netherlands

Mr. Hans Van den Berg
Applikon BV
Brauwweg 13
Schiedam 3125
The Netherlands

Mrs Danielle Vandenbergh
Pharos
Parc Scientific du Sart Tilman
Seraing 4102
Belgium

Dr. David Venables
Covance Laboratories Ltd.
Otley Road
Harrogate, N.Yorkshire HG3 2XH
UK

Prof. Angelika Viviani
Hochschule Wädenswil
Gruntal
Wädenswil 8820
Switzerland

Dr. Francis Verhoeye
Catholic University of Louvain
Place Croix du Sud 2/19
Louvain-la-Neuve B-1348
Belgium

Dr. Horst Vogel
EPFL Institute of Chemistry
Lausanne 1015
Switzerland

Mr. Stanislas Vermeire
Van der Heyden
Biotech Div.
49, Rue du Marais
Brussels 1000
Belgium

Mr. Ellwood Vogt
Sigma Aldrich
Grünwalder Weg 30
Diesenhofen 82041
Germany

Dr. Allison Vernon
Lonza Biologics
228 Bath Road
Slough, Berkshire SLI 4DY
UK

Dr. Michael Von Pein
BioWhittaker Europe
Parc Industriel de Petit Rechain
Verviers B 4800
Belgium

Mr. Thomas Viertel
Mavag Verfahrenstechnik AG
Zuricherstr. 94
Altendorf 8852
Switzerland

Dr. Bénédicte Vonach
Novartis Pharma AG
K-681.1.02
Basel 4002
Switzerland

Dr. Giuseppe Claudio Viscomi
Alfa Wassermann S.p.A.
Via Ragazzi del '99, 5
Bologna 40133
Italy

Dr. Juergen Vorlop
Chiron Behring GmbH & Co.
Emil-von-Behring-Str. 76
Marburg D-35041
Germany

Mr. Paolo Vismara
Sigma-Aldrich
Via Gallarate 154
Milano 20151
Italy

Dr. Nienke Vriezen
Centocor BV
P.O. Box 251
Leiden 2300 AG
The Netherlands

Dr. Roland Wagner
Institution Gesellschaft für
Biotechnologische
Mascheroder Weg 1
Braunschweig 38124
Germany

Dr. Johanna Wahlberg
Astra Biotech Laboratory
Byggnad 329
Södertälje S-151 85
Sweden

Dr. Anton Walser
Rentschler Biotechnologie GmbH
Etwin-Rentschler-Str. 21
Laupheim 88471
Germany

Mr. Chwan-Heng Wang
Hintere gasse 3
Tübingen 72070
Germany

Mr. Guozheng Wang
New Brunswick Scientific co. Inc.
44 Talmadge Road
Edison, NJ 08818
USA

Dr. Shue-Yuan Wang
AMGEN Inc.
1 Amgen Center Dr.
Thousand Oaks, CA 91320
USA

Mrs. Sally Warburton
ECACC
Porton Down
Salisbury, Wiltshire SP4 0JG
UK

Mrs. Angelika Weber
Greiner Labortechnik
Goethestrasse 6-8
Hirschberg 69493
Germany

Mr. Max Weber
IG Instrumenten-Gesellschaft
Raffelstrasse
Zurich 8045
Switzerland

Dr. Max Weber
IG Instrumenten Gesellschaft
Raffelstr. 32
Zurich 8045
Switzerland

Dr. Ludwig Weibel-Furer
Amersham Pharmacia Biotech
Lagerstrasse 14
Dübendorf 8600
Switzerland

Mr. Ezra Weisman
New Brunswick Scientific
Kerkenbos 11-01
Nijmegen 6546 BC
The Netherlands

Dr. Stefan Weiss
Biotechnology Consulting
1612 Beechwood Drive
Martinez, CA 94553
USA

Prof. John Wérenne
Universite Libre de Bruxelles
Laboratory of Animal Cell
Biotechnology
Faculty of Sciences - CP 160/17 Av.
F.D. Roosvelt
Bruxelles 1050

Prof. Rolf Werner
Boehringer Ingelheim GmbH
Birkendorferstrasse 65
Biberach/ Riss 88397
Germany

Miss Tanja When
University of Bielefeld
Universitaetsstrasse 25
Bielefeld 33615
Germany

Mr. Rich Whitehead
New Brunswick Scientific
Kerkenbos 1101
Njmegen 6546 BC
The Netherlands

Mrs. Jutta Winter
BI Pharma
Birkendorfer Str. 65
Biberach/Riss 88397
Germany

Dr. Marc Wintgens
Hyclone Europe NV - Belgium
Friedenstrasse 34
Asiar 35614
Germany

Dr. Manfred Wirth
GBF
Mascheroder Weg 1
Braunschweig 38124
Germany

Dr. Wilfred Wöhrer
Immuno AG
Uferstrasse 15
Orth/Donau 2304
Austria

Mr. E. Woizenko
Sarstedt AG
Bahnweg Sud 36
Sevelen 9475
Switzerland

Dr. Dieter Wolf
Boheringer Ingelheim Pharma KG
Birkendorf Strasse 65
Biberach/Riss 88397
Germany

Dr. Kathy Wong
University of Singapore
Bioprocessing Technology Centre
10 Kent Ridge Crescent
119260
Singapore

Mr. Michael Worbin
Hotel Tylösand
Box 643
Halmstad 30116
Sweden

Mr. Jens Christian Wortmann
Novo Nordisk A/S
Novo Alle
Bagsvaerd 2880
Denmark

Miss Gordana Wozniak
Blood Transfusion Center
Slajmerjeva 6
Ljubljana 1000
Slovenia

Dr. Jason Wright
EPFL - CBUE
Lausanne CH-1015
Switzerland

Dr. Paul Wu
Bayer Corporation
800 Dwight Way
Berkeley, CA 94701
USA

Prof. Florian Wurm
Institut de Génie Chimique de l'EPFL
Laboratoire de Biotechnologie
Cellulaire
Ecole Politechnique Fédérale de
Lausanne
Lausanne 1015

Prof. Diane E. Wyatt
University of Kansas
16012 W. 124th Circle
Olathe, KS 66062
USA

Mr. Christopher Adam Yallop
Novo Nordisk A/S
Novo Alle
Bagsvaerd 2880
Denmark

Mr. Shigeru Yasutake
Asahi Chemical Ind. Co. Ltd.
9-1 Kanda Mitoshirocho, Chiyoda-ku
Tokyo 101 8481
Japan

Dr. Monique Zahn
Laboratoire de Biotechnologie
Moléculaire, CBUE
Department de Chimie (DC-IGC)
EPFL
Lausanne 1015
Switzerland

Mr Michael O. Zang-Gandor
Eugenex Biotechnologies
Lauchefeld 31
Matzingen 9548
Switzerland

Dr. James Zanghi
Scios Inc.
2450 Bayshore Parkway
Mountain View, CA 94043
USA

Mrs. Tracey Zecchini
Cantab Pharmaceuticals
Bio Cambridge Science Park
Cambridge CB4 1LH
UK

Dr. S. Zeng
University of Zurich
Winterthurerstr. 190
Zurich 8057
Switzerland

Dr. Weichang Zhou
Merck & Co. , Inc.
Cell Culture Dev., Bioprocess R & D
126 Lincoln Avenue - P.O. Box 2000,
R810-121
Rahway, NJ 07065
USA

Mr. Thierry Ziegler
Ares-Serono
Corsier sur Vevey 1804
Switzerland

Mr. Patrick Ziemeck
Innovatis GmbH
Hauptstrasse 72
Berlin 12159
Germany

Dr. Frank Zimmermann
Bio West
Rue de la Caille
Nucille 49340
France

Dr. Heiner Zindel
Werthenstein Chemie AG
Biotechnology
Schachen, Lausanne 6105
Switzerland

Mr. Henk Zwirr
Quest International
P.O. Box 2
Bussum 1400
The Netherlands

Introduction

Products from Cells – Cells as Products

This book is the "lasting" product, a resource of up-to-date information in the scientific literature for the field of animal cell technology, as it was presented during a pleasant and stimulating meeting that was held in Lugano Switzerland in April 1999. "Products" appear twice in the title of the conference. This clearly indicates the fact that the focus of the papers presented during this meeting was really the application of new technologies (novel reactors or novel vectors, for example - for the preparation and/or the more efficient generation of products) that could be used, mainly, in the medical field. Classical approaches for the use of animal cells, for example for the production of virus vaccines for human and animal health, still remain an important technology and still have, surprisingly, quite significant potential for further development and improvement. However, it appears that major technological advances and major growth from an economical point of view are occurring in other areas. Most importantly, protein production on the basis of recombinant DNA molecules transferred into animal cells, appears to be an ever-increasing field of interest and innovation, even though the first production scheme with this technology was approved more than 15 years ago. Part of this boom may be based on the realization, by more and more companies and research institutions, that animal cells are very versatile in bridging the gap between a DNA concept and the reality of a protein product. Also, wide-spread availability and choice *of* catalogue-based tissue culture equipment, small to lab scale plastic wares for one-time use, sophisticated media for a large number of standard and specialized cell culture applications, provide a solid basis for doing work in this field - even for new-comers with little prior background. A reluctance to use immortalized animal cells, that may have been around in the early years, for production purposes has disappeared completely, since experiences with many therapeutic proteins now on the market demonstrated that this technology is both safe and reliable. Finally, and probably still to come with even more emphasis in the next 5 to 10 years, the "avalanche" of DNA sequences derived from the international genome sequencing efforts, will bring animal cell technology even more into the forefront of sciences, since they are most appropriate to unravel possible function behind these sequences.

Apart from the typical production systems in which CHO or other popular cell lines are used to express recombinant proteins, animal cell culture technology is growing rapidly into other and more diverse fields and it becomes difficult, even for "experts" to stay informed on the development in those diverse areas. It is fair to say that this growth is not only occurring in the pharmaceutical area (large and mid-size companies that discover the utility of animal cell technology) but also in the academic world. Increasingly animal cells are being used as substrates for the study of gene activation and repression, and also for the more rapid production of small and moderate quantities of interesting proteins. Transient expression technology, i.e. the production of protein from transiently provided nucleic acid templates, is being used successfully. Tissue engineering, somatic gene/cell therapy, organ-replacement technologies, and cell based bio-sensors, all these fields contribute to a considerable widening of interest and research activity, based on animal cell technology.

The conference in Lugano attracted the largest number of participants ever in the history of ESACT. We believe this to be a reflection of the continuing growth in the number of scientists and engineers working with animal cells, and the incorporation of the new fields of research that had not previously been considered to have an animal cell application. This book captures, in the form of concise papers of limited length, the essence of the latest developments in those fields of animal cell technology which were so well represented at the meeting in Lugano. We hope it will become a useful resource of the most up to date information in Animal Cell Technology, at least until the next meeting which will be held in the spring of the year 2001 in Sweden.

Florian Wurm
Alain Bernard
Wolfgang Noé
Bryan Griffiths

ACKNOWLEDGEMENTS

This meeting would not have been possible without the help of a large number of people, some of whom you will find listed below. In addition to the members of the Organising Committee, and most important to the two of us, were two wonderful ladies, our administrative assistants, Mrs. Giovanna Labemano and Mrs. Monika Loperiol. We wish to particularly acknowledge the very dedicated and expert organisational work related to the trade exhibition by Dr. Wolfgang Noé and his administrative assistant, Mrs. Jutta Winter. Fund raising was always important for providing the financial backbone for a good scientific meeting. Dr. Georg Schmid, our Fund-Raising Manager, has done an admirable job and has spent many hours that assured us finally a solid flow of funds. Dr. Ferruccio Messi bears the responsibility for having us lured to Lugano. This wonderful city is his hometown and it was not a difficult task for him to convince us. However, he helped us in many ways, last not least by speaking for us with local authorities and providing excellent ideas about the venue for the meeting. The flow and go of all the funds necessary was handled by two expert and kind gentlemen, Mr. Jean-Daniel Baki and Mr. Philippe Buon, of the Ares-Serono Company in Geneva.

Last but not least, the Serono Symposia Organisation became, already in 1997, a very important partner in setting up the Lugano meeting. We thank Mrs. Maria-Grazia Calí and her collaborators, especially Mrs. Roberta Cenci for providing to us expert and steadfast support. Finally, the Travel-Agency Meridiano helped most of you and us in a very efficient way and here special thanks to Mrs. Simona Germoni.

We are looking forward to a stimulating and enriching ESACT 1999 meeting. Your participation is already a success and we feel very privileged and honored that you decided to come to Lugano. We assure you that the team with whom we were lucky enough to work on this meeting has done the utmost to make this meeting a memorable and satisfying experience to you.

The Meeting Chairmen
Florian Wurm   Alain Bernard

IMPROVEMENT AND INDUCTION OF HIGH PRODUCTIVITY

Chapter I

RECOMBINANT PHARMACEUTICAL PROTEIN OVEREXPRESSION IN AN IRF-1 PROLIFERATION CONTROLLED PRODUCTION SYSTEM

C. GESERICK[1,2], K. SCHROEDER[1], H. BONARIUS[2], L. KONGERSLEV[2], P. SCHLENKE[1], H. HAUSER[1] and P. P. MUELLER[1]*

[1] *Department of Gene Regulation and Differentiation, GBF - National Research Center for Biotechnology, Braunschweig, Germany;* [2] *Novo Nordisk, Gentofte, Denmark*

* *Corresponding author, fax ++49-531-6181-262, e-mail PMU@GBF.DE*

1. Introduction

Mammalian producer cell lines have been selected for rapid and indefinite proliferation capacity due to the requirement to obtain large numbers of cells needed in industrial production processes. These cell lines are transformed and override natural growth control systems. However, unrestricted growth is associated with disadvantages. After reaching an optimal cell density, further growth leads to changes in the production conditions associated with decreased quality and consistency of the product. Excess cells lead to nutrient and oxygen depletion, rapid accumulation of toxic products, cell lysis, clogging of cell retention and product purification devices, product contamination with cellular debris and product deterioration due to glycosidases and proteases (Fussenegger et al., 1999). Prolonged proliferation periods are also associated with genetic instability.

In technical applications regulated cell growth could permit rapid proliferation initially until an optimal cell density is reached. Then reduced growth would extend the productive period and keep production conditions constant by lowering medium consumption and waste product accumulation. The reduced cell division rate is expected to reduce genetic drift, and by that stabilize the productivity. Therefore, growth regulation could increase production, product quality and consistency (Fussenegger et al., 1999).

Reduced growth has been achieved with various approaches, by starving cells for an essential energy source, by using DNA-synthesis inhibitors such as thymidine, hydroxyurea, TGF-β or genotoxic agents such as adriamycin, or by incubating temperature-sensitive mutant cells at the nonpermissive temperature (Al-Rubeai et al., 1992; Suzuki and Ollis, 1990; Jenkins and Hovey, 1993). All these procedures lowered the growth rate, and in some cases productivity was increased. However, the adaption to applied conditions is hampered by reduced cell viability or low productivity soon after the onset of growth arrest.

Genetic growth control systems are flexible and allow the stepwise improvement of the recombinant regulatory system, optimization of the producer cell and the production

A. Bernard et al. (eds.), Animal Cell Technology: Products from Cells, Cells as Products, 3–9.

conditions (Fussenegger et al., 1998; Müller et al., 1998). An advantage is that the expression of recombinant genes can be induced during growth arrest by using dedicated promoters. Specific requirements must be met concerning stability, productivity, cell viability, quality and product consistency and industrial applicability.

We have genetically engineered BHK-21 cells to express IRF-1, a transcriptional activator of genes which lead to growth inhibition. To allow sufficient growth for the formation of stable clones, the activiy of recombinant IRF-1 must be regulated. For this purpose, the IRF-1 open reading frame was fused to the regulatory domain of the human estrogen receptor (IRF-1-hER) to control growth in a ligand-dependent manner (Kirchhoff et al., 1996). The addition of ß-estradiol activates IRF-1-hER, leading to a reduced growth rate (Kirchhoff et al., 1996; Köster et al., 1995). Normal growth was observed in all cases in the absence of ß-estradiol. Upon estrogen addition to the growth medium, proliferation is reduced depending on several parameters, such as the estrogen concentration, the cell density and the duration of ligand exposure (Carvalhal et al., 1998).

We investigated properties that are important to control cell growth in production processes, that is cell viability, production, productivity and product quality.

2. Cell viability

After extended periods of IRF-1 activation cell viability decreases. The decrease in viability of BHK-21 cells begins at about day three after IRF-1 activation without the typical hallmarks of apoptosis. However, at higher cell densities the loss of viability is reduced. To use IRF-1 in technical processes, it cannot remain activated permanently. Either the activity of IRF-1-hER has to be reduced by low doses of estrogen or the inducer of IRF-1 may be applied in intervals, alternating with recovery periods.

3. Product formation

A critical factor for productivity is the recombinant gene promoter activity. For constitutive gene expression highly active recombinant promoters are available. For many applications constitutive expression is neither necessary nor desirable. In a biotechnological production process, high production levels in the initial phases of a fermenter process may be a metabolic burden, leading to a negative selection pressure of the most productive cells and decrease the performance of the system, while the contribution to the overall production is minor at low cell densities. Furthermore, product quality may change during the production process since it depends on the environment of the cells. Media compositions, in particular ammonia and glucose concentration, change during the course of fermentations, product glycan structures synthesized early and late in the process may differ as well. In addition, secreted products are subject to degradation processes. A long exposure time to degrading enzymes of product synthesized early lead to a lower quality and homogeneity of the product. With the exception of the production of highly cytotoxic products, maximal expression levels but not a very high induction rate is necessary. Despite these advantages, the drawback of regulated gene expression is the frequently far less efficient expression when compared to the best constitutively active promoters. Since IRF-1 acts as a transcriptional activator (Kirchhoff et al., 1993), a dedicated promoter was constructed by inserting IRF-1 binding sites into a strong constitutive promoter (Fig. 1).

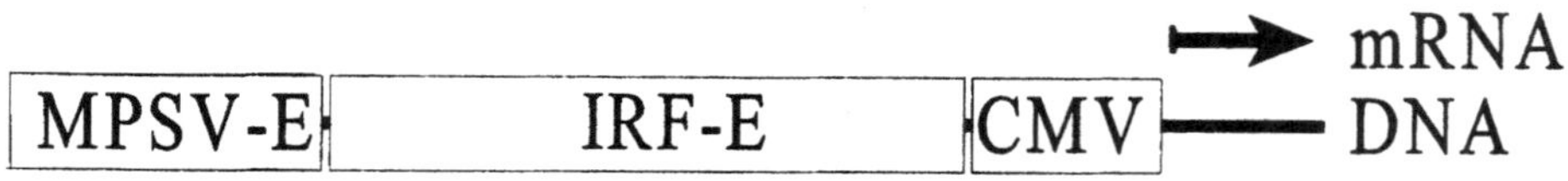

Figure 1. Composite promoter for high-level inducible expression. The promoter consists of constitutive viral enhancer elements derived from MPSV, IRF-1 binding sequences (IRF-E) and a minimal promoter sequence from CMV that determines the RNA start site.

This composite promoter has a high basal activity equivalent to the strong MPSV promoter (Artelt et al., 1988) in transient transfection experiments (Table 1). In IRF-1 proliferation controlled cells, the promoter can be induced to even higher levels (Fig. 2).

TABLE 1. Basal level activity of the MPSV/IRFE and MPSV promoter

Promoter	IgG [mg/ml]
IRFE	1.8 +/- 0.1
MPSV/IRFE	1.7 +/- 0.1
MPSV	1.7 +/- 0.1

4. Product quality

In addition to the expression level, a uniformly high product quality is a principal aim for biotechnological process applications. Protein quality refers to a number of parameters including protein folding, processing, post-translational modifications and protein integrity. Glycosylation is common for secreted proteins and shows the highest variability among the posttranslational modifications. Glycosylation is influenced by the producer cell and its environment; it can change in the course of cultivation. Pharmaceutical protein glycosylation can be important for biological activity, antigenicity and clearing time from the blood circulation. Towards the end of production processes high cell densities contribute to the ammonia accumulation in the medium that can lead to drastic changes in the glycosylation pattern (Gawlitzek et al., 1998; Gawlitzek et al., 1995; Grammatikos et al., 1998; Jenkins et al., 1996). Similarly, secreted glycosidases and enzymes released by lysed cells can degrade the carbohydrates. We have determined human erythropoietin (Epo) protein integrity and glycosylation pattern, a sensitive and relevant quality criteria, by SDS acrylamide gel electrophoresis and Western blot analysis (Fig. 3) by mass

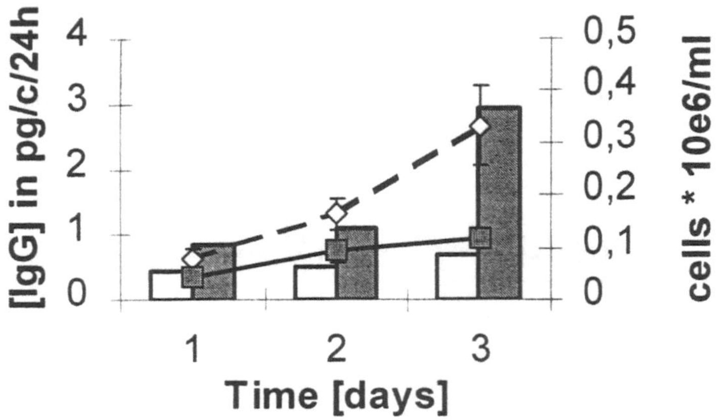

Figure2. Inducible productivity in proliferation controlled BHK cells. The compositeMPSV/IRFE promoter was used to express IgG in IRF-1-hER growth regulated BHK-21 single cell clones. The amount of IgG secreted into the supernatant was determined by a sandwich ELISA procedure. Open bars; basal productivity level. Hatched bars; productivity of growth arrested cells in the presence of 100 nM ß-estradiol. Interupted line; uncontrolled growth. Continuous curve; controlled growth in the presence of 100 nM ß-estradiol.

spectroscopic methods and HPLAE chromatography (Mueller et al., submitted). The Epo quality from the proliferation controlled culture was at least equivalent to that from the growing culture. IRF-1 activation did not influence the Epo protein integrity and there were no signs of proteolytic degradation. Therefore, the IRF-1 system can reduce excess cell growth and yield a consistent and high product quality.

5. Conclusions

The properties of the genetic IRF-1 mediated proliferation control system has been demonstrated in the biotechnologically relevant producer cell line BHK-21. Growth can be effectively regulated by pulsed ß-estradiol addition for periods of more than 50 days in a perfusion fermenter (not shown here). Productivity from IRF-1 inducible promoters, is enhanced while the product quality remains similar or is perhaps even superior to the product from uncontrolled proliferating cells. By avoiding excess growth, ammonia production and cell death, the environment of producer cells should remain more constant and is expected to result in improved product quality and consistency in applied production processes.

Acknowledgments: We thank Simon Klibisch, Meike Tümmler and Martina Grasshoff for experimental help and for providing results prior to publication, and Rosemary Avram for typing the manuscript.

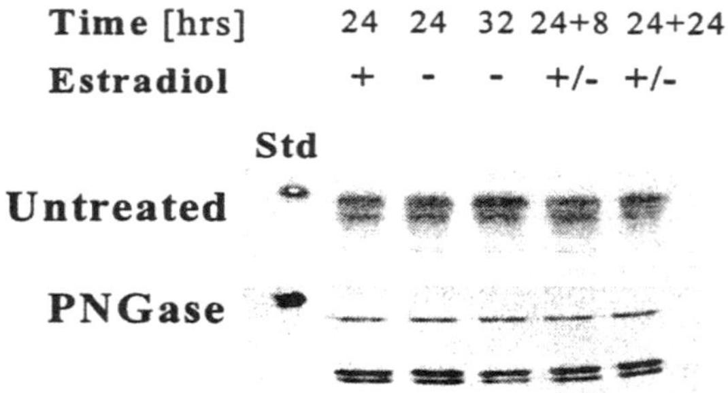

Figure 3. Secreted pharmaceutical product quality is constant in IRF-1 proliferation controlled cells. An IRF-1 proliferation controlled EPO secreting BHK single clone was grown in the presence (+) or absence (-) of a 24 hr ß-estradiol pulse to simulate conditions in a proliferation controlled perfusion system. EPO glycoforms from serum-free culture supernatant were separated by SDS gel electrophoresis, blotted to a nitrocellulose membrane and detected with anti EPO antibodies. Std, highly sialylated purified EPO isoform; PNGase, N-linked glycans were removed by PNGaseF treatment, leaving the O-glycosylated and the unglycosylated EPO isoforms (lower bands). The upper band is due to a crossreaction of antibodies with PNGase F (Schlenke, Ph.D. Thesis, 1999).

6. References

Al-Rubeai, M., Emery, A.N., Chalder, S. and Jan, D.C. (1992) Specific monoclonal antibody productivity and the cell cycle-comparisons of batch, continuous and perfusion cultures, *Cytotechnology* **9**, 85-97.

Artelt, P., Morelle, C., Ausmeier, M., Fitzek, M. and Hauser, H. (1988) Vectors for efficient expression in mammalian fibroblastoid, myeloid and lymphoid cells via transfection or infection. *Gene* **68**, 213-219.

Carvalhal, A.V., Moreira, J.L., Müller, P.P., Hauser, H., and Carrondo, M.J.T. (1998) Cell growth inhibition by the IRF-1 system, in New developments and new applications in animal cell technology (Merten, O.-W., Perrin, P. and Griffiths, B., eds.) Kluwer Academic Publishers, pp. 215-217.

Fussenegger, M., Schlatter, S., Dätwyler, D., Mazur, X. and Bailey, J. E. (1998) Controlled proliferation by multigene metabolic engineering enhances the productivity of CHO cells, *Nat. Biotechnol.* **16**, 468-472.

Fussenegger, M., Bailey, J., Hauser, H. and Mueller, P.P. (1999) Genetic Optimization of Recombinant Protein Production by Mammalian Cells, *TIBTECH* **17**, 43-50.

Gawlitzek, M., Conradt, H. S. and Wagner, R. (1995) Effect of different cell culture conditions on the polypeptide integrity and N-glycosylation of a recombinant model glycoprotein, *Biotechnol. Bioeng.* **46**, 536-544.

Gawlitzek, M., Valley, U., Wagner, R. (1998) Ammonium ion/glucosamine dependent increase of oligosaccharide complexity in recombinant glycoproteins secreted from cultivated BHK-21 cells, *Biotechnol. Bioeng.* **57**, 518-528.

Grammatikos, S. I., Valley, U., Nimtz, M., Conrad, H. S., Wagner, R. (1998) Intracellular UDP-N-acetyl-hexosamine pool affects N-glycan complexity: A mechanism of ammonium action on protein glycosylation, *Biotechnol. Progr.* **14**, 410-419.

Jenkins, N., Parekh, R. B. and James, D. C. (1996) Getting the glyosylation right: implications for the biotechnological industry, *Nat. Biotechnol.* **14**, 975-981.

Jenkins, N. and Hovey, A. (1993) Temperature control of growth and productivity in mutantChinese hamster ovary cells synthesizing a recombinant protein . *Biotechnol. Bioeng.* **42**, 1029-1036.

Kirchhoff, S., Schaper, F. and Hauser, H. (1993) Interferon regulatory factor 1 (IRF-1) mediates cell growth inhibition by transactivation of downstream target genes, *Nucleic Acids Res.* **21**, 2881- 2889.

Kirchhoff, S., Kröger, A., Cruz, H., Tümmler, M., Schaper, F., Köster, M. and Hauser, H. (1996) Regulation of cell growth by IRF-1 in BHK-21 cells, *Cytotechnology* **22**, 147-156.

Köster, M., Kirchhoff, S., Schaper, F. and Hauser, H. (1995) Proliferation control of mammalian cells by the tumor suppressor IRF-1, in Animal Cell Technology: Developments towards the 21st Century (Beuvery, Griffiths, Zeijlemaker, eds.) Kluwer Academic Publishers, pp. 33-44.

Mueller, P. P., Kirchhoff, S. and Hauser, H. (1998) in *New Developments and New Applications in Animal Cell Technology* (Merten, O.W., Perrin, P. and Griffiths, J.B., eds) Kluwer Academic Publishers, pp. 209-213.

Suzuki, E. and Ollis, D.F. (1990) Enhanced antibody production at slowed growth rates: experimental demonstration and a simple structured model. *Biotechnol. Prog.* **6**, 231-236.

Discussion (Mueller)

Piret: It is a shame you need to induce with estradiol periodically as you do not want to introduce that variability in a long-te rm process. Can you tell me more about the difficulty of induction with estradiol - is it a direct proliferation or is it an effect of estradiol itself?

Mueller: If you expose the cells longer than 3 days to estradiol one gets a decrease in viability. By adding estradiol as a pulse we can keep the viability over 90%.

Piret: Do you know the mechanism?

Mueller: No, it is not known. There are several individual proteins known that are induced by interferon and do reduce growth, but so far the initial one has not been identified.

Ozturk: Can you tell me if it is a full Factor VIII molecule or a truncated form?

Mueller: We produce Factor VII.

Nieba: Is IRF-1 alone affecting cell growth? Can you use other estradiols which do not bind that tightly to the receptor so you can continuously add the analogue?

Mueller: So far we have not used any estradiol analogues. We are in the process of screening compounds for estradiol-like activity and also for negative regulators of estradiol so that we can add an additional compound, rather than addition and removal. It is IRF-1 alone which reduces growth.

Kost: What concentrations of estradiol were you using, and have you looked at a dose response?

Mueller: We do have dose response data for estradiol. Lower doses reduce growth less drastically than high doses but, unfortunately, you still induce cell death with low doses. We use 100n Molar estradiol which is extremely high compared to the native estrogen receptor. We are using a deletion of the estrogen receptor which has deleted the last carboxy terminal alpha helix which acts on its own as the transcriptional activator. Since IRF-1 already has a transcriptional activation domain, and we did not want to change the function of IRF-1, this domain was deleted with the result that this fusion construct is much less sensitive to estrogen than the original estrogen receptor.

THE REAL MEANING OF *HIGH EXPRESSION*

S. I. GRAMMATIKOS, K. BERGEMANN, W. WERZ, I. BRAX,
R. BUX, P. EBERHARDT, J. FIEDER, W. NOÉ
Process Development Group
Department of Biopharmaceutical Manufacture
Boehringer Ingelheim Pharma KG
Birkendorfer Str. 65, D-88397 Biberach an der Riss, Germany

Introduction

There exist some common misconceptions about the application of high expression systems (HES) to an industrial setting. This paper intends to sensitize the developer and even more importantly the end-user of HES towards generic complications encountered during the industrial application of such systems, complications which, if overlooked, can render the term *high expression* meaningless.

Due to the large industrial interest in HES, the term „high expression" is used readily and amply in academic research and is almost always associated with attempts at commercialization. If one looks at the industrial perspective, however, what counts for the biopharmaceutical industry is not necessarily the HES but the high titer. All other things being equal, the titer dictates the production scale and the number of production runs required per year in order to produce the required quantity of recombinant protein. Clearly what is an acceptable or desired titer in batch and fed-batch processes depends on the dosage and the size of the market. Most of the time, however, and with only few exceptions, dosages are high and high titers (preferably >500 mg/l) are needed in suspension, serum-free cultures. Everyone hopes of course that the application of a HES automatically leads to high titer processes. This is where the misconceptions begin.

High Expression is a useful pre-condition

At BI Pharma we have dealt and are dealing with HES and with conventional expression systems and our experience has taught us that a HES is a useful and important pre-condition, but it is neither a necessary nor a sufficient condition for a high titer process. For example, in one case where a conventional expression system is used we have seen cells in serial seedstock cultures exhibiting specific productivities of about 1.5 pg/day delivering later in a thoroughly developed production process titers far exceeding 500 mg/l. In another case where a HES is used the specific productivity in serial seedstock culture is nearly 20-fold higher (20-30 pg/day) but still the process delivers titers well under 500 mg/l.

This curious discrepancy brings to the spotlight two important „facts of life" which are contributing to the successful application of a HES and can endanger, if overlooked, the presumed equation *high expression=high titer*.

A. Bernard et al. (eds.), Animal Cell Technology: Products from Cells, Cells as Products, 11–17.

Fact of Life #1: Every Cell is Different

First of all it is very important to realize that every cell, even within one transfection, is different. The transfection process and the ensuing selection and amplification lead to populations which differ even genotypically such that after a single-cell recloning of the culture, a variety of cell „entities" is obtained. „Entity" is probably a better term than „clone" for what we have in the 96-well plate. When one says „clone" one thinks of „Dolly" who is presumably identical to her mother. Here we have cell entities which react differently to culture conditions, which lend themselves more or less to optimization and which might require different media and process strategies in order to reach a maximum or near-maximum output.

An interesting example stems from our efforts in establishing at BI a HES licensed from IDEC Pharmaceuticals. The essence of this system in comparison to a conventional random-integration DHFR system is that it features 2 selectable markers: Neomycin resistance with impaired expression for primary selection of „clones" and intact DHFR. The idea behind this random-integration DHFR system is that only those transfectants which integrate the plasmid in a high expression region of the genome can survive in the presence of G418.

Using this system for the production of a proprietary antibody a CHO-DG44 host cell was transfected (Figure 1). After selection in the 96-well plate one „clone" is picked

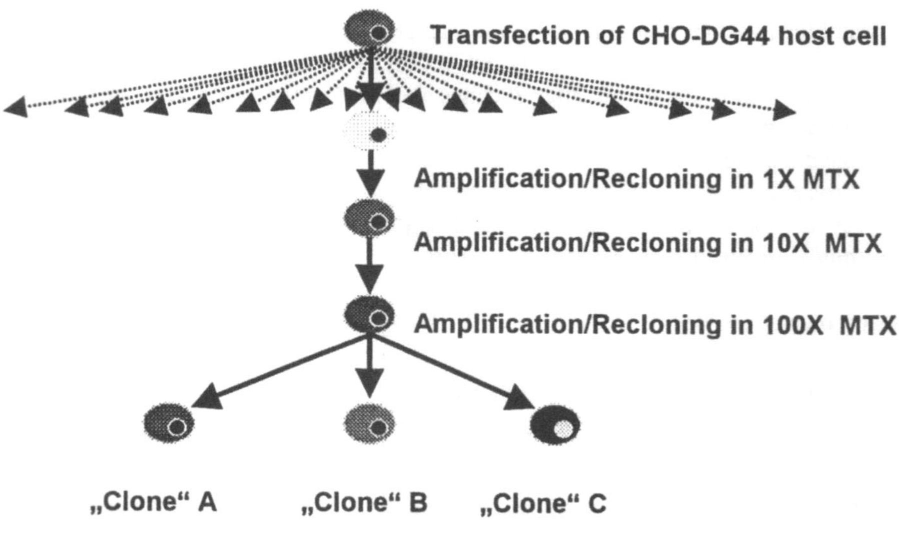

Figure 1. Parental lineage of 3 „best clones" coded A, B and C

on the basis of high product concentration for amplification in methotrexate (MTX). Every round of amplification involves single-cell recloning and from the last recloning in 100X MTX, 3 „clones" are selected for further study. Considering their parental lineage and since they originated from the same „clone" prior to amplification, the 3 „clones" coded A, B, and C should theoretically be very much related to each other.

Probing the potential of these three „clones" to deliver high titers, however, we note curious differences in their behavior. In the experiment shown in Figure 2 we compare the reaction of the 3 „clones" to feeding a nutrient mixture with or without a productivity enhancer. Without the productivity enhancer A and B perform similarly whereas C is clearly inferior. However, C reacts the most in the presence of the productivity enhancer and behaves similarly to A, whereas B remains unaffected. Here we see three types of responses to the productivity enhancer: no response (B), intermediate response (A) and dramatic increase in productivity (C).

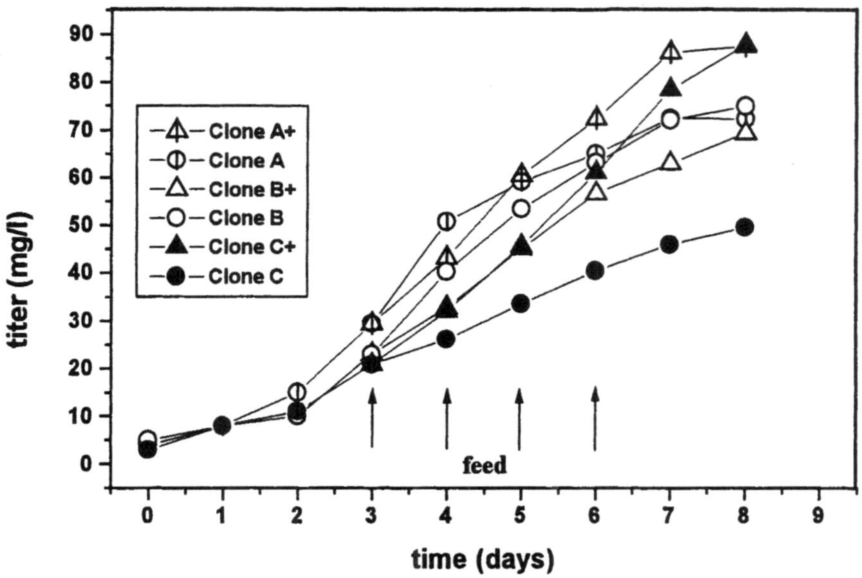

Figure 2. Spinner flask experiment probing the potential of the three „best clones" to deliver high titers in fed-batch processes. On the days indicated by arrows the same feeding mixture is added either containing (+) a productivity enhancer or not. The points represent the average of duplicate experiments which produced very similar results

Based on these observations and on the fact that A also shows the highest specific productivity in serial seedstock culture, one might select this „clone" (A) for further development. Studying A in serial culture, however, we note after several passages that it performs unstably perhaps due to a genetic instability (Fig. 3). The problem is manifested by a dramatic drop in specific productivity down to 10 pg/cell/day in 15 passages. On the other hand, neither B nor C show this instability (data not shown). All these differences have been observed within one „family" (Fig. 1), starting from the identical „clone" after transfection and recloning in the presence of G418. In fact there were more than 100 clones identified after the selection process. A few were chosen on the basis of high product concentration in the 96-well plate and all others were discarded.

The example shown here of variability in performance potential of transfected cell lines together with scattered but unambiguous evidence from the literature particularly with regard to stability[1,2] but also with regard to variable product quality[3] lead to the realization that every cell, even within one transfection, is different.

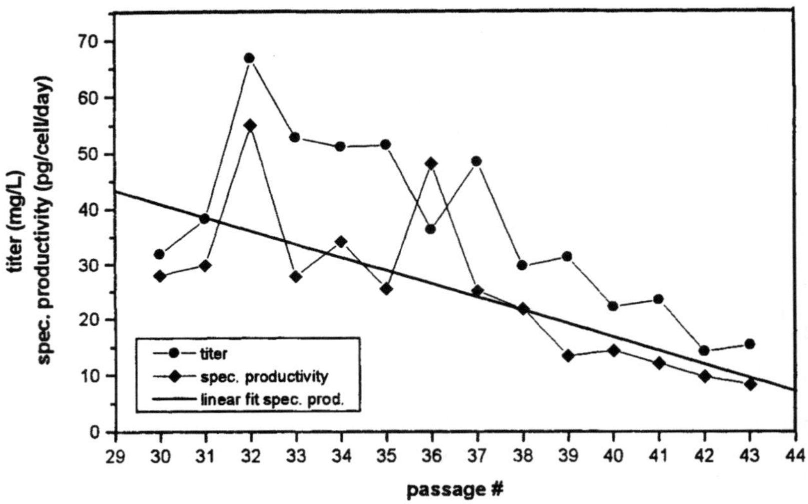

Figure 3. Stability of „clone“ A in serial seedstock culture. The points indicate measurements made at subcultivation

How do you pick the best cell?

Given the variability in behavior and potential of all these so-called „clones," it is far from certain that the „best" candidates manifest themselves with high product concentrations in the 96-well plates. This means that when a few „clones" out of 100 or 200 are selected on the basis of product concentration and the rest are discarded, it is very likely that a truly „best" cell goes „down the drain." The question that is raised immediately is: How can one pick then the best cell?

The answer is not easy because it must come from an as yet uncharted research territory, one that most certainly will involve the combination of molecular biology and cell culture engineering with cell culture applications of high throughput screening, nanotechnology and a large degree of automatization.

Fact of Life #2: Process Development

The importance and contribution of process development to the success of a HES cannot be overlooked. An interesting example that proves this point stems from an accumulation of data regarding the development of a process to produce sICAM-1, a soluble form of the Intercellular Adhesion Molecule 1 thought to be a docking station for rhinoviruses. The titer for this process in suspension conditions on the basis of a CHO cell and a conventional expression system started at about 20-40 mg/L and after two years of development it reached about 100 mg/L. Following further medium and process development efforts the process was thought to have matured at about 180 mg/L. Using the IDEC HES a new cell was created which surprisingly under the conditions of the old process delivered the same titer as the old cell (Fig. 4). As it turns out, the new high-expression cell (assuming it is the „best" cell out of the 96-well plate) has its own media and process requirements and must be subjected to customized process development before it can perform optimally. The choice of expression system gives the cell high potential but this potential has to be explored with a systematic process development. As is shown in Figure 4, a doubling in titer can be achieved after a period of process development.

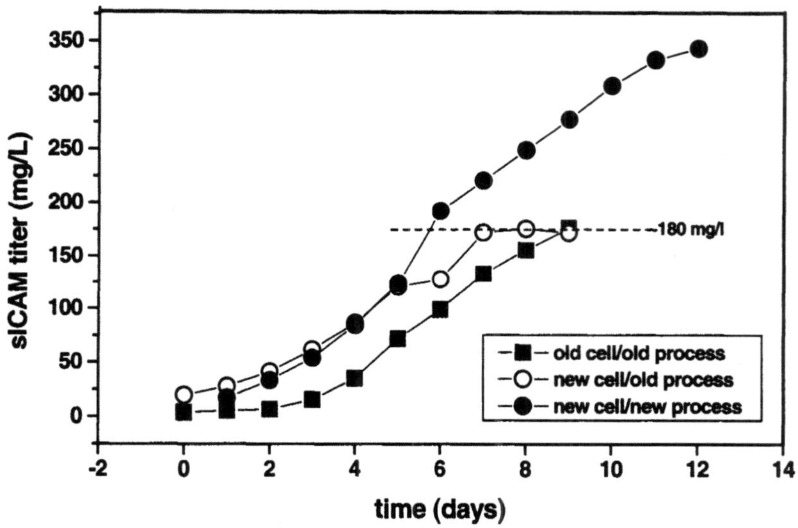

Figure 4. The benefit of using a high expression system (new cell) for the production of sICAM becomes apparent only after a period of process development

Conclusions

•Every transfected cell, even within one transfection is a different entity.

•Research in the area of „clone" selection is lacking. Application of methods of high-throughput-screening and nanotechnology to cell culture should afford a wide range of criteria to facilitate selection of cells with the highest production potential.

•Promise of high titers when applying high expression systems can be realized only after thorough „clone" selection and a period of process development.

•There is no such thing as a „quick fix" when it comes to the search for highly productive cells and high titer processes.

References and Notes

[1] Wurm, F.M., Pallavicini, M.G., Arathoon, R. 1992 Integration and stability of CHO amplicons containing plasmid sequences. *Dev. Biol. Stand.* **76:** 69-82.

[2] Kim, N.S., Kim, S.J., Lee, G.M. 1998. Clonal variability within dihydrofolate reductase-mediated gene amplified chinese hamster ovary cells: Stability in the absence of selective pressure. *Biotechnol. Bioeng.* **60:** 679-688.

[3] Oral presentation O1.04 at the 16th ESACT Meeting by M.S. Sinacore : Postranslational modifications limit high level expression of functionally active chimeric P-selectin glycoprotein ligand-1 in CHO cells.

Discussion (Grammatikos)

Zhou: Have you looked at NS0 cells as the expression system?

Grammatikos: We have looked at NS0 cells but these results were obtained with CHO cells.

Zhou: Where do you stop during the selection process for obtaining the best cells for further process development?

Grammatikos: Although we are trying to improve this shallow process, we are currently not doing anything different to everyone else for clone selection.

Cho: Is the integration site important, or the amplification process?

Grammatikos: Both - in this clever expression system from HiTech you probably select those clones that integrate in the high expression region, but the amplification certainly helps. A lot of variability can arise just from the amplification process.

A Novel Mammalian Gene Regulation System Responsive to Streptogramin Antibiotics

M. FUSSENEGGER, B. VON STOCKAR, C. FUX, M. RIMANN, R. MORRIS, C.J. THOMPSON AND J. E. BAILEY

CISTRONICS CELL TECHNOLOGY GmbH, Einsteinstrasse 1-5, P.O. Box 145, CH-8093 Zurich,
Tel.: +41 79 336 95 00; Fax: +41 86079 336 95 00;
e-mail: fussenegger@bluewin.ch

Abstract

We designed a new strategy for efficient regulation of cloned gene expression in mammalian cells. The new gene regulation system is responsive to streptogramins, a class of human therapy-proven antibiotics such as pristinamycin and virginiamycin, which are naturally produced by *Streptomyces* spp. The pristinamycin-responsive interaction between the repressor (Pip) of the pristinamycin resistance gene (*ptr*) and its target sequence the ptr promoter (P_{PTR}) formed the basis for construction of two chimeric genetic determinants: Pip was fused to a eukaryotic transactivator domain (PIT, pristinamycin-responsive transactivator) and the Pip binding DNA sequence (P_{PTR}) was cloned 5' of a minimal insect promoter resulting in P_{PIR} (pristinamycin I-responsive promoter). PIT retains its streptogramin-dependent P_{PTR} binding capacity in mammalian cells which enables sufficiently close contact between the transactivation domain and the minimal promoter to achieve transcription initiation in absence of streptogramin antibiotics. In the presence of streptogramins, PIT is released from P_{PIR} and gene expression is abolished. The novel PIT/P_{PIR} system shows excellent regulation characteristics (high expression levels and low basal activity) in a variety of mammalian cell lines including CHO-K1, BHK-21 and HeLa cells. In addition, the streptogramin regulation system is compatible with the well-established tetracycline-responsive expression concept which enables the combined use of these two distinct systems in a single cell for independent control of different transgenes.

Introduction

The ability to adjust gene expression in mammalian cells, tissues and organisms will be essential for future progress in many areas of biology, biotechnology, and medicine. Besides enabling functional genomic studies in an isogenic background and production of difficult-to-express protein pharmaceuticals in a desired production phase, regulated

A. Bernard et al. (eds.), Animal Cell Technology: Products from Cells, Cells as Products, 19–22.

gene expression technology is currently expected to revolutionize somatic gene therapy by enabling *in vivo* adjustment of gene product levels into the therapeutic range, for varying daily dosing regimes and to react to the changing nature of the patients' disease. The streptogramin-responsive expression technology is compatible with human therapeutic use since streptogramins show excellent pharmacokinetic behavior and have a successful track record in human antibiotic chemotherapy which is currently continued by Synercid, the only injectable antibiotic effective against most multidrug-resistant bacterial pathogens. Streptogramins consist of two structurally dissimilar molecules which are synergistically bactericidal. Only the the group B component (pristinamycin I (PI), virginiamycin I and quinupristin (Synercid group B component)) is able to regulate the streptogramin resistance operon in *Streptomyces* and consequently also the PIT/P_{PIR} system, a fact which renders the streptogramin-based antibiotic therapy compatible with almost independent use of the streptogramin-based regulation concept and reduces the emergence of streptogramin resistance.
In order to tolerate their own antibiotics, streptogramin synthesis in *Streptomyces* is strictly coregulated with streptogramin resistance which is thought to be mediated by antibiotic-dependent binding of a repressor protein (Pip) to its cognate sequence overlapping the promoter of the streptogramin resistance operon (P_{PTR}) (Fig. 1).

Fig. 1: Gel retardation assay showing Pip/P_{PIR} interaction. All three Pip binding sites (A in the absence of Pip) of P_{PTR} are saturated at high Pip concentrations resulting in a single shifted band (B). However, smaller amounts of Pip (0.02pmol) result in all possible permutations of site occupancy represented by three retarded bands (C). Pip was titrated with a range of PI (lanes D-H; 50pmol, 10pmol, 1pmol, 0.1pmol, 0.01pmol). Pip was completely released from P_{PTR} by excess molar concentrations of PI.

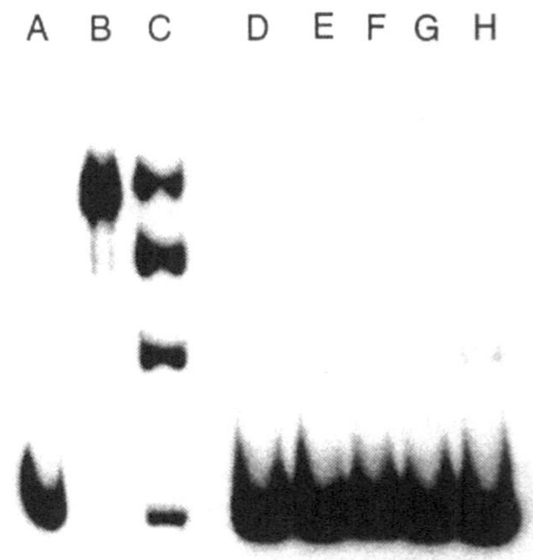

Fig. 1

Results and Discussion

1. PI-responsive mammalian gene expression system

We adapted the *Streptomyces* regulatory elements Pip and P_{PTR} for use in a eukaryotic context by construction of two chimeric determinants: The PI-responsive transactivator (PIT, Pip fused to the VP16 transactivation domain of *Herpes simplex*) and the PI-responsive promoter (P_{PIR}, P_{PTR} fused to a minimal insect promoter) (Fig. 2). Following cotransfection of a P_{PIR}-driven SEAP (human secreted alkaline phosphatase) construct and a PIT-encoding plasmid into CHO-K1, BHK-21 and HeLa cells, high SEAP expression could be observed which was repressed by addition of 2μg/ml PI to the cell culture medium (Fig. 3).

2. Stable expression of PIT in CHO cells

Two stable PIT-expressing cell lines were generated (CHO-PIT1 and CHO-PIT2). Constitutive overexpression of PIT seems to exert no adverse physiological effects on CHO cells since CHO-PIT1 and CHO-PIT2 show similar growth behavior comparable to wild-type CHO-K1 cells. A P_{PIR}-driven SEAP expression contruct which is transiently transfected into CHO-PIT1 showed reversible PI-responsive SEAP expression, reaching up to 13-fold higher expression levels in the absence of pristinamycin compared to the repressed state in the presence of 2μg/ml PI (Fig. 3). Unlike PI, the group A component pristinamycin II (PII) showed no regulatory potential on the streptogramin-based expression system. The PIT/P_{PIR} system is dose-responsive over 5 orders of magnitude of PI concentration which enables precise adjustment of gene expression over a wide range of expression levels. Such a characteristic is particularly important for certain metabolic engineering applications in cell culture and for emerging somatic gene therapies which require titration of the desired molecules into a specific therapeutic window.

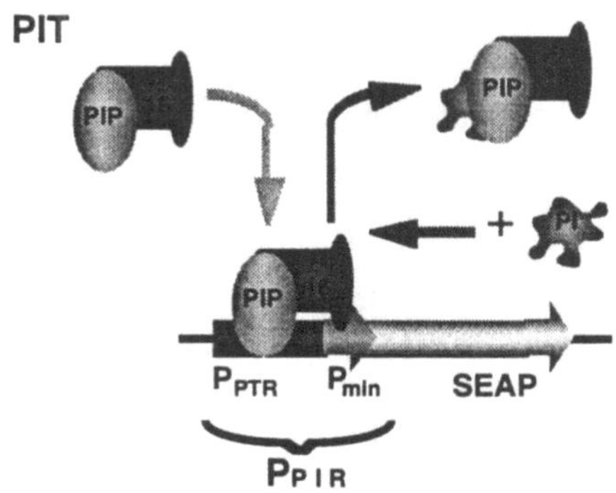

Fig. 1: PI-responsive mammalian gene expression

3. Regulation efficiency of different streptogramins

Besides PI several commercially available streptogramin sources were tested for their potential to regulate the PIT/P_{PIR} system (Fig. 3). Among available streptogramins, virginiamycin proved to be the most efficient regulating agent, showing an induction factor of up to 32. Least efficient was Synercid, displaying almost 4-fold lower regulation potential than Pyostacin. Since this mammalian regulation system is responsive to a wide varitey of streptogramins in the ng/ml range, this regulation concept represents also a very sensitive tool for the detection of streptogramin antibiotics and is likely to detect novel streptogramin antibiotics in the culture supernatants of *Actinomycetes* and fungi. Besides being at least one order of magnitude more sensitive than classical antibiotic detection concepts using indicator bacteria, the PIT/P_{PIR} system detects due to its mammalian setup predominantly non-cytotoxic and bioavailable antibiotics.

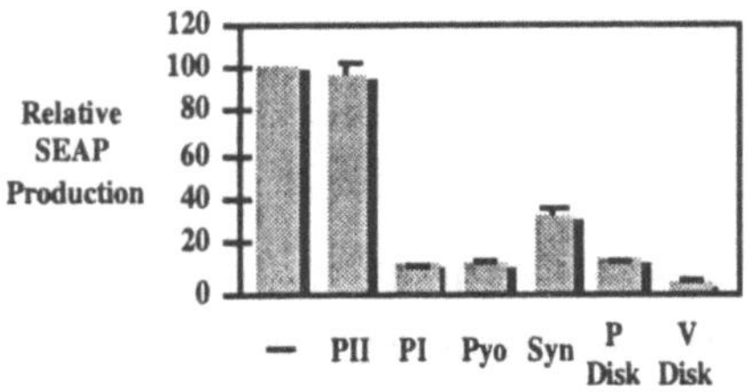

Fig. 3

Fig. 3: Regulation potential of different streptogramins. CHO-PIT1 was transiently transfected with the SEAP-encoding plasmid pMF172 and grown for 48 h in the absence or presence of different streptogramins at a concentration of 2 µg/ml of their group B component. (Pyo, Pyostacin; Syn, Synercid; P-Disk, PI eluted from antibiogram discs; V-disk, Virginiamycin eluted from antibiogram discs).

4. Comparison and compatibility of streptogramin- and tetracycline-based gene regulation systems

The regulation performance of the PI-based regulation system was compared directly with the widely used TET-responsive expression concept. pTWIN1 encoding both PIT and the tet-dependent transactivator in a single consitutive expression unit was stably established in CHO-K1 cells. Two clones CHO-TWIN1$_{95}$ and CHO-TWIN1$_{108}$ were transfected with P_{PIR}- and P_{TET}-driven SEAP expression constructs and assessed for individual regulation potential of these systems. As shown in Table 1, the regulatory characteristics of the streptogramin system are superior to the TET system in CHO cells using virginiamycin as regulating agent. However, more important than the individual performance of these systems is the observation that both of them function independently in the same cell. This compatibility of the PIT/P_{PIR} and the TET systems is expected to enable sophisticated metabolic engineering strategies in which two sets of transgenes can be regulated separately and to advance future tissue engineering and gene therapy concepts which require divergent regulation of proliferation control.

		SEAP Production (%)	
		CHO-TWIN1$_{95}$	CHO-TWIN1$_{108}$
P_{PIR}-SEAP	No inducer	100	100
	+ PI	8.84 ± 0.01	2.2 ± 0.4
	+ Vir	3.75 ± 0.39	1.2 ± 0.41
P_{TET}-SEAP	No inducer	100	100
	+ TET	8.57 ± 0.37	7.56 ± 0.01

Table 1: Comparison of the tetracycline- and streptogramin gene regulation systems

FEEDBACK CONTROL OF REDOX POTENTIAL IN HYBRIDOMA CELL CULTURE

A. MENESES, A. GOMEZ, AND O.T. RAMIREZ*
Instituto de Biotecnología
Universidad Nacional Autónoma de México
A.P. 510-3, Cuernavaca, Morelos, 62250 México

Abstract

Culture redox potential (CRP) is an easily measured variable and its utility in hybridoma culture has been recently shown. Nevertheless CRP is scarcely measured as a routine variable in cell culture and very few reports exist of its proper control. In this work we describe, for the first time, a proportional feedback control algorithm based on manipulation of oxygen partial pressure for maintaining CRP at a constant and predetermined value in hybridoma cultures. A broad spectrum of reducing and oxidizing conditions, in the range of -140 mV to 100 mV with respect to the initial CRP, was evaluated by maintaining CRP constant. Cultures were performed in 1-Lt instrumented and computerized stirred bioreactors operated in batch mode. It was observed that cell and MAb concentrations, and specific growth and thiol production rates increased with decreasing CRP. Maximum concentration of ammonium and lactate remained constant at all CRP values tested. Apoptotic cell death was present in the early stage of cultures at oxidative conditions, whereas at a reduced environment it was triggered only until nutrient depletion during late culture stages. Accordingly, oxidative stress was found to be an inducer of apoptosis. The results of this study show that CRP control in animal cell cultures can be exploited to increase productivity, opening a new way to optimize hybridoma cultures.

1. Introduction

The necessity to improve productivity in animal cell cultures has led to the development and design of new strategies, including the improvement of cell lines by genetic modification, design of new bioreactors and agitation systems, establishment of novel control systems, and the analysis of different culture parameters. Culture redox potential (CRP) is an easily measured parameter [2] whose total potential in cell culture has not been fully exploited. As reviewed by Kjaergaard [11], most reports of CRP in biotechnology have focused on microbial fermentations and soil science. CRP can provide valuable information of the status of various cultures and should be considered as a control parameter. Accordingly, strategies based on glucose addition, electric pulses, or agitation manipulation have been proposed to control CRP in microbial cultures [12,14,18].

A. Bernard et al. (eds.), Animal Cell Technology: Products from Cells, Cells as Products, 23–29.

One of the first reports of CRP in animal cell cultures [3] showed that viable cell concentration and growth rate depended on redox potential of the medium before inoculation. Optimum CRP was found to be in the range of -100 to -60 mV, whereas an oxidant medium (+140 mV to +170 mV) yielded the lowest growth rate. Accordingly, adjustment of medium redox potential was suggested as a suitable strategy prior to inoculation. Daniels et al. [4] also observed a decrease in CRP during exponential growth phase of Earle´s "L" cells, but no relationship between these two variables was established.

Various authors have recently shown that hybridoma cell concentration, during exponential growth phase is inversely related to CRP [6, 7, 9, 15]. Therefore, CRP can be used to estimate on-line viable cell concentration. It has been suggested that reduction of cell culture medium by animal cells is due to thiols generation, particularly cysteine, during metabolism [9]. An additional application of CRP measurements in cell culture has been suggested by Higareda et al. [8], who showed that CRP can be used, in combination with oxygen uptake rate (OUR) measurements, to discriminate between metabolic events and operational eventualities. Accordingly, simultaneous measurement of both variables can be used to differentiate glucose or glutamine depletion in hybridoma cultures from mechanical, electrical or instrumentation failures.

In this work we demonstrate the possibility of controlling CRP, at a constant and predetermined value, in hybridoma culture by manipulating oxygen partial pressure through a feed-back control algorithm. To our knowledge, the only attempt to control CRP in cell culture has been by Hwang and Sinskey [9], which was done through the dilution of thiol concentration by the addition of fresh medium to cultures. In the present study, we show the effect of constant CRP on induction of apoptotic death, as well as kinetic and stoichiometric parameters of an hybridoma cell culture.

2. Materials and Methods.

3.1. CELL LINE, CULTURE MEDIUM, AND ANALYTICAL METHODS.

A murine hybridoma (BCF2) cell line that secretes a neutralizing specific monoclonal antibody (MAb) to toxin 2 of scorpion *Centruroides noxius* Hoffman was used in this study [8]. The medium used was Dulbecco's modified Eagle's supplemented with 4 g/L glucose, 4 mM L-glutamine, 3.7 g/L $NaHCO_3$, 0.8 mg/L crystalline insulin, 5.5 mg/L sodium pyruvate, 1% nonessential amino acid solution, 1% antibiotic antimycotic solution and 10% (v/v) fetal bovine serum.

Viable and total cell concentration and cell size were determined by Trypan blue staining and Coulter Counter measurements, respectively. MAb production was determined by an ELISA sandwich technique. Glucose, glutamine and lactate concentrations were determined in a YSI 2700 analyzer. Ammonium was measured by Kaplan technique [10]. Thiols were determined by Ellman reaction [5] Apoptotic cell death was identified by acridine orange/ethidium bromide staining [13], flow cytometry [1], and the typical ladder pattern in agarose gels [16].

3.2. BIOREACTOR SYSTEM.

A Virtis 1L bioreactor, with a working volume of 700 ml and equipped with a magnetic bar stirrer, was used in this study. Temperature was maintained at 37° C and agitation at 150 rpm. Sterilizable electrodes for dissolved oxygen (DO) (polarographic), pH and CRP (combined platinum-reference) were used. Constant CRP and pH were maintained by an automatic proportional feed-back algorithm that manipulated the individual flows of oxygen, nitrogen and CO_2 through mass flow controllers, while keeping the total gas flow constant at 600 ml/min. Only superficial aeration was employed.

3. Discussion of Results.

A logarithmic relation between CRP and dissolved oxygen concentration, in agreement with Nernst equation [8], was observed. Accordingly, different controller gains (proportional constants) at the various working DO ranges were needed to maintain a suitable CRP control. These constants were calculated using the semi-empirical methods of Zieghler-Nichols and Cohen-Coon [17], and are summarized in Table I. It can be seen that for controlling CRP at reduced values it was necessary to use lower gains due to the stronger influence of DO on CRP. Namely, a small change on DO caused a large variation on CRP, resulting in a very sensitive response even to small variations in oxygen flow rate. In contrast, controlling CRP at oxidative ranges required 10-fold higher gains than those at reducing conditions. In this case, the logarithmic relationship between DO and CRP resulted in the need for higher oxygen flow changes in order to modify CRP.

TABLE I. Proportional control constants for controlling CRP at a fixed and predetermined value.

% DO range	Kc (Ziegler and Nichols)	Kc (Cohen and Coon)
0% - 20%	0.0318	0.0237
0% - 50%	0.0805	0.0626
100% - 200%	1.0013	0.7079

Typical results of the proposed CRP control system for hybridoma culture are shown in Figure 1. In the culture shown, it was possible to maintain CRP at –130 mV ± 5 mV for 106 h. Afterwards, CRP control was lost due to a drastic decrease in OUR upon glutamine depletion and pH control failure. The observed behavior is in agreement with previous reports by Higareda et al. [8]. A similar tight control in CRP using the proposed algorithm was also possible for a broad spectrum of reducing and oxidizing conditions between the range of -140 mV to 100 mV with respect to the initial CRP. A summary of culture parameters at such conditions is shown in Table II. In contrast to the CRP control strategy proposed by Hwang and Sinskey [9], in this case constant CRP was achieved without the need for diluting the culture medium. This allows the possibility of controlling CRP in batch cultures as well as in perfusion or continuous operation modes. In the latter cases, a CRP control independent of dilution rates, is possible.

As shown in Table II, viable cell concentration, MAb concentration and specific growth rate increased with a reducing environment. This is indicative of a deleterious oxidative stress. In addition, maximum thiol production was observed at reduced values, probably due to lower thiol autooxidation rates and low concentration of oxidative species under such conditions. Maximum concentration of lactate and ammonium were not affected by the various CRP values tested. Specific oxygen uptake rate remained relatively constant at around 3×10^{-10} mmol/cel-h for CRP in the range of -140 to 40 mV, but increased by more than 3-fold at higher CRP values.

TABLE II. Effect of CRP on cell culture parameters.

CRP (mV)	Cell conc. (10^6) (Viable Cells/ml)	MAb (mg/L)	Thiols (μM)	μ max (1/h)
-133	1.98	43.46	35.86	0.028
-7	1.16	31.06	29.54	0.026
90	0.43	23.52	20.65	0.021

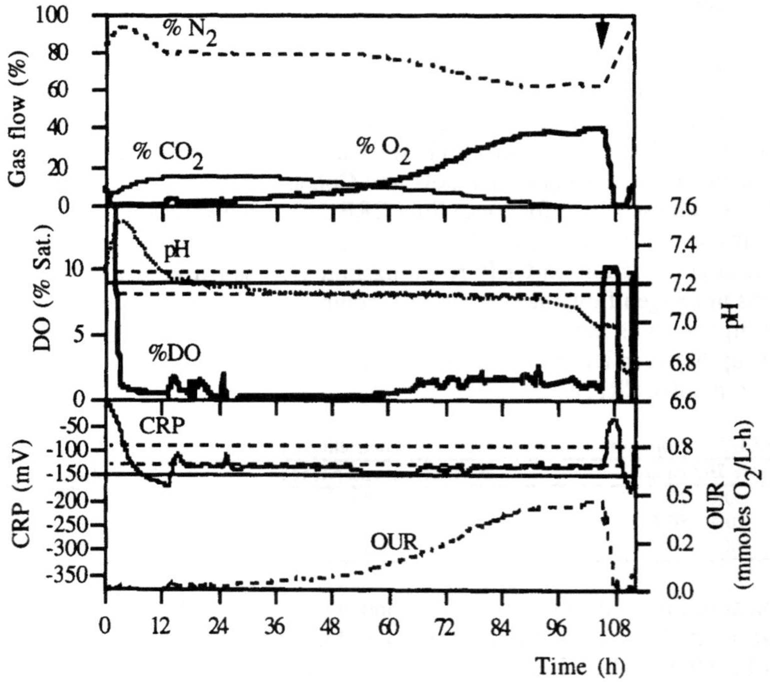

FIGURE 1. – Typical behavior of CRP control . Arrow indicates time of nutrient depletion.

In cultures maintained at reduced conditions, programmed cell death (apoptosis) occurred during the late culture stages (after 96 h). In these cases, apoptosis induction correlated with depletion of glutamine. In contrast, oxidative environments induced the development of apoptotic cell death in the early stages of the culture (before 60 h), even in the presence of non-limiting glucose and glutamine concentration and at non-inhibiting concentrations of lactate and ammonia. In these cases, a very long lag phase was observed (120 hrs). Surprisingly, after such long lag periods, cell growth resumed, although lower growth rates and lower maximum cell concentrations were observed. Such a behavior is indicative of a possible adaptation phenomena of hybridoma cells to the highly oxidative conditions.

4. Conclusions

It was shown that it is possible to control CRP at a predetermined and constant value in a range of -140 mV to 100 mV by manipulating the oxygen partial pressure. The proposed algorithm is based on a simple control scheme that can be easily applied to any mammalian cell culture under different operation modes. It was found that CRP strongly affects the main hybridoma cell culture parameters, including induction of apoptotic death. These results show that CRP control in animal cell cultures can be exploited to increase productivity and opens novel possibilities to improve hybridoma cultures.

5. Acknowledgments

Support by CONACyT 25164-B. A. M. thanks support CONACyT 93820, DGEP-UNAM and PAEP-202349.

6. References

1. Coligan, J., Krulsbeek, A. M., Marguiles, D.H. Ed. Related isolation procedures and functional assays. Morphological, biochemical and flow cytometric assays of apoptosis. Current Protocols of Immunology. John Wiley and Sons Inc. USA, supplement 16 CPI.
2. Dahod, S.K. (1982). Redox potential as a better substitute for dissolved oxygen in fermentation process control. Biotechnology and Bioengineering, 24, 55-56.
3. Daniels, W.F., Garcia, H., Rosensteel, J.F. (1970a). The relationship of oxidation-reduction potential to the growth performance of tissue culture media poised prior incubation. Biotechnology and Bioengineering, 12, 409-417.
4. Daniels, W.F., Garcia, H., Rosensteel, J.F. (1970b). Oxidation-reduction potential and concomitant growth patterns of cultures of Earle´s "L" cells in centrifuge bottle spinners. Biotechnology and Bioengineering, 12, 419-428.
5. Ellman, G.L. (1959). Tissue sulfhydryl groups. Archives of Biochemistry and physics, 82, 70-77.
6. Eyer, K., Hienzle, E. (1996). On-line estimation of viable cells in a hybridoma culture at various DO levels using ATP balancing and redox potential measurement. Biotechnology and Bioengineering, 49, 277-283.
7. Griffiths, B. (1984). The use of oxidation-reduction potential (ORP) to monitor growth during a cell culture. Develop. Biol. Standard. 55, 113-116.
8. Higareda, A.E., Possani, L.D., Ramírez, O.T. (1997). The use of culture redox potential and oxygen uptake rate for assessing glucose and glutamine depletion in hybridoma cultures. Biotechnology and Bioengineering, 56, 554-563.
9. Hwang, Ch., Sinskey, A.J. (1991). The role of oxidation-reduction potential in monitoring growth of cultured mammalian cells in Spier, R.E., Griffiths, J.B. and Meignier, (eds.), Production of biologicals from animal cells in culture. B. Halley Court, Oxford, pp. 548-567.
10. Kaplan, A. (1965). Urea, nitrogen and urinary ammonia. Stand. Methods Clin. 5, 245 - 256.
11. Kjaergaard, L. (1977). The redox potential: its use and control in biotechnology. Advances in Biochemical Engineering, 7, 131-150.
12. Kjaergaard, L., Joergensen, B.B. (1979). Redox potential as a state variable in fermentation systems. Biotechnology and Bioengineering Symp., 9, 85-94.
13. Mercille, S., Massie, B. (1994). Induction of apoptosis in nutrient-deprived cultures of hybridoma and myeloma cells. Biotechnology and Bioengineering, 44, 1140-1154.
14. Oh, D.K., Kim, S.Y., Kim, J.H. (1998). Increase of xylitol production rate by controlling redox potential in Candida parapsilosis. Biotechnology and Bioengineering, 58, 440-444.
15. Plushkell, S.B., Flickinger, M.C. (1996). Improved methods for investigating the external redox potential in hybridoma cell culture. Cytotechnology, 19, 11-26.
16. Smith, Ch., Williams, G.T. (1989). Antibodies to CD3/T-cell receptor complex induce death by apoptosis in immature T cells in thymic cultures. Nature, 337, 181-184.
17. Stephanopoulos, G. (1984). Chemical process control. Pretince Hall Inc. New Jersey. Pp. 352 - 355.
18. Thompson, B.G., Gerson, D.F. (1985). Electrochemical control of redox potential in batch cultures of Escherichia coli. Biotechnology and Bioengineering, 27, 1512-1515.

Discussion (Ramirez)

Naveh: How did you measure redox potential?

Ramirez: With a combined platinum/silver chloride electrode. The important factor is how you calibrate it, which is tricky. We use the initial redox potential of the medium as a reference for calibration. The probe equilibrates for several hours in the medium at a steady state at the desired temperature and initial pH.

Piret: Most of us, I expect, are used to controlling pH and O_2, but letting the redox run free. I wonder how much you felt yourself constrained by controlling the redox and having to lower, for example, your oxygen tension to such low levels? What were the penalties of this approach?

Ramirez: In the most highly reducing culture we had zero dissolved oxygen in the medium, however we still had a very high partial pressure of oxygen in the overlay gas phase. It was interesting that even at this very low dissolved oxygen, our cultures performed very similarly to the control cultures maintained at 50% dissolved oxygen. So, in this case, there was no negative price. On the other hand, in highly oxidising cultures we paid a price as we observed a drastic decrease. We are now looking to see if we can have a low dissolved oxygen concentration, and manipulate it in higher oxidising environments with various agents which I showed in my talk.

Noé: I did not see any beneficial effects on titre in your presentation.

Ramirez: We had small variations in monoclonal antibody production - a 2-fold increase from the very highly oxidising to the highly reducing conditions. Compared to the control, there was a slight increase in antibody which could be within the limits of the ELISA technique. We were not interested in optimising the culture, but rather in studying the effect of redox potential with regard to kinetic and stoichometric parameters.

Genetic Manipulation of the Protein Synthetic Capacity of Mammalian Cells

Michèle Heaton[1], *John Birch*[2], *Alison Hovey*[2], *Robert Kallmeier*[2], *Chris Proud*[3], *David James*[1]

[1] *Research School of Biosciences, University of Kent, Canterbury CT2 7NJ, U.K.*
[2] *Lonza Biologics plc, 228 Bath Road Slough, Berkshire SL1 4DY, U.K.*
[3] *Dept. of Anatomy & Physiology, Medical Sciences Institute, University of Dundee, DD1 4HN, U.K.*

1. Introduction

Major improvements in the yield of recombinant proteins produced by animal cell cultures are likely to result from an extension of productive cell lifetime. Genetic engineering of cell cycle and death mechanisms and /or improved fed-batch processes have been used to enhance cell productivity. In contrast, by direct genetic manipulation of the rate of protein synthesis in animal cells we intend to augment both cell specific production, and productive cell lifetime in batch culture. This approach also provides a means to increase the transient expression of recombinant proteins by 'transient' host cell engineering. mRNA translation initiation factors control the overall rate of protein synthesis. The phosphorylation state of a subset of these factors; 4E-BP1, eIF2B, eIF4G, eIF2 and eIF4E are important in the regulation of protein synthesis (Figure 1.). An increase in the phosphorylation state of the α subunit of eIF2 is associated with an inhibition of protein synthesis in response to stimuli such as heat shock (Duncan & Hershey, 1987), amino acid, glucose, or serum deprivation (Scorsone *et al.*, 1987). The impairment of translation seen in response to heat shock is partially overcome by expression of a non-phosphorylatable mutant of eIF2α, Ser51Ala (Murtha-Riel *et al.*, 1993). Thus, reduced rates of protein synthesis seen when cells encounter other such conditions are also expected to be attenuated by expression of this mutant, which is otherwise functional in translation.

A. Bernard et al. (eds.), Animal Cell Technology: Products from Cells, Cells as Products, 31–33.

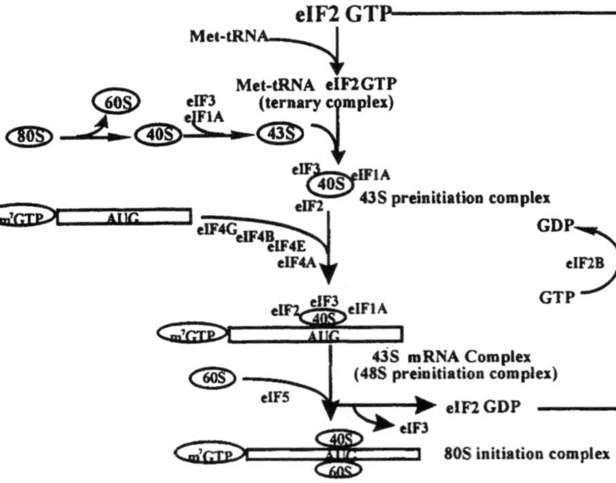

Fig. 1: Mechanism of Initiation of Protein Synthesis. (Adapted from Pain, 1996)

2. Results and Discussion

2.1 CONSTRUCTION OF DOUBLE GENE CONSTRUCTS

eIF2α Ser[51] was mutated to an alanine residue by site-directed mutagenesis. Double gene constructs were created by cloning eIF2αSer[51] and eIF2αAla[51] cDNA into the mammalian expression vector pEE14.1 (Lonza Biologics). The luciferase reporter gene (Promega) was cloned into the mammalain expresion vector pEE6.1 (Lonza Biologics). A Not I / Bam HI restriction digest was performed on both vectors and the fragment containing the hCMV-MIE promoter and luciferase gene was ligated into the pEE14.1 vector backbone (Figure 2.).

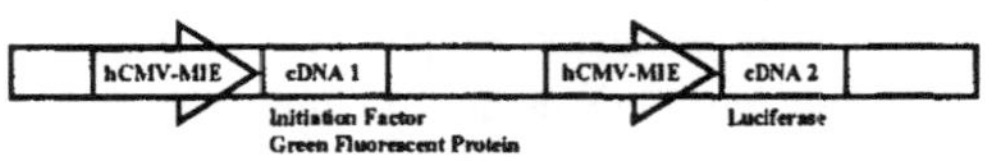

Fig. 2: Schematic diagram of cDNA configuration used for double gene expression.

2.2 PHOSPHORYLATION STATE OF eIF2α DURING BATCH CULTURE

The phosphorylation state of eIF2α was investigated during a 500ml spinner batch culture of CHO cells (Figure 3.).

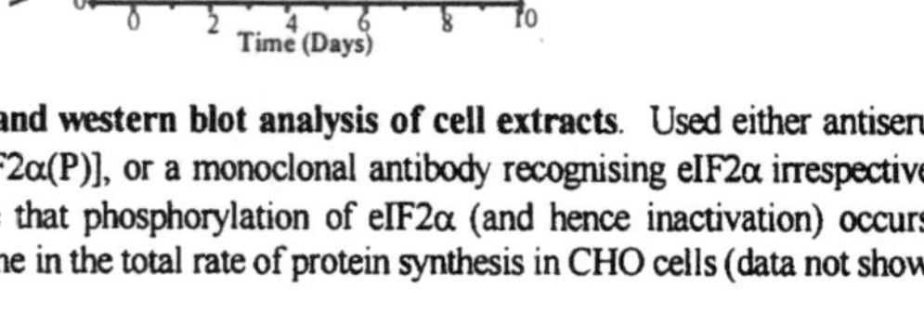

Fig. 3: Batch grwoth of CHO cells and western blot analysis of cell extracts. Used either antiserum specifically recognising phosphorylated eIF2α [eIF2α(P)], or a monoclonal antibody recognising eIF2α irrespective of its state of phosphorylation. These data indicate that phosphorylation of eIF2α (and hence inactivation) occurs late in batch culture. This is associated with a decline in the total rate of protein synthesis in CHO cells (data not shown).

2.3 LUCIFERASE EXPRESSION TIME COURSE

A mammalian expression vector containing luciferase was transfected into CHO cells at various concentrations using electroporation in Optimix® buffer (Flowgen). Luciferase expression was determined at different time points after transfection using Luclite™ kit (Packard) (Figure 4.).

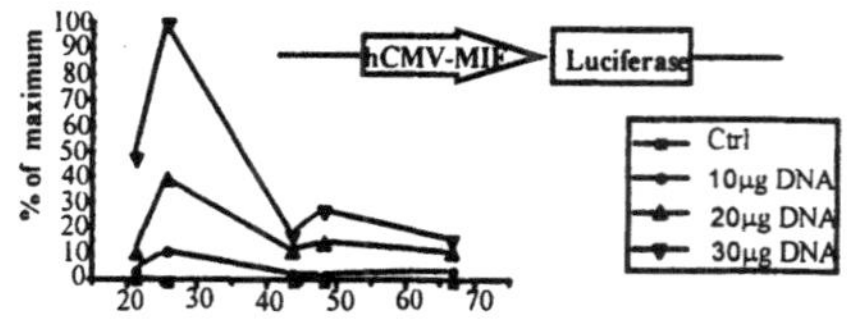

Fig. 4: Dose dependent luciferase activity. Maximum reporter gene expression occurring at 24 h post transfection

2.4 TRANSIENT CO-EXPRESSION OF LUCIFERASE REPORTER WITH WILD-TYPE AND MUTATED eIF2α IN CHO CELLS

CHO cells were transfected with 10μg of each of the following DNA: pEE14 Luc or pEE14.1 2α WT/Luc or pEE14.1 2α Ala/Luc (Figure 5.). The experiment was then repeated using mammalian expression vector containing pEE14 EGFP/Luc or pEE14.1 2α WT/Luc or pEE14.1 2α Ala/Luc (Figure 6.).

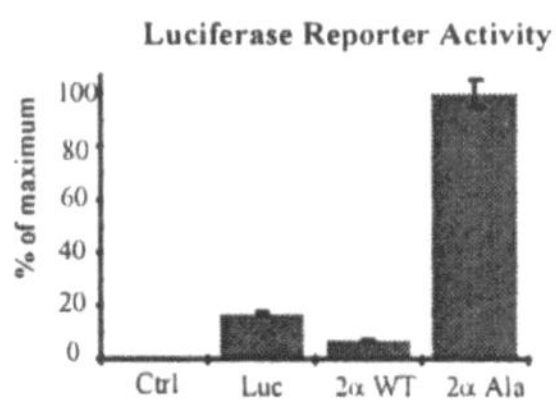

Fig. 5: % of maximum luciferase activity 24 h post transfection. The data was obtained from 9 separate experiments.

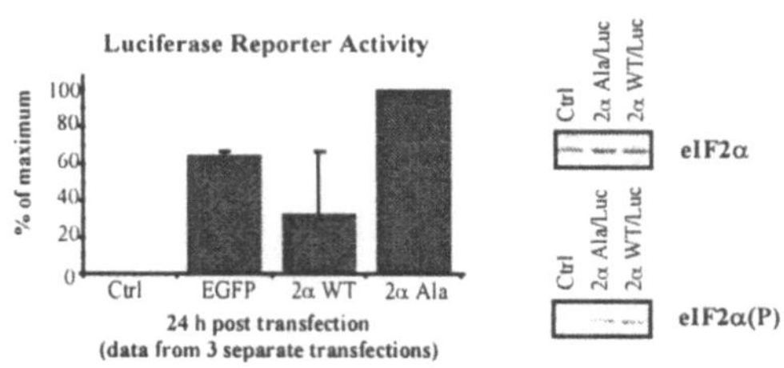

Fig. 6: % of maximum luciferase activity 24 h post transfection. The data was obtained from 3 separate Experiments. Western blot analysis of eIF2α expression and eIF2α(P) expression in cell extracts.

3. Conclusion

This technique provides an effective means of cell engineering for improved stable productivity **and** transient expression.

4. References and Acknowledgements

Duncan R. & Hershey J. W. (1984) Heat-shock-induced translational alterations in HeLa cells. *J. Biol. Chem.* **259**, 11882-11889.

Murtha-Riel P., Davies M. V., Scherer B. J., Choi S-Y, Hershey J. W. B. & Kaufman R. J. (1993) Expression of a phosphorylation resistant eukaryotic initiation factor 2α subunit mitigates heat shock inhibition of protein synthesis. *J. Biol. Chem.* **268**, 12946-12951.

Pain V. M. (1996) Initiation of protein synthesis in eukaryotic cells. *Eur. J. Biochem.* **236**, 747-771.

Pause A., Belsham G. J., Gingras A. C., Donze O., Lin T-A., Lawrence J. & Sonenberg N. (1994) Insulin-dependent stimulation of protein synthesis by phosphorylation of a regulator of 5'-cap function. *Nature.* **371**, 762-767.

Scorsone K. A., Panniers R., Rowlands A. G. & Henshaw E. Ç. (1987) Phosphorylation of eukaryotic initiation factor 2 during physiological stresses which affect protein synthesis. *J. Biol. Chem.* **262**, 14538-14543.

I would like to thank Lonza Biologics for my case award and the BBSRC for sponsoring this project.

RETINOIC ACID ENHANCES MONOCLONAL ANTIBODY PRODUCTION OF HUMAN-HUMAN HYBRIDOMA BD9

Y. INOUE[1], M. FUJISAWA[2], M. SHOJI[3], S. HASHIZUME[3], Y. KATAKURA[2] AND S. SHIRAHATA[2]

[1]*Department of Biochemical Science and Technology, Faculty of Agriculture, Kagoshima University, 1-21-24 Korimoto, Kagoshima 890-0065, Japan;* [2]*Graduate School of Genetic Resources Technology, Kyushu University, 6-10-1 Hakozaki, Higashi-ku, Fukuoka 812-0053, Japan;* [3]*Morinaga Institute of Biological Science, 2-2-1 Shimosueyoshi, Tsurumi-ku, Yokohama 230-0012, Japan*

Abstract The enhancement of human monoclonal antibody (hMAb) production by retinoic acid (RA) was evaluated using the human-human hybridoma cell line BD9 cultured in serum-free medium. The hMAb production of BD9 hybridomas was enhanced up to eightfold by stimulation with 10^{-7} M of RA. Northern blot analysis showed that mRNA levels of hMAb were significantly increased by RA when compared with control without RA. The comparison between intracellular and extracellular hMAb contents by immunoblot analysis showed that hMAb secretion of BD9 hybridomas would be accelerated by RA. These results provide direct evidence for enhancement of hMAb production by RA, suggesting that RA may be effective for mass production of hMAb using BD9 hybridomas.

1. Introduction

Lung cancer specific human monoclonal antibody (hMAb) BD9D12 produced by the human-human hybridoma cell line BD9 has been shown to be useful for the immunocytochemical detection of lung adenocarcinomas (Shoji *et al.*, 1994). However, low productivity of human-human hybridomas remains a major obstacle for mass production of hMAbs. Aotsuka *et al.* (1991) reported that retinol and retinoic acid (RA) enhanced antibody production of some human-human hybridoma cell lines. We have also shown that retinyl acetate (RAc), a storage form of retinol, would be effective for middle scale production of hMAbs by human-human hybridomas (Inoue *et al.*, in press). However, there have been few reports describing how retinoids such as RA and RAc enhance hMAb production of human-human hybridomas. In the present study, we gave attention to the enhancement process of hMAb production by RA using BD9 hybridomas.

A. Bernard et al. (eds.), Animal Cell Technology: Products from Cells, Cells as Products, 35–37.

2. Materials and Methods

2.1. CELLS AND CELL CULTURE

A human-human hybridoma cell line BD9 was generated by fusing peripheral blood lymphocytes from a healthy adult with the T lymphoblastoid cell line A_4H_{12} derived from Molt4 cells using an *in vitro* immunization method (Kawahara *et al.*, 1992). Cells were maintained in ERDF medium (Kyokuto Pharmaceutical Industrial. Co., Tokyo, Japan) supplemented with 5 μg/ml insulin, 20 μg/ml human transferrin, 20 μM ethanolamine and 25 nM sodium selenite (ITES-ERDF), at 37°C in humidified 5% CO_2/95% air.

2.2. TREATMENT OF CELLS WITH RA

all-*trans*-RA (Wako Pure Chemical Industries Co., Osaka, Japan) was dissolved in ethanol, diluted in the culture medium and added to the cell cultures immediately after plating. After culture for 4 days, antibody concentration in the spent medium was measured by an enzyme-linked immunosorbent assay as described previously (Shoji *et al.*, 1994), using anti-human IgG antibody (#4100; TAGO, Burlingame, USA) as the first antibody, and anti-human immunoglobulin peroxide conjugate antibody (#2390; TAGO) as the second antibody. Cell number was counted by using a cell counter, and viability was determined by the trypan blue dye exclusion method.

2.3. NORTHERN BLOT ANALYSIS

Total RNA was recovered from cells using TRIZOL reagent (GIBCO BRL, Tokyo, Japan). Two micrograms of RNA were loaded on 1.5% agarose-formaldehyde gel and the separated RNA was transferred onto Hybond N+ membrane (Amersham, Buckinghamshire, UK). Northern blot analysis was performed by using the *Gene Images* random prime labelling module (Amersham) with the variable region specific probe for BD9D12 antibody κ or γ chain.

3. Results and Discussion

The hMAb production of BD9 hybridomas was enhanced up to eightfold by stimulation with 10^{-7} M of RA (Fig. 1a). The stimulation with RA was more efficient for hMAb production than that with RAc, which enhances about fivefold hMAb production of BD9 hybridomas at the concentration of 10^{-5} M. Northern blot analysis showed that mRNA levels of hMAb were significantly increased by RA when compared with control without RA (Fig. 1b). RA have been shown to enhance the expression of many genes (Vu Dac *et al.*, 1996; Minucci *et al.*, 1997). Our findings further expanded these facts. On the other hand, the comparison between intracellular and extracellular hMAb contents by immunoblot analysis showed that hMAb secretion of BD9 hybridomas would be accelerated by RA (data not shown). These results could provide direct evidence for enhancement of hMAb production by RA. Ballow *et al.* (1996) reported that the enhancing effects of RA on hMAb synthesis by Epstein-Barr virus transformed B cells were mediated, at least in part, by production of interleukin-6 (IL-6). In our experiment system using human-human hybridomas, however, the mediation of IL-6 was still not confirmed. This study suggest that RA may be effective as a culture additive to obtain a large amount of hMAb using BD9 hybridomas. Further study on the practical use of RA for hMAb production is needed.

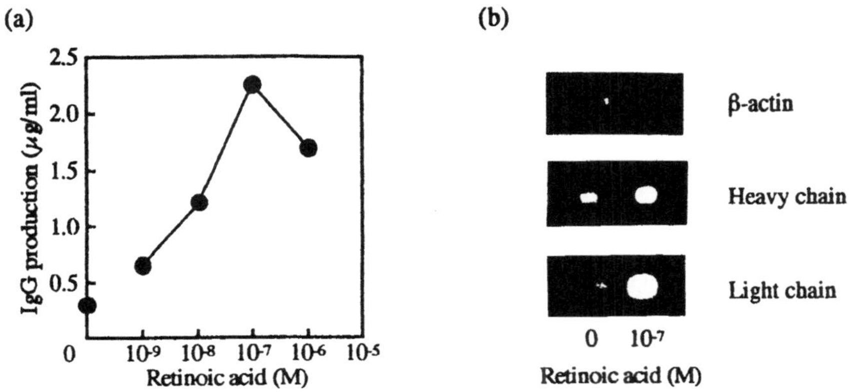

Fig. 1 Enhancing effect of RA on hMAb production of BD9 hybridomas. (a) dose response curve, (b) Northern blot analysis.

4. References

Aotsuka, Y. and Naito, M. (1991) Enhancing effects of retinoic acid on monoclonal antibody production of human-human hybridomas, *Cell. Immunol.* **133**, 498-505.

Ballow, M., Xiang, S., Wang, W. and Brodsky, L. (1996) The effects of retinoic acid on immunoglobulin synthesis: role of interleukin 6, *J. Clin. Immunol.* **16**, 171-179.

Inoue, Y., Fujisawa, M., Kawamoto, S., Shoji, M., Hashizume, S., Fujii, M., Katakura, Y. and Shirahata, S. (in press) Effectiveness of vitamin A acetate for enhancing the production of lung cancer specific monoclonal antibodies, *Cytotechnology*.

Kawahara, H., Shirahata, S., Tachibana, H. and Murakami, H. (1992) *In vitro* immunization of human lymphocytes with human lung cancer cell line A549, *Hum. Antibod. Hybridomas* **3**, 8-13.

Minucci, S., Leid, M., Toyama, R., Saint Jeannet J.P., Peterson, V.J., Horn, V., Ishmael, J.E., Bhattacharyya, N., Dey, A., Dawid, I.B. and Ozato, K. (1997) Retinoid X receptor (RXR) within the RXR-retinoic acid receptor heterodimer binds its ligand and enhances retinoid-dependent gene expression, *Mol. Cell. Biol.* **17**, 644-655.

Shoji, M., Kawamoto, S., Sato, S., Kamei, M., Kato, M., Hashizume, S., Seki, K., Yasumoto, K., Nagashima, A., Nakahashi H, Suzuki T, Imai T, Nomoto K and Murakami H (1994) Specific reactivity of human antibody AE6F4 against cancer cells in tissues and sputa from lung cancer patients, *Hum. Antibod. Hybridomas* **5**, 116-122.

Vu Dac, N., Schoonjans, K., Koskyh, V., Dallongeville, J., Heyman, R.A., Stael, B. and Auwerx, J. (1996) Retinoids increase human apolipoprotein A-II expression through activation of the retinoid X receptor but not the retinoic acid receptor, *Mol. Cell. Biol.* **16**, 3350-3360.

EXTRACELLULAR ADENOSINE 5' MONOPHOSPHATE CONTROLS PROLIFERATION AND STIMULATES PROTEIN PRODUCTION OF rCHO CELLS

Hendrik P.J. Bonarius*, Karen Hansen, and Leif Kongerslev.

Biologics Development, Novo Nordisk A/S, Gentofte, Denmark
* *Corresponding author, fax ++45-4443 9210, e-mail bona@novo.dk*

Extracellular AMP inhibits cell growth while at the same time stimulates protein and RNA synthesis (Hugo et al. , 1992, *J. Cell. Phys.* 153: 539). This principle is applied to establish a two-phase production process for recombinant prethrombin. It is shown that the addition extracellular AMP doubles prethrombin yields and triples cell-specific productivity compared to control cells. Intracellular nucleotide and cyclic AMP measurements together with cell-cycle analyses suggest that growth control by AMP is firstly a result of pyrimidine starvation, and subsequently a result of cyclic-AMP mediated G1 arrest.

1. AMP controls proliferation and stimulates protein production in a concentration-dependent manner

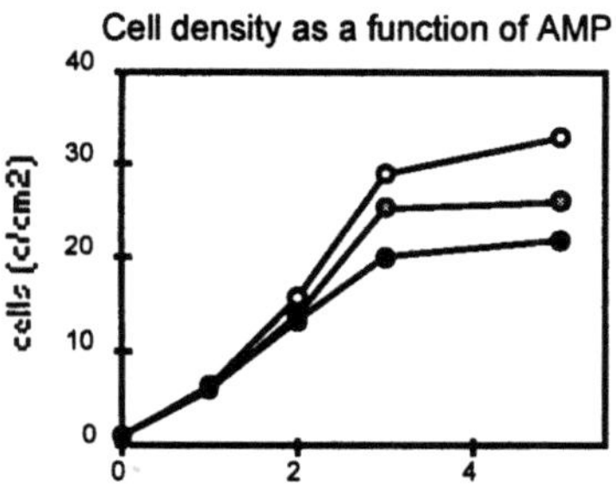

Fig. 1. CHO cells were grown in T-flasks in serum-containing medium and AMP was added at day 1 in different concentrations. After 2 days, the cell densities in t-flasks with higher AMP levels is significantly lower. White, grey, and black symbols indicate 0, 2 and 5 g/L AMP respectively.

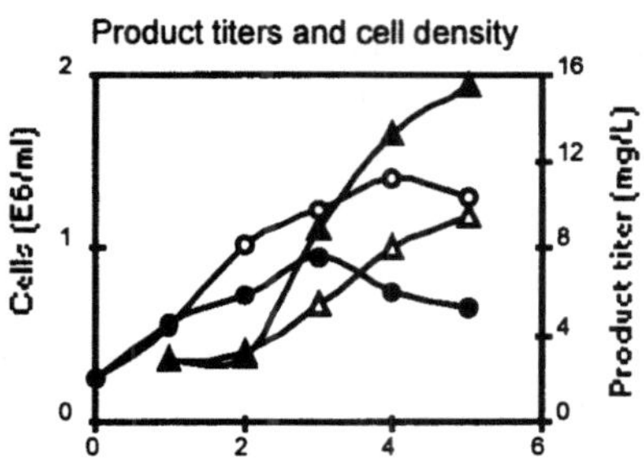

Fig. 2. CHO cells were grown in spinner flasks in serum-free culture medium and AMP was added at day 1. Compared to the control culture, the cell density remained low. In contrast, the prethrombin titer is significantly higher. The cell-specific productivity increases > 3-fold. Circles: Cell Density; Triangles: Product concentration. Closed symbols: AMP addition, Open symbols: control.

A. Bernard et al. (eds.), Animal Cell Technology: Products from Cells, Cells as Products, 39–41.

2. Intracellular AMP levels increase rapidly, followed by a delayed accumulation of intracellular cyclic AMP

CHO THR 101 cells were grown in spinner flasks. At day 1 AMP was added and samples were taken for the analysis of intracellular nucleotides. Four hours after the addition, the intracellular AMP concentration is > 50 fold compared to the control cells (Figure 3). At day 2, most of the extra AMP disappears. In contrast, the intracellular *cyclic* AMP increase is delayed. Cyclic AMP accumulates only at day 2 and 3.

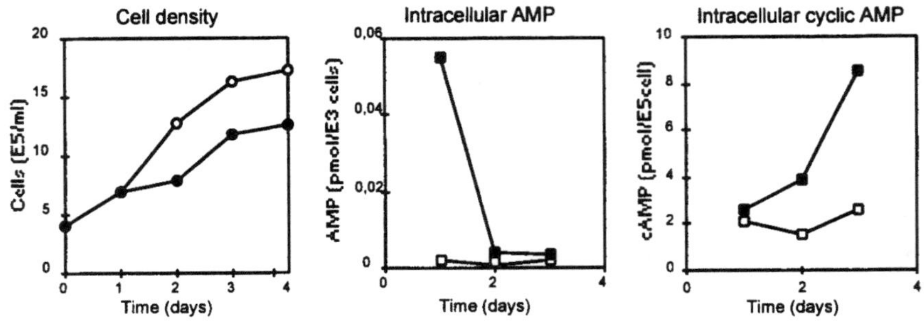

Fig. 3. Cell density, AMP and cyclic AMP pools. Open symbols: Control, Closed symbols: AMP addition.

In addition to AMP and cyclic AMP, other nucleotides were measured (data not shown here). Four hours after AMP addition, intracellular UTP and CTP pools decreased compared to control values. At day 2, the concentrations of these pyrimidines were similar to the control. Together with the intracellular AMP data shown above, these data suggest that during the first day after addition of AMP, growth is inhibited by pyrimidine starvation.

3. Cell-cycle analysis shows that CHO cells are arrested in G1 phase 24 hours after AMP addition

Figures 4 and 5 show the result of cell-cycle distribution as determined by FACS. AMP was added at day 1 and samples for cell-cycle analysis were taken 0, 2, 6 and 24 hours after AMP addition. 2 and 6 hours after the addition of AMP there is no difference in cell cycle phase between the control AMP-stressed cells (Fig. 4). Only after 24 hours the cell-cycle distribution of the AMP-stressed cells is different from the control as shown here by the FACS diagram (Fig. 5). The increase in G1 occurs simultaneously to the increase in intracellular cyclic AMP.

These data suggest that growth control by AMP occurs via different mechanisms: Within the first 24 hours after AMP addition, an increase in intracellular

AMP results in a general growth inhibition, which is not specific for any phase of the cell cycle and most likely a result of pyrimidine starvation. This is followed by G1-arrest associated with a sharp increase of the cyclic AMP pool.

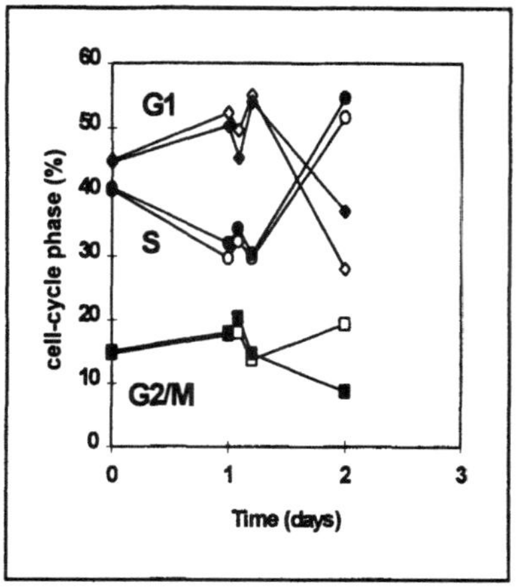

Fig. 4. Cell-cycle distribution. Open symbols: Control, Closed symbols: AMP addition.

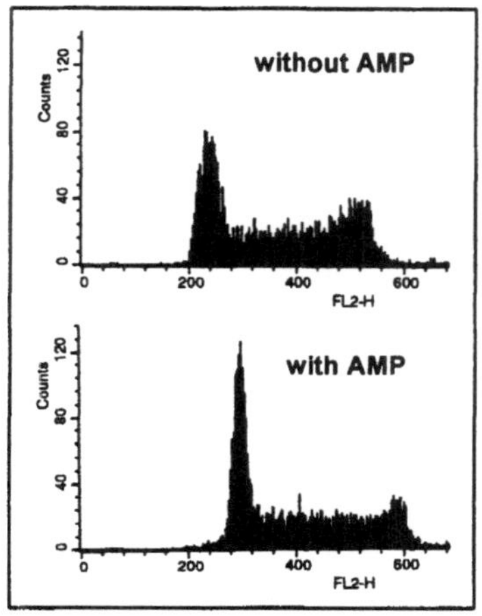

Fig. 5. FACS diagram taken at day 2: 24 hours after addition of AMP.

Conclusions

1. Prethrombin yields produced by rCHO cells increase 2-fold in batch culture after addition of the growth inhibitor AMP. The cell-specific productivity of increases 300 %.

2. Cell-cycle analyses and nucleotide measurements suggest that growth is first inhibited as a result of pyrimidine starvation. Approximately 24 hours after AMP addition, proliferation is controlled by a cylic-AMP mediated G1 arrest.

Acknowledgments

We acknowledge Charlotte Bertelsen and Louise Jepsen (Dept. of Cell Biology, Novo Nordisk) for technical assistance, Dr. Roland Wagner (GBF Braunschweig, Germany) for the analyses of nucleotide pools, and Dr. Carsten Stidsen (Dept. of Molecular Pharmacology, Novo Nordisk) for the analyses of cyclic AMP.

THE MODE OF ACTIONS OF LYSOZYME AS AN IMMUNOGLOBULIN PRODUCTION STIMULATING FACTOR (IPSF)

T. SUGAHARA, F. MURAKAMI, AND T. SASAKI
College of Agriculture, Ehime University.
3-5-7 Tarumi, Matsuyama, Ehime 790-8566, Japan.

1. Abstract

Lysozyme derived from hen egg white stimulated immunoglobulin (Ig) production by human-human hybridoma, HB4C5 cells producing monoclonal IgM. IgM production by HB4C5 cells was enhanced more than 13-fold by lysozyme at 380 μg/ml in serum-free medium. The mode of actions of lysozyme as an Ig production stimulating factor (IPSF) was investigated. Lysozyme enhanced Ig production by transcription-suppressed HB4C5 cells. However, the enzyme was ineffective to accelerate IgM production by translation-suppressed HB4C5 cells. In addition, the intracellular IgM content of HB4C5 cells, which were suppressed post-transcription process, was obviously increased by the addition of lysozyme. These findings suggest that lysozyme accelerates translation process to enhance Ig productivity.

2. Introduction

We have attempted to stimulate Ig production of hybridomas and lymphocytes by modification of culture media supplements under serum-free conditions for effective mass production. Then, we screened the Ig production stimulating factor (IPSF) [1, 2]. As the result of screening, it was revealed that several basic proteins and poly-basic amino acids facilitated Ig production by human-human hybridoma, HB4C5 cells [3]. These findings urged us to inquire IPSF activity of other basic proteins. Finally, we found out that lysozyme had the IPSF activity against human-human hybridoma and human peripheral blood lymphocytes [4]. Lysozyme is a very simple and stable protein, and easily separated from antibodies in culture medium by gel filtration. Moreover, this enzyme is so cheap that we can use it for mass production of monoclonal antibodies. Therefore, we investigated the IPSF activity of lysozyme.

3. Materials and method

Human-human hybridoma HB4C5 cells producing monoclonal IgM were used for the assay of the IPSF activity of lysozyme from hen egg white. HB4C5 cells were fusion

A. Bernard et al. (eds.), Animal Cell Technology: Products from Cells, Cells as Products, 43–45.

product of a human lymphocyte from lung cancer patient and a human fusion partner, NAT-30 cells. HB4C5 cells were cultured in ERDF medium (Kyokuto Pharmaceutical, Japan) supplemented with 10 μg/ml of insulin, 20 μg/ml of transferrin, 20 μM ethanolamine and 25 nM selenite (ITES-ERDF) at 37 °C under humidified 5% CO_2-95% air. The IPSF activity was determined by measuring the amount of IgM secreted by HB4C5 cells in culture media. HB4C5 cells were inoculated in ITES-ERDF medium containing 380 μg/ml of lysozyme at $1x10^5$ cells/ml. For determination of the IPSF activity of lysozyme, the amount of IgM secreted in each culture medium was measured by enzyme-linked immunosorbent assay (ELISA).

4. Results and discussion

4.1. Effect of lysozyme on IgM production

Human-human hybridoma HB4C5 cells were cultured in ITES-ERDF medium supplemented with lysozyme at various concentrations for 6 h to investigate a dose-response effect of lysozyme. IgM production by HB4C5 cells was stimulated dose-dependently by the addition of lysozyme. Lysozyme facilitates IgM production more than 13-fold at 380 μg/ml in ITES-ERDF medium. The enzyme immediately started to enhance IgM production soon after inoculation, and the effect was maintained for 5 days. Lysozyme, however, showed no significant growth promoting. This suggests that lysozyme stimulates specific IgM productivity of HB4C5 cells.

4.2. Correlation between IPSF and enzymatic activities of lysozyme

The first point for discussion regarding the mode of actions of lysozyme as an IPSF is whether the enzymatic activity takes part in its IPSF activity, or not. The time-course effect of trypsin digestion on the IPSF and enzymatic activities of lysozyme was investigated. Lysozyme was treated with 250 unit/ml of trypsin, and the digestion was terminated by the addition of 5000 units/ml of soybean trypsin inhibitor. As the result of that, the IPSF activity was lost in consequence of fragmentation. On the other hand, the enzymatic activity was stable against trypsin digestion and the fragments fully retained the enzymatic activity. The cleavage sites of trypsin on lysozyme do not affect the active center of the enzyme. Hence, trypsin treatment had no influence on lysozyme activity. Our previous data also indicated that lysozyme, which lost the enzymatic activity by boiling for 30 min, stimulated IgM the same as native enzyme [4]. These results mean that the IPSF activity is not derived from the enzymatic activity. Moreover, this fact suggests that the IPSF and enzymatic activities are independent, and the IPSF activity is a novel function of lysozyme.

4.3. The IPSF effect of lysozyme on transcription-suppressed HB4C5 cells

The IPSF effect of transcription-suppressed HB4C5 cells was investigated. Following actinomycin D (Act D) treatment, HB4C5 cells were cultured in ITES-ERDF medium supplemented with 380 μg/ml of lysozyme, and the amount of IgM in the medium was measured. As the result of that, IgM production of transcription-suppressed HB4C5 cells was enhanced as much as that of control cells by lysozyme. This result indicates that the transcriptional suppression does not influence the IPSF effect of lysozyme at all.

4.4. The IPSF effect of lysozyme on translation-suppressed HB4C5 cells

Then, the IPSF effect of lysozyme on translation-suppressed HB4C5 cells was examined to confirm whether the enzyme stimulates post-transcriptional process, or not. The IPSF effect on HB4C5 cells treated with cycloheximide was investigated. After pre-cultivation with 10 μg/ml cycloheximide for 12 h, the IPSF effect on HB4C5 cells was examined in ITES-ERDF medium supplemented with or without 10 μg/ml of cycloheximide and 380 μg/ml of lysozyme. The result showed that lysozyme did not facilitate translation-suppressed HB4C5 cells.

4.5. The IPSF effect of lysozyme on HB4C5 cells which suppressed the post-translation process.

Monensin was adopted to suppress the post-translational process. After pre-incubation with monensin, HB4C5 cells were inoculated with lysozyme under successive inhibition by monensin. The IPSF effect of lysozyme on HB4C5 cells was obviously suppressed by monensin-treatment. IgM production in the culture medium by monensin-treated cells was stimulated only 2-fold by lysozyme. This means that the IPSF effect was reduced by 1/5. It seems like that lysozyme is ineffective to enhance IgM production by monensin-treated HB4C5 cells. The intracellular IgM was analyzed by staining with FITC conjugated anti-human IgM. HB4C5 cells were stained with FITC-anti human IgM after monensin-treatment. The laser confocal microscopic analysis revealed that the intracellular IgM of HB4C5 cells treated with monensin was obviously increased by lysozyme. These findings mean that IgM synthesis of monensin-treated HB4C5 cells was facilitated by lysozyme, even though IgM secretion was reduced by the drug. According to these results, the fact is that lysozyme stimulates IgM synthesis by accelerating the translation process of hybridomas.

5. References

1. Sugahara, T., Shirahata, S., Akiyoshi, K., Isobe, T., Okuyama, T., and Murakami, H. (1991) Immunoglobulin production stimulating factor-IIα (IPSF-IIα) is glyceraldehyde-3-phosphate dehydrogenase like protein, *Cytotechnology* **6**, 115-120.
2. Sugahara, T., and Sasaki, T. (1998) A novel function of enolase from rabbit muscle; an immunoglobulin production stimulating factor, *Biochim, Biophys. Acta* **1380**, 163-176.
3. Sugahara, T., Sasaki, T., and Murakami, H. (1994) Enhancement of immunoglobulin productivity of human-human hybridoma HB4C5 cells by basic proteins and poly-basic amino acids, *Biosci. Biotech. Biochem.* **58**: 2212-2214.
4. Murakami, F., Sasaki, T., and Sugahara, T. (1997) Lysozyme stimulates immunoglobulin production by human-human hybridoma and human peripheral blood lymphocytes, *Cytotechnology* **24**, 177-182.

A DISSECTION OF RECOMBINANT PROTEIN EXPRESSION IN CHO CELLS INDICATING PROTEIN-SPECIFIC, NOT CELL SPECIFIC, LIMITATION.

K. ALTOBELLO, Y. CHUNG, J. IOVINO, C. LUCHETTE[1], S.OVERTON and MARK CUNNINGHAM
Ares Advanced Technology, 280 Pond Street, Randolph, MA 03268
[1]Millennium Pharmaceuticals, 640 Memorial Drive, Cambridge, MA 02139

1. Abstract

The expression of recombinant Component B (rCB) in Chinese hamster ovary (CHO) cells was investigated. No factor (s) limiting expression were identified when the **cell-specific** parameters of transcription, translation and post-translational modification were investigated. Following this, recombinant growth hormone (rGH) was used as a surrogate marker to demonstrate that the CHO cellular machinery was not limiting the rCB expression. The **protein-specific** nature of rCB expression was then confirmed when it was found that CHO cells expressing rCB had elevated levels of GRP78 suggesting cellular stress from protein misfolding.

2. Introduction

Component B (CB) is a small non-glycosylated protein (8 kDa) containing 5 disulfide bonds, and a tyrosine sulfation site. To evaluate CB as a potential therapeutic, expression in CHO cells was evaluated. However, the highest rCB specific expression level observed, following amplification to 5 µM MTX and cell cloning, was 3 pg/cell-day. This poster details the strategies taken to investigate this low rCB expression.

3. Methods

3.1 DISSECTION OF rCB EXPRESSION

3.1.1 mRNA Steady-state Level

Total RNA was extracted from CHO clones expressing rCB and rHEP (Highly Expressed Protein). Northern blot analysis was performed using a hybrid oligonucleotide probe containing both rCB *and* rHEP sequences (see Fig. 1).

A. Bernard et al. (eds.), Animal Cell Technology: Products from Cells, Cells as Products, 47–49.

3.1.2 Tyrosine Sulfation

CHO cells expressing rCB were grown in a microcarrier-based continuous perfusion system, 5 l scale, until a confluent culture was obtained, the perfusion medium was then supplemented with sodium chlorate after 6 days the inhibitor was removed (see Fig. 1).

3.2 PROTEIN-SPECIFIC EXPRESSION

A clonal CHO cell line already expressing rGH (G418 selection) was supertransfected with an expression vector encoding rCB (DHFR selection). The rCB amplicon was preferentially amplified using MTX until a level of 5μM was reached, during which time the effect of increasing rCB expression upon rGH expression was evaluated (see Fig. 2).

3.3 rCB AND GRP78

CHO cells expressing rCB were sacrificed for total RNA extraction and intracellular protein analysis. GRP78 mRNA and protein levels were investigated using Northern and western blot analysis (blots loaded with equal RNA and protein, see Fig. 3).

4. Results

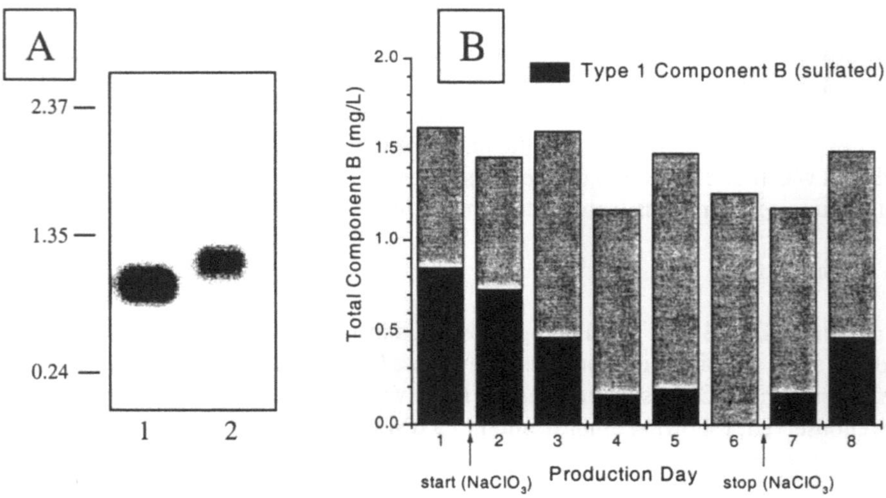

Figure 1. Data from transcription, translation and post-translational modification analysis

Panel A: Northern blot analysis of rCB mRNA and rHEP mRNA (see 3.1.1). *Lane 1* – CHO rCB (1.2 pg/cell-day), *Lane 2* – CHO rHEP (15.0 pg/cell-day)

Panel B: Volumetric productivity of rCB following inhibition of tyrosine sulfation (see 3.1.2).

Figure 1 indicates that the steady state level of rCB mRNA is high and probably not limiting in the expression of rCB protein. In addition, inhibiting tyrosine sulfation did not result in improved expression.

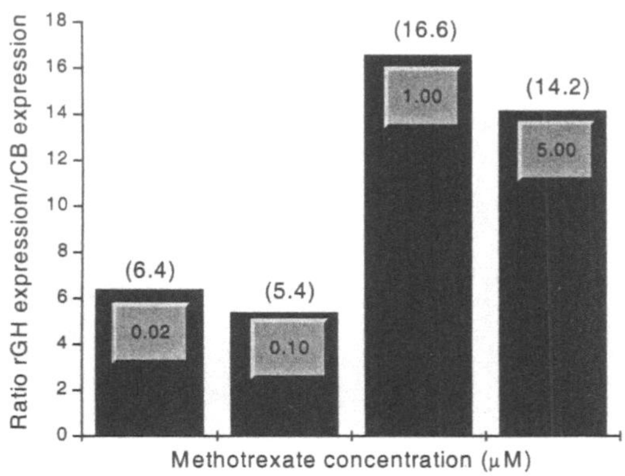

Figure 2. Ratio of rGH expression to rCB expression during amplification of the rCB expression cassette using MTX. Ratio is indicated in parenthesis and MTX level is indicated in the bar.

Figure 2 illustrates that rGH expression is not compromised by increasing rCB expression i.e. the ratio of rGH/rCB did not **decrease** as rCB expression **increased**.

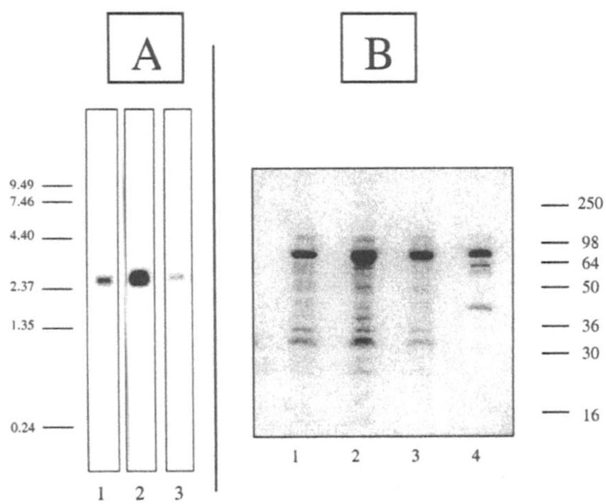

Figure 3. Northern blot (Panel A) and western blot (Panel B) on CHO cells expressing rCB probed with an oligo for GRP78 mRNA and an anti-GRP78 polyclonal antibody, respectively.

Panel A: 1 – Control CHO
2 – rCB CHO
3 – rGH CHO

Panel B: 1 – rGH CHO
2 – rCB CHO
3 – Control CHO
4 – GRP78 (purified)

Figure 3 shows elevated levels of both GRP78 mRNA *and* protein in cells expressing rCB compared to control CHO or CHO expressing rGH.

5. Interpretations

- CHO 'cellular-machinery' for protein expression is not limiting rCB expression.
- Expression of rCB in CHO induces a stress response probably a result of interaction with GRP78 following misfolding.
- Recombinant Component B expression is a function of protein-specific determinants ***i.e. ULTIMATELY CONTROLLED BY THE AMINO ACID PRIMARY SEQUENCE.***

EFFECT OF BIOREACTOR PROCESS CONTROL PARAMETERS ON APOPTOSIS AND MONOCLONAL ANTIBODY PRODUCTION DURING THE PROTEIN-FREE FED-BATCH CULTURE OF A MURINE MYELOMA.

ENDA MORAN, STEVE MCGOWAN, NIAMH REYNOLDS, CLAIRE WILSON

Glaxo Wellcome Research & Development, Biotechnology Development Laboratories, South Eden Park Road, Beckenham, Kent, BR3 3BS, UK

1. Introduction

Apoptosis is now well known to occur in cell culture systems using commercially important cell lines, for example Chinese hamster ovary (CHO), SP2/0 and NS0 myeloma cell lines. Apoptosis is undesirable in commercial production systems - culture productivity becomes diminished, the evolution of fragmented DNA may complicate downstream processing operations, and the proteases activated early in the apoptotic cascade may compromise the quality of any desired peptide or protein product. This study aimed to extend current knowledge by systematically investigating the effect of important process control parameters such as pH and temperature on the growth, death and productivity of a commercially important murine myeloma cell line producing a humanised monoclonal antibody during protein-free fed-batch culture.

2. Experimental Methods

2.1 FED-BATCH CULTURE & EXPERIMENTAL DESIGN

A GS-NS0 myeloma cell line (3622W94) producing a humanised IgG1 monoclonal antibody was used in this study. Fed-batch cultures were initiated in a proprietary protein-free growth medium in FT Applikon 3 L bench scale reactors in a scaled-down model of a commercial production process. Cultures were fed with a concentrated protein-free feed medium at 10% v/v (volume of feed/culture volume). An experimental design was created using a half-fractional factorial design for fed-batch culture incorporating half of the 32 possible combinations of five selected control parameters at high and low levels. These high and low levels of control parameters were as follows: pH 7.1 & 7.5, dissolved oxygen tension with respect to medium saturation with air (5% & 30%), seeding density (0.27×10^6 cells/mL & 0.43×10^6 cells/mL), temperature (34°C & 38°C) and a pair of culture feeding regimes involving early or late feed medium addition (Feed Regime A/ 44 h, 92 h, 140 h; Feed Regime B / 52 h, 100 h, 148 h; all times from culture inoculation).

2.3 ANALYTICAL METHODS

Antibody concentrations were measured by nephelometry. Cell numbers and viability were determined using a Neubaeur haemacytometer and erythrosin B staining. Cell size distributions were measured using a Coulter Multisizer. DNA was extracted from whole cells and electrophoresed on a 1.2% agarose gel. The fluorescent DNA-intercalating dyes acridine orange and ethidium bromide were used to allow UV fluorescent microscopic identification of viable and non-viable apoptotic and non-apoptotic cells based on DNA structure. A commercially available assay kit (Oncogene Research Products, US) and a FACS-Coulter Epics XL-MCS flow cytometer were used to enable the identification of viable apoptotic cells exposing phosphatidylserine (PS) at the outer leaflet of the cell membrane.

A. Bernard et al. (eds.), Animal Cell Technology: Products from Cells, Cells as Products, 51–53.

3. Results and Discussion

3.1 MULTI-ASSAY ANALYSIS OF APOPTOSIS IN 3622W94 GS-NS0 FED-BATCH CULTURE

We used the full complement of analytical techniques in a preliminary experiment to determine which were most suitable for the monitoring of the occurrence of apoptosis during fed-batch culture of the 3622W94 cell line. A 7 L fed-batch culture was analysed daily to quantify the populations of apoptotic, necrotic and non-apoptotic cell populations using dual dye staining and UV microscopy, Figure 1(a). Cell size distributions correlated with the observed phenomenon of cell shrinkage and fragmentation during apoptosis, Figure 1(b). Endonucleases were clearly at work digesting DNA from the onset of apoptosis at day 5, Figure 1(c). Numbers of viable apoptotic cells measured using annexin V-FITC/flow cytometry were compared to those levels as measured by UV microscopy, Figure 1(d). The flow cytometric assay failed to detect viable apoptotic cells at culture initiation (day 0) and day 4. On induction of 3622W94 to apoptosis using etoposide, the percentage viable apoptotic cells as measured by UV microscopy was not matched by a concomitant increase in the number of viable apoptotic cells as measured by flow cytometry, Figure 1(e). We therefore conclude that annexin V-FITC labelling is not a useful method for the detection of apoptosis in this particular GS-NS0 line.

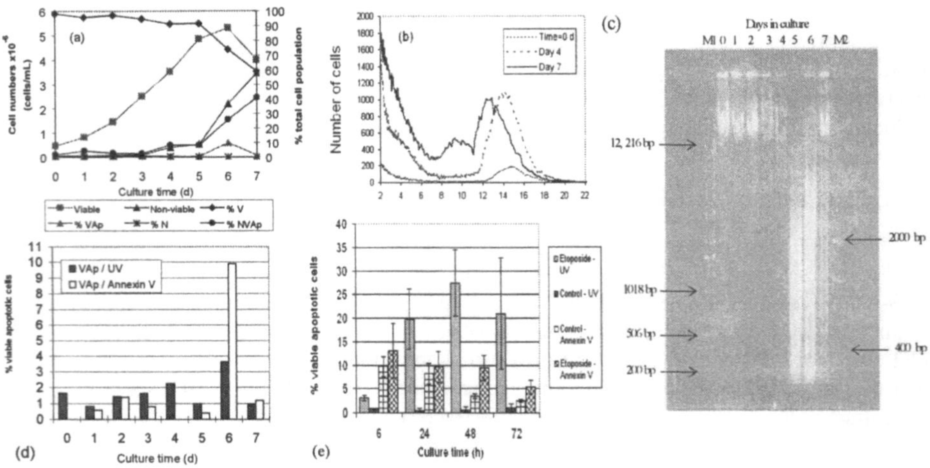

Figure 1 Identification of apoptosis in a 7 L 3622W94 GS-NS0 fed-batch culture. (a) cell growth and death profiles determined using UV fluorescent microscopy. V-Viable cells, VAp-Viable apoptotic cells, N-necrotic cells, NVAp-Non-viable apoptotic cells. (b) cell size distributions, (c) agarose gel electrophoresis, (d) viable apoptotic cells (VAp) as a percentage of the total cell population determined using both annexin V-FITC labelling / flow cytometry and UV microscopy. (e) comparison of annexin V-FITC labelling / flow cytometry and UV microscopy for the detection of viable apoptotic cells.

3.2 EFFECT OF PROCESS CONTROL PARAMETERS ON APOPTOSIS IN FED-BATCH CULTURE

The average specific growth rate μ, Figure 2(a), the total integral of non-viable apoptotic cells (INVAp cells, Figure 2(b)) and the total integral of viable cells (IVC) were selected as suitable culture performance measures of cell growth and death. Only temperature and culture seeding density of the five control parameters studied had significant causative effects on apoptosis in this culture. At the higher limits of temperature and seeding density, cultures reached maximum cell densities at days 5 or 6 (high specific growth rate cultures) and rapidly entered the decline phase with the majority of cells dying through apoptotic mechanisms - necrotic cell populations were very rarely observed throughout all fed-batch cultures. In this work, data suggest that apoptosis was not induced by nutrient depletion (glucose or primary amino acids, data not shown) in the cultures of high specific growth rate. Possibly the most useful method of assessing apoptosis in a culture system such as that described here is to examine factors impacting on the relative proportions of the total integral of apoptotic cells (viable apoptotic and non-viable apoptotic) to the total integral of viable & non-viable cells (%IAp of Total Int.Cells) - clearly, it is desirable to keep this culture response to a minimum. The analysis of these data shows a significant interaction effect between seeding density and temperature on this response, Figure 2(c). The approach to culture optimisation from an apoptosis point of view may be demonstrated by selecting a few examples of control parameter combinations and examining their effect on culture responses such as growth rate, productivity and occurrence of apoptosis, Table 1. When the INVAp was minimised (34°C /Seeded at 0.27×10^6 cells/mL), it was as a result of poor growth rate - culture antibody productivity, q_p, was additionally diminished when the growth rate was low. However, at the minimum of the %IAp of Total Int.Cells (38°C /Seeded at 0.27×10^6 cells/mL), growth rate and productivity were at acceptable levels. In this study, the cultures in which apoptosis was most prevalent were also the cultures in which cell growth (IVC) and specific productivity were superior e.g. 38°C /Seeded at 0.43×10^6 cells/mL, Table 1.

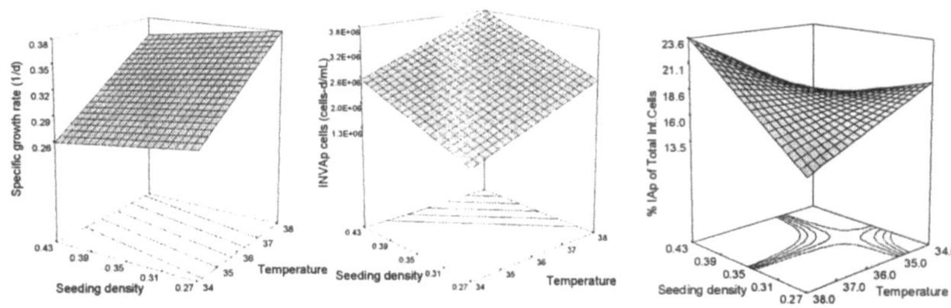

(a) (b) (c)

Figure 2 Effect of seeding density ($\times10^6$ cells/mL) and temperature (°C) on the average specific growth rate (a) & the total integral of non-viable apoptotic cells (b) and the interaction effect (c) of seeding density ($\times10^6$ cells/mL) and temperature (°C) on the total integral of apoptotic cells (IAp) as a percentage of the total integral of cells (both viable and non-viable; Total Int.Cells).

Culture Response	Temperature (°C) / Seeding Density ($\times10^{-6}$ cells/mL)			
	34 / 0.27	38 / 0.27	38 / 0.43	37 / 0.35
INVAp cells (10^6 cells-day /mL)	1.2	2.78	4.3	3.5
% IAp of Total Int. Cells	19.5	13.5	23.6	18.7
IVC (10^6 cells-day/mL)	7.4	14.1	18.2	14.5
μ (d^{-1})	0.29	0.38	0.36	0.35
q_p (μg MAb/10^6 cells/d)	8.3	23.5	25.2	20.5

Table 1 Statistical model values of 3622W94 fed-batch culture responses relating to apoptosis and culture production performance at different levels of the two control parameters having significant effects on these responses. Symbols are described in the text.

4. Conclusions

The principal cause of cell death in this GS-NS0 myeloma culture system is apoptosis. This has been identified by a combination of agarose gel electrophoresis, cell sizing and fluorescent microscopic techniques. Apoptosis was particularly prevalent in certain of the 16 fed-batch cultures of this study. However, culture control conditions have been identified at which the occurrence of apoptosis is minimal without compromising the specific growth rate, specific antibody productivity and final antibody titre at harvest.

ADENOVIRUS VECTOR PRODUCTION IN 293 FED-BATCH CULTURES

KATHY WONG[1], MARIA JESUS GUARDIA[2], STANFORD LEE[1], AND WEI-SHOU HU[3]
[1]*Bioprocessing Technology Centre, The National University of Singapore, Singapore 119260.* [2]*Department of Chemical Engineering, University of Alcala, 28871 Alcala de Henares, Spain.* [3]*Department of Chemical Engineering and Materials Science, University of Minnesota, MN 55455, USA*

Keywords: adenovirus, 293 cells, GFP, fed-batch, metabolic shift, lactate, glucose and glutamine control

1. Abstract

Adenovirus vector production kinetics using 293 cells were investigated in serum-free fed-batch cultures. The monitoring of infection progress and vector production was facilitated by the use of a recombinant adenovirus expressing green fluorescence protein (GFP). The metabolic activities at different stages of the culture were monitored with on-line measurements of oxygen uptake rate (OUR), which allowed for the controlling of glucose and glutamine levels in a fed-batch mode. The metabolism of 293 cells shifted from a high lactate producing state to a low lactate producing state gradually. As a result the maximum viable cell density reached was higher than that achieved in a batch culture. Comparing cell metabolism before and after the metabolic shift, the specific consumption rates of glucose and all amino acids in cells in low lactate producing state were significantly reduced. The extension of such strategy enabled us to perform adenovirus infection at a high cell density without medium replacement. It is envisaged that the above studies will aid the development of production process that delivers the required level of viral vectors for gene therapy.

2. Introduction

The use of recombinant adenovirus as vectors for gene therapy is increasing rapidly. Currently, 17% of the total protocols undergoing clinical trials worldwide involves the use of adenovirus vectors (http://www.wiley.co.uk/genetherapy/clinical/vectors.html). The average vector dosage per patient required range from 10^{11} to 10^{13} PFU (particles)/dose. At present, 293 cells (Human embryonic kidney cells) are most widely used for the production of E1 defective adenovirus vectors due to the cells' intrinsic expression of E1 polypeptides (Graham 1977). In this study, we address our aim to improve the production of adenovirus vectors in 293 serum-free fed-batch cultures.

3. Material and Methods

293 cells and culture medium. The 293 cells were a gift from Dr. Scott McIvor (University of Minnesota, USA) and were adapted to serum-free suspension cultures in our laboratory. The basal medium used was calcium free DMEM/Ham's F12 (Gibco). It was supplemented with 100 µM $CaCl_2$, ethanolamine, chemically-defined lipid concentrate, bovine insulin, transferrin, and Kraft Primatone RL. Cells were routinely maintained in siliconised glass Erlenmeyer shake-flasks. The concentrated feed medium used for the fed-batch cultures contained 10x amino acids and 1x salts and s upplements of the maintenance

A. Bernard et al. (eds.), Animal Cell Technology: Products from Cells, Cells as Products, 55–57.

medium. The glucose and glutamine levels in the feed concentrate were 133 mM and 12 mM respectively.

Ad-GFP. The GFP-adenovirus vector was obtained from Bernard Massie (Mosser et. al., 1997). Under the control of the early promoter of cytomegalovirus, the S65T GFP mutant gene is expressed in infected cells. The determination of viral infectivity in viral stocks and culture samples was performed by the end-point dilution assay.

Bioreactor set-up and off-line analysis. The instrumentation, reactor set-up and off-line analysis used had been described previously (Zhou et al., 1997).

Cell enumeration. Cell concentration was estimated by microscopic counting with a hemocytometer. Prior to counting, samples were treated with equal volume of 0.25% trypsin and 1 mM EDTA in PBS for 10 minutes at 37 °C. This serves to disperse the cell aggregates which grow up to 150 μm in diameter when the cell density is above $2x10^6$ cells/ml.

4. Results and Discussions

Figure 1 shows the results of a fed-batch culture of 293 cells. A maximum density of $6.5x10^9$ cells/L was achieved in this culture, which was higher than that from a batch culture (Table 2). Lactate production steadily increased for approximately 90 hours and beyond which no further accumulation was observed. Coinciding with this shift of metabolism, glutamine in the medium declined to a very low level, while glucose concentration was still sufficiently high as compared to that used previously to trigger metabolic shift in hybridoma cells (Zhou et. al., 1997).

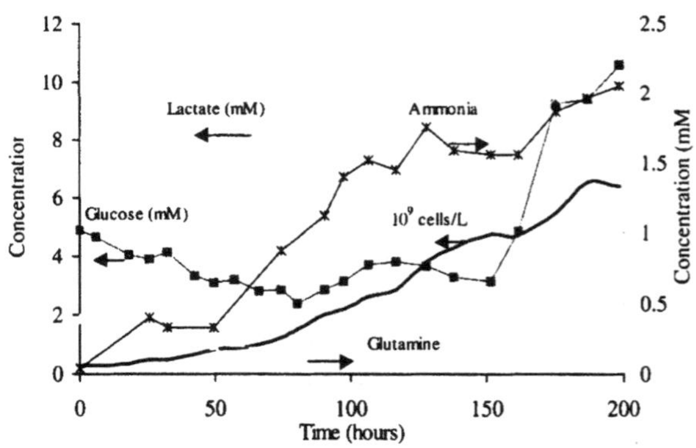

Figure 1. Profiles of cell density, nutrients and waste metabolites in a 293 fed-batch culture.

As clearly shown in Table 1, the consumption of nutrients and production of waste metabolites were reduced significantly after the shift of metabolism at 90 hours. The reduction levels ranged from 50-90%. The suppression of the pathway flux from pyruvate to lactate implies that glucose and glutamine metabolism in 293 cells can be made more efficient under a nutrient controlled environment, which also appeared to benefit the efficiency in the consumption of all other major amino acids. The above fed-batch strategy was further tested for the production of Ad-GFP. The cell density at which infection was initiated was $3.7x10^9$ cells/L and the final Ad-GFP yield was $1x10^{12}$ PFU/L (Table 2).

Table 1. Specific rates of consumption of selected nutrients and production of waste metabolites in 293 a fed-batch culture.

Nutrient/ metabolite	Specific consumption rate (µmole/10^9 cells/hour)	
	Before 90 hours	After 90 hours
Glucose	116	41
Glutamine	1.2	0.3
Lactate	-131	-5.5
Ammonia	-15	-2
Serine	6.1	0.4
Arginine	7.4	1.9
methionine	2.4	0.4
Leucine	8.9	1.9
Glycine	2.1	1.0
Asparagine	2.1	1.0
Glutamate	0.7	0.3

Table 2. Comparisons of batch (B) and fed-batch (F-B) 293 cells in growth only and infection cultures.

	CELL ONLY		INFECTION	
	B	F-B	B	F-B
Initial glucose concentration (mM)	32.5	4.9	29.4	3.3
Initial glutamine concentration (mM)	3.1	0.4	2.8	0.2
Cell concentration at infection (10^9 cells/L)			0.7	3.7
Maximal cell concentration (10^9 cells/L)	3.9	6.5	1.2	4.3
Final glucose consumption (mM)	17.5	22.6	21.5	17.8
Lactate Production (mM)	21.5	9.2	40.9	23.6
Ad-GFP Production at 4 days post infection (10^{11} PFU/L)			4.4	9.9

Also shown in Table 2 are comparisons of the results obtained from batch and fed-batch cultures. The specific production of lactate in infected cells increased significantly from that of uninfected cells in both batch and fed-batch cultures. However, although the total infected cell density in the fed-batch culture was nearly four times higher than that of the batch culture, the low specific production of lactate in cells cultured under nutrient-controlled environment did not result in its accumulation to inhibitory level. Base was not needed any time during the fed-batch culture. The specific productivity of Ad-GFP at 4 d.p.i. in the fed-batch culture was fractionally lower than that in the batch culture. This was likely due to the decreasing efficiency in infection as cell density increased, since the cell aggregate size in the fed-batch culture reached 150-200 µm.

4. Conclusions

In this study we demonstrated that 293 cell metabolism could be altered in fed-batch cultures. The resulting high cell density culture was effectively used for the production of large quantities of adenovirus vectors. The reduction of lactate accumulation, especially after infection, was significantly advantageous to the process. Further improvement may be expected by manipulating the cell aggregate size, which will enhance the infection kinetics of the system.

5. References

Zhou W.; Rehm, J.; Europa, A.; & Hu, W-S (1997) Alteration of mammalian cell metabolism by dynamic nutrient feeding. Cytotechnolgy, 24:99-108.

Graham, F.L.; Simley, J.; Russel, W.C.; and Nairn, R (1977) Characterisation of human cell line transformed by DNA from human adenovirus 5. J. Gen. Virol. 36:59-72

Mosser, D.D.; Caron, A.W; Bourget, L.; Jolicoeur, P.; & Massie, B (1997) Use of a dicistronic expression cassette encoding the green fluorescent protein for the screening and selection of cells expressiong inducible gene products. BioTechniques, 22:150*-161.

INFLUENCE OF BCL-2 OVER-EXPRESSION ON NS0 AND CHO CULTURE VIABILITY AND CHIMERIC ANTIBODY PRODUCTIVITY

B.T.TEY, R.P.SINGH, M.AL-RUBEAI
Animal Cell Technology Group, School of Chemical Engineering, University of Birmingham, Edgbaston, Birmingham, B15 2TT, UK.

INTRODUCTION

Cell death during the cultivation of mammalian cell lines for biopharmaceutical production occurs by one of two morphologically and biochemically distinct mechanisms. Apoptosis is an active and highly regulated cell death pathway induced by a range of environmental and physiological factors. Necrosis is a passive accidental form of cell death that results from extreme environmental stress. Factors such as nutrient deprivation, growth/survival factors, oxygen deprivation as well as exposure to excess toxic metabolites can all induce high levels of apoptosis (Franek and Dolnikova, 1991; Mercille and Massie, 1994; Singh et al., 1994).

Apoptotic pathway can be modulated by over-expression of anti-apoptotic gene such as bcl-2 gene, and/or the addition of anti-apoptotic chemical such as protease (Caspase) inhibitors into the culture medium. Modulation of the apoptotic pathway by the over-expression of the bcl-2 gene provides highly effective approach to the reduction of cell death in the bioreactor. This has been demonstrated in several industrially important cell lines such as burkits lymphoma, hybridoma, myeloma and insect cell line (Mitchell-Logean and Murhammer, 1997; Simpson et al., 1997; Singh et al., 1996; Suzuki et al., 1997).

In the present study, we have investigated the influence of bcl-2 on the death rate of NSO and CHO cell lines that were previously developed for the industrial scale production of a chimeric antibody. We report that bcl-2 expression significantly reduces the rate of cell death under batch and nutrient limited conditions and leads to an increase in maximum cell numbers. However, due to the exhaustion of biosynthetic precursors, the enhanced survival of cells under batch conditions did not translate into an increase in maximum product titres.

MATERIALS AND METHODS

The parental cell line CHO 22H11 and NS0 6A1 were kindly provided by Lonza Biologics (Slough, UK) and had previously been transfected with glutamine synthetase (GS) expression system carrying a gene for human-mouse chimeric antibody (cB72.3). The CHO 22H11 and NS0 6A1were further transfected with the expression vector pEF bcl-2 the control vector pEFneo respectively. Cells were maintained in glutamine free GMEM (Gibco, UK), supplemented with 5%(v/v) foetal bovine serum (FBS) (Gibco, UK), MEM non-essential amino acids (Gibco, UK), sodium pyruvate (Gibco, UK), nucleosides, glutamic acid, asparagine, sodium bicarbonate and methionine sulphoximine (MSX) (all chemicals from Sigma, UK).

A. Bernard et al. (eds.), Animal Cell Technology: Products from Cells, Cells as Products, 59–61.

RESULTS AND DISCUSSION

Growth comparison between NS0 and CHO control and bcl-2 cell line is shown in Table 1. For both cell lines, the over-expression of bcl-2 resulted in a significant increase in maximum cell number. The culture duration of CHO bcl-2 was extended to more than 48 days compared to 35 days in control culture. In the NS0 cultures, culture duration was extended from 7 days in control culture to 14 days in bcl-2 culture. The antibody titre in bcl-2 cultures of both NS0 and CHO were only slightly higher compared to the control cultures despite having a considerable beneficial effect on cellular viability. However, this was not surprising as key nutrients would be a limiting concentrations by this stage of the culture. Although bcl-2 expression may allow survival under these conditions, the cells may enter a metabolically inactive or quiescent state at least as regard to cell proliferation and production of heterologous protein.

The effect of exposure of cells to various nutrient limited conditions were shown in Table 2. In all cases, the viable cell number of the bcl-2 cultures was significantly higher than the control cells. As expected, the effect of deprivation of all amino acids was particularly marked in the NSO control cells, with the complete loss of viable cell after 3 days. However, what was surprising was that the bcl-2 transfectants only exhibited a 25% fall in viable cell number. This indicated the high levels of robustness of bcl-2 transfectant even extreme starvation conditions. Interestingly, the rate of fall in viable cell number in the CHO was much lower. Clearly the CHO cell line was more robust that the NS0 cell line.

CONCLUSSION

The over-expression of bcl-2 significantly reduces the rate of cell death in CHO and NS0 cell cultures. There was only 19% and 25% increase in antibody productivity under bath culture conditions of CHO and NS0 respectively.

ACKNOWLEDMENTS

This work was funded by the EC Framework IV programme. BTT was funded by Universiti Putra Malaysia. We would like to thank ESACT for the award of a bursary to BTT, and Dr John Birch (Lonza Biologics plc, U.K.) for the CHO 22H11 and NS0 6A1 cell lines used in this study.

REFERENCES

Franek , F. and Dolnikova, J. (1991) Nucleosomes occurring in protein-free hybridoma cell cultures. Evidence for programmed cell death. FEBS Lett. 248: 285-287.

Mercille, S. and Massie, B. (1994) Induction of apoptosis in nutrient-deprived cultures of hybridoma and myeloma cells. Biotechnology and Bioengineering. 44:1140-1154.

Mitchell-Logean, C. and Murhammer, D.W. (1997) Bcl-2 expression in Spodoptera frugiperda Sf9 and Trichoplusia ni BT-Tn-5B1-4 insect cell: effect on recombinant protein expression and cell viability. Biotechnology and Bioengineering. 56:380-388.

Simpson, N.H., Milner, A. N., Al-Rubeai, M. (1997) Prevention of hybridoma cell death by bcl-2 during sub-optimal culture conditions. Biotechnology and Bioengineering. 54:1-46.

Singh, R.P., Al-Rubeai, M., Gregory, C.D., Emery, A.N. (1994) Cell death in bioreactors: A role for apoptosis. Biotechnology and Bioengineering. 44:720-726.

Singh, R.P., Finka, Emery, A.N., Al-Rubeai, M. (1996) Enhancement of survivability of mammalian cells by overexpression of the apoptosis suppressor gene bcl-2. Biotechnology and Bioengineering. 52:166-175.

Suzuki, E., Terada, S., Ueda, H., Fujita, T., Komatsu, T., Takayama, S., Reed, J.C. (1997) Establishing apoptosis resistant cell lines for improving protein productivity of cell culture. Cytotechnology. 23: 55-59.

Table 1 : Batch culture

	CHO 22H11		NSO 6A1	
	Bcl-2	Control	Bcl-2	Control
Max. viable cell number, (X 10E5 cell/ml)	6.9	4.34	6.43	5.15
Culture duration, (Day)	48	35	14	7
Antibody titre, (ug/ml)	43	36	15	12

Table 2 : Influence of nutrient deprivation

Medium	Viable cell number, (X 10E5 ell/ml)				
	Day 0	Day 3 (NSO 6A1)		Day 7 (CHO 22H11)	
		Bcl-2	Control	Bcl-2	Control
Serum (0.5%)	2	3.39	2.81	3.06	1.77
Amino acids free	2	1.5	0.03	2.13	0.85
Glutamine & asparagine free	2	4.42	3.69	3.5	2.33
Glucose free	2	2.69	0.92	2.74	1.54

A REGULATABLE SELECTIVE SYSTEM FACILITATING ISOLATION OF HIGH EXPRESSION MAMMALIAN CELLS

Kiichiro Teruya, Ying-Pei Zhang, Yoshinori Katakura and Sanetaka Shirahata
Graduate School of Genetic Resources Technology, Kyushu University, 6-10-1 Hakozaki, Higashi-ku, Fukuoka 812-8581, Japan.

ABSTRACT: We describe a regulatable selective system which facilitates the isolation of high expression mammalian cells. The system is based on a modified tetracycline inducible expression system where the binding of the tetracycline-controlled rtTA transactivator to the tetracycline responsive promoter PhCMV*-1 depends on the presence of doxycycline (Dox). When the cells that produce functional tetracycline-controlled transactivator rtTA and express the dominant selectable marker *bsr* gene under the control of the tetracycline responsive promoter PhCMV*-1 are cultured in the presence of sufficient amount of the inducer (1-2 μg/ml Dox), the rtTA transactivators bind to the PhCMV*-1 promoter and drive the transcription of the *bsr* gene. The cells therefore are resistant to the selective agent blasticidin S (5 μg/ml). However, when the cells are cultured in a medium containing the concentration of blasticidin S but only very low concentration of the inducer Dox (1 ng/ml or lower) to repress the rtTA-dependent transcription of the *bsr* gene, only the cells capable of expression of the *bsr* gene relatively independent of Dox can survive. These cells, when cultured in the presence of sufficient amount of inducer (1-2 μg/ml Dox), express much higher level of *bsr* gene than the parent population. By co-transfecting a target gene with the *bsr* gene using calcium phosphate co-precipitation, high expression cells can be isolated with the same selective procedures. As the regulatable selective system uses a dominant selectable marker, it can be used in prototrophic mammalian cells. Theoretically, the same strategy can be extended to other eukaryotic expression systems.

1. Introduction

Many genetic variations can increase the expression of a foreign gene in mammalian cells. Cells carrying such variations can be isolated if there are appropriate selective or screening methods. A well-known example is the DHFR-dependent gene amplification system [1]. The DHFR enzyme can be inhibited by methotrexate (MTX). Besides DHFR system, several other such selective systems are available for mammalian cells. However, all of them require autotrophic hosts for optimal results [1].

We designed a regulatable selective system which allows to do this in prototrophic cells. The system is based on a modified tetracycline inducible expression system where the binding of the tetracycline-controlled rtTA transactivator to the tetracycline responsive promoter PhCMV*-1 depends on the presence of Doxycycline (Dox)[2]. When the cells that produce functional tetracycline-controlled transactivator rtTA and express the dominant selectable marker *bsr* gene [3] under the control of the tetracycline responsive promoter PhCMV*-1 are cultured in the presence of sufficient amount of the inducer (1-2 μg/ml Dox), the rtTA transactivators bind to the PhCMV*-1 promoter and drive the transcription of the *bsr* gene. The cells therefore are resistant to the selective agent blasticidin S (BS)(5 μg/ml). When the cells are cultured in a medium containing the same concentration of the BS but only very low concentration of Dox (0-1 ng/ml), most cells loss the resistance to BS due to the suppression of the rtTA-dependent transcription of the *bsr* gene. The cells survived in this medium are capable of expressing the *bsr* gene relatively independent of the inducer. When cul-

A. Bernard et al. (eds.), Animal Cell Technology: Products from Cells, Cells as Products, 63–65.

tured in the presence of sufficient inducer Dox, these cells should express higher level of the *bsr* gene from Dox-dependent and independent transcription.

2. Results and Discussion

We tested the regulatable selective system in several cell lines (populations). CHO (DUKBX-11) cells maintained in 10% FBS/MEM-α medium were co-transfected with pUHD172-1neo, which carries an rtTA gene driven by CMV promoter [2], and pUBsr9, which carries a *bsr* gene driven by the PhCMV*-1 promoter. TA5 is a clones isolated in the medium containing 5 μg/ml BS and 1 μg/ml Dox. TA5 cells were then cultured in a medium containing 5 μg/ml BS but only 1 ng/ml Dox, a clone capable of growing in this medium was isolated and named D1A. TA5 cells were co-transfected with pTD4 which carries a *dhfr* gene driven by the PhCMV*-1 promoter and pUL5 which carries a β-lactoglobulin (β-LG) gene driven by the PhCMV*-1 promoter. A clone expressing β-LG and *dhfr* gene in the presence of 1 μg/ml Dox was named DL1. DL1 cells were then cultured in a Dox-free medium containing 5 μg/ml BS, a pool of the cells capable of grow in this medium was named ER/B. On the other hand, CHO cells were co-transfected with pUHD172-1neo and pTD4. The transfectants displaying *dhfr+* in 1 μg/ml Dox and resistance to G418 were then further co-transfected with pUL5 and pUBsr9. A clone16A is isolated after limiting dilution cloning. 16A cells express β-LG gene and display resistance to 5 ng/ml BS in the presence of 1 μg/ml Dox. 16A cells were then cultured in a Dox-free medium containing 5 μg/ml BS, a pool of the cells capable of growing in this medium was named 16B. Calcium phosphate co-precipitation method was used in all transfections [4]. According to the selective procedure, these cells can be divided into two groups. Group A includes D1A (derived from TA5), ER/B (derived from DL1) and 16B (derived from 16A), all of which are capable of expressing *bsr* gene relatively independent of Dox. Group B includes TA5, DL1 and 16A, which are parents of group A.

The expressions of the bsr gene in these cells were first analyzed by Northern blotting. In the same medium containing sufficient amount of Dox (1 or 2 μg/ml), The expression level of *bsr* gene in group B is significantly higher than in group A (Data not shown). The expression of the *bsr* gene in ER/B and 16B cells was still inducible by Dox. These results demonstrated that the regulatable selective system was efficient in isolation of the cells expressing high level of the *bsr* gene. We then further examined whether the system can be used if the target gene does not encode a selectable marker. In DHFR-dependent gene amplification system, a foreign gene can be co-amplified if it is co-transfected with *dhfr* gene, because the genes co-transfected with calcium phosphate co-precipitation tend to integrate into the same chromosome site and the gene amplification often involves a large chromosome region [1]. To examine whether the co-transfection strategy works in the regulatable selective system, we analyzed the expression of the gene co-transfected with the *bsr* gene in these cells. In TA5, D1A, DL1 and ER/B cells, the gene co-transfected with the *bsr* gene is rtTA, which is under the control of CMV promoter. The mRNA level of rtTA gene in D1A and ER/B cells is significantly higher than in TA5 and DL1 cells. The high rtTA level in ER/B cells is also reflected by the enhanced expression of β-LG gene which is under the control of the PhCMV*-1 promoter. In 16A and 16B cells, the gene co-transfected with *bsr* is β-LG, the mRNA level of β-LG gene in 16B is also much higher than in 16A. The transcription of β-LG gene which is driven by the PhCMV*-1 promoter is inducible in 16B and 16A cells. However, there is no significant difference between 16A and 16B cells in the expression of rtTA, gene not co-transfected with *bsr*.

The mechanism whereby the expression of the *bsr* and co-transfected gene is enhanced remains to be identified. We had suspected that this was resulted from gene amplification. However, the results of Southern blotting demonstrated that there was no significant change in the copy number of the β-LG, *bsr*, and rtTA gene (Data not shown). If the co-transfected genes have been integrated into the same chromosome site as frequently occurs in calcium phosphate co-precipitation transfection, the transcription of the two genes may be simultaneously enhanced if there is a genetic change facilitating the assembly of the transcriptional machinery at this site or improving the transcriptional efficiency of the machinery. Besides co-transfection, a foreign gene can be linked to the *bsr* gene with other methods such as using a bi-directional tetracycline responsive promoter to drive the transcription of the *bsr* and the foreign gene or positioning the foreign gene downstream of the *bsr* gene in a bicistronic expression unit [5, 6].

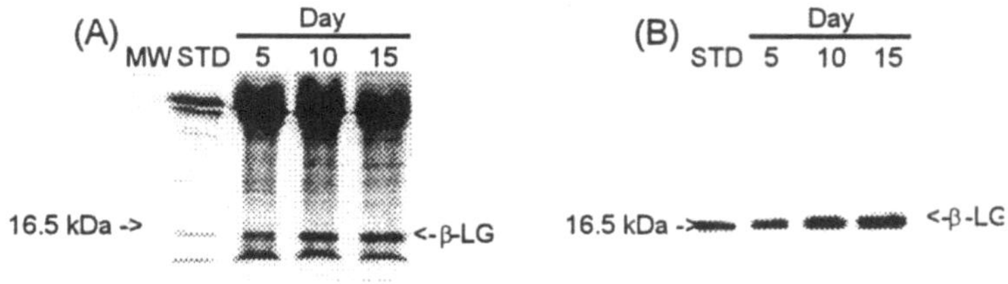

Fig. 1. β-LG production in Tecnomouse™. 7.8x10^7 of 16B cells were inoculated into a 5 ml incubation cassette and cultured at 37°C in 5% CO_2. FBS-free ERDF medium containing 2 μg/ml Dox was supplemented in a recycling manner. Samples were harvested on day 5, 10 and 15. MW, molecular weight standard. STD, 1 μg β-LG protein dissolved in 0.1% bovine albumin. (A). SDS-PAGE stained with Coomassie brilliant blue R-250. (B) β-LG protein detected by Western blotting as described [7].

So far, the tetracycline inducible expression system has been mainly used as a tool in studying gene regulation. Since the regulatable selective system allows to isolate high expression cells rapidly, a combination of them may provide a powerful tool for recombinant protein production. To evaluate the usefulness of the system, we used 16B cells to produce recombinant β-LG protein in a hollow-fiber bioreactor Tecnomouse system (Integra Biosciences, Germany). As shown in Fig. 1, large amounts of recombinant β-LG protein were accumulated in the medium. The β-LG concentration is over 300 μg/ml in the supernatant harvested after 15-day's culture as determined by ELISA. This level is comparable to the most successful cases achieved with other mammalian expression system.

In conclusion, we established a regulatable selective system which facilitates the isolation of high expression cells. This system is based on a dominant selectable marker, therefore, can be used in prototrophic cells. Theoretically, the same strategy can be extended to other eukaryotic expression systems.

3. Acknowledgment

We would like to thank Dr. H. Bujard for the plasmids pUHD172-1neo and pUHD10-3, and Dr. S. Kaminogawa for the β-LG cDNA.

4. References

1. Kaufman, R.J. (1990) *in* Goeddel, D.V. (ed.), *Methods in Enzymology*. Academic Press. San Diego, Vol. 185, pp.537-566.
2. Gossen, M. *et al.* (1995). *Science*, **268**, 1766-1769.
3. Izumi, M. *et al.* (1991) *Exp. Cell Res.*, **197**, 229-233.
4. Chen, C. and Okayama, H. (1987) *Mol. Cell. Biol.*, *7*, 2745-2752.
5. Baron, D., Freundieb, S., Gossen, M. and Bujard, H. (1995) *Nucleic Acids Res.*, **23**, 3605-3606.
6. Dirks, W., Wirth, M. and Hauser, H. (1993) *Gene*, **128**, 247-249.
7. Totsuka, M. *et al.* (1990) *Agric. Biol. Chem.*, **54**, 3111-3116.

COMPARATIVE ANALYSIS OF IRES EFFICIENCY OF DICISTRONIC EXPRESSION VECTORS IN PRIMARY CELLS AND PERMANENT CELL LINES

LORIN SCHUMACHER[1], PAULA M. ALVES[2], AND MANFRED WIRTH[1]
[1)] *Dept. of Regulation and Differentiation, GBF, Braunschweig, Germany*
[2)] *Animal Cell Technology/NMR Lab., IBET/ITQB, Oeiras, Portugal*

Summary

Internal ribosome entry sites (IRES) are used in eukaryotic vectors for correlated co-expression of genes in eukaryotic cells. To investigate the IRESs' tissue and species specificity we have compared the translational efficiencies in dicistronic expression vectors mediated by the IRES of the bovine viral diarrhea virus (BVDV), the poliovirus type I IRES and by a vector harboring a random intercistronic region in certain primary cells and permanent cell lines derived from the liver, kidney, brain, cervix and endothelium. We found that IRES mediated expression differs considerably between the IRES elements used in one given cell necessitating the careful choice of IRES elements for a given cell type. BVDV IRES was superior to the strong poliovirus IRES in rodent neuronal and liver cells. The polio-IRES is stronger than the BVDV IRES in human neuronal, liver and kidney cells. The BVDV IRES tends to function optimal in the mouse (rodent) context, the poliovirus IRES in the human context. Interestingly, dicistronic expression differs considerably between primary cells and permanent cell lines. This indicates that for gene therapy of primary cells using dicistronic vectors, permanent cell lines cannot be recommended as models for gene expression.

1. Introduction

Internal ribosome entry sites (IRES) allow the efficient translation of multiple genes on one single mRNA in cells of higher eukaryotes. The advantage of placing individual cistrons on one single mRNA is based on the tight coupling of expression of the individual cistrons from such constructs. Coupled expression allows e.g. equimolar production of antibody light and heavy chain, selection for animal cell clones highly expressing the gene of interest and guarantees defined and uniform expression ratios of genes introduced by retroviral vectors into target cells. The knowledge of IRES strength in individual cell lines and primary cells is an important prerequisite for biotechnological and gene therapy applications. Recent investigations have adressed the question of translational efficiency of IRES elements *in vitro* in retic lysates or in animal cell lines after transfection (1 , 2). The IRESs of viral origin differ in strength in individual cell lines. Little is known about IRES mediated

A. Bernard et al. (eds.), Animal Cell Technology: Products from Cells, Cells as Products, 67–69.

translational efficiency in primary cells.

2. Results and discussion

To investigate IRES efficiency in permanent cell lines and primary cells we have transfected different mono- and dicistronic expression plasmids into different cell lines (Fig. 1). Expression plasmids carried either the poliovirus type I IRES, the recently characterised IRES of the bovine viral diarrhea virus (BVDV) strain SD1 or were devoid of any IRES.

Plasmids	HeLa human cervix	293 human kidney	HGBM1 human glioma	HT1080 human fibro-sarcoma	HepG2 human hepatoma	C6 mouse neuro-blastoma	AS-30D rat hepatoma
	Luciferase expression %						
SEAP – Polio type I 5'UTR – Luciferase AA	100	100	100	100	100	100	100
SEAP – BVDV-SD1 5'UTR – Luciferase AA	241	77	27	78	67	502	563
Stemloop – SEAP – BVDV-SD1 5'UTR – Luciferase AA	172	70	23	38	50	273	468

Fig. 1 IRES mediated translation efficiencies in human and rodent cell lines. Luciferase expression values of dicistronic expression plasmids transiently transfected into human, mouse and rat cell lines derived from different tissues. The polio IRES and the BVDV 5'UTR were used as intercistronic region. A plasmid with inhibitory stemloop upstream the first cistron was also included. Translation mediated by the polio IRES served as a standard for comparison.

The BVDV IRES functions well in the mouse background and translation efficiency was prominent in rat hepatoma and mouse glioma derived cell lines. Poliovirus IRES worked well in most human cells where it exhibited 3-5 fold higher levels of translation from the second cistron compared to the BVDV (with the exception of HeLa cells).

Plasmids	HUVEC SEAP l.u./OD_{562}	HUVEC Luciferase l.u./OD_{562}
SEAP – BVDV-SD1 5'UTR – Luciferase AAAA	23.194	15.969
SEAP – BVDV-SD1 5'UTR+NH2 – Luciferase AAAA	18.788	4.900
Stemloop – SEAP – BVDV-SD1 5'UTR – Luciferase AAAA	7.222	12.544
SEAP – Δ – Luciferase AAAA	31.547	3.553
SEAP – Polio type I 5'UTR – Luciferase AAAA	20.325	7.524
Luciferase AAAA	3.315	234.381
SEAP AAAA	2.250.138	4.218

Fig. 2 Expression in HUVEC cells mediated by cap-dependent and cap-independent mechanisms. SEAP and firefly luciferase expression values of bicistronic and monocistronic expression plasmids transiently transfected into the primary endothelial cells derived from human umbilical cord. Expression is given in light units and is normalized to protein levels via BCA assay.

To assess translational efficiency in primary cells as the final target for e.g. gene therapy approaches several rat brain cell types and human endothelial cells were transfected using optimised lipofection methods (Figs. 2, 3). Interestingly, expression from dicistronic vectors was weak in primary cortical neurons (Fig 3) and astrocytes (data not shown) as well as primary endothelial cells if compared to the respective monocistronic equivalent and permanent cell lines derived from the same tissue (C6 glioma, Fig. 3 or ECV304, a human endothel derived cell line (data not shown). This may be caused by inefficient internal initiation of translation in primary cells compared to permanent cell lines. Expression from monocistronic IRES-minus plasmids was e.g. approximately 100 fold higher from a monocistronic plasmid compared to a BVDV-IRES dicistronic plasmid (Fig. 3, plasmid1 and 6) in cortex neurons, but only 5fold increased in C6 cell line. This may be explained by decreased steady state levels of long dicistronic mRNAs in primary cells as a result of insufficient nuclear export or lower mRNA stability of such RNAs in primary cells. Taken together, vectors used in gene therapy should therefore be carefully tested in both permanent and in the primary target cells.

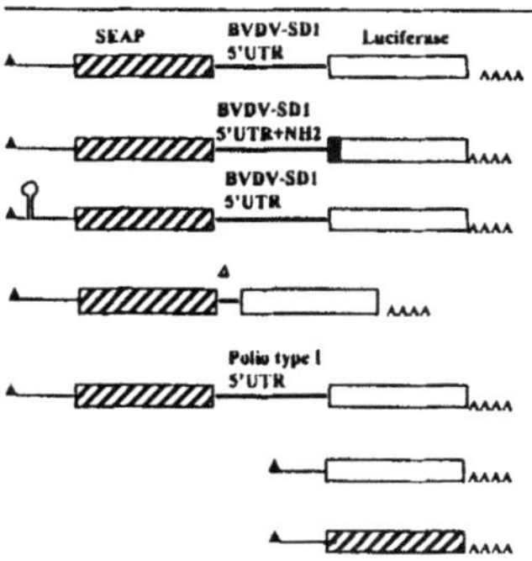

	Rat C6 glioma		Neurons from rat cortex	
	SEAP l.u./OD_{562}	Luciferase l.u./OD_{562}	SEAP l.u./OD_{562}	Luciferase l.u./OD_{562}
	2 027 625	6 142 076	343 391	22 149
	2 169 692	7 644 564	355 395	45 925
	87 537	1 562 227	36 095	20 025
	3 407 562	216 269	571 457	1 750
	2 783 175	542 602	370 219	7 209
	2 710	31 366 932	14 344	2 592 311
	13 692 893	10 484	-	-

Fig. 3 Expression in rat brain tumor cells and primary neuronal cells mediated by cap-dependent and cap-independent mechanisms.
SEAP and firefly luciferase expression values of bicistronic and monocistronic expression plasmids transiently transfected into the glioma cell line C6 and in primary neuron cells derived from the cortex of 15 day old rat embryos. Expression is given in light units and is normalized to protein levels via BCA assay.

3. References

Borman, A.M. Bailly-J.L., Girard, M. and Kean, K.M. (1995). Picornavirus internal ribosome entry segments: comparison of translation efficiency and the requirements for optimal initiation of translation *in vitro*. *Nucl. Acids Res.* 23, 3656-63.

Borman, Le-Mercier, P. , Girard, M. and Kean, K.M. (1997). Comparison of picornaviral IRES-driven internal initiation in cultured cells of different origins. *Nucl. Acids Res.* 25, 925-32.

USE OF GEL MICRODROPS SELECTION SYSTEM FOR RECOMBINANT CELLS.

Karen Hansen*, Charlotte Bertelsen, and Leif Kongerslev.

Novo Nordisk A/S, Biologics Development, DK-2820 Gentofte, Denmark
**Corresponding author, fax +45 4443 9210, e-mail kha@novo.dk*

1. Introduction

The gel microdrop system is a unique tool for selecting cells expressing a specific protein either naturally or transfected. The technique provides an assay based on expression and secretion from single cells. By using a FACS sorter the cells can be selected and propagated for further study, enabling a faster selection of producing cells and ensuring a more suitable pool for subsequent cloning.

The Gel Microdrop selection assay system can be licenced from One Cell Systems Inc., Cambridge, MA. The technique has been implemented in our lab using cells expressing recombinant coagulation factor VII.

2. Encapsulating cells

The cell suspension was mixed with melted CelBioGel^{TM}, a biotinylated agarose agar. The mixture was added to CelMix^{TM} emulsion matrix and microdrops of a 25 - 30 μm were made by choosing the right propeller speed of the CelSys microdrop maker.

By looking at the preparation in a FACS window showing forward scatter and side scatter, it is possible to select a gate where almost all cells have been encapsulated as single cells.

Fig. 1a shows a phase contrast photo of the gate with single cells in microdrops and Fig. 1b is from a gate where more than one cell has been encapsulated per bead.

A. Bernard et al. (eds.), Animal Cell Technology: Products from Cells, Cells as Products, 71–73.

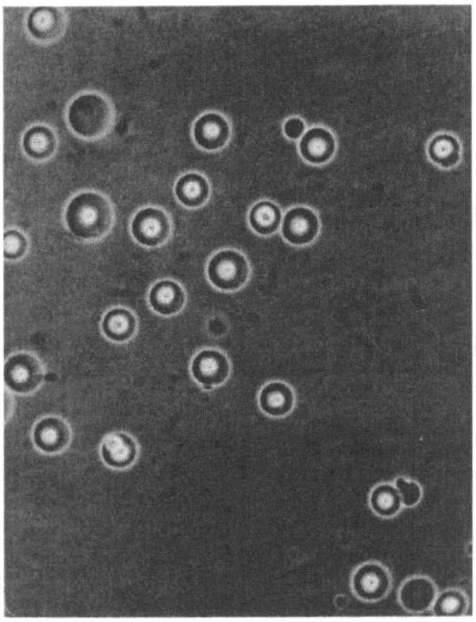

Fig. 1a

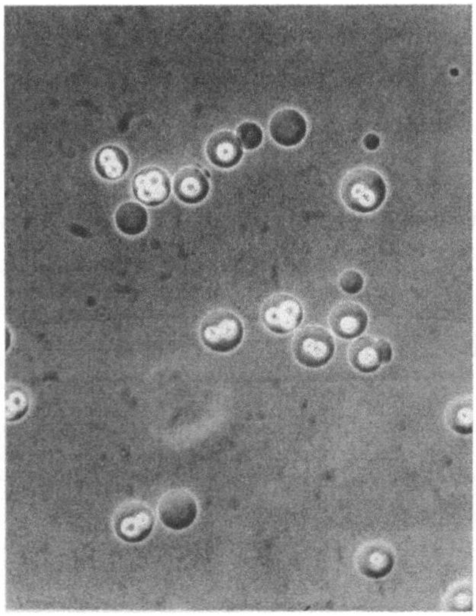

Fig. 1b

3. Single cell ELISA - staining of secreted protein.

After encapsulation of the cells, streptavidin and then biotinylated FVII antibody (Ab1) was added to the microdrops. The preparation was incubated in production medium for the number of hours that would ensure secretion above negative control but less than saturation of the microdrops. After incubation the secreted FVII was detected by FITC-conjugated FVII antibody (Ab2).

4. Recovery of FVII producing cells.

A mixture of 90% non-producing cells and 10% FVII producing cells was encapsulated, stained and sorted on a FACS Calibur after a 2 hour incubation. A fraction of single cells in beads was identified and from this fraction the highest 10% fluorescent beads were isolated.

The sorted fraction of cells in microdrops was seeded in growth medium. One day later, cells (of adherent type) started to leave the beads. As the beads created a favourable micro-environment for initial growth, cells seeded at a low density had a higher survival rate than normally seen.

After propagation in t-flasks, FVII analysis showed that the sorted fraction had the same cell specific FVII expression as the cells used for the initial 10%.

Thus, it is possible to select a producing fraction of the cell population without significant contamination with the non-producing. The transfected and non-transfected cells had roughly the same growth rate.

5. Evaluation of the gel microdrop technique.

a) The cells are selected not only for expression but also for secretion of the protein.

b) As no marker proteins are needed, cells transfected previously can be tested.

c) Suitable antibodies against the secreted protein of interest is a must, and for new proteins they may not be available.

d) Compared to using marker proteins, the technique is complex and time consuming to set up. Antibodies have to be conjugated and the appropriate incubation time needs to be determined.

BOVINE SERUM-ALBUMIN WITH GRAFTED RGD TAILED CYCLIC PEPTIDES AS AN ENGINEERED PRO-ADHESIVE PROTEIN

JANIQUE DEWELLE, DOMINIQUE DELFORGE, SABINE PIROTTON, JOSE REMACLE, MARTINE RAES
Laboratory of Cellular Biochemistry, Facultés Universitaires Notre-Dame de la Paix, 61, rue de Bruxelles, B-5000 Namur, Belgium

A synthetic adhesion protein was designed by chemical grafting of the RGD tailed cyclic peptide c[-$_D$VRGDE-(εAhx-YCNH$_2$)-] on the carrier protein bovine serum albumin (BSA). Native BSA displays poor pro-adhesive properties. However polystyrene (PS) modified with BSA-RGD peptide conjugates, promotes adhesion in a way similar to PS coated with the pro-adhesive extracellular matrix protein fibronectin (FN). Our data also confirm that the RGD motif is required for conferring pro-adhesive properties to the BSA-peptide conjugates. Interestingly, similar results were obtained either with the RGD peptides covently grafted or simply adsorbed on BSA. Adhesion on PS modified with BSA-RGD peptide conjugates, was able to induce integrin downstream signalling such as tyrosine phosphorylation of proteins in focal adhesion sites to the same extent than FN. In summary, PS can be reconditionned with BSA-peptide conjugates bearing constrained RGD motifs, since BSA has been turned into a pro-adhesive protein mimicking the proteins of the extracellular matrix (MEC) BSA has been turned.

1. Introduction

The Arg-Gly-Asp (RGD) sequence is one of the minimal motifs for cell adhesion present on extracellular matrix proteins such as FN, collagens, laminin,.... The RGD motif is often located in a loop in these proteins [1]. Conformationnally constrained RGD peptides have been used by several authors to study the importance of the three-dimensional structure of this motif for the recognition by the integrins [2]. The aim of this work was to turn BSA, a protein with poor pro-adhesive properties into a FN-like protein, by engineering BSA-RGD peptide conjugates, used to recondition PS. The tailed cyclic peptide cyclo[-$_D$VRGDE-(εAhx-YCNH$_2$)-] was grafted chemically or adsorbed on BSA according to the method described by Delforge [3]. Firstly, we compared the chemical coupling and the simple adsorption of the RGD peptides on the pro-adhesive properties of the BSA-RGD peptide conjugates. Secondly, we checked if the BSA-RGD peptide conjugates were able to interact with cellular integrins and to induce downstream signalling, and in particular protein tyrosine phosphorylation in focal adhesion sites, as observed for FN.

A. Bernard et al. (eds.), Animal Cell Technology: Products from Cells, Cells as Products, 75–77.

2. Methods

2.1. CELL ADHESION ASSAYS

Peptide synthesis and grafting on BSA was performed as described by Delforge et al. [3]. HUVEC (Human Umbilical Vein Endothelial Cells) were cultured and harvested as described by Michiels et al.[4]. The cells, in serum free medium were plated in 96-well plates. PS was beforehand treated with FN, native BSA or different BSA-peptide conjugates for 1 hour at 37°C and then washed twice with PBS. Cellular adhesion was estimated after 15 minutes by the hexosaminidase colorimetric method [5].

2.2. FOCAL CONTACT SITE FORMATION

The cells were incubated 2 hours (in serum free medium) on glass reconditionned with FN, BSA, BSA-cystein and BSA-peptide conjugates and a primary antibody directed to phosphorylated tyrosine residues of proteins was added. The focal adhesion complexes were located thanks to a secondary FITC-labelled antibody.

3. Results and Discussion

Three methods of coupling were used to graft RGD peptides on BSA: (A) chemical grafting of the peptide was performed with BSA in solution before coating; (B) chemical grafting of the peptide was performed with BSA already adsorbed on wells; (C) simple adsorption of the peptides on BSA already adsorbed on PS. The three methods gave similar results (Table 1). Adhesion on PS coated with BSA-RGD peptide conjugates was even better than on FN used as a positive control. The RGD linear peptides had a positive effect, but less pronounced than the cyclic peptide as described by Gurrath [2]. The non-RGD peptide didn't promote adhesion of HUVECs, which confirms that the interaction of the cells with the modified PS is mediated by the RGD motif. We observed a very low adhesion on BSA modified only with a cystein residue.

The activation of intracellular events (phosphorylation of tyrosine on focal adhesion sites) was verified with cells incubated on glass modified with the BSA-RGD peptide conjugates. The cells spread on FN, which is an adhesive protein of the MEC, and presented several focal adhesion sites with phosphorylated proteins. On the other hand, when plated on an anti-adhesive protein such as BSA, the cells presented a round morphology without focal adhesion sites. The results obtained with the BSA-RGD peptide (linear and cyclic) conjugates, were comparable to those obtained with FN, with clear focal adhesion sites (data not shown). So, the RGD motif in the BSA conjugates is clearly recognized by the integrins and functional.

TABLE 1 : Effects of the modifications of PS on cell adhesion (%). Three methods for preparing BSA-RGD peptide conjugates were compared.

Modifications of PS	A*	B*	C*
PS	-	30	30
FN	100	100	100
BSA	10	35	35
BSA-Cys	50	50	-
BSA-RGD linear peptide	-	115	65
BSA-RGD tailed cyclic peptide	120	140	125
BSA-non RGD tailed cyclic peptide	-	50	30

Cell adhesion was assayed on PS coated with FN alone (set to 100%), with BSA alone, with BSA-Cys and BSA-RGD peptides
* see text.

4. Conclusions

In conclusion, PS can be reconditionned with conformationnally constrained RGD peptides to promote cell adhesion and to induce integrin downstream signalling. Moreover, we have converted the non-adhesive protein BSA into a pro-adhesive protein, that behaved even better than FN, in the conjugate with the cyclic RGD peptides. This approach could be extended to both other peptide motifs and carriers (protein, polymers, ...).

5. Acknowledgements

This work was supported by the Walloon Region.

6. References

1. Main, A. L., Harvey, T. S., Baron, M., Boyd, J. and Campbell, I. D. (1992) The three dimensional structure of the thenth type III module of fibronectin : an insight into RGD-mediated interactions, *Cell* **71**, 671-678.

2. Gurrath, M., Mller, G., Kessler, H., Aumailley, M. and Timpl, R. (1992) Conformation/activity studies of rationale designed potent anti-adhesive RGD peptides, *Eur. J. Biochem.* **210**, 911-921.

3. Delforge, D., Gillon, B., Art, M., Dewelle, J., Raes, M. and Remacle, J. (1998) Design of a synthetic adhesion protein by grafting RGD tailed cyclic peptides on bovine serum albumin, *Lett. Pept. Sci.* **5**, 87-91.

4. Michiels, C., Arnould, T., Houbion, A. and Remacle, J. (1992) Human umbilical vein endothelial cells submitted to hypoxia-reoxygenation in vitro : implication of free radicals, xanthine oxidase, and energy deficiency, *J. Cell Physiol.* **153**, 53-61.

5. Givens, K ; T., Kitado, S., Chen, A. K., Rothschiller, J. and Lee, D. A. (1990) Proliferation of human ocular fibroblasts : an assessement of in vitro colorimetric assays, *Invest. Ophtalmol. Visual Sci.* **31**, 1856-1862.

MATRIX ATTACHMENT REGIONS IN STABLE CHO CELL LINE DEVELOPMENT

MONIQUE M. ZAHN, M. KOBR, N. MERMOD
Laboratory of Molecular Biotechnology
CBUE, DC-IGC, EPFL, 1015 Lausanne-Switzerland

Abstract

We have tested several Locus Control Regions (LCRs), Scaffold/Matrix Attachment Regions (S/MARs) and Boundary Elements (BEs) for their effect on transgene expression in CHO cells. The chromatin elements had little or no effect on transient expression levels. A modest 2- to 4-fold increase in stable expression level was seen for all but a MAR that showed a 20-fold increase. Analysis of individual clones indicates that the MAR generally increases stable transgene expression and reduces the occurrence of poorly expressing clones. Hence stable CHO cell line development with MAR-bearing constructs is clearly promising.

Introduction

The generation of highly productive CHO stable cell lines is a lengthy and costly process. One of the bottlenecks to the establishment of efficiently producing cell lines is the drastic and uncontrollable variation in the specific productivity of different clones. This variation is thought to result from the effect of the local chromatin structure at the site of integration of the transgenic DNA within the chromosome.

One approach to abolish this effect is to make use of chromatin elements capable of shielding the transgene locus from the repressive effects of flanking chromatin, such as S/MARs and BEs. S/MARs may act as a topological barrier, whereas BEs may act as a functional barrier, as they block the action of an enhancer when placed between the enhancer and a promoter driving reporter gene expression. Alternatively, LCRs may possibly be used to overcome a repressive chromatin context.

A. Bernard et al. (eds.), Animal Cell Technology: Products from Cells, Cells as Products, 79–81.

Results and Discussion

Single or combinations of chromatin elements were cloned on either, or both sides, of the luciferase expression unit of pGL3-Control (Promega) (Fig. 1, left pannel). The rat 3' liver activated protein LCR (LAP LCR) and the T-cell receptor alpha LCR (TCRα LCR) were positioned as in their original expression locus. The chicken lysozyme 5' MAR was tested flanking the luciferase expression unit as in Stief *et al.* (1989). Various combinations of the *D. melanogaster* histone spacer SAR (his SAR), heat shock protein 70 locus SAR (hsp SAR), and the special chromatin structure (scs and scs') BEs were also tested.

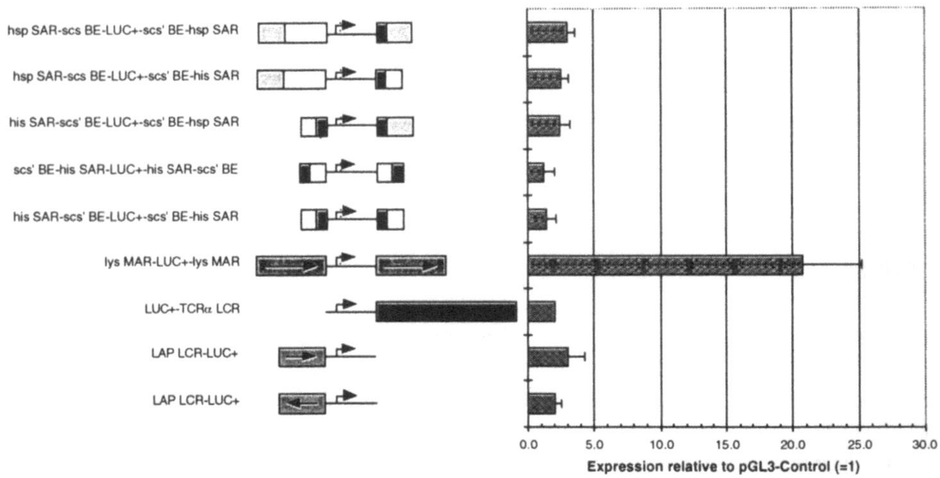

Figure 1. Effect of various chromatin elements on stable transgene expression. Left pannel – Schematic diagram of test constructs. S/MARs are depicted as green boxes, BEs as red boxes, and LCRs as blue boxes. The reporter expression unit (in black) consists of the SV40 promoter, LUC+ reporter gene and SV40 enhancer. Right pannel – CHO DG44 cells were co-transfected by the polyethylenimine (PEI) method (Boussif et al., 1995) with the linearized test constructs and pSV2neo (10:1 molar ratio). The luciferase activity of pools of stable clones was measured and normalized with respect to the protein content.

The chromatin elements had little or no effect on transient expression levels (data not shown) where chromatin structure does not come into play. Pools of stable clones were analyzed as a first approximation of the effect of the various elements. A modest 2- to 4-fold increase in expression levels was seen (Fig 1, right pannel) for all but the chicken lysozyme 5' MAR, which shows a 20-fold increase in stable reporter expression. The orientation of the chicken lysozyme MAR has no effect on transgene expression (data not shown). However, two flanking MARs have a greater effect than a single MAR (Fig. 2 and data not shown). Analysis of individual clones shows that high transgene expression is more prevalent in clones with MAR-bearing constructs than in clones with one or no MARs (Fig. 2).

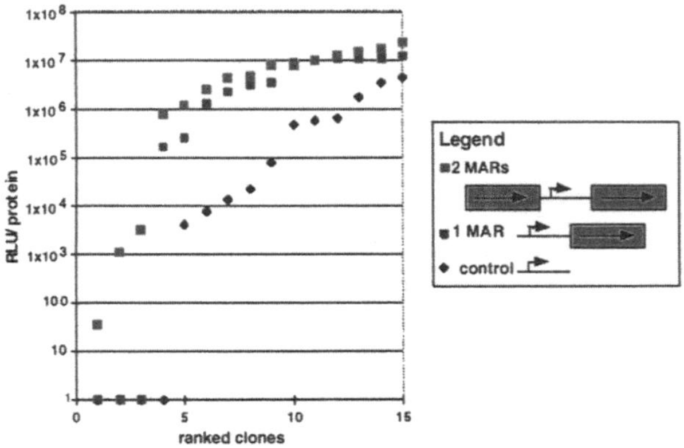

Figure 2. Expression levels for stable clones. The luciferase activity of individual clones was measured and normalized with respect to protein content.

Conclusion

The chicken lysozyme 5' MAR is the only element tested which significantly increases stable transgene expression in CHO cells. As it also appears to reduce the occurrence of poorly expressing clones, the MAR has been used in the development of efficiently producing CHO cell lines (de Jesus *et al.*, 1999).

Acknowledgements

This work was financed by the Swiss National Science Foundation, as part of the Swiss Priority Program in Biotechnology Module 1.

References

Boussif, O., Lezoualc'h, F., Zanta, M.-A., Mergny, M.D., Scherman, D., Demeneix, B., and Behr, J.-P. (1995) A Versatile Vector for Gene and Oligonucleotide Transfer into Cells in Culture and *in vivo*: Polyethylenimine, Proc. Natl. Acad. Sci. USA **92**, 7297-7301.

De Jesus, M.J., Bourgeois, M., Jordan, M., Zahn, M., Mermod, N., Amstutz, H., and Wurm, F. (1999) Establishing and Developing CHO Cell Lines for the Commercial Production of Human Anti-Rhesus D IgG, this volume.

Stief, A., Winter, D.M., Strätling, W.H., and Sippel, A.E. (1989) A Nuclear DNA Attachment Element Mediates Elevated and Position-independent Gene Activity, Nature **341**, 343-345.

HIGH CELL DENSITY CULTIVATION OF HYBRIDOMA CELLS: SPIN FILTER VS IMMOBILIZED CULTURE

H. Heine*, M. Biselli, C. Wandrey
Institut für Biotechnologie 2, Forschungszentrum Jülich GmbH,
52425 Jülich, Germany
*Serono Pharmaceutical Research Institute, Ch. des Aulx 14,
1228 Plan-les-Ouates, Switzerland

Introduction

For long term production of monoclonal antibodies, standard fermentation procedures like batch or open chemostat cultures of hybridoma cells are not sufficient in terms of product quality and volumetric productivity (space time yield). Two approaches for high density cultivations are compared in this study: Single cell suspension cultures using a Spin-Filter[1] as cell retention device and fluidized bed immobilized cultures[2]. Both use perfusion techniques to decouple media and cell residence time in the fermenter. The comparison is carried out with respect to growth rate, specific productivity and space-time yield.

Material and Methods

This study was done using a mouse-mouse hybridoma cell-line secreting a monoclonal antibody of the subclass IgG_{2A} which was kindly provided by Merck, Germany. The culture medium consisted of a 3:1 mixture of DMEM and Ham's F-12 supplemented with: 1 mg/l insulin, 100 mg/l BSA, and 4 mg/l transferrin. Additional fatty acids, vitamins, amino acids and trace elements (all Serva/Sigma, Germany) were added. To determine the suspended cell density the cells were stained with Erythrosin B and counted using a hemocytometer. The immobilized cells were lysed, the nuclei were stained with Crystalviolet and counted using a hemocytometer. The quantification of the product titer was done by a HPLC method, using a PerSeptive ID-Protein-G-Cartridge with UV detection at 280 nm.

For the Spin-Filter experiments a stirred tank fermenter (Applikon, Germany) with 1,2 l working volume was used. The Spin-Filter, which had a pore size of 75 µm and a surface area of 100 cm^2, was mounted on the stirrer shaft. Reinforced silicon tubing with a wall size of 0.8 mm and a surface area of 700 cm^2 was used for the aeration. The pO_2 was controlled at 30% air saturation using an air/O_2-mixture at 0.2 bar overpressure. In the steady state the perfusion rate was 0.085 h^{-1} using 15% culture bleed and 85 % filtrate of the Spin-Filter.

For the immobilized cultures a lab-scale fluidized bed fermenter with 60 ml carrier and 280 ml total volume was used. Siran-beads (Schott, Germany) with a size of 400-710 µm coated with gelatine[3] were used as carrier. Bubble free aeration by thin walled (0.3 mm) silicon tubing with a surface area of 2 m^2 was used to maintain the pO_2 at 75 % air

A. Bernard et al. (eds.), Animal Cell Technology: Products from Cells, Cells as Products, 83–85.

saturation at the fermenter inlet to prevent the immobilized cells from being oxygen limited. During the steady state a perfusion rate of 0.4 h^{-1} was used.

Results and Discussion

IMMOBILIZED CULTURE:

Our results show a major impact of the immobilized cell density on the growth rate and cell specific productivity (Fig. 1 phase 1 and 2; process data given in tab. 1). Due to the very short residence time in this fermenter products can be protected from being degraded. Also the space time yield is about five times higher than in the Spin-Filter system. For the aeration the silicon membranes are absolutely sufficient to prevent the cultures form being oxygen limited.

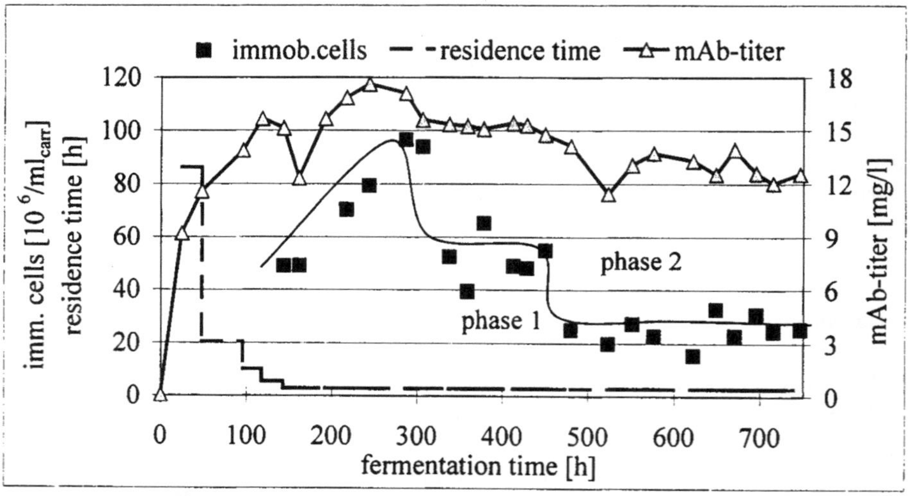

Figure 1: Immobilized cells and mAb titer in fluidized bed fermenter

SPIN-FILTER CULTURE:

During the suspension cultures nearly 500 hours of fermentation were necessary to achieve a steady state (Fig. 2). This results from about 400 hours of time which are necessary to partly foul (coat) the Spin Filter[1]. After this first fouling has taken place no further fouling or clogging of the spin filter could be observed during the next 400 hours of fermentation. Due to the longer residence time the mAb-titer is four times higher compared to the fluidized bed fermenter (Tab. 1).

In table 1 the results of our experiments are shown in comparison to standard batch and chemostat cultures performed in the same stirred tank fermenter. While growth rate and cell specific productivity are in the same order of magnitude, the mAb titer of the Spin-Filter system is comparable with batch fermentations but four times higher than in the other perfusion processes. Nevertheless the space time yield of the fluidized bed fermenter is five times higher than the Spin-Filter system reflecting the very high immobilized cell density and the small volume of carriers used.

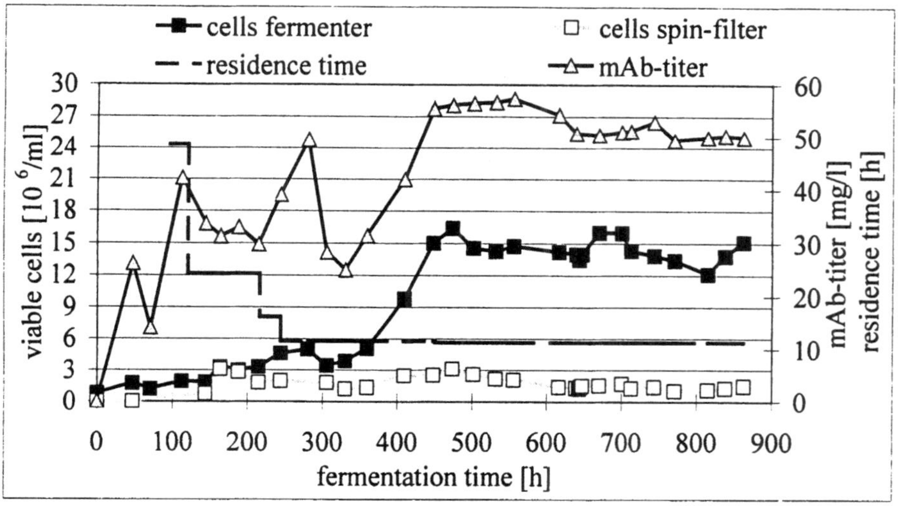

Figure 2: Viable cells and mAb-titer during Spin-Filter fermentation

		Batch Typical	Spin-Filter steady state	Fluidized bed Steady state	Chemostat steady state
growth rate	h^{-1}	0.042	0,033	0,032-0,085#	0.035
specific productivity	mg/(10^9c*h)	0.65	0,29	0,31 / 0,9#	0.45
cell density	10^7/ml##	0.19	1,5	5,8 / 3,0#	0.12
mAb titer	mg/l	44	51	15 / 13#	15
residence time	H	144	11.5	2.8	19.8
space time yield	mg/(l_{reac}*d)##	7.3	105	625 / 541#	18.1
## $ml_{Carrier}$ or $l_{Carrier}$ for fluidized bed fermenter			# Phase 1 / Phase 2		

Table 1: Comparison of process data for the different fermentation systems used

Summary

Both systems are superior to batch or chemostat cultures. Long time fermentations of more than 900 hours could be carried out. The space time yield of the Spin-Filter fermenter was 6 times higher compared to chemostat cultures. The space time yield of the fluidized bed fermenter was even 5 times higher than the Spin-Filter fermenter meaning a 30-fold increase compared to the chemostat culture. The higher mAb-titer of the Spin-Filter systems makes it valuable if the product concentration is the main focus. For unstable and easy degradable products the fluidized bed fermenter is the better choice because of its possibility to be operated at very low residence times.

References

[1] Deo, Y.M.; Mahadevan, M.D.; Fuchs, R.: "Practical considerations in operation and scale-up of spin-filter based bioreactors for monoclonal antibody production"; Biotechnol. Prog., 12, (1996), 57-64

[2] Heine, H.: "Inhibierungen bei hochzelldichten Hybridomakulturen"; Dissertation Universität Bonn, 1998

[3] Lüllau, E.; Dreisbach, C.; Grogg, A.; Biselli, M.; Wandrey, C: Immobilization of animal cells on chemically modified carrier, in: Spier, Griffith (eds.): Animal cell technology, Butterworth-Heinemann,(1992), 469-475

INFLUENCE OF THE INOCULUM ON THE PRODUCTION OF ANTIBODIES WITH MAMMALIAN CELLS

M. INGLIN[1], U. GRAF-HAUSNER[1], B. SONNLEITNER[1],
R. SPINNLER[1], CH. LEIST[2]
[1]*University of Applied Sciences, Winterthur, Switzerland*
[2]*Novartis Pharma AG, Basel, Switzerland*

1 Abstract

The mouse mouse hybridoma cell line BT 376 was cultivated in T-flasks, spinner flasks (30 rpm) and in a surface-aerated loop reactor (5% CO_2 in air, pH and pO_2 unregulated, 300 rpm, tip speed 1.0 m s^{-1}). The aim was to determine the influence of the inoculum on the growth kinetics, viability and antibody production. For this purpose young and old cells, defined by the number of subcultivations, were used. Different procedures for the inoculation of the bioreactor, using T-flasks or a spinner flask, were applied. In addition to the online monitoring of pH and pO_2, several offline measurements were carried out (cell count, viability and antibody concentration). In the bioreactor the specific growth rate of the young cells was lower (μ_6 = 0.93 d^{-1}, μ_{24} = 1.18 d^{-1}), but without a difference in the maximum antibody concentration (c_6, c_{24} = 100 mg l^{-1}) and the volumetric production rate (Qp_6 = 17 mg $l^{-1}d^{-1}$, Qp_{24} = 15 mg l^{-1} d^{-1}). The comparison of the preparation of the inoculum in T-flasks and in spinner flasks showed no significant difference.

2 Introduction

The productivity of monoclonal antibodies (MAb) depends on a variety of parameters. One important step is the preparation of the inoculum. Besides the age of the inoculum (Martial et al. 1991) and the initial cell density (Ozturk et al. 1990) the procedure from the frozen cryovial to the needed inoculum volume has not to be neglected. In this work the influence of the way to prepare the inoculum as well as of the numbers of subcultivation was investigated.

Martial, A., Dardenne M., Engasser J.M., Marc A. (1991) Influence of inoculum age on hybridoma culture kinetics, *Cytotechnology* **5**, 165 – 171.
Ozturk S., Palsson B. (1990) Effect of initial cell density on hybridoma growth, metabolism and monoclonal antibody production, *Journal of Biotechnology* **16**, 259 – 278.

A. Bernard et al. (eds.), Animal Cell Technology: Products from Cells, Cells as Products, 87–89.

3 Materials and Methodes

3.1 CELL LINE AND MEDIUM COMPOSITION

The cell line used in the experimental part of this work was a mouse mouse hybridoma BT 376 secreting an antibody which could be applied as an immunotherapeutic active substance. The cryovials (-196°C, liquid nitrogen) were reconstituted according to a standardized procedure. The commercially available serum-free PFHM-II medium with additions (RPMI 1640 amino acids, Pluronic F68, insulin, 2-mercaptoethanol, ascorbic acid, $NaHCO_3$ and lipids) was used for all the cultivations. Every 2 – 3 days the T-flasks cultures were subculitivated to $8*10^4$ cells ml^{-1}.

3.2 CELL CULTURE SYSTEMS

There were three different systems in use: T-flasks, spinner flasks (500 ml) and a surface-aerated loop reactor (2.5 l) with a four-bladed marine propeller and draft tube. The cultivation conditions were 5% CO_2 in air, pH and pO_2 unregulated, 37 °C and gentle agitation in the spinner flasks (30 rpm) and the bioreactor (300 rpm, tip speed 1.0 m s^{-1}). The cultivations were aborted nine days at the latest.

As in figure 1 shown, the inoculation of the bioreactor was realized in two different ways: from T-Flasks directly and after one split in spinner flasks respectively. The inocula were taken in the exponential phase of growth and inoculated to an inital cell density of $8(\pm1)\cdot10^4$ cells ml^{-1}.

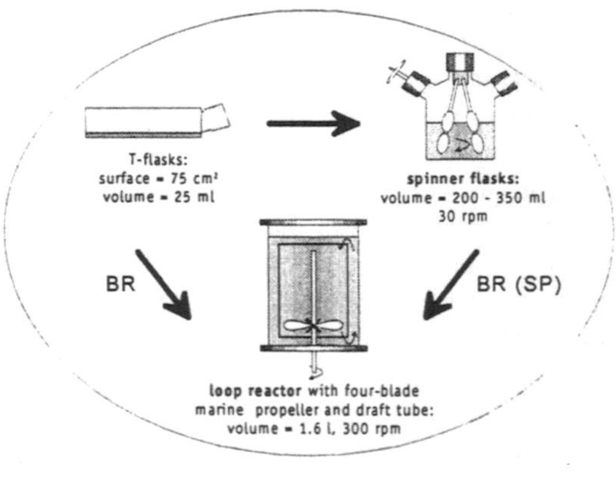

Figure 1. Different ways to inoculate the bioreactor

3.3 ANALYSES

The total cell count and the viability was determined with a haemocytometer (Neubauer improved) and the classical trypan blue assay. Furthermore, the antibody concentration was quantified using an enzyme-linked immunosorbent assay (ELISA). In this sandwich realization horseradish peroxidase (HRP) and 3,3',5,5'-tetramethylbenzidine (TMB) were used for the conjugate-substrate-system. During the cultivations in the bioreactor, pH and pO_2 were monitored inline.

4 Results

4.1 COMPARISON OF YOUNG AND OLD CELLS

As observed in the spinner flasks, the young cells (< split 8) showed a slightly decreased growth kinetics (μ) than the older ones (> split 24) in the loop-reactor as well (fig. 2), but they reached the identical total cell count ($2.3*10^6$ cells ml^{-1}). With respect to the maximal concentrations of MAb and the volumetric production rates (Qp), there was no significant difference (fig. 3).

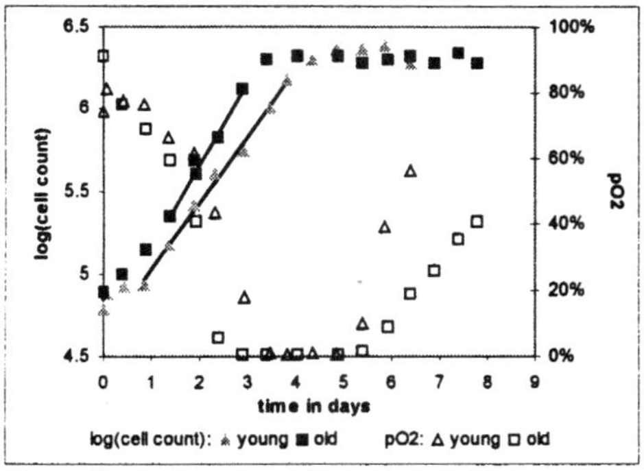

Figure 2. time-courses of growth and pO_2
$\mu_{old} = 1.18\ d^{-1}$, $\mu_{young} = 0.93\ d^{-1}$

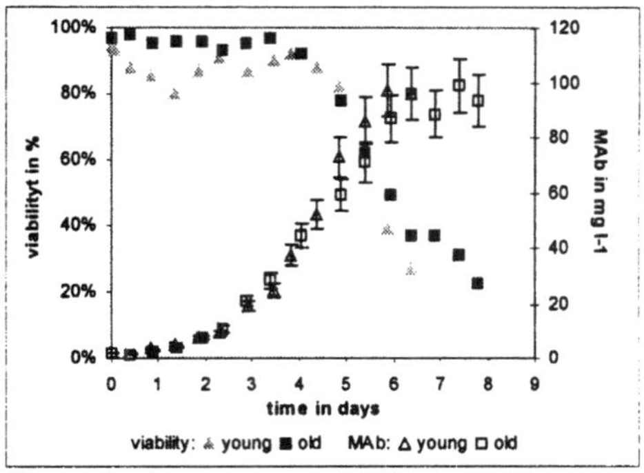

Figure 3. time-courses of viability and MAb
$Qp_{old} = 15$ mg l^{-1} d^{-1}, $Qp_{young} = 17$ mg l^{-1} d^{-1}

After 6 to 8 days there were agglomerations of detritus and viable cells observed in the cultures of the young cells (spinner flasks and bioreactor). Released free nucleic acids are supposed to be responsible for this.

4.2 COMPARISON OF THE WAY FOR PREPARING THE INOCULUM

Except the total cell counts (BR: $2.3\cdot10^6$ cells ml^{-1}, BR(SP): $1.7\cdot10^6$ cells ml^{-1}), the courses of growth kinetics, viability and the antibody concentrations developed similarly (fig.4). So, it doesn't matter whether the needed volume of the inoculum is prepared with spinner flasks or T-Flasks.

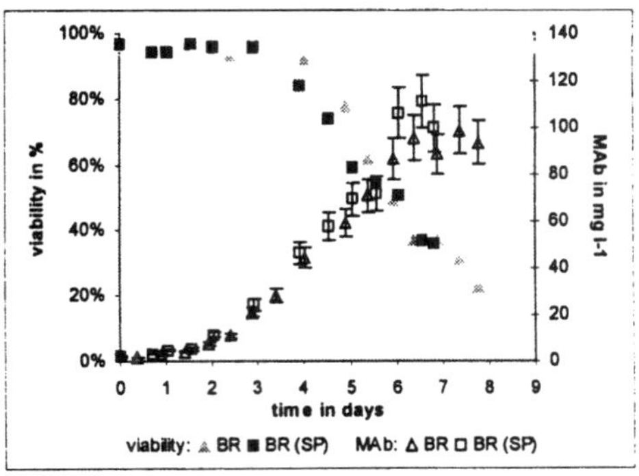

Figure 4. Comparison BR – BR(SP)

5 Conclusions

The observation of the slightly increased growth kinetics of the old cells in spinner flasks was corroborated in the bioreactor. With respect to the maximal concentrations of MAb and the volumetric production rates (Qp) there was no difference. Because there is no significant influence of preparing the inoculum with T-flasks or spinner flasks, practical aspects like consumption of disposable materials or general handling will govern the choice of either procedure.

ESTABLISHING AND DEVELOPING CHO CELL LINES FOR THE COMMERCIAL PRODUCTION OF HUMAN ANTI-RHESUS D IgG

M. J. DE JESUS, M. Bourgeois, M. Jordan, M. Zahn, N. Mermod, H. Amstutz*, F. M. Wurm
Laboratories of the Center of Biotechnology UNIL-EPFL,
1015 Lausanne-Switzerland
ZLB Central Laboratory*
Blood Transfusion Service Swiss Red Cross, 3000 Bern-Switzerland

1. Introduction

Hemolytic Disease of the Newborn (HDN) can be a life threatening condition in Rhesus D positive babies born from Rhesus D negative mothers. Successful prophylaxis in mothers is commonly based on the use of plasma-derived anti-D immunoglobulin preparations, which are rapidly becoming a scarce resource. In consequence, the substitution of the human derived polyclonal anti-D is both a health and an ethical concern. Several groups are involved in the expression of immunoglobulins anti Rh D in different hosts, including fused EBV-transformed human lymphocytes with murine myeloma (only IgG1) (Olovnikova, Belkina et al. 1997) and EBV-transformed B-lymphoblastoid cell lines (IgG1 and IgG3 subclasses) (Kumpel 1997). Among all the eukaryotic expression systems, CHO cells are still the most popular cell host for the production of recombinant proteins as human therapeutic agents (Wurm 1997), because of safety considerations, as well as practical issues related to scale up.
Hoping to provide an alternative method for pharmaceutical manufacturing, we present here various aspects such as expressing vectors, transfection technology and selection strategies as approaches for the establishment of suitable CHO cell lines. We also tried to address the time and material constraints of our multidisciplinary team by using simple (rollerbottle) and more complex (2-20 L bioreactor) systems for production of human anti-Rhesus D antibodies.

2. Results and Discussion

2.1. VECTORS, TRANSFECTION AND SCREENING OF HIGH PRODUCERS

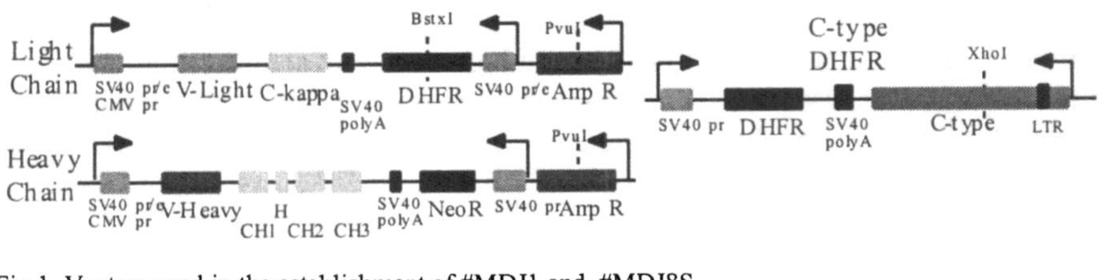

Fig 1. Vectors used in the establishment of #MDJ1 and #MDJ8S.

A. Bernard et al. (eds.), Animal Cell Technology: Products from Cells, Cells as Products, 91–93.

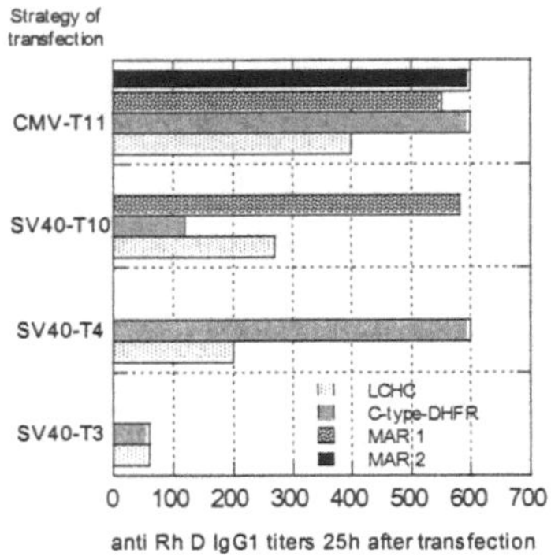

Transient IgG titers subsequent to transfection, provided leads for high productivity stable clones obtained weeks later. From all the transfections done, two "clones" were selected for the purpose of production of anti Rh-D IgG1 to answer the needs of this project, namely related to process development and downstream processing. They were selected from transfection SV40-T3/LCHC (#MDJ1) and SV40-T4/C-type-DHFR (#MDJ8S)(1pg/cell/day) and more recently, #AMW.1 (10 pg/cell/d) derived from SV40-T10 (MAR 1).

Figure 2. Transient titers 25h after transfection. Linearized plasmids encoding for the light and heavy chain constructs were co-transfected into CHO DG44 (dhfr-/dhfr-) cells with a modified calcium phosphate precipitation method (Jordan, et al. 1996). Transfection was performed in 10 cm Petridishes using 25 µg DNA per plate. Colonies were isolated after 10-14 days, with a cotton swap. High producers were screened by an ELISA based assay of the IgG content of their supernatants.

Interestingly, the productivities of all clones selected from transfection CMV-T11 dropped with time. Evidently, none of these clones were further investigated.

2.2. PROCESS DEVELOPMENT AND PRODUCTION OF ANTI RH D IgG1

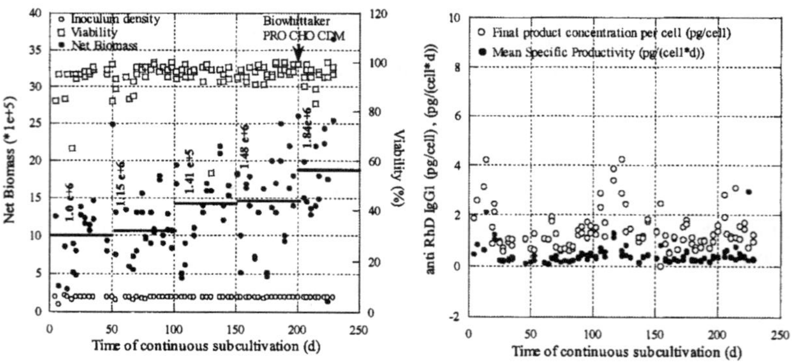

Figure 3. Stability of cell culture parameters and productivities during the period of medium optimization and production of anti Rh-D in 1998, for # MDJ8S.

Clone #MDJ8S was grown in chemical defined medium (CHO FA4) (Biowhittaker) and was maintained in spinner cultures for an extended period of 9 months. During this period of time, medium optimization resulted in higher cell densities, but specific productivities were maintained which indicates a high degree of stability (Fig. 3).
Under batch conditions, #MDJ8S was characterized by a rapid decrease in viability after approximately 110h which closely correlated with the depletion of glucose in the medium (Fig. 4-BR2).

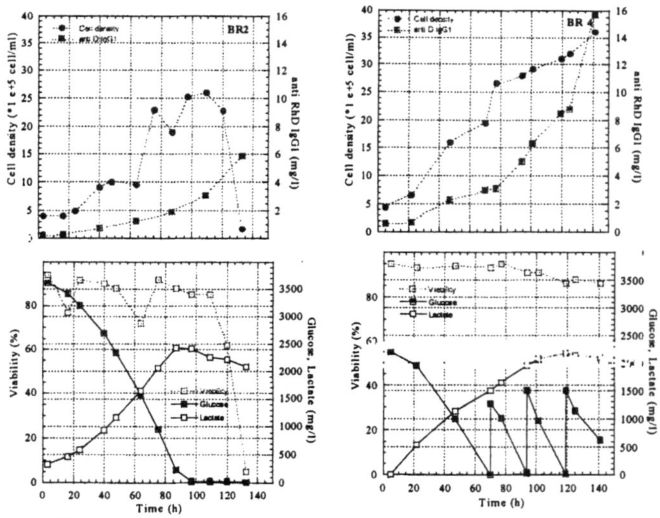

Figure 4. Kinetic profiles of #MDJ8S cultivated in a15L Bioengineering bioreactor with serum free medium (Biowhittaker-CHO Formula A4), under the following culture conditions: T 37°C, pH 7.21, DO 30-50% air saturation and agitation speed 150-200 rpm. The harvest was done before the viability had droped below 50%.

The adopted strategy of glucose feeding (Fig.4-BR4) prolonged viability and increased antibody titers as well as maximum cell density.

	Clone	Period (months)	Nb runs (performed/failed)	Total purified IgG (mg)
15L Bioeng Bioreactor	MDJ8S	10	14/3	320
Roller bottle	MDJ1	4	8/0	55

Table 1. Production of IgG1 in roller bottles and bioreactor in 1998. The production of anti-Rh D IgG1 in rollerbottles (850 cm^2) was performed in DMEM/F12 medium using the #MDJ1 as production host.

CONCLUSIONS

The development of a suitable stable producing cell line is a dynamic and multifactorial process that involves many aspects such as transfection, selection, screening, adaptation in serum free media as well as the behaviour of cell lines under suitable bioreactor conditions. While the work towards a final producer host and an industrially acceptable process is progressing, simple production systems, like rollerbottles can be used to provide supernatants for downstream processing and analysis of product characteristics

References

Jordan, M., A. Schallhorn, et al. (1996). "Transfection of mammalian cells: optimization of critical parameters affecting calcium-phosphate precipitate formation." Nucleic Acids Research **24**(4).

Kumpel, B. M. (1997). "Monoclonal anti-D for the prophylaxis of RhD haemolytic disease of the newborn." Transfus Clin Biol **4**(4): 351-356.

Olovnikova, N. I., E. V. Belkina, et al. (1997). "Immunoglobulin G monoclonal human anti-rhesus Rho(D) to prevent rhesus-incompatibility." Klin Med (Mosk) **75**(7): 39-43.

Wurm, F. M. (1997). Aspects of Gene Transfer and Gene Amplification in Recombinant Mammalian Cells. Mammalian Cell Biotechnology in Protein production. R. W. Hansjorg Hauser. Berlin, Walter de Gruyter & Co: 87-120.

GLUCOSE AND GLUTAMINE REPLACEMENT FOR THE ENHANCEMENT OF FED-BATCH PROCESSES USING CHO CELLS

C. ALTAMIRANO, C. PAREDES, J.J. CAIRÓ AND GÒDIA, F.
Departament d'Enginyeria Química. Escola Tècnica Superior d'Enginyeria. Universitat Autònoma de Barcelona. 08193. Bellaterra. Barcelona (Spain).

1. Abstract

The formulation of the culture medium for a CHO cell line has been investigated in terms of the simultaneous replacement of glucose and glutamine, the most commonly employed carbon and nitrogen sources, pursuing the objective of achieving a more efficient use of these compounds in fed-batch processes, simultaneously avoiding the accumulation of lactate and ammonium in the medium. Despite this substitution, no enhancement of fed-batch culture profiles was observed. Alternatively, preliminary perfusion operation showed better results.

2. Introduction

It has been generally accepted that the compounds to use in cell culture the medium as main carbon, energy and nitrogen sources are glucose and glutamine. However, the rapid glucose and glutamine metabolism that animal cells exhibit when they are cultured *in vitro* leads to a very inefficient use, causing their rapid depletion from the culture medium and eventually the accumulation of lactate and ammonium ions in the medium, that often have inhibiting effects for the cells. In the present work, the substitution of glucose and glutamine in a low-protein medium for the culture of a CHO cell line producing t-PA is analysed. This substitution has been adressed by selecting carbon and nitrogen souces slowly metabolizable to replace the original ones. Once identified, these compounds would be tested in a fed-batch strategy in order to check the viability of this process.

3. Material and Methods

Cell culture: Cell line CHO TF 70R (Pharmacia & Upjohn, Stockholm, Sweeden) producing t-PA. Cells culture were carried out in spinner flasks (Techne) 125 mL, 50 rpm, in a CO_2 incubator, at 37 °C, with 96% relative humidity in an atmosphere of 5% CO_2 in air. Fed-batch was started at 80 h of culture time and feeding every 24 h. Perfusion culture was performed replacing 20% of medium every 60 h.
Culture medium: The basal medium was a propietary serum-free and low protein medium BIOPRO1 (Bio Whitaker Europe, Verviers, Belgium). BIOPRO1 was supplemented with vitamins (Sigma), lipids and cholesterol (GibcoBRL), proline, serine and aspartic acid

A. Bernard et al. (eds.), Animal Cell Technology: Products from Cells, Cells as Products, 95–97.

(Sigma). This medium was also supplemented with 20 mM glucose, fructose, mannose or galactose and 7 mM glutamine or glutamic acid in the different experiments.

Viable and cell concentration: Cell viability was determined by the trypan blue exclusion method, using a haemacytometer.

Metabolite determinations: Glucose and lactate concentrations were determined with a YSI 2700 automated glucose and L-lactate analyser. Ammonium concentration by a flow injection analysis system Amino acids concentrations were measured by HPLC. t-PA concentration by ELISA.

4. Results and Discussion

4.1. SUBSTITUTION OF GLUCOSE BY AN HEXOSE

The glucose substitution in the formulation of the culture medium was investigated, in all cases maintaining the glutamine as nitrogen source. Fructose, galactose and mannose were tested. The results are presented in *Figure* 1.a. Glucose and mannose are consumed at a high rate, whereas fructose and galactose are consumed much more slowly. As a consequence of these different consumption rates, the generation of lactate also differs greatly. For glucose and mannose, the high consumption rate leads to high lactate generation. When cells are grown on fructose and galactose, the lactate production decreases drastically. The glutamine consumption shows a similar profile for all the compounds; however, taking into account the different level of cells generated, the specific consumption of glutamine is notably higher when galactose and fructose are used. These profiles are linked to those corresponding to the ammonium ion generation in the medium, that is higher for the compounds providing a higher specific consumption of glutamine, that is, fructose and galactose.

4.2. REPLACEMENT OF GLUTAMINE BY GLUTAMATE

In the second series of experiments, glutamine was substituted by glutamate in the culture medium, with the main objective of obtaining a more efficient use of the nitrogen source. Galactose, as a slowly metabolizable sugar, and glucose, as the control system were selected for this set of experiments. The results are presented in *Figure* 1.b. It can be observed that the final cell density attained growing on glucose-glutamate is the highest, also notably higher that the final cell concentration for the control culture, glucose-glutamine. The cell profile for the growth on galactose-glutamate and galactose-glutamine is quite similar. It can be observed that galactose is consumed more slowly than glucose, as discussed in the previous section, and also that in presence of glutamate, the consumption of the carbon source is always lower that in presence of glutamine, both for glucose and galactose. With respect to the generation of lactate, it is minimal for the media formulated with galactose. A clear difference can be observed between the control culture, glucose-glutamine and the experiment with glucose-glutamate, as a consequence of the lower consumption for glucose when glutamine is replaced by glutamate.

The most relevant consequence of this different up-take profile is the ammonium generation. Indeed, it is markedly lower when glutamate is used instead of glutamine. It is also interesting to appreciate that the carbon source also affects, although to a minor extent, the ammonium generation, the general trend being that for any of the two nitrogen sources, ammonium concentration always reaches higher values for galactose than for glucose.

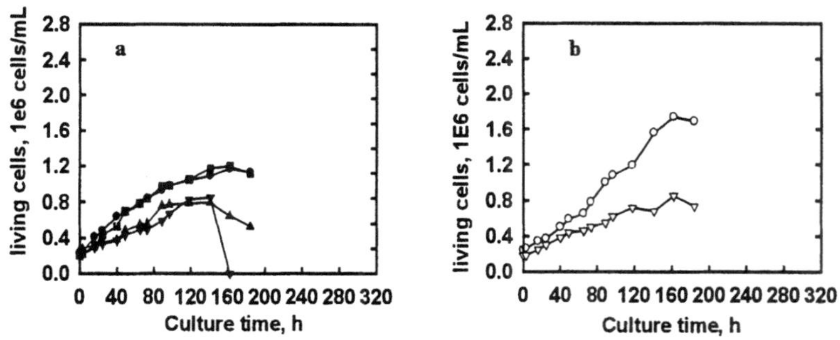

Figure 1. Results of the batch cultures. **a.** glutamine, **b.** glutamate. (●) glucose, (■) mannose, (▲) fructose and (▼) galactose.

4.3. FED-BATCH EXPERIMENTS

Using the information obtained from the previous experiments, a set of fed-batch cultures was carried out. The best results of growth were obtained using glucose and glutamate as subtrates. In any case all the cultures stoped the growth after 120 hours of cultivation as show in *Figure 2.a.* This fact can be explained in terms of cell inhibition due to a toxic compound generated in the media different than lactate or ammonium or due to an increace of the osmotic pressure. When a perfussion strategy was used, this problem could be overcomed as it can seen from the growth of *Figure* 2.b.

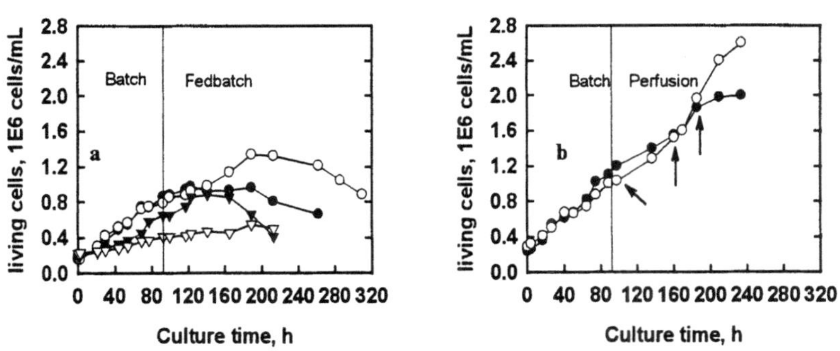

Figure 1. Results of the fedbatch and perfusion cultures. **a.** fed-batch and **b.** perfusion. (●) glucose/glutamine, (▼) galactose/glutamine, (O) glucose/glutamate and (∇) galactose/glutamate.

PILOT SCALE PRODUCTION OF HUMAN PROTHROMBIN USING A RECOMBINANT CHO CELL LINE

D. LÜTKEMEYER, R. HEIDEMANN, H. TEBBE, F. GUDERMANN J. LEHMANN*, M. SCHMIDT, M. RADITSCH[+]
**Institute of Cell Culture Technology, University of Bielefeld, Germany*
[+]BASF AG, Pharmaceutical Research, Ludwigshafen, Germany

1 Introduction

The production of gram quantities of human Prothrombin using a recombinant CHO cell line was investigated.
Hirudin is used as therapeutic agent to treat coagulation disorders. For clinical purposes it is necessary to have an antidote for Hirudin. After activating the rProthrombin an inactive Thrombin is formed which selectively neutralizes Hirudin. This neutralization leads to the restored activity of natural Thrombin the main protease of the coagulation cascade.
Starting with an adaptation towards serum free medium conditions and further optimization of key amino acids and glucose, the process development was started in a 1 L SuperSpinner and a 5 L bioreactor. Further scale up steps included 20 and 100 L Bioreactors. In 20 L scale an internal perfusion system was used to achieve high cell densities as well as high product titers. The cell free harvest was directly concentrated via ultrafiltration 10 to 20 fold. In 100 L scale both batch and continuous cultures were performed. For the continuous cultures a sedimentation device was used. Cell separation was carried out by continuous centrifugation. The cell free supernatant was also concentrated by ultrafiltration (1:10 to 1:20 fold). Prior to freezing the concentrated Prothrombin was diafiltrated and remaining particles were removed by centrifugation.
In total 20 g of recombinant Prothrombin was produced. The Prothrombin from the concentrates was activated and the rThrombin further purified by a number of chromatographic steps.

2 Material and Methods

The cultivated cell line was a recombinant CHO cell line with the genetically modified gene for human Prothrombin (Degen et al. 1983). The culture medium consisted of a mixture of DMEM/F12 plus 1 - 5 % FCS. In the serum free medium 10 mg/L human Transferrin, 10 mg/L bovine Insulin and 0.1 - 2 g/ L Albumin (BSA) were included as protein supplements. 23 nmol/L Na-Selenit was added. Glucose and amino acids were supplemented as needed.
Nuclei were counted via crystal violet stain, free dead cells were stained using Trypan blue dye exclusion. Glucose and Lactic acid were analysed using the YSI-Analyser (Yellow Springs Inst.). The activity of rThrombin was measured after activation of rProthrombin.
For the cultivation 1 L SuperSpinner, 5 and 100 L bioreactors (B. Braun Biotech International, Melsungen) and a 20 L bioreactor (Diessel, Hildesheim) were used.

A. Bernard et al. (eds.), Animal Cell Technology: Products from Cells, Cells as Products, 99–101.

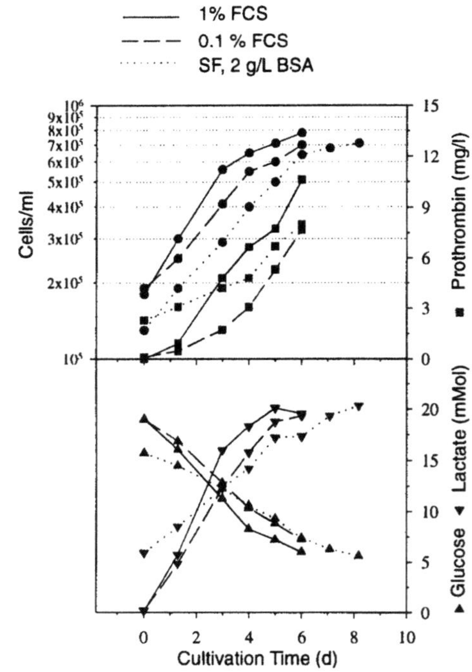

Fig. 1: Adaptation of rCHO cells to serum free conditions

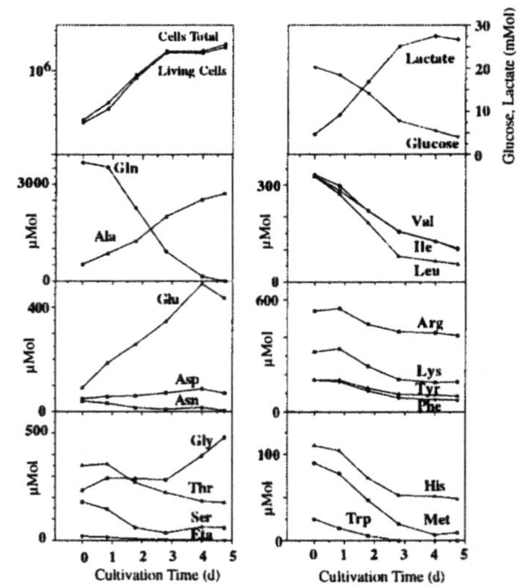

Fig. 2: Batch cultivation in a 5 L bioreactor. According to the amino acid uptake the medium was fortified.

All bioreactors were aerated bubble free. For continuous cultivation an internal microfiltration system based on the Accurel membrane (Akzo) was used in 20 L scale. In 100 L scale a self made sedimentation device was implemented. The cell separation was done with a continuous centrifuge (Contifuge 300 MD, Heraeus Sepatech). The cell free supernatants were concentrated using a ultrafiltration system and the conductivity was reduced via diafiltration (SP20 Ultrafiltration system with S10Y10 modules (cut off: 10 kdal, Amicon)). Remaining particles were removed by another centrifugation step (Cryofuge 6000 (Heraeus Sepatech)).

Results

The first step was the adaptation of the rCHO cells from serum containing medium to serum free (sf) conditions in T-flasks. An adaptation of the cells to the serum free standard medium with a high concentration of albumin was possible (Fig. 1). The cell and product concentration reached similar values compared to a culture with low serum concentration. The anchorage dependent cells changed in sf medium to a suspension culture which allowed the further cultivation in stirred tank bioreactors without carriers.
A further optimization in 1 L SuperSpinner and 5 L bioreactor was performed (Fig. 2). According to the amino acid uptake the medium was fortified. A reduction of Albumin from 2 g/L to 0.1 g/L was possible. The maximum cell density increased to $2 \cdot 10^6$ cells/ml, an increase of the growth rate from $\mu = 0.3 - 0.6$ at the end of the adaptation phase was reached. During the production phase continuous cultures in 20 L scale were performed, an example is given in Fig. 3. The perfusion with total cell retention was started at day 5 and was set to $D = 0.4 - 0.6\ d^{-1}$. The cell density reached $7 \cdot 10^6$ viable cells/ml, viability remained above 90%. The product concentration increased to 33 mg/L.

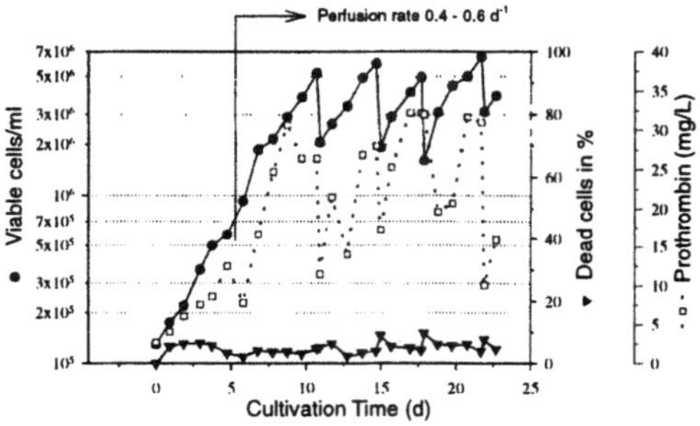

Fig. 3: Continuous cultivation in 20 L scale.

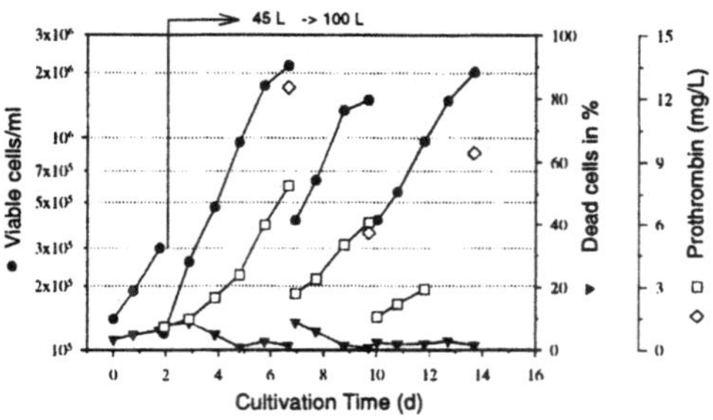

Fig. 4: Repeated batch cultivation in 100 L scale. The diamonds shows the Prothrombin concentrations calculated from the concentrates.

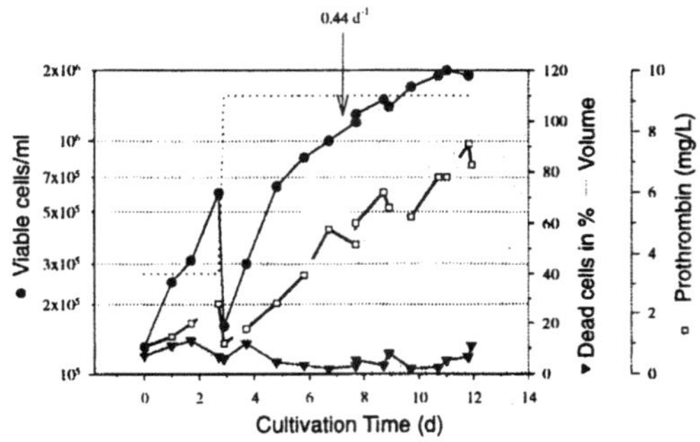

Fig. 5: Continuous cultivation in 100 L scale with partial cell retention using a sedimentation device.

	Volumetric productivity mg/(L•d)	Productivity mg/d
20 L continuous	1.01	247
100 L repeated batch	0.66	176
100 L continuous	0.41	208

Tab. 1: Comparison of different production systems. The continuous 20 L system showed the highest volumetric productivity and productivity per day.

To harvest the product from the cell broth as well, every fourth day 8-12 L culture were removed from the bioreactor and processed.

For the Prothrombin production in 100 L scale batch experiments and a continuous cultivation was done. Three repeated batch cultivations were performed (Fig. 4). The viable cell densities reached up to $2 \cdot 10^6$ cells/ml with product titres of maximum 13 mg/L.

The continuous cultivation in 100 L scale with partial cell retention using a sedimentation device yielded in 496 L supernatant and cell broth with 2.5 g of Prothrombin (Fig. 5).

In table 1 a comparison of the production runs is given. The continuous cultivation in 20 L scale showed the highest volumetric productivity and productivity per day. The cell retention in the continuous 100 L system was not good enough to achieve high cell densities and therefore the productivity was not quite as good.

Conclusion

In total 20 g of recombinant Prothrombin was produced in different scales. An adaptation to a serum free standard medium was possible and allowed the cultivation in suspension mode. Cell densities and specific growth rates in a batch mode were comparable to other in house CHO cell lines. The best volumetric productivity was reached in a continuous 20 L experiment with high cell densities and total cell retention. In 8 pilot scale runs 2000 L of supernatant were produced.

References

Degen et al.1983, Biochem 1983, 22, 2087-2097

MAINTENANCE OF HIGH VIABILITY DURING DIRECT TRANSITION OF CHO-K1 CELL GROWTH FROM 10% TO 0% SERUM: NO ADAPTATION REQUIRED

M.C. TSAO; J.M. JOHNSON; C. BACHELDER; M.A.WOOD; J.A. BOLINE; R.N. BERZOFSKY
BioWhittaker, Inc.
Walkersville, Maryland 21793-0127, United States

Introduction

Historically, serum or serum-fractions have had a universal applications in cell culture. It is an indispensable supplement for classical cell culture media (CCCM) because CCCM were developed containing serum. Utilizing these media to grow cells without the addition of serum cannot be easily achieved. In addition, the CCCM were developed primarily to support the growth of low density adherent cells in t-flasks where cell growth is limited by surface area. The usual method to grow cell without serum is to adapt the cells to the CCCM by decreasing serum levels in a step-wise fashion. To decrease the serum levels, these cells must maintain a reasonable cell viability and density. It is very important to keep in mind the ultimate objectives when adapting cells to a serum-free environment. If the secretion of a large amount of a specific protein is the final goal, monitoring the level of secreted protein while adapting the cells to serum-free medium becomes very critical. With decreasing serum levels, the number of viable cells tends to be low, making the adaptation process laborious and time consuming. Furthermore, the adapted cells are in CCCM that cannot effectively support the desired high density suspension culture. Basically, there is no universal serum-free CCCM applicable to all cell lines. Customized media that has been specifically tailored to each cell line is required. The goal was to develop a defined serum-free medium that would support the direct transition of *serum-supplemented adherent* CHO-K1 culture to *serum-free suspension* (spinner) culture. An optimized chemically defined medium (CDM), ProCHO4-CDM, that achieves high density cell growth for the Chinese Hamster Ovary (CHO) cell line, was compared to two commercially available serum-free CHO media.

Material and Methods

Media Preparation:

ProCHO4-CDM is a selective medium and was made complete before each experiment by the addition of hypoxanthine, thymidine, and L-glutamine. The Commercial Suppliers' (Comm.Sup.1 and 2) media were completed by following supplier's instructions. The Commercial Media were supplemented with additional recombinant (r-) insulin (denoted with +I) which was required to achieve the transition from serum-supplemented adherent culture to serum-free suspension cell growth. DMEM/F12 (D/F) and DMEM (CCCM) media were supplemented with 10mg/L r-Insulin and 0.1% F68.

Culturing Protocol:

The CHO-K1 cells were cultured in DMEM/F12 supplemented with 10% FBS. These cells were subcultured and used as the inoculum to start the serum-free and suspension adaptation in spinners and t-flasks. The starting densities for spinners and t-flasks were 2-3 X 10^5 cells/ml and 10 million cells per flask in a total volume of 45 ml. T-flasks were maintained in a 37°C, 5% CO_2 environment. Both 250 ml and 500 ml unsiliconized spinners were utilized in suspension experiments. When feeding was included in the spinner experiment, 25 to 50 ml of fresh media were added to culture on specified days. For 250 ml and 500 ml spinners, 120 and 200 ml volumes were initially used respectively. Spinners were maintained at approximately 80 rpm in either 37°C incubators supplied with either non-CO_2 or 5% CO_2 environment.

A. Bernard et al. (eds.), Animal Cell Technology: Products from Cells, Cells as Products, 103–106.

Counting Protocol:
To ensure that accurate cell counts were obtained, daily aliquots of suspension cultures were aseptically removed from the spinners and t-flasks and 1/10 volume of 2.5% Trypsin (10X) was added. After incubation at 37°C for 15 to 20 minutes, the suspension was mixed briefly, and cells were counted by hemacytometer with trypan blue.

Results:

Concurrent suspension and serum-free CHO culture adaptation in ProCHO4-CDM

5% CO_2 incubator environment

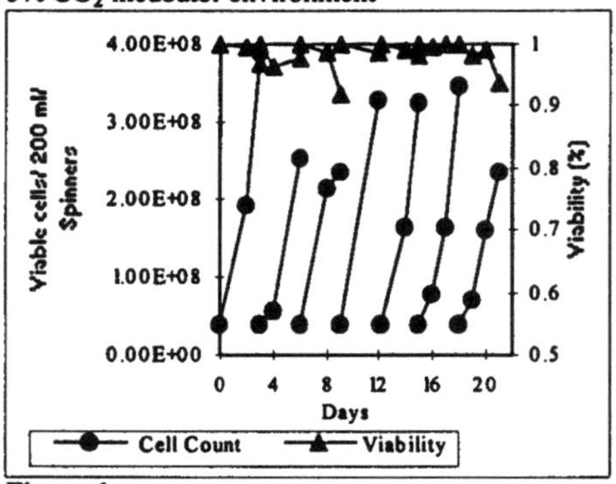

Figure 1

Non-CO_2 incubator environment

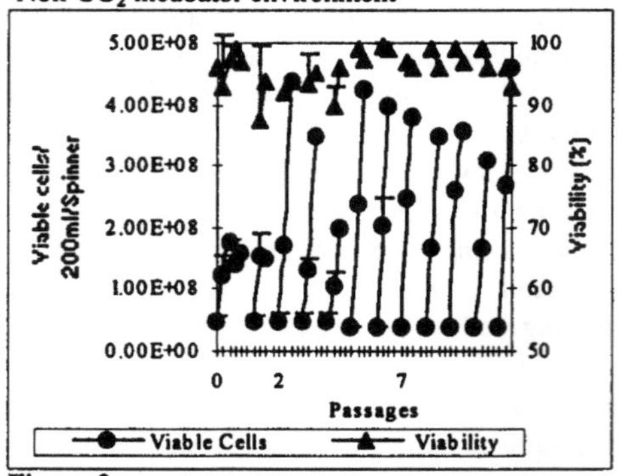

Figure 2

The 500 ml spinners were filled with 200 ml of ProCHO4-CDM and inoculated at 2 X 10^5 cells/ ml. CHO-K1 inoculum cells were subcultured from cells grown with DMEM/F12 + 10% FBS in t-flasks. The adherent cells in t-flasks were detached by trypsin, neutralized by serum containing medium, and spun to remove the trypsin and medium. Every three to four days, these cells were passaged in ProCHO4-CDM. The data supports the direct transition from adherent culture in 10% serum to suspension culture in 0% serum. Both cell growth and viability were maintained in ProCHO4-CDM for multiple passages. The spinners for this experimenrt were maintained at 37°C and 5% CO_2 and non-CO_2 environments.

Ability of serum-free media to support CHO cell adaptation from cells grown previously in 10% attached serum medium

In to T-flasks static cultures

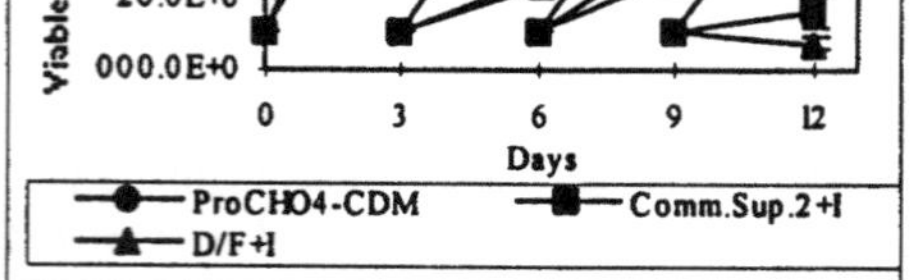

Figure 3

In to suspension spinner cultures

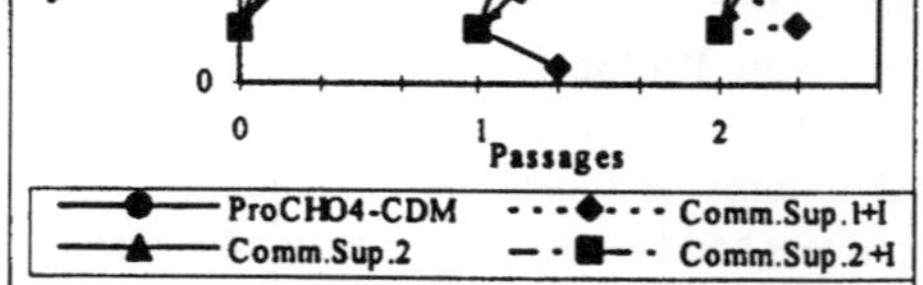

Figure 4

The goal was to develop a defined serum-free medium that would support the direct transition of ***serum-supplemented adherent*** CHO-K1 culture to ***serum-free suspension*** (spinner) culture in one step. The experiment was initiated with CHO-K1 cells grown attached in DMEM/F12 + 10% FBS. Three t-150 flasks were initiated with each media, cells were seeded at 10 million cells per flask in a total volume of 45 ml. Every three to four days, these cells were passaged in the appropriate media. By the fourth passage, both the Comm.Sup.2 and D/F supplemented with r-Insulin (+I) were not able to sustain the high growth rate supported by ProCHO4-CDM.

SF adapted CHO cells subcultured into five different defined media

Conditioned in Commercial Supplier medium 2+I

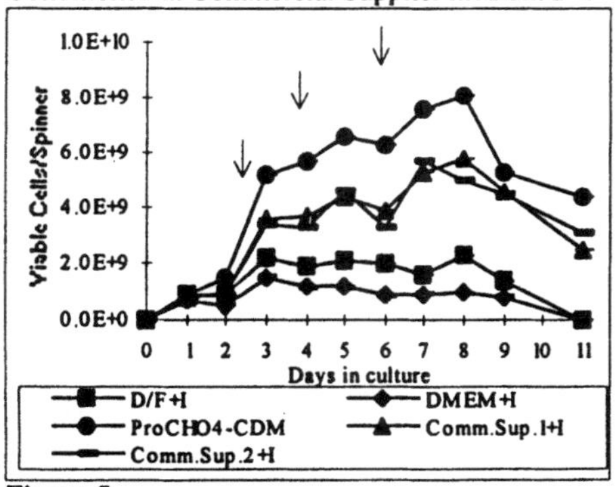

Conditioned in ProCHO4-CDM

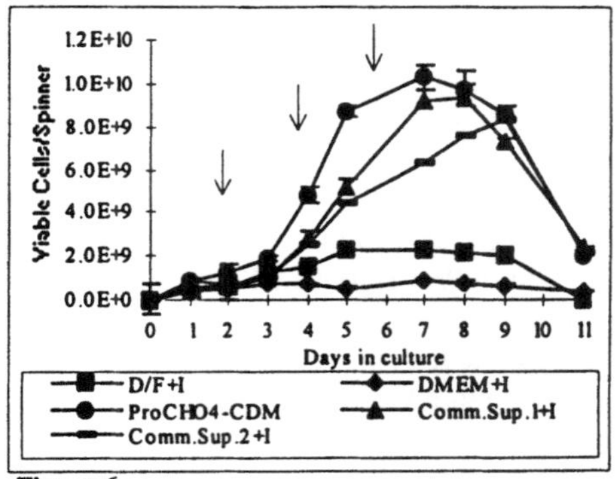

Figure 5 **Figure 6**

These growth kinetic experiments were initiated with pre-adapted SF suspension cells conditioned in both Comm.Sup. 2 medium supplemented with additional r-Insulin and ProCHO4-CDM These conditioned cells were subcultured into five different defined medium listed in the legend. Media denoted with '+I' had r-Insulin added (10mg/L). The classical media, DMEM and D/F were supplemented with additional 0.1% F68. Each medium had triplicate spinners with starting seed density at 3 X 10^5 cells/ml in 250 ml spinners with 120 ml fill, and were fed on days 2, 4, and 6 with 50 ml to all spinners. Cells counts were performed daily from days 1 to 11. The ProCHO4-CDM supported the most rapid growth rate with the highest viability.

SF-CHO cells pre-conditioned in Commercial Supplier 2 medium and ProCHO4-CDM

Conditioned in Commercial Supplier medium 2+I

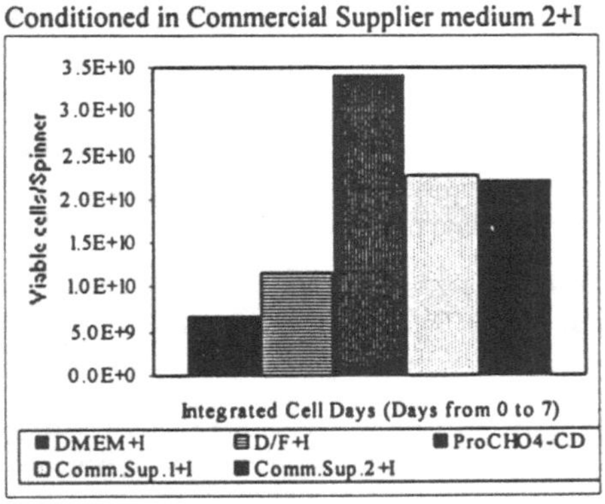

Conditioned in ProCHO4-CDM

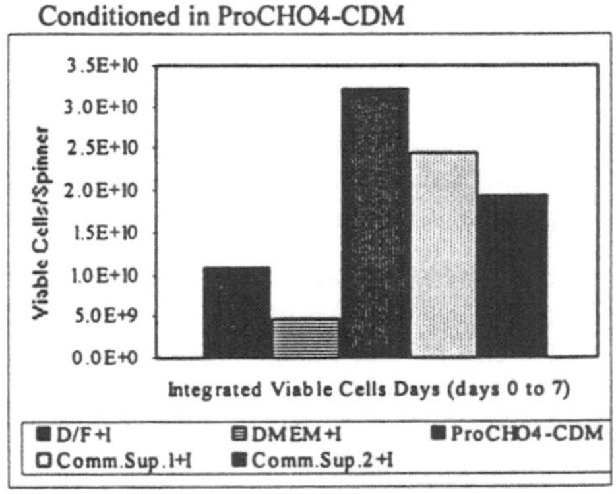

Figure 7a **Figure 7b**

CHO-K1 cells used in the above experiments were pre-conditioned with multiple passages to serum-free growth using two different defined media (Panel a, b). The accumulated viable cells per spinner were calculated from the average of triplicate spinners for each media formulation. The medium that consistently supported the highest accumulated viable cells per medium was ProCHO4-CDM.

Discussion:

The current biotechnological approach to CHO cell culture is to avoid the use of serum. Some therapeutic proteins can be economically produced in the CHO cell line without the need to supplement the culture with defined natural animal proteins. The transition process to non-animal supplemented culture medium is difficult. These difficulties are further compounded by using CCCM and sub-optimal CDM. The gradual step wise reduction of serum was developed by the need to minimize the repercussion to poor growth rate, viability and potential loss of secreted products. The step wise serum reduction part of process development can consume developmental time by one to three additional months. Sometimes the transition to chemically defined medium must be abandoned due to failure of the cells to adequately adapt to the medium. This time consuming and difficult process can be minimized by starting the serum withdrawal process with an optimized defined CHO medium.

A one-step transition from serum to serum-free growth requires a medium that can support active cell growth and prolonged high cell viability. By starting with ProCHO4-CDM, the need for gradual reduction of serum was eliminated. This medium can save months in process development time by circumventing the gradual adaptation process because ProCHO4-CDM provides the optimized micro-nutrient environment required. No matter how easy it was to transition to ProCHO4-CDM, it is important to verify that the product yield is not compromised during the transition process. There is an abundance of evidence in the literature to support the direct correlation between viable biomass and the amount of secreted proteins. ProCHO4-CDM supported more viable cell mass over time (fig. 7a and b) when compared to other commercial supplier of CHO media.

There is no one universally defined CHO medium (ie.without natural animal protein) that will support the production of all types of therapeutic proteins. However, ProCHO4-CDM promotes easy transition to defined medium conditions, has been shown to support high cell density concurrent with prolonged high viability, and may be a good basal medium to select to further optimize for maximum protein yield.

Conclusions:

- The ProCHO4-CDM bridges the gap for adapting CHO cells from 10% serum containing attached t-flask CHO culture to serum-free suspension spinner culture in both CO_2 and non-CO_2 culture environments.
- An optimized micro-nutrient environment medium minimizes the 'lag' period in each new subculture of cells. The accumulated 'lag' time saved, during a protein production scale up process can be translated to a shorter production process, and potentially more product yield.
- ProCHO4-CDM supports the active respiration of cells in non-CO_2 culture environments, at low seeding density, enabling sufficient release of CO_2 into the new subculture medium thereby, minimizing the pH of the medium to drift.
- ProCHO4-CDM supports high cell density with extended high viable days in fed-batch uncontrolled spinner cultures when compared to CCCM and other commercial suppliers' of CHO media.

References:

1. Man Bock Gu, et al (1996) Metabolic burden in recombinant CHO cells: effect of *dhfr* gene amplication and *LacZ* expression. Cytotechnology 18: 159-166.
2. Mitsua Satoh, et al (1990) Chinese Hamster Overy cells continuously secrete a cysteine endopeptidase. In vitro Cell Dev. Biol. 26:1101-1104.
3. Ralph E. Parchment et al (1992) A free-radical hypothesis for the instability and evolution of genotype and phenotype *in vitro*. Cytotechnology 10: 93-124.
4. David T. Berg, et al (1993) High -level expression of secreted proteins from cells adapted to serum-free suspension culture. Biotechniques: vol 14,No 6.
5. Seamans, et al (1994) Use of lipids emulsions as nutritional supplements in mammalian cell culture. Ann N.Y. Acad. Sci. 245: 240-243.

ANALYSIS AND SIMULTANEOUS ISOLATION OF RECOMBINANT t-PA USING HIGH PERFORMANCE MEMBRANE AFFINITY CHROMATOGRAPHY

G. RENEMANN[1], G. KRETZMER[1], T. B. TENNIKOVA[2]

[1]*Institut für Technische Chemie, Universität Hannover, Hannover, Germany*
[2]*Institute of Macromolecular Compounds, Russian Academy of Sciences, St. Petersburg, Russia*

INTRODUCTION

Recently found serine protease called as tissue plasminogen activator (t-PA) is able to dissolve efficiently the blood clots. Thus this protein seems to be extremely useful in clinical practice in the cases of heart attack victims.

The real process of fibrinolysis in human blood system represents very complicated network of simultaneous biological events. It is clear that t-PA has a branched set of functional complements with their own, and probably different, affinity to this enzyme. It seemed to be possible and quite interesting to investigate all these pairs separately *in vitro*. At the same time, it is clear that the affinity chromatography approach can become as the most convenient way to study such artificially created biological pairs.

The recently developed *high performance membrane chromatography* (HPMC) is quite promising in this regards because of its high capacity and selectivity, combined with low back pressure and short operation times. Due to the inherent speed of the isolation it facilitates the recovery of a biologically active product, since the exposure to putative denaturing influences such as solvents, temperature, and contact time is reduced.

Affinity HPMC using the specially designed discs with a macroporous structure identical to effective particle sorbents is likely to overcome many critical disadvantages. Most importantly, the better mass transfer mechanism allows to consider only the biospecific reaction at time limit.

A. Bernard et al. (eds.), Animal Cell Technology: Products from Cells, Cells as Products, 107–109.

This fact seemed to be used effectively not only in affinity separation processes but also at the in vitro modeling of different biological events following the forming of complementary functional pairs.
The development of the fast analysis of t-PA directly from supernatants by High Performance Membrane Affinity Chromatography using monoclonal antibodies as ligands is shown.

METHOD

In this method for the analysis of t-PA by High Performance Membrane Affinity Chromatography macroporous discs instead of columns are used as solid phase. This means the advantage of high flow rates, no backpressure and short analysis times (2 min). The only time limiting factor is the biospecific reaction because mass transport takes place by convection instead of diffusion.
After immobilisation of ligands (antibodies) on the surface of the disc material affinity chromatography of t-PA is carried out. As loading buffer PBS (10 mM, 150 mM NaCl) is used and for desorption HCl (2n) is used.

RESULTS

Figure 1 shows a typical chromatogram of the analysis of t-PA directly from supernatants.

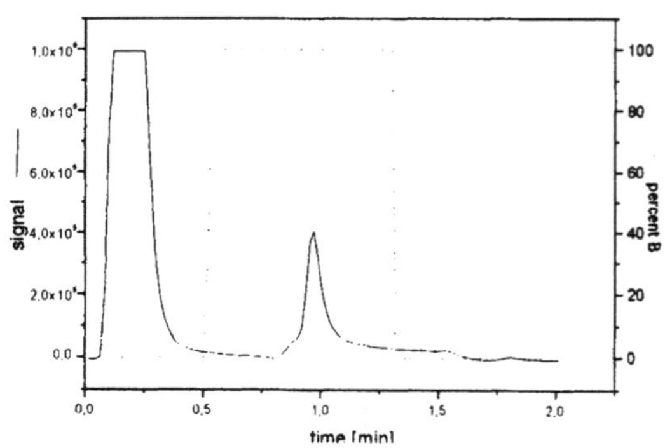

Fig. 1: Typical chromatogram of the analysis of t-PA from supernatants. Dot line: gradient shape (percentage B, buffer A: PBS, buffer B:HCl (2 n))

Figure 2 shows the calibration of the HPMC.

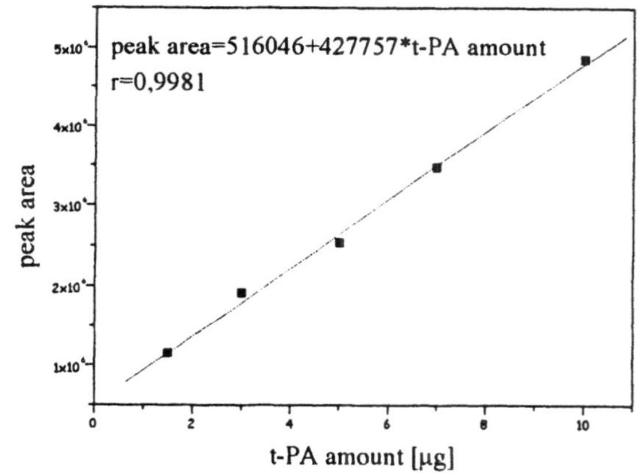

Figure 2: Calibration of the affinity HPMC

In figure 2 a good correlation between peak area and t-PA amount loaded onto the disc can be seen. The lowest traceable t-PA amount is 1.5 µg t-PA.

The affinity constant of the pair t-PA--mAB was determined by frontal analysis.

Frontal analysis of the disc with immobilized antibodies leads to an affinity constant of $2{,}83*10^{-7}$ M.

OUTLOOK

In the future it is planned to immobilise other proteins like activators, inhibitors, substrates and artificial peptides on the macroporous disc material.

By determination of the affinity constants of these different ligands by frontal analysis the most „economic" ligand for the analysis and purification of t-PA by affinity HPMC will be chosen.

After establishing the off-line analysis and purification of t-PA, devices for on-line analysis and purification of t-PA directly from cultivations shall be developed.

METABOLIC AND PROCESS ENGINEERING

Chapter II

IMPROVEMENT OF THE PRIMARY METABOLISM OF CELL CULTURES INTRODUCING A PYRUVATE CARBOXYLASE REACTION PATHWAY

N. IRANI[1], M. WIRTH[2], J.v.d. HEUVEL[3], A.J. BECCARIA[4], R. WAGNER[1]
[1]Cell Culture Technology Dept., [2]Regulation and Differentiation Dept., [3]Production and Expression Systems Dept., Gesellschaft für Biotechnologische Forschung mbH, Mascheroder Weg 1, D-38124 Braunschweig, Germany
[4]Instituto de Tecnologia Biologica, Facultad de Bioquimica y Ciencias Biologicas, Universidad Nacional del Litoral, C.C. 242 - 3000 SANTA FE - Pcia. Santa Fe, Argentina

I. Introduction

Mammalian production cell lines are unable to completely oxidize Glucose to CO_2 and H_2O [1-4]. This leads to a high throughput of the substrates glucose and glutamine giving a low energy yield and ample toxic side products such as lactate and ammonia[5]. Low activity of the enzymes connecting glycolysis with the TCA cycle are shown to be the cause for these metabolic disorders[6]. We show that introducing a cytosolic yeast pyruvate carboxylase (PYC) gene into a BHK cell line reduces this problem by reconstituting this connection and channeling pyruvate and NADH derived from glycolysis into the TCA cycle. As a result substrates are better exploited and higher product yields and concentrations are achieved.

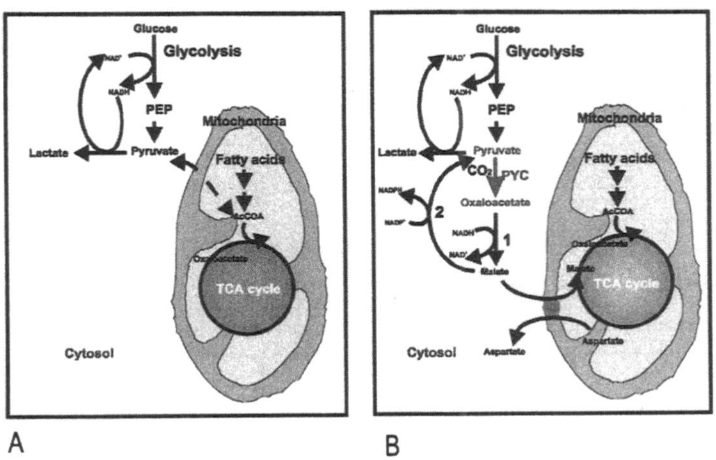

Fig. 1 Strategy of Pyruvat Carboxylase. (a) Usual pathway of glucose oxidation in cell lines. (b) After introduction of pyruvate carboxylase pyruvate derived from glycolysis is converted to oxaloacetate in the cytosol and enters the mitochondria after conversion to malate via different transporter shuttles. The backward reaction of malate to pyruvate supplies extra NADPH for reductive processes.

A. Bernard et al. (eds.), Animal Cell Technology: Products from Cells, Cells as Products, 107–109.

II. Results

Cell growth. After transfection stable clones were tested for growth, viability and nutrient consumption in batch experiments. It could be shown that the yeast PYC enzyme did not impair cell proliferation since growth rates were almost identical for the clones and the control. The life-span of the PYC expressing cultures in batch mode was extended for 2 to 3 days versus the control. Subclones of the control all showed shorter viability periods suggesting that our results were indepent of cloning effects. In perfused bioreactor experiments cell concentrations more than 2 fold of the control, could be achieved.

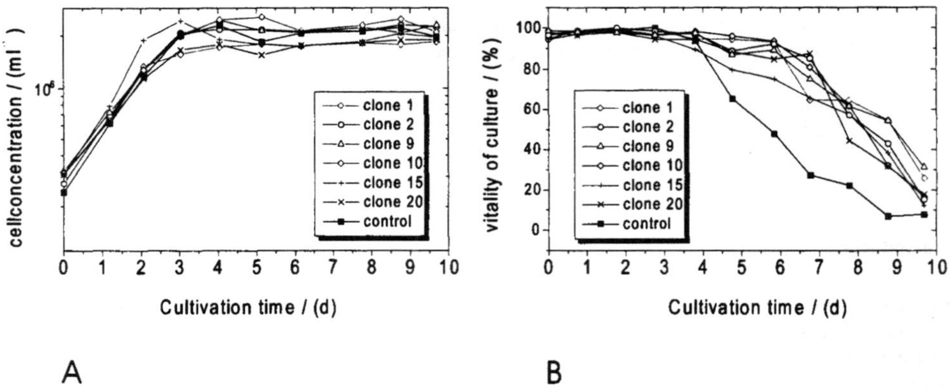

Fig. 2 (A) Cell nuclei concentrations in batch cultures. (B) Viability in cultures assessed by viable cell counts. In all cases cells were cultured in ZKT-1 medium supplemented with serum and geneticin under identical conditions.

Nutrient consumption. In batch cultures PYC expressing cells revealed a reduced glucose and glutamine consumption up to 4 fold and 1.8 fold, respectively. Moreover, the rate of lactate production was reduced up to 2.5 fold, resulting in lower lactate concentrations in culture media. In a chemostat reactor lower consumption rates could be reproduced (table 1).

Table 1 Metabolic data for control and PYC-transfected cells cultivated in batch and in a chemostat bioreactor

Cell specific metabolic data	Batch cultivation control	Batch cultivation clones	Continuous cultivation control	Continuous cultivation clone	Fold improvement batch	Fold improvement continuous
glc consump rate ($nmol*s^{-1}*10^{-6}$)	0.205	0.051 - 0.140	0.062	0.036	1.4 - 4.0	1.78
gln consump rate ($nmol*s^{-1}*10^{-6}$)	0.048	0.027 - 0.044	0.0103	0.0072	1.1 - 1.8	1.4
lac prod rate ($nmol*s^{-1}*10^{-6}$)	0.332	0.13 - 0.21	0.076	0.032	1.6 - 2.5	2.3

Flux analysis and O_2 consumption. The flux of labelled D-[6-^{14}C]-glucose and pyruvate was increased indicating a higher rate of oxidative glucose degradation (table 2). Moreover O_2 consumption measured throughout a chemostat fermentation was elevated up to 2.9 fold, suggesting an increase in the oxidative phosphorylation rate.

Table 2 Oxygen consumption and flux of metabolites into the TCA cycle

Cell specific metabolic data	control	clones
O_2 consumption ($pmol*s^{-1}*10^{-6}$)	35	102
D-[6-^{14}C]-Glucose flux ($nmol*h^{-1}*10^{-6}$)	0.27	0.39
^{14}C-Pyruvate flux ($nmol*h^{-1}*10^{-6}$)	9.57	21.01

Productivity. To evaluate the production capacity the PYC bearing cells and the control were transfected once again with a model glycoprotein. Identical productivity of the cells was ensured using eGFP as reporter protein. PYC expressing cells cultivated in perfusion mode showed almost twice the concentration of product in the culture harvest and also a better exploitation of glucose for production demonstrated by a higher value of product produced per glucose consumed.

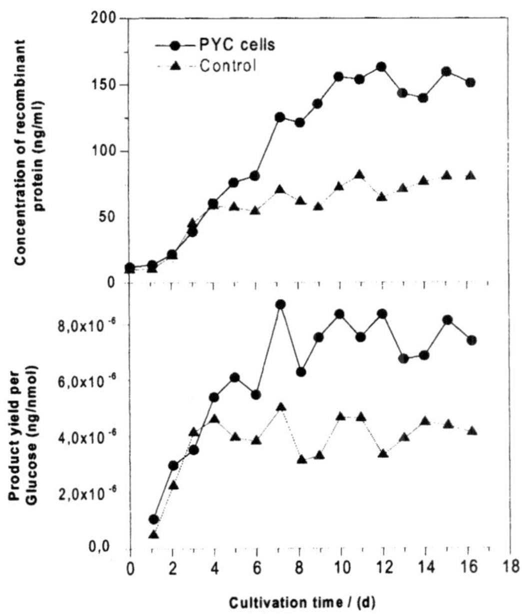

Fig. 3 Concentration (top) and yield of product per consumed glucose (down) in a perfused bioreactor. Perfusion rates in both reactors were almost identical. The mean product concentration achieved is 138,6 ng/ml for the PYC expressing culture and 69,7 ng/ml in the control.

III. Conclusion

Introducing a yeast pyruvate carboxylase enzyme into a production cell line enables us to extend life-span and production capacity in batch as well as in perfusion mode, leading to reduced production costs of protein pharmaceuticals. With the PYC metabolic engineering experiment the following biochemical goals could be achieved:

1. Gucose carbon is completely oxidized in the TCA cycle.
2. The cytosolic pyruvate carboxylase reaction competes with lactate dehydrogenase for the substrate pyruvate leaving less substrates for lactate formation.
3. The malate dehydrogenase reaction replenishes NAD^+ which is needed in the glycerinaldehyde-3-phosphate reaction during glycolysis. With this the importance of the LDH reaction for the regeneration of NAD^+ is reduced. Moreover, reconversion of additional malate produced in the MDH reaction can serve as a useful cycle for the production of NADPH.

IV. References

1. Lanks, K.W., and Li, P.W. 1988. End products of glucose and glutamine metabolism by cultured cell lines. *J. Cell. Physiol.* **135**:151-155

2. Fitzpatrick, L., Jenkins, H.A, and Butler, M. 1993. Glucose and glutamine metabolism of a murine B-Lymphocyte hybridoma grown in batch culture. *Appl. Biochem. Biotech.* **43**:93-116

3. Glacken, M.W. 1988. Catabolic control of mammalian cell culture. *Bio/Technology* **6**:1041-1050

4. Petch D., and Butler M. 1994. Profile of energy metabolism in a murine hybridoma: Glucose and glutamine utilization. *J. Cell. Physiol.* **161**:71-76

5. Ozturk, S.S., Riley, M.R., and Palsson, B.O. 1992. Effects of ammonia and lactate on hybridoma growth, metabolism, and antibody production. *Biotechnol. Bioeng.* **39**:418-431

6. Neermann, J., and Wagner, R. 1996. Comparative analysis of glucose and glutamine metabolism in transformed mammalian cell lines, insect and primary liver cells. *J. Cell. Physiol.* **166**:152-169

Discussion (Irani)

Bailey: I notices in your perfusion culture that the biggest difference between your construct and the control occurred very late, towards the end of the perfusion culture. What was happening there- was there a special condition in the culture at that time where the cells had an advantage?

Irani: We started to drive up the perfusion rate and when it gets high then the advantage becomes more pronounced. However, in the production perfusion we chose a smaller rate of perfusion increase and there you can see it already at the beginning - you do not need to go as high as 3 vols/reactor volume to see that difference..

Zhou: Did you see changes in the acetyl-CoA pathway in the TCA cycle after you had expressed the pyruvate carboxylase gene?

Irani: We did not look at acetyl-CoA - we assumed that it is built through better oxidation from fatty acid oxidation. The flux of radioactive labelled glucose to CO_2 was measured and it was enhanced.

Moran: Have you a measure of the stability of your expression of *pyc* transfected clone - are you confident that it maintains expression in long-term culture?

Irani: Yes, we used a bi-cistronic construct for the product - later we used an enhanced GFP and EPO in one construct and with that we could see that the EGFP expression was still there after weeks of cultivation.

GROWTH KINETIC STUDIES (IN HOURS) BASED ON THE FLUORESCENCE OF STABLE, GFP-EXPRESSING CHO CELLS

L. Hunt, M. Jordan, M. De Jesus, F. M. Wurm
Laboratoire de Biotechnologie Cellulaire,
Swiss Federal Institute of Technology, 1015 Lausanne, Switzerland

Keywords: GFP, CHO, growth kinetics, microplate assay

ABSTRACT

This study correlates the fluorescent signal from a stable CHO cell line expressing the **G**reen **F**luorescent **P**rotein (calcium phosphate cotransfection of CHO DG44 cells with 1:20w/w linearized pEGFP-N1 from Clontech and a DHFR vector) with viable cell number, extending the use of fluorescent proteins to kinetic applications. This cell line has maintained its GFP expression over months without selective pressure (data not shown). Using a standard fluorometer, growth of these cells can be quantified non-invasively in multiwell plates, and since signals are obtained without preparation, the same culture samples can be measured repeatedly. In this way, the dynamics of cell growth can be studied with high sensitivity, low error rate and minimum sample preparation and on a time scale practically impossible with traditional methods. It is thus possible to identify and to follow growth trends without counting cells.

INTRODUCTION

Mammalian cells can be quantified with a variety of methods. Manual counting, with a haemocytometer for example, is tedious and time-consuming. Microplate assays, such as LDH and MTT, that exist for high-throughput applications are prone to errors and high variability due to the sample preparation and enzymatic reactions involved. We propose an alternative method to measure cell growth kinetics that is based on the fluorescence signal of constitutively expressed, intracellular GFP.[1,2,3] It is more sensitive, less disruptive and requires less manipulation than other available methods. Many features of GFP, such as its excellent stability and the non-invasive measurement contribute to the high reproducibility observed. Furthermore, the dramatic reduction in operator interaction reduces variations introduced by handling and therefore improves the accuracy of the data obtained. Reliably measurable effects of modifications of culture conditions from individual data points that may be only a few hours apart can thus be detected.

A. Bernard et al. (eds.), Animal Cell Technology: Products from Cells, Cells as Products, 119–121.

RESULTS AND DISCUSSION

Two important conclusions can be drawn from the first two figures:

- Fluorescence intensity varies linearly with cell number over a dynamic range.
- The kinetics of GFP fluorescence can be used to calculate the specific growth rate.

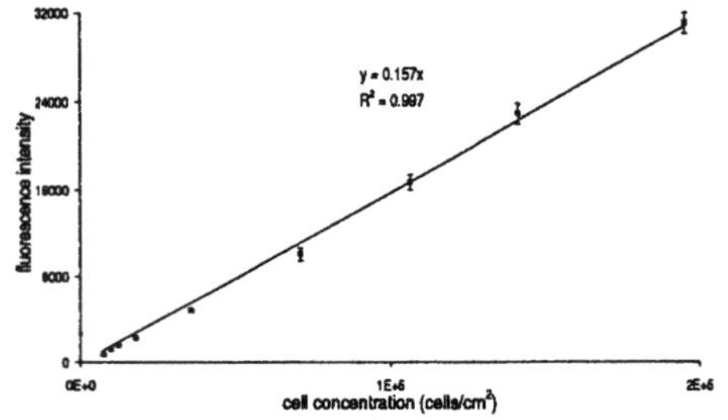

Figure 1. Correlation between fluorescent signal and cell number. n=5, in PBS. Fluorescence in multiwell plates was quantified in relative fluorescence units (RFU) using a Cytofluor™ 4000 (PerSeptive Biosystems, Farmingham, MA), gain 80, excitation filter: 485nm, bandwidth 20nm, emission filter: 530nm, bandwidth 25nm. The non-specific signal of wells containing cell-free PBS was subtracted.

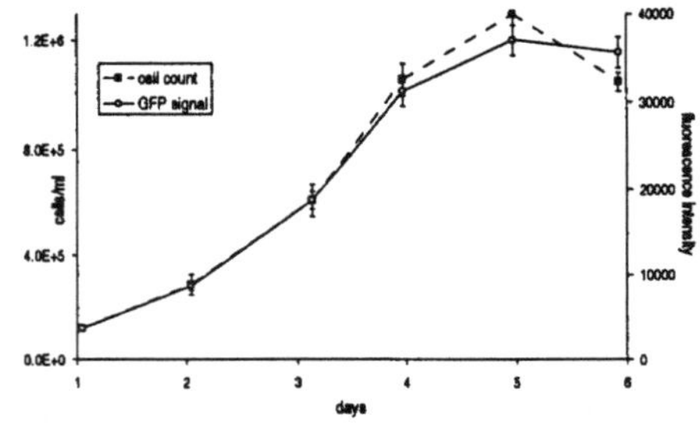

Figure 2. Growth curves based on GFP signal and viable cell number over time. Cell samples were counted in duplicate and viability was determined using trypan blue exclusion (haemocytometer). Using average (n=3) cell numbers from 1-3d, the maximum specific growth rate was calculated to be 0.78 $days^{-1}$ (R^2=0.994). The same growth rate of 0.78 $days^{-1}$ was calculated using the average (n=6, blanked) GFP signal (R^2=0.996).

To demonstrate that growth rates can be precisely determined and to test the limits of detection we applied the method to optimize the concentration of fetal calf serum (FCS), an essential supplement for the DMEM/F12 medium used here. GFP-expressing cells routinely grown at 5%FCS were cultivated for 5 days in medium supplemented with 0-20% serum. As shown in Fig. 3, medium without serum does not support growth, 1% FCS moderately stimulates growth, and 3% FCS or more results in maximal growth.

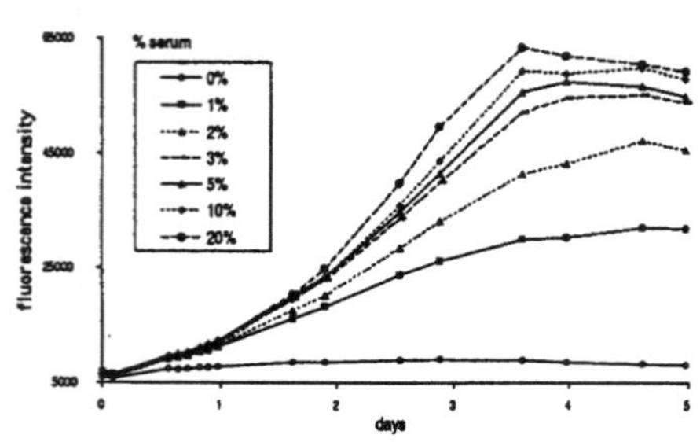

Figure 3. Fluorescence intensity as a function of serum concentration at different time points in the culture. n=6. All samples and blanks in the same 96-well plate.

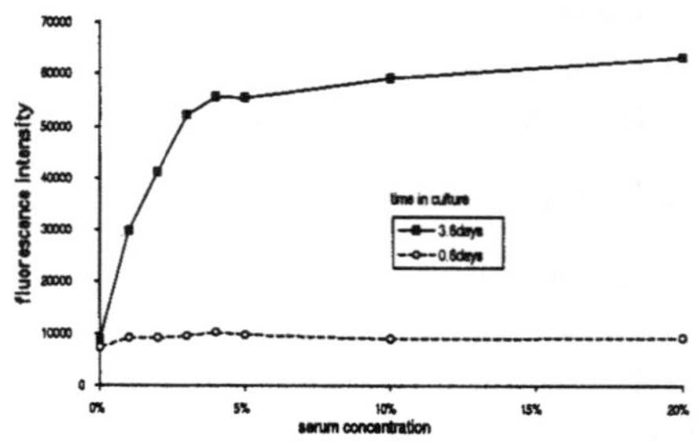

Figure 4. Fluorescence signal versus %FCS in the early exponential phase (t=0.6days) and at the maximum signal obtained (t=3.6days).

The fluorescence measurements as a function of %FCS at a given point in time are shown in Figure 4. The effect of serum concentration on cell growth is clear at 3.6d, when the maximum fluorescence intensities were measured. However, the effect of serum is not apparent early in the exponential growth phase, after 0.6 days in culture. For this particular experiment, the overall increase in signal for cells in media with 0-3% FCS was significant, whereas above 3% FCS little influence on the signal was

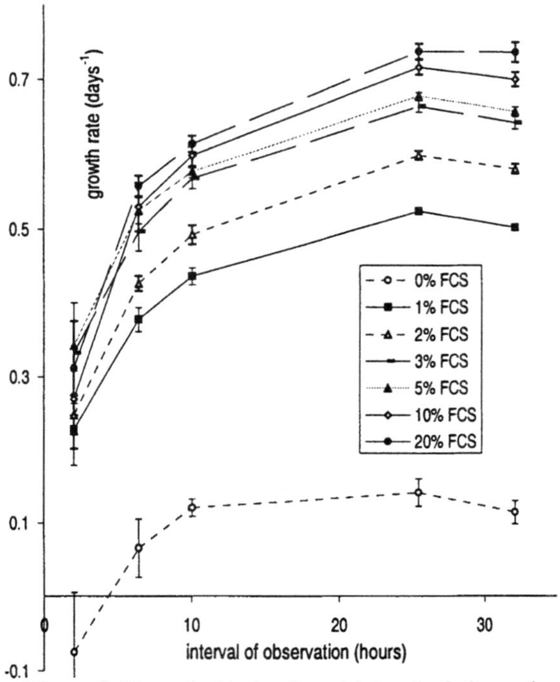

Figure 5. Recognizable trends and interval of observation. Growth rate calculations were based on initial (0.6d) and final fluorescent signals (up to 1.9 days after seeding).

found. Thus, in the evaluation of factors affecting cell growth, total fluorescence indicates trends. Differences in total fluorescence reveal these trends even faster.

Standard errors for samples from individual wells were routinely below 6%. A low standard deviation is needed to detect small differences in biomass values. The excellent reproducibility of fluorescence measurements makes it possible to detect kinetic differences in growth rates very early in the culture. A 2h period from 14-16h after seeding (first interval in Fig. 5) already reveals significant differences between the 0% and 1% samples. For the other serum concentrations, whose corresponding growth rates are more similar, differences can be reliably detected over the 10h interval from 14-24h. Extended observation, as in Fig. 4, confirms these early kinetic trends.

CONCLUSIONS

Kinetic analysis of two fluorescence measurements indicates trends within hours. Fluorescence measured at 1 point in time gives similar indications over days, as well as indicating other growth characteristics such as maxima.

This method of evaluating growth kinetics using such stable, GFP-expressing cells deserves to be studied in more detail due to its potential as an optimization tool; reducing the time and labor intensive analysis of protein production would dramatically impact the efficiency of process development by quickly and qualitatively indicating trends.

This work is supported by research funds from the Swiss National Science Foundation, Biotechnology Priority Program.

[1] Green Fluorescent Protein: Properties, Applications, and Protocols. Ed. M.Chalfie and S.Kain. Wiley-Liss. 1998.

[2] Hunt, L., Dejesus, M., Jordan, M., Wurm, F.M. GFP expressing mammalian cells for fast, sensitive, non-invasive cell growth assessment in a kinetic mode. Biotech & Bioeng. In press.

[3] Subramanian, S., Srienc, F. Quantitative analysis of transient gene expression in mammalian cells using the green fluorescent protein. Journal of Biotechnology 49 (1996) 137-151.

INTRACELLULAR LOCALIZATION OF NON-SECRETED RECOMBINANT β-TRACE PROTEIN IN VESICLE-LIKE STRUCTURES OF BACULOVIRUS-INFECTED INSECT CELLS

N. SCHULZE[1], U. ALBERS[1], E. GRABENHORST[2], H. S. CONRADT[2], M. ROHDE[3], M. NIMTZ[4] and V. JÄGER[1]
[1]Cell Culture Technology Dept., [2]Protein Glycosylation Dept., [3]Div. of Microbiology, [4]Dept. of Structure Research, Gesellschaft für Biotechnologische Forschung mbH, Mascheroder Weg 1, D-38124 Braunschweig, Germany

Introduction

The baculovirus expression vector system (BEVS) has become an important tool for the production of high levels of recombinant proteins within short time. It possesses the capability to ensure all co- and posttranslational modifications like glycosylation, phosphorylation, signal-peptide cleavage, cellular targeting, secretion etc. In spite of that it is widely known that during the expression of recombinant proteins by BEVS substantial amounts of these proteins are accumulated intracellularly even in the presence of correct secretion signals [1] suggesting a bottleneck in the secretion pathway using this expression system. Using β-Trace Protein (β-TP), a glycoprotein bearing two N-glycosylation sites at Asn_{29} and Asn_{56} as a model, this complex phenomenon was investigated by high level expression in infected High Five™ cells. SDS-PAGE/western blotting indicates that after 72 hours about 55 % of the produced recombinant β-TP is accumulated within these cells. In the present work intracellular β-TP has been localized in vesicle-like structures and the accumulation process was analyzed. Furthermore, the N-glycan structures of the intracellular β-TP have been investigated proving that the recombinant protein is localized in the endoplasmic reticulum and the cis-golgi network.

Results

Localization of intracellularly accumulated β-TP

High Five™ cells were *Ac*MNPV-β-TP infected with a MOI=5, fixed after 2 days and observed under a laser scanning microscope (Figure. 1). Immunolabelling with a rabbit anti-β-TP shows a cytoplasmic staining of β-TP within vesicle-like structures in these cells.
In order to investigate the emergence of these vesicles and to study the kinetics of the accumulation process the cells were analysed in more detail at different times of infection using immunogold electron microscopy (Fig. 2).

A. Bernard et al. (eds.), Animal Cell Technology: Products from Cells, Cells as Products, 123–125.

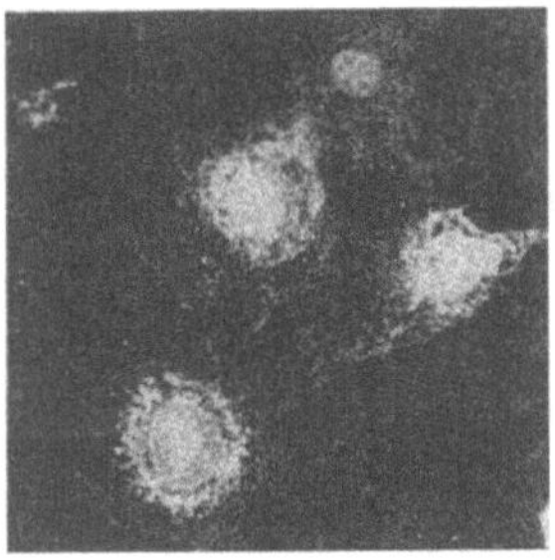

Figure 1: Immunolocalization of β-TP in infected High Five™ cells 2 days post infection by confocal laser scanning microscopy.

Even in the early phase of recombinant protein production mainly golgi-derived structures are immunogold labelled (Fig. 2a.). With increasing time the β-TP accumulates in two different vesicles: in small electron dense-(DV) and electron lucent-(LV) vesicles (Fig. 2b) Subsequently, an increasing number of these vesicles emerges and seems to fuse to larger vesicles (Fig. 2c.)

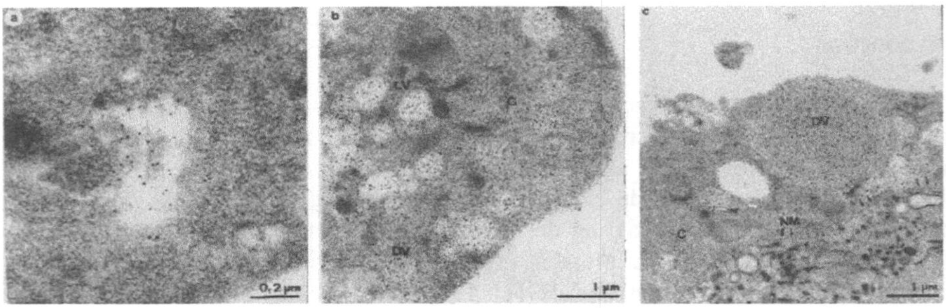

Figure 2: Time course study of the accumulation process of recombinant β-TP in High Five™ cells 24,- 48,- and 72 h post infection respectively (a)-(c). C, Cytoplasm; DV, electron dense vesicle; LV, electron lucent vesicle; N, nucleus; NM, nuclear membrane

Purification and characterization of intracellular β-TP

For the analysis of the N-glycan structure the intracellular recombinant protein β-TP was purified as presented in Fig. 3.

intracellular ß-TP (3 d.p.i.)

↓

cell disruption

↓

ultracentrifugation

↓

gel permeation chromatography

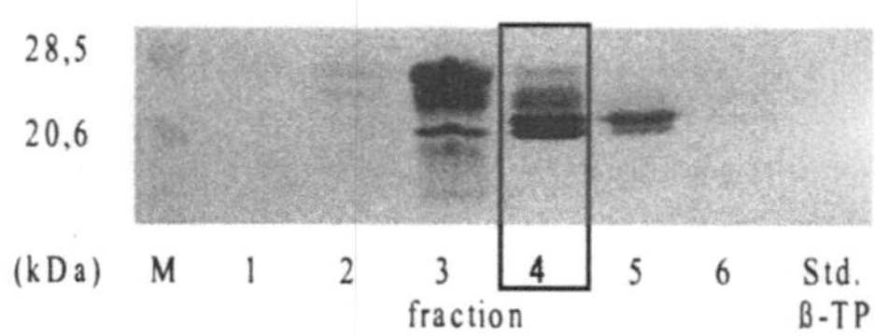

Figure 3: Western blot analysis after gel permeation chromatography and concentration of the collected peak fraction

The results of oligosaccharide analysis were schematically shown for fraction 4 (Fig. 3) representing mainly non -and monoglycosylated forms of intracellularly accumulated β-TP. The analysis of fraction 3 containing the non-, mono- and biglycosylated forms were carried out in the same way.

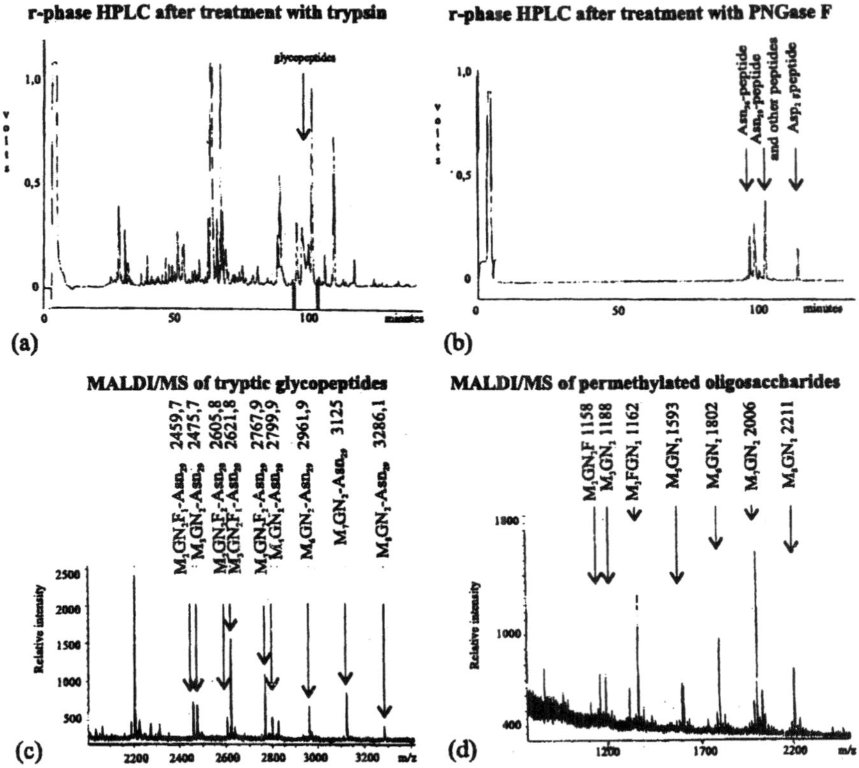

Figure 4: Oligosaccharide structure analysis of fraction 4.M, Mannose; F, Fucose; GN, N-Acethylglucosamine

After reduction and carboxamidomethylation the fraction was digested with trypsin and rechromatographed on a C_{18}-resin (Fig. 4a). An aliquot of the obtained glycopeptides was analysed by MALDI/MS (Fig. 4c) and sequence analysis. The remainder glycopeptides were digested with PNGase F and rechromatographed on the same resin (Fig. 4b). The peak fractions were identified by MALDI/MS and sequence analysis. N-glycans were analysed by MALDI/MS and HPAEC-PAD (not shown). Our experiments clearly demonstrate that the majority of this fraction is high mannose type glycosylated with mainly $Man_7GlcNAc_2$ structures. Monoglycosylated species are only occupied at the first N-glycosylation site at Asn_{29}. This was confirmed by sequence analysis of the peak fractions after the second reversed phase HPLC (Fig. 4b). After PNGase F treatment no change from Asn_{56} to Asp_{56} has occured. Fraction 3 also reveals the presence of high mannose type oligosaccharides.

Reference

1. Hsu, T.A., Eiden. J.J., Bourgarel, P., Meo, T. and Betenbaugh, M.J. (1994) Effect of co-expressing chaperone BIP on funtional antibody production in the baculovirus system. Protein Expr. Purif. **5**, 595-603

INDUCTION KINETICS OF APOPTOSIS IN MAMMALIAN CELL CULTURES

M. LINZ, A.-P. ZENG, W.-D. DECKWER
GBF - Division of Biochemical Engineering
Mascheroder Weg 1, D-38124 Braunschweig, Germany

1. Summary

The morphological and biochemical features of cell death in three production cell lines and one human carcinoma cell line were characterised in order to investigate the kinetics of apoptosis induction, apoptotic and necrotic death.

In BHK and hybridoma cultures apoptosis induction was found to be the limiting step of apoptosis. Lack of glutamine and elevated concentrations of ammonium alone had no obvious effect on the induction of cell death, whereas lack of both glucose and glutamine resulted in significantly enhanced apoptosis induction.

2. Methods

1.1. CULTURE CONDITIONS

Cell lines BHK-21 C13, CHO K1 and HyGPD YK-1-1 and a human carcinoma cell line (HeLa) were cultivated using serum-free DIF1000 medium. Glucose- and glutamine-free cultivations were done in a 3-component medium otherwise equal to DIF1000. All cultivations were performed in 125 mL or 500 mL Techne spinner flasks at 37°C, 12,5% CO_2-content and 90% humidity. Cell number and viability, glucose, lactate, ammonium and amino acid concentrations were determined at different intervals.

1.2. DETECTION OF APOPTOSIS

Three different methods based on special features of apoptosis were used for its detection:

- Agarose gel electrophoresis detecting the ladder pattern of fragmented DNA
- Fluorescence microscopy with acridine orange/ethidium bromide (AO/EB) staining of DNA revealing morphological changes
- FACS analysis with Annexin-V-FITC/propidium iodide staining indicating alterations of the plasma membrane during apoptosis

A. Bernard et al. (eds.), Animal Cell Technology: Products from Cells, Cells as Products, 127–129.

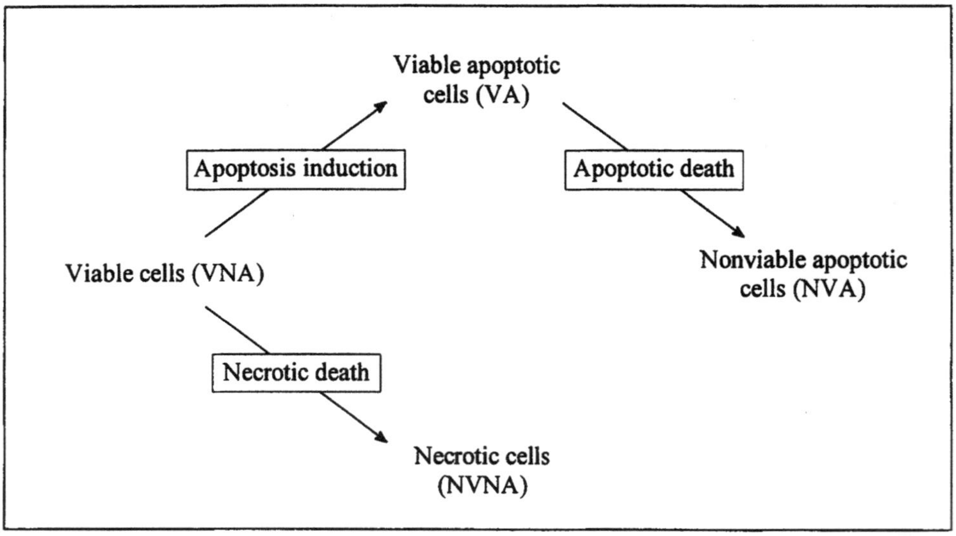

Figure 1. Routes of cell death

3. Characterisation and Quantification of Cell Death

For the hybridoma cell line HyGPD YK-1-1 DNA-laddering, morphological and cell membrane changes showed apoptosis to be the prevalent mode of death. For the other cell lines DNA-fragmentation so far could not be shown and varying portions of dead cells could neither be classified apoptotic nor necrotic by means of morphology. Therefore a combination of total cell number, percentage of nonviable apoptotic and necrotic cells from AO/EB-staining and percentage of viable, viable apoptotic and dead cells from FACS-analysis was used for quantification of cell death in BHK and hybridoma cultures.

4. Kinetics of Apoptosis

The transition of viable (VNA) to viable apoptotic cells (VA) and of viable apoptotic to nonviable apoptotic cells (NVA) is described by the apoptosis induction rate k_{ApInd} and the apoptotic death rate $k_{ApDeath}$:

$$k_{ApInd} = \left(\frac{dNVA}{dt} + \frac{dVA}{dt}\right) \times \frac{1}{VNA} \tag{1}$$

$$k_{ApDeath} = \frac{dNVA}{dt} \times \frac{1}{VA} \tag{2}$$

TABLE 1. Maximum percentage of viable apoptotic cells (VA) and average apoptosis induction and apoptotic death rates during death phases of different BHK-21 C13 and HyGPD YK-1-1 cultures (FACS-data). The cultures with + 5 mM NaCl serve as standard cultures. w/o = without

	BHK-21 C13			HyGPD YK-1-1		
	max. VA [%]	k_{ApInd} [d^{-1}]	$k_{ApDeath}$ [d^{-1}]	max. VA [%]	k_{ApInd} [d^{-1}]	$k_{ApDeath}$ [d^{-1}]
+ 5 mM NaCl	22,0	0,51	1,14	7,4	0,90	5,4
+ 5 mM NH_4Cl	24,3	0,53	1,18	7,4	0,63	5,6
w/o Glucose and Glutamine	10,3	0,81	4,17	18,6	3,20	13,8
w/o Glucose	15,5	0,46	1,77	10,1	0,87	10,3
w/o Glutamine	6,9	0,16	1,38	6,0	0,34	6,4

The kinetics of apoptosis induction and apoptotic death of BHK and hybridoma cells were studied in normal batch cultures and under stress conditions (lack of substrates and elevated ammonium concentrations).

For both cell lines:

- The apoptotic death rate $k_{ApDeath}$ was 2 - 19times higher than the apoptosis induction rate k_{ApInd} (depending on cell line and culture conditions), resulting in usually low amounts of living apoptotic cells in the culture (0,1 - 5%).
- Apoptosis induction was enhanced by lack of both glucose and glutamine.
- No enhanced induction of apoptosis by lack of glutamine or elevated ammonium concentrations was observed.

5. Conclusion

The kinetic of apoptosis strongly influences quantitative studies of cell death. This is due to the fast changes of features used for the detection of apoptosis as could be shown for different cell lines. These results must be taken into account when cell death is studied for optimising production processes in mammalian cell cultures.

A MODEL FOR THE ON-LINE SCRUTINY OF METABOLISM: ITS APPLICATION TO THE CHANGING NUTRITIONAL DEMANDS OF CULTURED ANIMAL CELLS

Y.H. GUAN and R.B. KEMP
Institute of Biological Sciences, University of Wales
Aberystwyth, SY23 3DA, UK

Abstract The development of a novel on-line metabolic probe, the combined heat flux biosensor, has facilitated the formation of a working metabolic model that can be integrated with an on-line strategy to feed nutrients. The methodology is based on the monotonic relationship between heat flux and the consumption flux of substrates. A theoretical demonstration of it is validated by data from continuous cultures on the relationship that is used to implement a feeding strategy in which heat flux is the control variable.

Introduction

A biosensor has been developed to measure on-line the heat flux of cells in a bioreactor by combining the calorimetric measurement of heat flow rate with the estimation of biomass from changes in capacitance using a dielectric spectrometer [1]. CHO320 cells genetically engineered to produce interferon-γ (IFN-γ) were grown in a batch culture as the model system and the results showed that heat flux was monotonically related to the consumption fluxes of the substrates. Based on principles established by Battley [2], a simplified growth reaction was constructed from the measured changes in the consumption of substrates and output of products. It was validated by the enthalpy balance approach in which the calculated overall molar enthalpy flux was compared with the heat flux [3]. The formulated reaction indicated that the demand by the rapidly growing cells for the major substrates, glucose and glutamine, was in the stoichiometric ratio 3:1 rather than in the ca. 5:1 mix found in the medium [4]. Stimulated by this finding, an improved medium was formulated aimed at providing more closely the cellular requirements for sustained cell growth and cytokine production while minimising the conversion of the major substrates into biosynthetic precursors with the formation of lactate as a toxic by-product [5]. The advantage of using this medium was revealed by the increase in both cell growth and the specific production of IFN-γ.

The classical method to sustain the growth and productivity of cells over long periods is to feed them nutrients during the culture. It had not escaped our attention that the heat flux might be the ideal variable to control the timing of feeding because it apparently measures the overall metabolic flux. The aims of this paper are to give the theoretical reasons for the monotonic relationship between heat flux and the material fluxes, to

A. Bernard et al. (eds.), Animal Cell Technology: Products from Cells, Cells as Products, 131–133.

prove the relationship experimentally using continuous cultures and to apply the heat flux probe as the control variable in fed-batch culture.

Experimental

CHO320 cells were grown in an improved medium based on RPMI 1640 [5] using an Applicon bioreactor system that included a Thermometric TAM customised flow calorimeter [6] and an Aber Instruments dielectric Viable Cell Monitor. The Applicon BioXpert software was employed for data acquisition, analysis and control. Off-line measurements of oxygen uptake rate (Paar Oroboros Respirometer) and materials have been detailed previously [1].

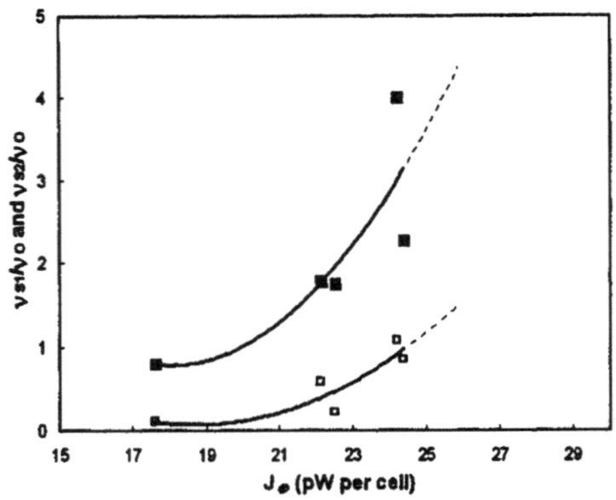

Figure 1. The heat flux over a specified set of values is compared against the stiochiometric ratios for the consumption of glucose (■) and glutamine (□) to oxygen.

Results and Discussion

Our previous work to characterise cell growth as a chemical reaction emphasised that the reactants and products including biomass (X) formed a set of stoichiometric coefficients (v_i for the ith species) with the heat production being regarded as their equivalent [1,4]. It was reasoned that this set represented the monotonic relationship between the metabolic flux and the stoichiometry of the growth reaction. It follows that

$$(1/X)\,d\xi/dt \leftrightarrow \vec{v} \tag{1}$$

where ξ is the advancement of the growth reaction. Since heat flux, J_{th}, is a form of metabolic flux, Eq. (1) can be converted as

$$v_{\text{i}} = f\left(J_{\text{th}}\right) \tag{2}$$

The strength of this relationship is illustrated from the data for a 120-h batch culture of cells utilising glucose (glc), glutamine (gln) and oxygen in relation to their heat flux for the following stoichiometric ratios,

$$v_{\text{Glc}}/v_{\text{O}} = 0.0581J_{\text{th}}^2 - 2.0949J_{\text{th}} + 19.663 \tag{3}$$

$$v_{\text{Gln}}/v_{\text{O}} = 0.0279J_{\text{th}}^2 - 1.045J_{\text{th}} + 9.8561 \tag{4}$$

From a graphical representation (Figure 1) of Eq (3) and Eq. (4), it is clear that the heat flux is a monotonically increasing function of the two ratios for the specified condition.

In order to broaden the validity of heat flux as a probe of metabolic activity to the steady state conditions that would accurately pinpoint the exact relationship, the cells were grown in continuous culture at different dilution rates. It is seen in Figure 2 that the monotonically increasing relationship is verified at the different steady states created in the culture.

The complete confidence in the validity of heat flux as the control variable allowed us to conduct the controlled fed-batch experiments. For them, the averaged decrease in on-line heat flux over a 1-h period was the biosensor signal to trigger the feeding of a nutrient cocktail (glucose, 50 mM, and glutamine, 16 mM) to the cells. As seen in Figure 3, biosensor-controlled nutrient feeding had the effect of restoring the level of metabolic activity at cell concentrations below about 10^6 cm^{-3}. Above this number, feeding slowed the decrease in metabolism to a degree that depended on the culture time.

The current study has amply validated the on-line heat flux probe as a biosensor of the overall metabolic activity of cells. Most importantly, we demonstrated its use as the control variable to prolong batch cultures by feeding a nutrient cocktail that has the major substrates in the stoichiometric ratio determined by cellular demand. Specific cell growth and IFN-γ production were superior and there should be improvements to the cytokine quality.

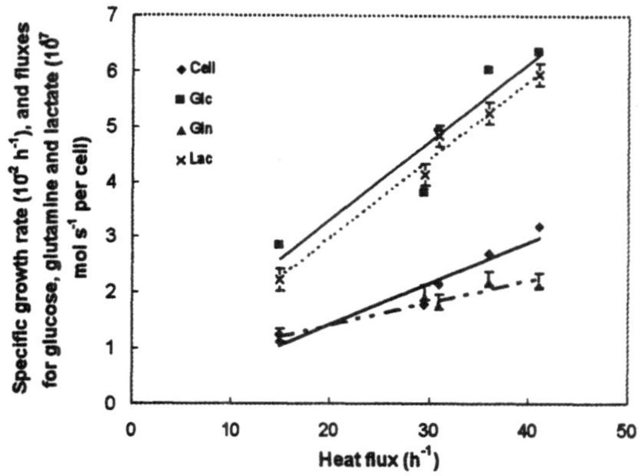

Figure 2. The cell specific growth rate, and fluxes for glucose, glutamine and lactate were correlated to heat flux at different dilution rates in a continuous culture.

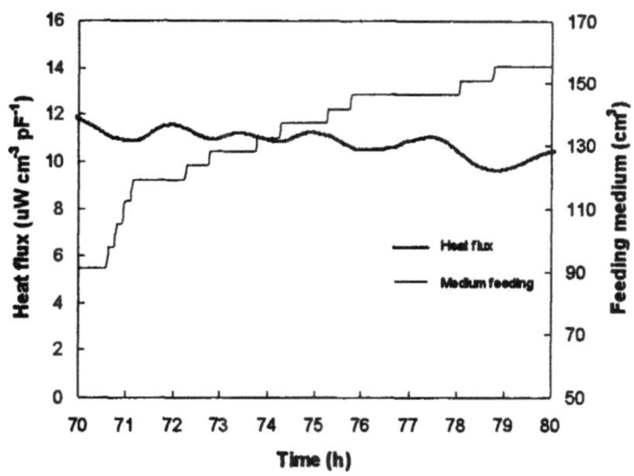

Figure 3. This shows a small section of the heat profile for a fed-batch culture (from 70 to 80h) to illustrate that the medium feeding was triggered by the declining heat flux values over 1-h assessment period. The heat flux was restored, to a varied extent. by this feeding strategy.

Acknowledgements: The research is funded by the Biotechnology and Biological Sciences Research Council (UK) with grants numbered 2/3680, 2/TO3789 and 2/E10985.

References:

(1) Guan, Y., Evans, P.M., and Kemp, R.B. (1998) *Biotechnol. Bioeng.* **58**, 464-477.
(2) Battley, E.H. (1987) *Energetics of Microbial Growth*, Wiley-Interscience, New York.
(3) Guan, Y.H. and Kemp, R.B. (1999) *J. Biotechnol.* **69**, 95-114.
(4) Guan, Y. and Kemp, R.B. (1998), in O.-W. Merten, P. Perrin, and B. Griffiths (eds.), *Animal Cell Technology: New Developments - New Applications*, Kluwer, Dordrecht, The Netherlands, pp. 355-357.
(5) Guan, Y.H. and Kemp, R.B. (1999) *Cytotechnol.* **30**, 107-120.
(6) Guan, Y.H., Lloyd, P.C., and Kemp, R.B. (1999) A calorimetric flow vessel optimised for measuring the metabolic activity of animal cells, *Thermochim. Acta* (in press).

CASPASES AS A TARGET FOR APOPTOSIS INHIBITION IN HYBRIDOMA CELLS IN CULTURE

C. GABERNET[1], A. TINTÓ[1], J. VIVES, E. PRATS [2], L. CORNUDELLA[2], J.J. CAIRÓ[1] AND F. GÒDIA[1]

1. Departament d'Enginyeria Química. Escola Tècnica Superior D'Enginyeria. Universitat Autònoma de Barcelona. 08193 Barcelona (Spain).

2. Departament de Biologia Molecular i Cel.lular. Centre de Investigació i Desenvolupament. CSIC. Jordi Girona, 18-26. 08034 Barcelona (Spain).

The importance of glutamine as the major apoptotic inductor in batch cultures of an hybridoma cell line has been investigated.
The importance of the caspase cascade, as a possible target for apoptosis inhibition, in the signaling of apoptosis in cultures deprived of glutamine has been also studied. The specific peptidic inhibitor of caspases Ac-DEVD-CHO partially inhibited (25%) the apoptotic process while cysteine protease inhibitor Z-VAD-FMK was a much more effective inhibitor of apoptosis (75%).
On the other hand, only with the combination of both inhibitors has been possible to make reversible the programed cell death process.

1. Introduction

It has been described that PCD in hybridoma cells is a critical factor which reduces productivity of the cells when cultured in vitro.Apoptosis occurs when a cell activates an internal suicide program, encoded in cellular genes, as a result of different stimuli leading to important changes in cell morphology such are chromatin condensation, DNA fragmentation and citoplasmatic colapse resulting in disassembly of the cell.

The regulation of the genetic machinery of apoptosis has raised considerable interest due to the possibility to prevent it and therefore extend the life-span of cultures improving both cell viability and productivity.The role of both inductor stimuli and effector molecules cooperating in the apoptotic process in batch cultures of an hybridoma cell line has been examined in order to define a strategy for controlling gene expression in cell death pathways.

2. Materials and methods

Cell line: murine hybridoma cell line KB 26.5 producing an IgG_3 antibody directed against antigen A_1 of red cells. Culture medium and culture conditions : DMEM supplemented with 2% foetal calf serum in 75cm^2 stationary flask, 37ºC, 5% CO_2.

A. Bernard et al. (eds.), Animal Cell Technology: Products from Cells, Cells as Products, 135–137.

Viable cell concentration: Tripan blue dye exclussion test and hemacytometer counts. Apoptosis identification: Morphological examination by fluorescence microscopy and confocal microscopy of cells stained with acridine orange and ethidium bromide. DNA fragmentation assay by agarose gel electrophoresis. Phosphatidylserine translocation by flow cytometry with Annexin-V-Fluos (Boehringer Mannheim).

3. Results and discussion

The contribution of glutamine in the PCD process is studied. The control culture died basically by apoptosis when glutamine was exhausted showing a 52% of apoptotic cells at 72h. Glutamine deprivation showed a instantaneous effect on cell viability with similar percentage (56%) of apoptotic cells at 72h.

In order to study the role of caspases in the transduction of the effector signal once the apoptotic machinery has been trigered by glutamine, inhibitors of this cysteine proteases were used.

The effect of the specific inhibitors Ac-DEVD-CHO and Z-VAD-FMK at different concentrations (0, 10, 50 and 100uM) on cell viability can be seen in *Figures 1a and 1b.*

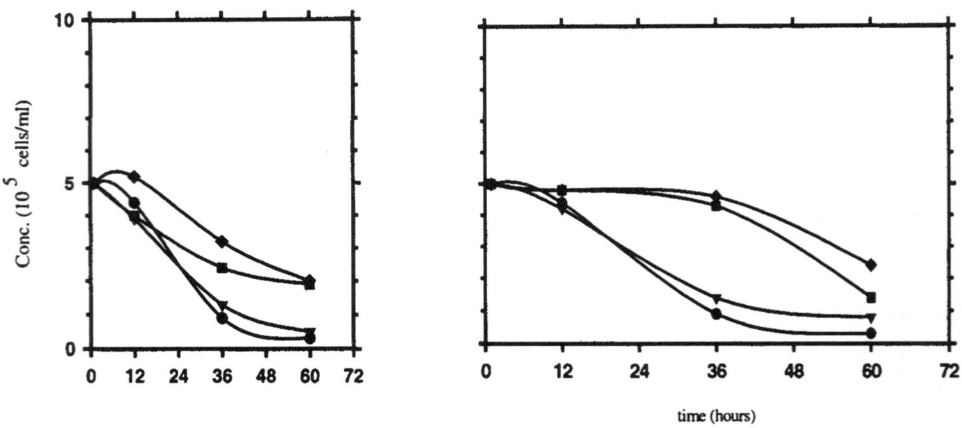

The effect inhibitor of caspases 2 and 9 has reached a great success in apoptosis inhibition, reducing the cell death and the phosphatidylserine translocation in a 75%. Also the DNA degradation pattern has been attenuated . The caspase 3 family inhibitor plays also a role in the inhibition of the PCD process but not so strong as the first inhibitor.

Finally, the effect of the different inhibitors alone or combined is presented in *Figure 2*. There is no significant diference between the viability of cultures with inhibitors Ac-DEVD-CHO + Z-VAD-FMK and the culture with Z-VAD-FMK alone.

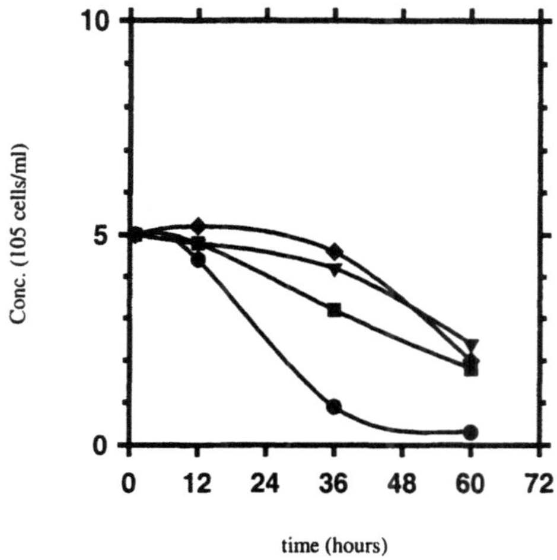

Figure 2. Results of cell viability obtained with different combinations of z-VAD and DEVD caspase inhibitors. Inhibitors combinations: Control – Gln ● – Gln + z-VAD ▼, – Gln + z-VAD ■, –Gln + z-VAD + DEVD ◆.

The main difference was observed in the reversibility of the apoptotic process. As can be seen in table 1, only with the two inhibitors of caspase is possible to get a reversible effect on the apoptotic process. This fact is very important for the selection of caspases as genetic targets for apoptosis inhibition.

TABLE 1. Reversible effect of the two caspase inhibitors on the apoptotic process.

	12 h.	24 h.	36 h.	60 h.
CONTROL – Gln	–	–	–	–
– Gln + z-VAD	+	–	–	–
– Gln + DEVD	+	–	–	–
– Gln + z-VAD + DEVD	+	+	+	–

z-VAD-FMK is an inhibitor of caspases 2 and 9, these caspases are responsible of the transmission of the apoptotic signal from mitochondria. Ac-DEVD-CHO is an specific inhibitor of caspase 3 family, also present in mitochondria. The results obtained suggest that hybridomas use more than one caspase pathway in the transmission of the apoptotic signal. The inhibition of caspases 2 , 9 and caspase 3 family are assential in order to keep the cells in a state where is possible to make reversible the programmed cell death process.

CELL ACTIVITY AFTER CELL GROWTH INHIBITION BY THE IRF-1 SYSTEM

A.S. COROADINHA[1], A.V. CARVALHAL[1], J.L. MOREIRA[1] & M.J.T. CARRONDO[1,2]

1- IBET/ITQB, Apartado 12, 2780 Oeiras, Portugal,
2- Lab. Eng. Bioq., FCT/UNL, 2825 Monte da Caparica, Portugal

1. Abstract

A genetic approach based on the activation of interferon-regulated-factor-1 (IRF-1) has been applied to regulate BHK cell growth. The presence of 100 nM estradiol in the culture medium in agitated flasks leads to the activation of the constitutively expressed IRF-1/estrogen receptor fusion protein (IRF-1-hER) and inhibits cell growth. Two days after estradiol addition cell concentration was still maintained but a significant decrease in cell viability was observed. The IRF-1 activation clearly interferes with the cell energetic metabolism, since there is an extra metabolic activity, with higher glucose, glutamine and oxygen consumption rates, while the yield lactate/glucose is maintained. Although there is a higher proteolytic activity, the protein content per cell increases significantly, suggesting an overall increase in the protein synthesis. A significant increase in the lactate dehydrogenase enzyme activity and a higher reduction of the MTT were also observed, indicating the increase in cell activity after cell growth inhibition. ATP, ADP and AMP contents were evaluated, and the cells were able to maintain the energy charge (EC) higher than 0.8. The significant increase in the total content on protein per cell can be the reason for the higher energetic needs. Thus, in order to maintain the EC the cells need to increase the ATP production, i.e., increase the catabolic metabolism. Nevertheless, 2 to 3 days after estradiol addition this equilibrium is no longer stable, leading to an increase in the ADP and AMP contents, and at that time the cell viability decreases and the other alterations previously referred start to be significant. The IRF-1 activation leads to a significant increase of the cell activity until the cell is no longer able to maintain its metabolic equilibrium and starts to die.

2. Materials and Methods

Cell line and medium: BHK 21A (ATCC CCL10) cells genetically modified to express IRF-1-hER were stably transfected by Dr Peter Muller (GBF, Braunschwig, Germany). The cells were grown in DMEM supplemented with 10% (v/v) of FBS, 4g l^{-1} of glucose and 5 mg l^{-1} of puromycin (except for the non-modified cells). FBS was supplied by Sigma (St. Louis, MO); all others were supplied by Life Technologies (Glasgow, UK).

A. Bernard et al. (eds.), Animal Cell Technology: Products from Cells, Cells as Products, 139–141.

Culture System: Cells were grown in 500 ml spinner flasks (Wheaton, at a agitation rate of 80 rpm and inoculated with a 2.5×10^5 cells ml^{-1} cell density.

Analytical Methods: Total cell concentration was determined by cell nuclei counting using crystal violet solution (0.1 % (p/v) in 0.1M citric acid). Cell viability was determined by the trypan blue exclusion method. Cell protein content was determined by a BCA kit (Pierce, Rockford, U.S.A). Cell proteolytic activity, MTT assay and LDH activity measurements were performed as described elsewhere[2,3,4]. Glucose and Lactate measurements were performed by exclusion and ion exchange HPLC (Sugar SH1011, Shodex-Waters, Tokyo, Japan). Glutamine and ammonia concentrations were determined by commercial enzyme assay kits, Sigma and Boehringer Mannheim GmbH (Mannheim, Germany) respectively. Oxygen measurements were performed using a Clark-type oxygen electrode. ATP, ADP and AMP measurements were done by ion pair reversed-phase HPLC (Supelcosil LC-18T, Supelco, Bellefont, U.S.A.)

3. Results and Discussion

For each test two experiments were always done in parallel using cells from the same origin and prepared exactly in the same way: in one experiment estradiol was added at inoculation time with a final concentration of 100 nM in the culture medium. In the other, used as a control, no estradiol was added, therefore without IRF-1 activation.

3.1. CELL GROWTH KINETICS AND CELLULAR VIABILITY

Cell concentration and cell viability were measured, showing that IRF-1 activation leads to an efficient cell growth inhibition, achieving a cell concentration 3 fold lower than that observed in the absence of estradiol, 72 hours after estradiol addition. Nevertheless, there is a more pronounced decrease in cell viability than in the control.

Similar cell growth inhibition and cell viability patterns were observed with different regulated clones expressing IRF-1-hER[1] showing that the IRF-1 effect is clone independent.

3.2. PROTEIN CONTENT

Total cellular protein determination is important since protein content per cell depends on growth phase. Although DNA content per cell was not altered after cell growth inhibition (data not shown), it was clear that IRF-1 activation led to an increase in protein content per cell (Figure 1), in agreement with an higher cell volume observed.

IRF-1 is a transcriptional activator that is known to modulate the expression of several genes, some of which interferes with protein synthesis pathways. Thus, the observed protein increase could be specific to a set of cellular proteins rather than an overall increase. Protein content per cell depends on a fine balance between protein synthesis and breakdown. The proteolytic activity was determined in order to assess if the increased cellular protein content observed after cell growth inhibition could be due to a decreased protein turnover. The observation of an increased protein content per cell simultaneous with an increased protein breakdown can represent a higher protein synthesis.

3.3. ENERGY METABOLISM

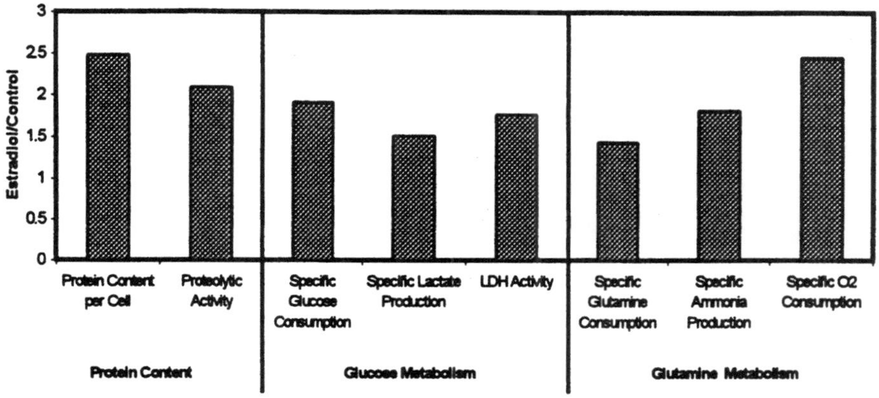

Figure 1: IRF-1 activation effect upon cell metabolism. The reasons between the values achieve in the presence of estradiol and in its absence (control) were calculated 72 hours after estradiol addition.

Glucose Metabolism: IRF-1 activation leads to an higher specific rate of glucose consumption and lactate production., with the lactate/glucose yields not being significantly altered (Figure 1). The observed increase in LDH intracellular activity and in MTT reduction (data not shown) is in agreement with a higher metabolic rate of glucose through lactate formation.

Glutamine Metabolism: IRF-1 activation led to higher specific consumption rate of glutamine, in agreement with a higher specific ammonia production and an higher oxygen consumption rate (Figure 1).

ATP, ADP and AMP content per cell: ATP, ADP and AMP are the primary energy units used by cells with a vast regulatory potential, indicating the cell status during growth. The ATP level was not significantly altered by the IRF-1 activation, but an increase in the levels of ADP and AMP after 72 hours indicated an higher energetic demand. The energy charge (EC) remained above 0.8

Control experiments using the parental cell line BHK 21A clearly confirmed that alterations in the cell activity due to IRF-1 activation were not due to the addition of estradiol *per se* or to its solvent, ethanol (data not shown).

4. Conclusions

The IRF-1 activation leads to a significant increase of the cell activity until the cell is no longer able to maintain its metabolic equilibrium and starts to die.

Acknowledgements
The authors are grateful to Ms. Maria do Rosário Clemente from IBET/ITQB for technical support. The authors acknowledge and appreciate the financial support received from the European Commission (BIO4-CT95-0291) and from Fundação para a Ciência e Tecnologia - Portugal (FMRH/BIC/1788/95 and BIO1117/95).

References

[1]-Carvalhal, A.V., Moreira, J.L., Cruz, H., Mueller, P., Hauser, H. and Carrondo, M.J.T (1999) Manipulation of culture conditions for BHK cell growth inhibition by IRF-1 activation.– submitted.

[2]- Twining, S.S. (1984) Isothiocyanate-labeled casein assay for proteolytic enzymes. Analytical Biochemistry 143:30-34.

[3]- Mosmann, T.(1983) Rapid colorimetric assay for cellular growth and survival: application to proliferation and cytotoxicity assays. J. Immunol Methods 65:55-63.

[4]- Racher, A.J., Moreira, J.M., Alves, P.M., Wirth, M., Weidle, U.H., Hauser, H., Carrondo, M.J.T., Griffiths, J.G. (1994) Appl. Microbiol. Biotechnol 40:851-856.

ESTABLISHMENT OF A PRIMARY LIVER CELL CULTURE FROM A TELEOST, *OREOCHROMIS MOSSAMBICUS*, THE TILAPIA: A VALID TOOL FOR PHYSIOLOGICAL STUDIES.

[1]A. C. SCHMID, [2]W. KLOAS, [1]M. REINECKE
[1]Institute of Anatomy, University of Zürich, Winterthurerstr. 190, CH-8057 Zürich, Switzerland
[2]Institute of Zoology, University of Karlsruhe, Kaiserstr. 12, D-76128 Karlsruhe, Germany

1. Introduction

It is generally accepted that the liver is the main source of circulating IGF-I in mammals as well as in lower vertebrates, such as in teleosts (Reinecke and Collet 1997). In mammals, the primary stimulus for the regulation of liver IGF-I is growth hormone (GH) (Daughaday and Rotwein, 1989). Although few studies have dealt with the potential GH-IGF-I axis in submammalians there is evidence that the expression of IGF-I in bony fish liver is also regulated by GH. Liver cell cultures are useful models for the investigation of physiological effectors on the hepatic system. Thus, we established a primary cell culture of hepatocytes from a teleost. Because the IGF-I cDNA sequence has been revealed IGF-I mRNA from the liver of the tilapia (*Oreochromis mossambicus*) (Reinecke et al., 1997) the tilapia was used as representative species.

2. Material and Methods

Adult individuals of tilapia were injected with heparin (3000 U) in a Ringer's solution and anesthetized in water containing MS 222 (1g/l). The liver was perfused retrogradely with 100 ml of a calcium-magnesium-free (CMF) buffer and digested with 50 ml calcium-magnesium-containing (CMC) buffer containing collagenase D (0.5 mg/ml). The liver cells were resuspended in CMF buffer and selected by filtering through nylon gazes with 250 and 50 µm meshes, collected by centrifugation at 70 g for 10 min and washed twice in CMF buffer. After resuspending in minimal essential medium, the cells

A. Bernard et al. (eds.), Animal Cell Technology: Products from Cells, Cells as Products, 143–145.

were seeded as monolayers (about 1.5 x 10^6 cells per ml) onto culture plates and cultivated at 20°C under high humidity. Under these conditions, hepatocytes could be kept alive up to 3 days. In order to study the effect of GH on bony fish liver, the cells were treated with recombinant tilapia (t) GH every 12 h over 2 days. To investigate the time dependence of GH action, cells were incubated with 10 nM tGH and frozen after 0, 6, 12, 18, 24, 36, and 42 h. Dose dependency was studied with cells incubated with tGH ranging from 0.1 nM to 1 µM. The cells were lysated in a buffer containing phenol (Ultraspec/ams) and RNA was extracted by phenol/chloroform extraction. The IGF-I signal was measured by semiquantitative RT-PCR with tilapia specific IGF-I primers. Primers specific for human actin served as internal standard. The PCR products were separated on a 2% agarose gel stained with ethidium bromide and the signal intensities were determined by scanning the optical density.

3. Results

RT-PCR revealed the expected IGF-I fragment of 208 bp and the actin fragment of 661 bp. Untreated cells showed the maximal IGF-I mRNA expression at the onset of the experiment and the signal decreased with time down to 16% of the control level (Fig. 1). The tGH-treated hepatocytes revealed higher IGF-I mRNA expression levels than the untreated cells throughout the experiment. The stimulatory effect of tGH on the IGF-I production in tilapia cultured hepatocytes was time dependent. After 6 h, the amount of IGF-I mRNA expression reached 180% of the control level. However, after 42 h the total IGF-I mRNA decreased to the control amount (Fig. 2). The increase of IGF-I mRNA expression was dose dependent reaching its maximum at 1 µM tGH. Even a concentration as low as 0.1 nM tGH resulted in IGF-I signals exceeding those in the controls (Fig. 3).

4. Discussion

In the newly established primary tilapia hepatocyte cell culture the liver cells could be kept alive up to 3 days. Thus, this experimental model represents an useful tool to study the physiological interactions and hormonal regulation in bony fish liver *in vitro*. The IGF-I signal of the controls decreased with time which may be due to the lack of growth factors such as GH in the serum free medium. This hypothesis is supported by the results obtained with tGH treated hepatocytes. The amount of hepatic IGF-I mRNA was increased after GH incubation in a time and concentration dependent manner. Therefore, the results support the presence of an GH-IGF-I-axis in teleosts.

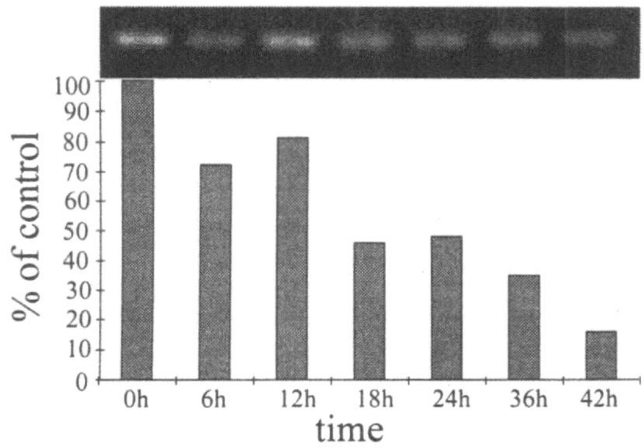

Fig. 1 RT-PCR of IGF-I mRNA expression in untreated hepatocytes and densitometric evaluation of time dependence (control values were set to 100 %)

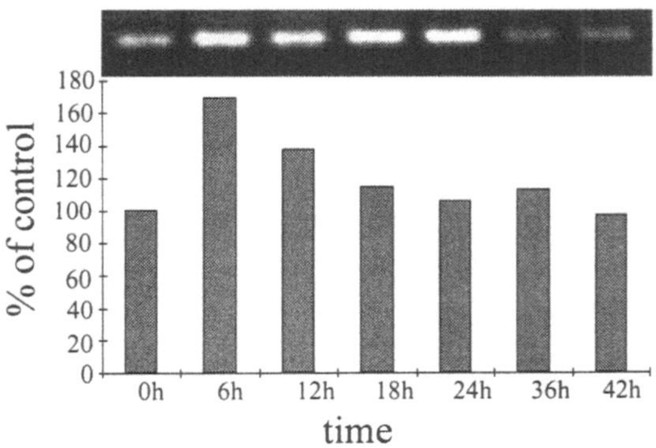

Fig. 2 RT-PCR of IGF-I mRNA expression in tGH- treated hepatocytes and densitometric evaluation of time dependence (control values were set to 100 %)

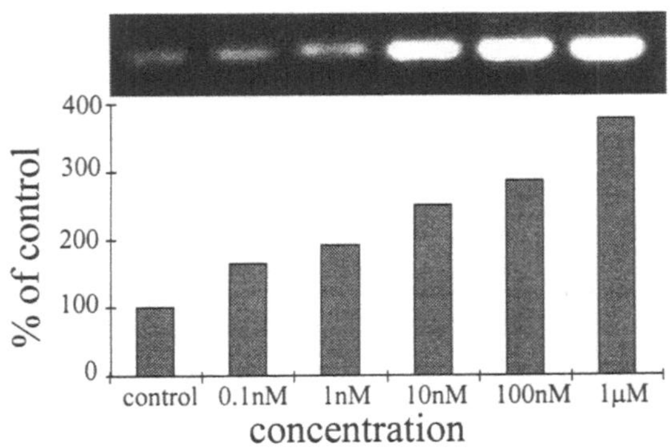

Fig. 3 RT-PCR of IGF-I mRNA expression in tGH-treated hepatocytes and densitometric evaluation of concentration dependence (control values were set to 100 %)

5. References

Daughaday, W. H. and Rotwein, P. (1989) Insulin-like growth factors I and II. Peptide, messenger ribonucleic acid and gene structures, serum, and tissue concentrations, *Endocrine Rev.* **10**, 68-91

Reinecke, M., Collet, C. (1998) The Phylogeny of the Insulin-like Growth Factors, *Int. Review Cytol.* **183**, 1-94

Reinecke, M., Schmid, A., Ermatinger, R., Loffing-Cueni, D. (1997) Insulin-like Growth Factor-I in the Teleost *Oreochromis mossambicus*, the Tilapia: Gene Sequence, Tissue Expression and Cellular Localization, *Endocrinology* **138**, 3613-3619

TGF-β INDUCES PREMATURE AND REPLICATIVE SENESCENCE IN CANCER CELLS

T. MIURA, Y. KATAKURA, E. NAKATA, N. UEHARA, AND S. SHIRAHATA

Graduate School of Genetic Resources Technology, Kyushu University 6-10-1 Hakozaki, Higashi-ku, Fukuoka 812-8581, JAPAN

1. Abstract

Transforming growth factor-β (TGF-β) has been demonstrated to regulate extracellular matrix formation, cell differentiation, and cell cycle in most cell types. Here we demonstrate that TGF-β induces a premature senescence in A549 cells (human lung adenocarcinoma cell line) independently of the telomere shortening, which was evidenced by the expression of senescence marker β-galactosidase and morphological changes. However, A549 cells did not completely inhibit for its growth by the treatment of TGF-β, which was thought to be one of key features of premature senescence. These results suggest that the induction of premature senescence and growth inhibition are independently regulated in the TGF-β treated A549 cells. On the other hand, telomerase activity was downregulated via transcriptional repression of hTERT (human reverse transcriptase) in the TGF-β treated A549 cells. Furthermore, when A549 cells was cultured with TGF-β for a long term, telomere length gradually shortened from 8 kb to 2 kb. These results indicate A549 cells treated with TGF-β entered into the replicative senescence state. This study provides a new correlation between TGF-β signals and a cellular senescence programs.

2. Introduction

Telomerase activity is absent in most of somatic cells and normal cells [1]. Therefore, as these cells proliferate, telomeres undergo gradual shortening [2]. On the other hand, in cancer cells, the shortening of telomeres dose not take place. Because the length of telomeres in cancer cells is maintained at a certain level by telomerase [3]. Telomerase activity seems to be necessary for the proliferation of malignant tumors. Therefore, inhibition of telomerase may become a target for anti-cancer therapies.

We have been interested in the molecular mechanism of the cellular senescence of A549

A. Bernard et al. (eds.), Animal Cell Technology: Products from Cells, Cells as Products, 147–149.

cells as cancer cells. We have previously shown that A5DC7 cells ,which were isolated from A549 cells by treating interferon-γ (IFN-γ), showed no telomerase activity and entered the replicative senescence after a limited number of cell division [4]. Therefore, A5DC7 cells may become a useful model cell for study of a cellular senescence. The aim of the present our work is to hunt new other factors inducing a cellular senescence in A549 cells.

In most cell types, transforming growth factor-β (TGF-β) has multifunctional roles including proliferation, differentiation, development, and extracellular matrix production [5]. TGF-β also inhibits growth by repression of the CDK activator Cdc25A in cells with $p15^{INK4B}$ deletion, whereas A549 cells are not growth-inhibited by TGF-β [6]. In our study, whereas TGF-β-treated A549 cells did not show downregulation of Cdc25A, its growth rate was slightly lower. Therefore, another molecular mechanisms except cell-cycle arrest may participate in the weak antiproliferation by TGF-β. In order to analyze this mechanism response to TGF-β in A549 cells, we approached from the viewpoint of cellular senescence.

3. Results and Discussion

To explore the ability of TGF-β to induce senescence in A549 cells, we treated A549 cells with 10 ng/ml TGF-β (AUSTRAL Biologicals, San Roman, CA). After TGF-β was treated for 7 days, we determined a senescence-associated acidic β-galactosidase (SA-β-galactosidase) activity that has been associated with senescence in human cells [7]. Addition of TGF-β into A549 cells elicited the expression of acidic β-galactosidase, which was apparent maximal by 7 days after treatment of TGF-β. Simultaneously, morphological changes like senescence also were detected in the TGF-β treated A549 cells (data not shown). We conclude that TGF-β induces a premature senescence which is independent of the telomere shortening.

A549 cells are capable of extended proliferation responsible for negation of tumor suppresser pathways. One of features in premature senescence is that growth inhibition is induced in a short time. We therefore investigated growth phenotypes for A549 cells cultured with TGF-β for a long term. As a result, its growth rate was slightly lower by treatment of TGF-β (Fig. 1). Next, it was of interest to examine A549 cells for changes in cell cycle progression associated with stimulation of TGF-β. Cell cycle analysis was performed on a FACS Calibur (Becton Dickinson Immunocytometry Systems, San Jose, CA). As shown in Table 1, cell cycle phase distribution of A549 cells was no greatly changed by the addition of TGF-β. Furthermore, when we tested functions of p53, $p15^{INK4B}$ and Cdc25A acting as cell cycle regulator in TGF-β-treated A549 cells, these all factors were lost functions (data not shown). Accordingly, these functional incompletion of regulators which take part in cell cycle arrest may be prevented growth inhibition in A549 cells by TGF-β. Furthermore, we determined the telomere length in A549 cells that were treated with TGF-β for a long term. As a result, accompanied with the transcriptional repression of hTERT in A549 cells, telomere length gradually shortened from 8 kb to 2 kb (data not shown). This result indicates that TGF-β-treated A549 cells entered into replicative senescence state.

In conclusion, TGF-β triggers two independent-senescence programs, which mean premature and replicative senescence, in cancer cells. This study provides a new correlation between TGF-β signals and a cellular senescence programs. However, it is not clear whether TGF-β signals directly links to a cellular senescence induction. We presently investigate to clarify the senescence induction mechanism by TGF-β.

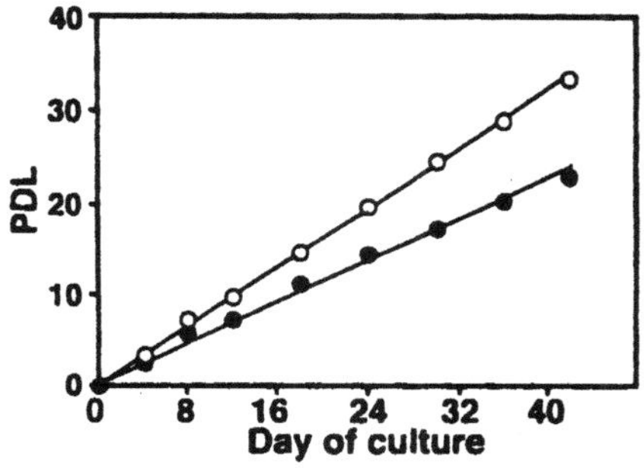

FIG. 1. Growth phenotypes of A549 cells grown in media (+) TGF-β (●) vs. (-) TGF-β (○).

TABLE 1. Cell cycle analysis in TGF-β-treated A549 cells

	-TGF-β	+TGF-β
G0/G1	72.9%	70.9%
G2/M	6.3%	3.3%
S	20.8%	25.8%

Cell cycle analysis of A549 cells which was cultured with or without TGF-β. The percentage of cells in the different phases of the cell cycle was assessed by flow cytometry.

References

1. Counter, C.M., Avilion, A.A., LeFeuvre, C.E., Stewart, N.G., Greider, C.W., Harley, C.B., and Bacchetti, S. (1992) Telomere shortening associated with chromosome instability is arrested in immortal cells which express telomerase activity, *EMBO J.* **11**, 1921-1929.
2. Hastie, N.D., Dempster, M., Dunlop, M.G., Thompson, A.M., Green, D.K., and Allshire, R.C. (1990) Telomere reduction in human colorectal carcinoma and with aging, *Nature* **346**, 866-868.
3. Shay, J.W., and Wright, W.E. (1996) Telomerase activity in human cancer, *Curr. Opin. Oncol.* **8**, 66-71.
4. Katakura, Y., Yamamoto, K., Miyake, O., Yasuda, T., Uehara, N., Nakata, E., Kawamoto, S., and Shirahata, S. (1997) Bidirectional regulation of telomerase activity in a subline derived from human lung adenocarcinoma, *Biochem. Biophys. Res. Commun.* **237**, 313-317.
5. Haldin, C.-H., Miyazono, K., and Dijke, P. (1997) TGF-β signalling from cell membrane to nucleus though SMAD proteins, *Nature* **390**, 465-471.
6. Iavarone, A., and Massague, J. (1997) Repression of the CDK activator Cdc25A and cell-cycle arrest by cytokine TGF-β in cells lacking the CDK inhibitor p15, *Nature* **387**, 417-422.
7. Dimri, G.P., Lee, X., Basile, G., Acosta, M., Scott, G., Roskelley, C., Medrano, E.E., Linskens, M., Rubelj, I., Pereira-Smith, O., and Peacocke, M. (1995) A biomarker that identifies senescent human cells in culture and in aging skin in vivo, *Proc. Natl. Acad. Sci.* **92**, 9363-9367.

DEVELOPMENT OF A FED-BATCH STRATEGY THAT UTILIZES MULTIPLE CARBON SOURCES TO MAXIMIZE RECOMBINANT PROTEIN EXPRESSION IN CHO CELLS

FRANK DEER AND MARK CUNNINGHAM
Ares Advanced Technology
280 Pond St., Randolph, MA, 02368, USA

1. Introduction

We faced the challenge of developing an upstream process to support preclinical production of sIBP2, a soluble receptor therapeutic, in recombinant CHO cells. We wanted to combine scalable suspension technology with the use of a regulatory-friendly animal source-free media. Our approach was to adapt an existing stable recombinant CHO clonal cell line to IS CHO-V™, a medium developed by Irvine Scientific to support high density CHO cell growth and is free of components derived from human, bovine or other mammalian sources.

2. Methods

A CHO cell line (clone 31), expressing sIBP2, was derived from a cell pool originally amplified to 5.0 μM MTX and cloned by limiting dilution in non-selective medium. Adaptation of clone 31 to IS CHO-V was accomplished in two stages. First, the serum component was removed by subculturing in an animal component-rich serum-free formulation (IS CHO™, Irvine Scientific) in suspension, following the suggested adaptation protocol. Then, the transition to animal component-free IS CHO-V was easily performed in 4 passages. A cell bank was cryopreserved in IS CHO-V and served as a source of cells for the studies reported here. Unless specified otherwise, all cultures were performed in 100 mL spinner flasks with magnetic stirrers (stir rate = 60 RPM) at 37°C in a humidified atmosphere containing 5% CO_2.

3. Results and Discussion

After successful adaptation of clone 31 to IS CHO-V, a single 100 mL spinner culture was run and monitored daily. Figure 1 shows that the cell density increased rapidly to 7.5E5/mL by day 3, and then more slowly to 2E6/mL by day 10. Maximum sIBP2 titer attained was ~30 mg/L, the majority of which was produced in the first 8 days of culture.

A. Bernard et al. (eds.), Animal Cell Technology: Products from Cells, Cells as Products, 151–153.

The metabolic profile shows that glucose was consumed over the first 5 days, while the lactate level rose to 10mM. When glucose became depleted, the culture began to consume the lactate that had been generated. The ammonia level rose steadily throughout the culture to a final concentration of approximately 8mM. Supernatant samples from the spinner culture were analyzed for amino acid levels by HPLC. Results showed that Gln, Glu, Asn, Asp, Ser and Leu had the highest consumption rates, whereas Ala and Gly were both produced during the culture. It is plausible to predict that the shift to lactate consumption helped to extend culture life by preventing the build-up of this known inhibitory agent. This form of metabolic activity in CHO cells has been reported previously (Lübben, H. and Kretzmer, G. 1998 Merten OW, Perrin P, Griffiths B (eds.), *New Developments and New Applications in Animal Cell Technology*, Kluwer Academic Publishers, Dordrecht, pp. 267-271).

A study was designed to determine which additives could be fed to the culture to improve productivity during the final four days of high viability. Five factors were examined, based on initial results and references from the literature: MEM non-essential amino acid solution, glutamine, fructose, KCl and Na butyrate. The only feed component to have a significantly positive effect on final titer was butyrate.

Next, we investigated strategies for reducing ammonia generation in IS CHO-V by decreasing the initial glutamine concentration and feeding with glutamine at various timepoints. Three conditions were run: 1) Glutamine-free IS CHO-V (control), 2) Glutamine-free IS CHO-V supplemented with 8mM glutamine on day 0 (normal IS CHO-V), 3) Glutamine-free IS CHO-V supplemented with 2mM glutamine on day 0, then fed with an additional 2mM glutamine on day 3 and 3 mM on days 5 and 7. Figure 2 shows that roughly equivalent cell growth was observed for all conditions. Only the glutamine-free control culture exhibited reduced ammonia levels. Low ammonia levels during the latter stage of culture improved productivity. Finally, we investigated optimal butyrate concentrations at 1mM and below in glutamine-free IS CHO-V. Figure 3 shows that the 1mM butyrate condition gave a two-fold induction of sIBP2 expression compared with the control (no butyrate). Feeding with butyrate in a low ammonia environment supported a final titer of 196 mg/L.

4. Conclusions

We developed a highly productive, suspension based recombinant CHO process that utilizes a regulatory-friendly animal component-free culture medium, IS CHO-V (Irvine Scientific). The use of this medium induces a metabolic shift in the culture, characterized by the consumption of lactate, resulting in an extended viable period. Several process improvements were made over the course of this study. First, the removal of glutamine from the process allowed us to control ammonia toxicity and resulted in a two- to three-fold titer increase. Second, induction of sIBP2 expression with sodium butyrate provided us with an additional two-fold titer increase. When combined, the improvements increased the process yield from 30 to 200 mg/L, an increase of greater than six-fold.

FIGURE 1. 100 mL suspension culture of clone 31 adapted to IS CHO-V medium. Left panel: cell density and productivity. Right panel: glucose, lactate and ammonia profiles.

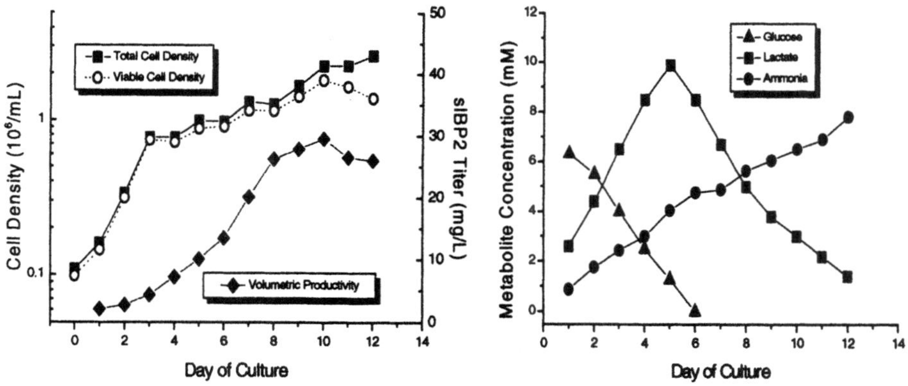

FIGURE 2. Results of the ammonia reduction study. Left panel: cell density and ammonia levels. Right panel: productivity.

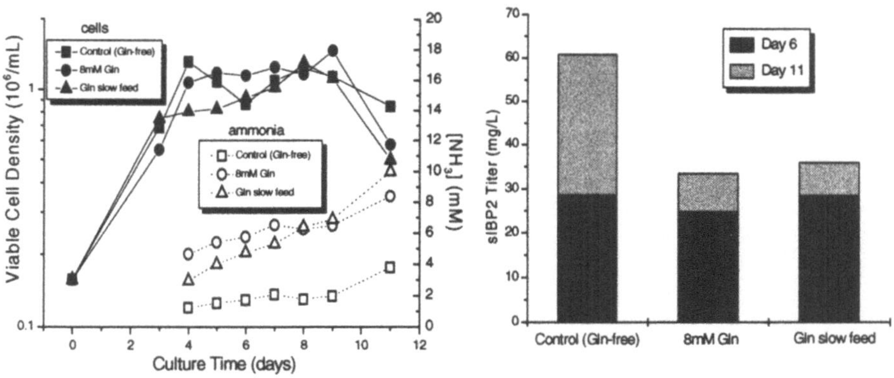

FIGURE 3. Results of the butyrate titration study. Left panel: cell viability. Right panel: productivity.

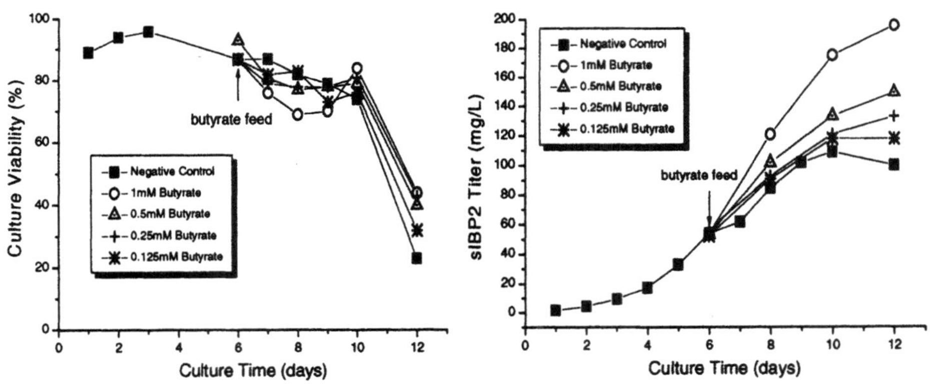

DEVELOPMENT OF A NEW, HIGHLY FLEXIBLE, FULLY COMPUTER INTEGRATED FERMENTOR CONTROL SYSTEM

H. BÜNTEMEYER, M. NERKAMP, S. BOCKHOLT, C. WEBER, F. GUDERMANN AND J. LEHMANN
Institute of Cell Culture Technology, University of Bielefeld
P.O. Box 10 01 31, 33501 Bielefeld, Germany

1 Introduction

Bioreactor control software is most often offered from bioreactor manufacturers. This software package is especially designed for the features of their fermentors and peripherals. In most cases it is not flexible enough to integrate additional devices of other vendors, such as pumps, balances, sampling devices or analytical online systems (FIA).
We developed a new highly flexible control software package. The software is completely modular and follows the principle of the client/server concept. It is written in C++ and runs on the QNX platform, an extremely stable, real time operating system for PCs.
An adaption to any existing bioreactor and additional hardware should easily be possible if those systems accept and/or provide remote control by analog or digital signals.

2 Software Concept

The software is structured completely modular and follows the principle of the client/server concept. It is written in C/C++ (Watcom 10.6). Depending on the purpose, some modules are only servers (e.g. hardware drivers), some are only clients (e.g. Graphical User Interface, data storage) and some are clients and servers as well (e.g. sensors, actors, controlling processes). The module hierarchy for the pH and pO_2-controllers can be seen from figure 1.
QNX (QNX Software Systems Ltd., Canada, www.qnx.com) was used in the version 4.24 as operating system (OS). Photon 1.13 was installed as a graphical platform.
The QNX OS is a stable, UNIX based, real-time operating system for PCs (Intel CPUs).
QNX was installed in the institute on 75 - 120 Mhz Intel Pentium PCs with 16 - 32 MB RAM. The computers were equipped either with ATI Mach32 or S3 Trio64 graphic adapters (2MB RAM) and 500-1000 MB SCSI (Adaptec) or EIDE hard disks. For networking SMC network adapters (ISA) were installed.

A. Bernard et al. (eds.), Animal Cell Technology: Products from Cells, Cells as Products, 155–157.

3 Implemented Features of the Fermentor Control Software

The fermentor control software has, among others, following main features: a graphical user interface; alarming, logging and auto startup procedures; short and long term trend displays; user configurable data storage and transformation (for calculation and plotting programs); process controller for pH (PID, fuzzy, adaptive fuzzy), pO_2 (PID, fuzzy, adaptive fuzzy), temperature (PID), 2 pressure (PID), agitation and liquid handling modes such as fed-batch, chemostat, perfusion, harvesting; sensor processes for pH, pO_2, 5 gas flow meters (O_2, N_2, CO_2, 2 Air), temp. (PT100), stirrer, 2 pressure sensors, serial and analog balances (Prezisa, Sartorius), serial and analog pumps (Watson Marlow); actor processes for 5 gas flow controller (O_2, N_2, CO_2, 2 Air), stirrer, heating system, pressure valves, serial and analog pumps (Watson Marlow).Hardware drivers for some AD adapter (Advantech PCL818h, ELV ADA16), DA adapter (Advantech PCL727, ELV ADA16) and TTL I/O adapter (ELV PIO32) are available. For other adapters new drivers can easily be programmed from modular routines.

Some other useful modules outside the fermentor control system are also available: automatic pump calibration routine, medium filtration control, automatic dosage of supplements or test specimen [1] and automatic sampling systems for bench-scale and pilot-scale fermentors [2]. At present the following modules are under development: online determination of OUR and OTR, connection to MFCS/win (B.Braun Biotech Intl.) and integration of online analysers.

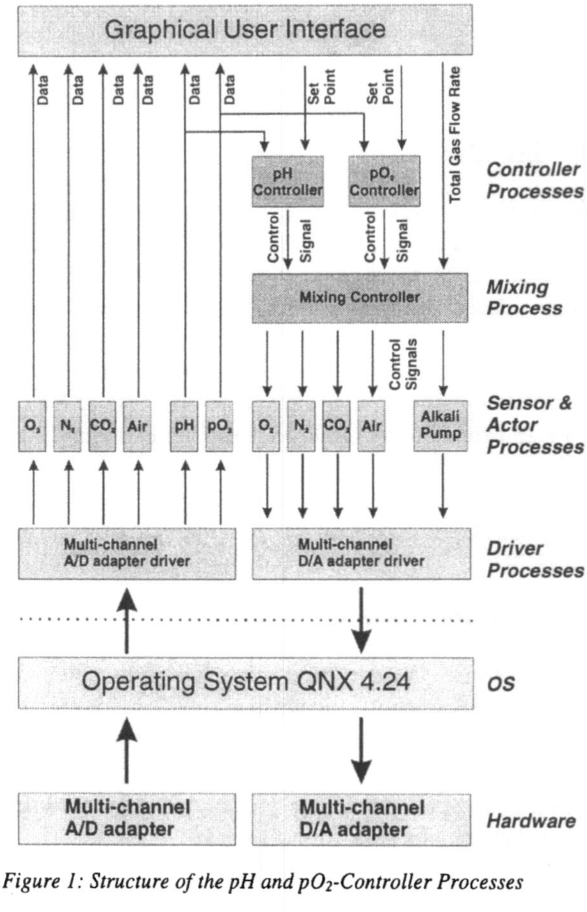

Figure 1: Structure of the pH and pO_2-Controller Processes

4 Results

The software was adapted to 2 L bench-scale and 20 L pilot-scale fermentors. Batch and continuous cultivations were performed in both systems. A time plot of some online parameters of a distinct cultivation is shown in figure 2. On the left hand side the pO_2 and pO_2 - controller output signal (pO_2-COS) are plotted. The progress of the cultivation can be easily seen from the pO_2-COS which can be regarded as an online signal from the metabolic activity of the cells. On the right hand side the gas flow rates of the gases are displayed. All four gases are mixed by the mixing controller process (see figure 1) depending on the pH and pO_2 controller output signals to a constant total flow rate (in this case: 10 L/h).

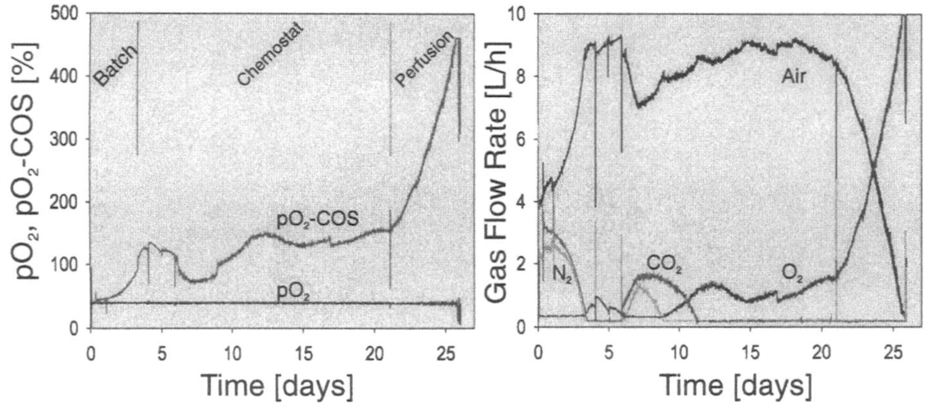

Figure 2: Cultivation of a human cell line in a 2 L bench-scale bioreactor controlled by the fermentor control system. Different dilution rates were used in chemostat mode. In perfusion mode an increasing perfusion rate was used.

5 Conclusion

The software presented is a powerful, but easy to use tool for fermentor control, which should easily be adaptable to many fermentor systems. A lot of peripheral equipment such as pumps, balances etc. can be integrated and controlled.

6 References

1. K. Iding, H. Büntemeyer, F. Gudermann, S. Deutschmann, C. Kionka, J. Lehmann (1998) An automatic system for the assessment of complex medium additives under cultivation conditions. Poster: Cell Culture Engineering VI, Feb. 1998, San Diego, USA
2. D. Lütkemeyer, S. Plahl, J. Lehmann (1998) Vollautomatische, sterilisierbare Probeentnahme und -abfüllung für einen Bioreaktor im Pilotmaßstab. Poster: DECHEMA-Jahrestagung, May 1998, Wiesbaden, Germany

CELL GROWTH RATE ESTIMATION IN PACKED-BED AND HOLLOW FIBERS BIOREACTORS: METABOLISM AND PRODUCTIVITY

M.T.A.RODRIGUES, A.GARBUIO, L.T.NAGAO, I.RAW & A.M. MORO
Centro de Biotecnologia, Instituto Butantan
05503-900, São Paulo, SP, Brasil. E-mail: anammoro@usp.br

Abstract: Hybridoma cells can be cultivated at high densities for long periods through the combination of immobilization devices and continuous perfusion of medium, providing an efficient oxygen and nutrients feeding without cell loss. The disadvantage of these systems is that the cell density cannot be directly monitored and the cellular health is followed by online measurements of pH and dissolved oxygen and off-line determination of residual metabolites. We have been cultivating hybridoma lines in two types of bioreactors: a packed-bed created by hollow glass cylinders inside an airlift bioreactor and a hollow fibers bioreactor. Through the glucose uptake and the lactate production rates we calculated the efficiency of glucose consumption and estimated the growth rate, allowing a comparison of hybridoma lines in the two types of bioreactor, with serum-supplemented and serum-free culture media. The metabolic data together with monoclonal antibody productivity in a per cell basis would make it possible to detect changes of metabolic processes in a perfusion steady-state culture and would allow to improve monoclonal antibody productivity.

Material and Methods: An IgG1 producing hybridoma line was used in this study. The medium culture was DME. The bioreactor was purchased from Bellco Glass, Inc. (fashioned in a packed-bed version (1)). The daily measurements of residual glucose, lactate and glutamine and ammonia (using commercial kits) were used as data for the following equations:

$$GUR = \frac{Glucose_{in} - r\ Glucose_{t2} - e'(Glucose_{in} - r\ Glucose_{t1})}{1-e'}$$

$$LPR = \frac{r\ (Lactate_{t2} - e'\ Lactate_{t1})}{1-e'}$$

$$GlnUR = \frac{Gln_{in} - r\ Gln_{t2} - e'\ (Gln_{in} - r\ Gln_{t1})}{1-e'}$$

A. Bernard et al. (eds.), Animal Cell Technology: Products from Cells, Cells as Products, 159–161.

$$APR = \frac{r(A_{t2} - e'A_{t1})}{1-e'}$$

$$MAbFR = \text{mg Ab/mL x harvest pump flow rate x } \Delta t$$

where $e' = e^{-rt/v}$; r (total inflow rate, mL/hr); t = Δt; v (system volume); GUR (glucose uptake rate, mg/hr); Glu_{in} (perfusion rate x glucose concentration); LPR (lactate production rate, mg/hr); GlnUR (glutamine uptake rate, mmol/hr); APR (ammonia production rate, mmol/hr), MAbFR (monoclonal antibody formation rate). Cellular density estimation was based on glucose balance and considering $Y_{X/GLU} = 4.4 \times 10^8$ cell/g (2).

Results: The run was kept for 94 days being sampled daily. Figure 1 shows the MAbFR in relation to perfusion rate along the 94 days of culture. Lactate balance was used to evaluate LPR and $Y_{LAC/GLU}$; the correlation (r = 0.999) can be seen in Fig.2. Fig. 3 shows the antibody yield ($Y_{MAb/X}$) on the estimated number of cells = 0.409×10^{-10} g/cell

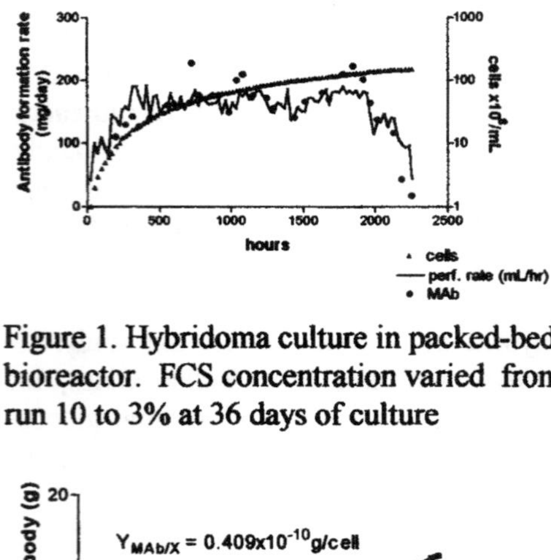

Figure 1. Hybridoma culture in packed-bed bioreactor. FCS concentration varied from run 10 to 3% at 36 days of culture

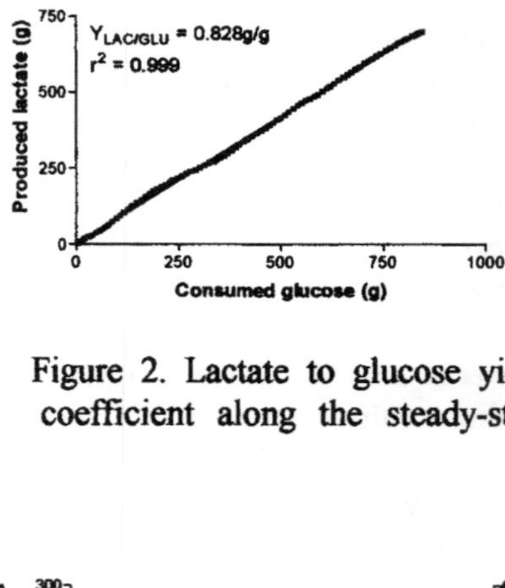

Figure 2. Lactate to glucose yield coefficient along the steady-state

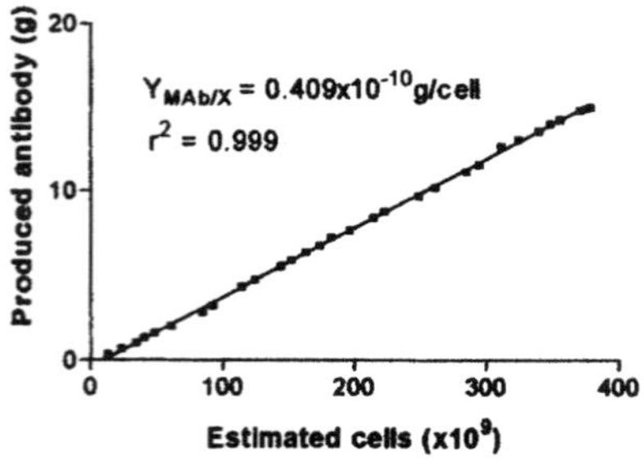

Figure 3. Antibody yield versus cell number

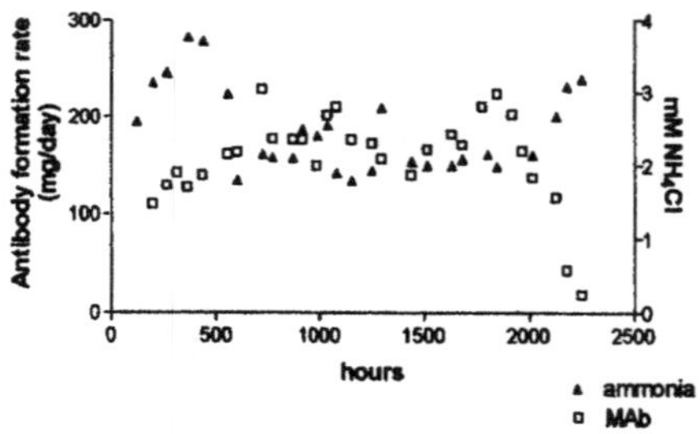

Figure 4. Effect of ammonia on cell growth and antibody production

along the run. In Fig. 5 is displayed the relation between MAbFR and APR.

Discussion: The results were concerned about a typical culture of hybridoma in packed-bed bioreactor with internal aeration. Neither oxygen or FCS or temperature was limiting, as observed previously for this hybridoma line. The run was kept at state-state conditions, the glucose and lactate and ammonia and antibody production were nearly the same along the time; so the perfusion was maintained at constant rate. The use of metabolic data, especially glucose uptake rates, to estimate cellular proliferation was based on the facts that only viable cells consume substrate and produce antibodies and specific rates of incorporation of nutrients into cell mass are proportional to the specific growth rate (3). In consequence to the steady-state discussed above, the cellular growth rate estimated was found to be gradual and constant. However, in contrast to other packed-bed and also hollow fiber system cultivations, the antibody productivity was low in the present case. The possible explanations are related to metabolite toxicity, as lactate and ammonia. Ammonia concentration higher than 2mM seemed to cause an inhibition of antibody yield/cell. The ammonia concentration reached these levels because of the medium perfusion rate, which was enough to maintain a culture for longer period but insufficient for a higher antibody productivity as obtained in other runs. Besides ammonia, also lactate concentration could be blamed for the poor antibody yield, being higher than 22mM several times along the run, in contrast to the concentrations measured in more productive runs (4).

In conclusion, for a higher antibody productivity, besides a good cellular growth, the medium perfusion must be maintained at higher rates, even if the spent medium is still capable of supporting cell growth. It is not worthwhile to save medium culture by maintaining slower perfusion rates or recycling in detriment of a reasonable and possible higher antibody production.

A.M.Moro received financial support from FAPESP.

References:

1. Moro AM, Rodrigues MTA, Gouvea MN, Silvestri MLZ, Kalil JE and Raw I (1994). Multiparametric analyses of hybridoma growth on glass cylinders in a packed-bed bioreactor system with internal aeration. Serum-supplemented and serum-free media comparison for MAb production. J. Immunol. Meth., 176:67-77.

2. Kurokawa H, Park YS, Iijima S and Kobayashi T (1994). Growth characteristics in fed-batch culture of hybridoma cells with control of glucose and glutamine concentrations. Biotechnol. Bioeng., 44:95-103.

3. DiMasi D and Swartz RW (1995). An energetically structured model of mammalian cell metabolism. 1.Model development and application to steady-state hybridoma cell growth in continuous culture. Biotechnol. Prog., 11:664-676.

4. Rodrigues MTA, Vilaça PR, Garbuio A, Takagi M, Barbosa Jr. S, Léo P, Laignier NS, Silva AAP and Moro AM (1999). Glucose uptake rate as a tool to estimate hybridoma growth in a packed bed bioreactor. Bioprocess Eng., in press.

INFLUENCE OF OXYGEN LIMITATION ON MYELOMA CELL CULTURES:
Various Methods of Inoculum Proliferation for Bioreactor Cultivation

R. SPINNLER[1], U.GRAF-HAUSNER[1], M. INGLIN[1], CH. LEIST[2]
[1] *Universitiy of Applied Sciences (ZHW), Winterthur, Switzerland*
[2] *Novartis Pharma AG, Basel, Switzerland*

1. Introduction

The way in which the inoculum of myeloma cells is prepared has an important influence on the antibody production in the bioreactor. The aim of this work was to study the influence of oxygen in the inoculum production on the volumetric production rate in the bioreactor.

2. Materials and methods

Figure 1

Step 1: working cell bank: –196°C (liquid N_2)
Step 2: splits in T75-flasks if cell conc. is $7\text{-}9 * 10^5$ cells ml^{-1}, incubation 37°C, 5% CO_2.
Step 3: 1-2 splits T150-flasks (inocula for bioreactor and spinner).
Step 4: surface aerated loop reactor: 2.5 litres, 37°C, 5% CO_2 in synthetic air, 150 rpm (propeller tip-speed: 0.5 m s^{-1}).
Step 5: 500 ml spinner-flasks with two magnetic stirrers, incubation: 30 rpm, 37°C, 5% CO_2.

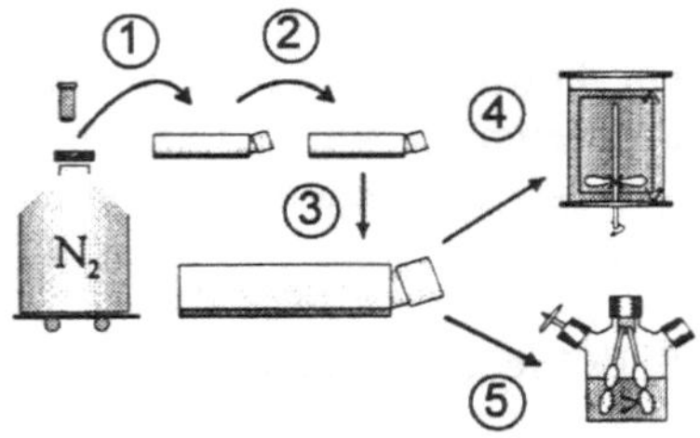

3. The influence of different volumes in spinner- and T-flasks:

Cells were cultivated in T75-flasks at five different volumes of cell culture medium. The global antibody production rates were determined (figure 2).

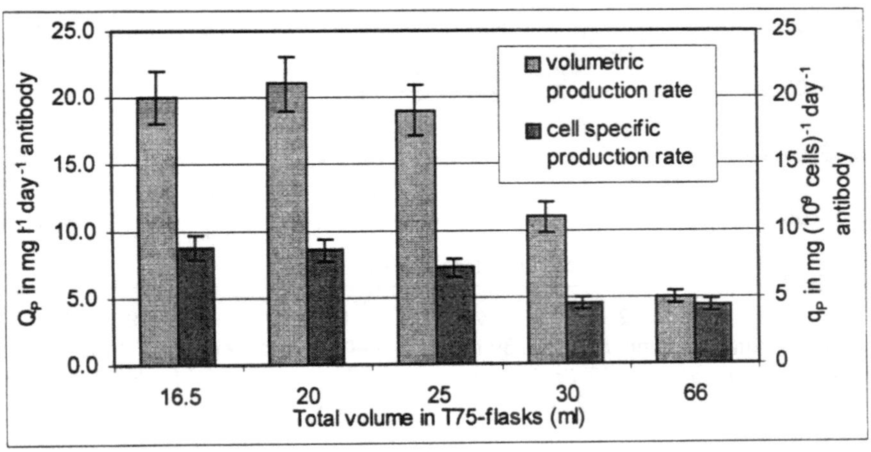

Figure 2: Antibody production rates in T-flasks filled with different volumes.

A. Bernard et al. (eds.), *Animal Cell Technology: Products from Cells, Cells as Products*, 163–165.

The lower the volume in a T75-flask the better were the specific and volumetric production rates. The optimal volume was 20 ml with a Q_P of 21 mg l^{-1} d^{-1} antibody. Cells in a low volume have a better oxygen-supply because there is a thinner medium layer between the surface and the cells.
The spinner showed similar results: cells incubated in spinner-flasks in a total volume of 110 ml had a three times higher Q_P than cells in a volume of 440 ml (data not shown).
Other cell lines of Novartis Pharma AG showed the same or similar results (unpublished).

4. Influence of sampling from spinner- and T-flasks

The influence handling the spinner- or T-flask for sampling was studied. To do this, one flask was opened repeatedly after 1, 2, 3, 4, 7, 9 and 11 days (7 samples). Another flask was opened just after 4, and at the end after 11 days (2 samples), because after 4 days the highest total cell-concentration, and after 11 days the highest antibody-concentration is normally reached.
In T-flasks viability, total cell concentration and antibody concentration between seven- and two-times sampling is the same. There was no significant influence of sampling.
In spinner-flasks there was an influence between seven- and two-times sampling (data not shown). After 4 days there was no significant difference. But after 11 days there was a difference: With a total volume of 110 ml, the handling procedure increases the antibody production by about 30%. With a total volume of 440 ml the results are vice versa.

5. Influence of inocula-proliferation with different volumes in the bioreactor

The inoculum was proliferated twice in T75-flasks with the total volume of 25 ml. Then the cells were proliferated for six splits in 20 or 30 ml in T75-flasks. The inoculum (last split in T150-flasks with 40 or 60 ml) was transferred into the bioreactor.

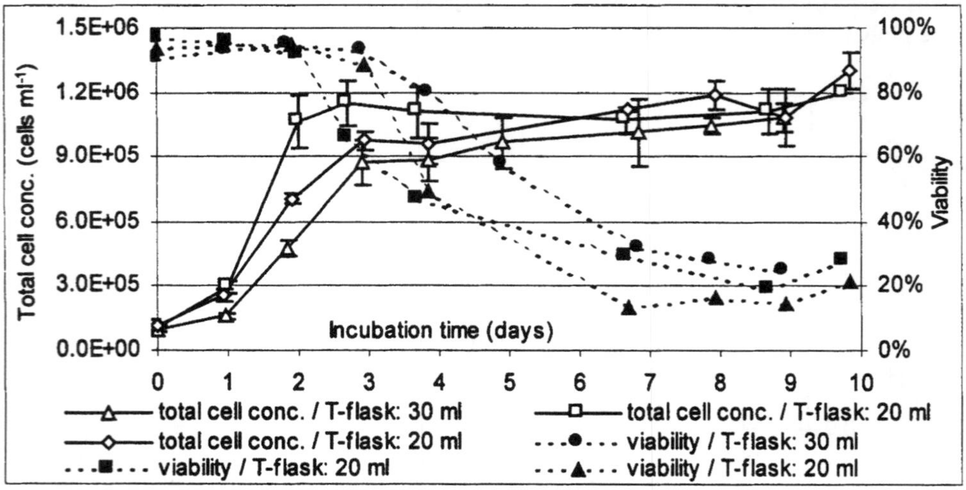

Figure 3: Viability and total cell concentration in the bioreactor with different inocula, proliferated in T75-flasks with 20 or 30 ml.

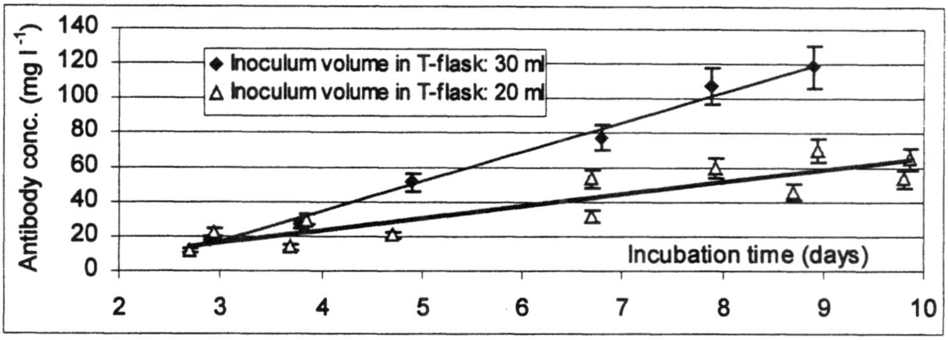

Figure 4: Linear regression of antibody concentrations in the bioreactor with different inocula, proliferated in T75-flasks with 20 or 30 ml.

The bioreactor-experiment with inocula proliferated in 20 ml volume showed a faster growth in the bioreactor (figure 3). But cell viability decreased sooner and faster than the inoculum proliferated in 30 ml. In all experiments the highest cell concentration was reached after 3-4 days. The inoculum proliferated in 30 ml had a volumetric production rate (Q_P) of 13.5 mg l^{-1} d^{-1} (figure 4). This was twice as high as with the Inocula proliferated in 20 ml (Q_P of 6.4 mg l^{-1} d^{-1}). This result contradicts the results of T-flask where the lower volume had the better production rate !

6. Conclusions

6.1 LOWER VOLUME (BETTER O_2-SUPPORT) → BETTER ANTIBODY PRODUCTION

Cells proliferated in a low volume such as 20 ml in T75-flasks, have a significantly better volumetric production rate (Q_P) than cells proliferated in larger volumes. A possible reason is the oxygen supply of the cells: A thinner layer between the surface and the bottom of the flask means an improved oxygen transfer. The results in spinner-flasks are similar. Different cell lines showed similar results (unpublished).

Whether sampling occurred more frequently or less frequently, had no effect on the antibody production in T-flasks. Therefore oxygen supply or other handling-effects of the sampling process have no significant effect on the antibody production in T-flasks.

In spinner flasks the sampling seems to be more difficult, with regard to antibody production. There seems to be a problem keeping cells in good antibody-production conditions. Possible reasons are: shear stress (stirrer) or a limitation of O_2-diffusion through the sterile gas filter.

6.2 INOCULUM: "STRESS" → TWOFOLD ANTIBODY PRODUCTION IN BIOREACTOR

There was a significant influence using different volumes during inocula proliferation for the bioreactor. A twofold antibody production was achieved. The production in the bioreactor may be enhanced due to diffusion limited oxygen in the inoculum. It seems that these cells are better adapted to oxygen-stress, or other stress factors in the bioreactor.

COMPARISON OF FED-BATCH STRATEGIES IN HYBRIDOMA CULTURES

J. O. SCHWABE, INES WILKENS, R. PÖRTNER
Technische Universität Hamburg-Harburg, Bioprozeß- und Bioverfahrenstechnik, D-21071 Hamburg, Germany

Abstract

The classical cell culture process in stirred tank or airlift reactors is still the mostly used large scale production system. Fed-batch processes with different feeding strategies have been investigated to improve the productivity of hybridoma cultures. The following approaches for substrate control are compared in this work: model-based control, on-line characterisation and the application of linear feed trajectories. The required knowledge and the effort for implementation is discussed.

1. Introduction

Cell culture processes are characterised by cell specific limits such as low growth and production rates, substrate limitation and inhibition by metabolite. Fed-batch processes can significantly improve cell concentration and productivity, if substrates, especially glucose and glutamine, are controlled at concentrations close to limitation to avoid enhanced metabolite formation. Fed-batch processes with different strategies have been applied by various authors. The choice of the method is dependent on the knowledge about the process and cell specific kinetics. The model based approaches require quite detailed knowledge and a model with high accuracy, otherwise they show only limited success when they are used for substrate control. If the trajectories are determined by 'a priori' simulation, changes in metabolism during the process are not considered and this can result in different behaviour than expected. The on-line characterisation requires less information but higher implementation effort is necessary. The feed trajectory strategy shows that a simple method is able to realise high cell concentration at low substrate concentrations. This however demands experience in growth behaviour and substrate uptake.

2. Cell line and Culture Conditions

The hybridoma cell line IV F 19.23 produces monoclonal antibodies against penicillium amidase. Cultures were initially started with standard medium (1:1 mixture IMDM/Ham's F12) and standard medium with reduced glutamine and glucose of 1 mmol l^{-1} and 5 mmol l^{-1} respectively (feed trajectories).

A. Bernard et al. (eds.), Animal Cell Technology: Products from Cells, Cells as Products, 167–169.

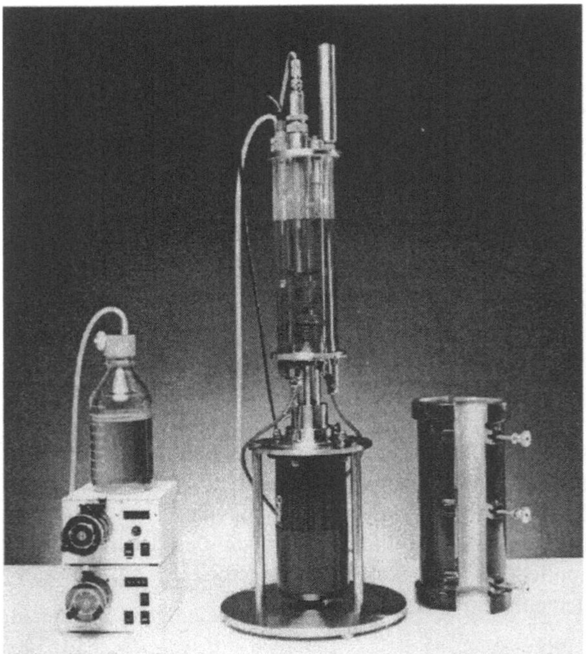

Fed-batch were performed in a 2L polyamid foil reactor (figure 1). For the fed-batch phase a 10-fold concentrated medium (CELLCONCEPT, D) containing glucose (50 mM/ 100 mM) and a concentrated glutamine solution (200 mM, Life Technologies, D) were supplied. In the example for model based control standard medium was used as feed.

Figure 1: VSF-Bioreactor (2L) , Bioengineering (CH): Bioreactor with transparent polyamid foil, in situ sterilisable with metal shell. Setup for fed-batch culture of suspended mammalian cells.

3. Feeding Strategies

a) Model Based Process Control

Application of a structured or unstructured model for "a priori" simulation/optimal control of substrate concentrations. Feed trajectories are defined by simulations and are optimised according to determined criteria before starting the process (figure 1). Model and experiment were previously described by Pörtner et al. (1996).

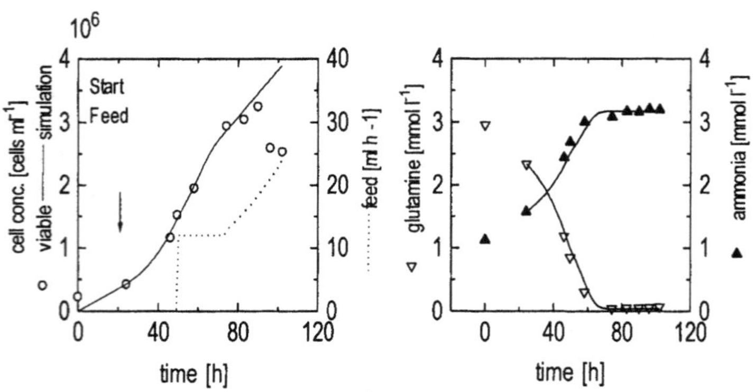

Figure 2: Modell based control: Fed-batch culture with "a priori" determination of the feed rate with a kinetic model, Pörtner et al. (1996).

b) On-line Characterisation

The cell metabolism is characterised by internal process variables, which are calculated on-line, such as oxygen uptake rate (OUR) or ATP production. Substrate feed is controlled via coupling of on-line data and off-line substrate measurements. This strategy is based on the ratio of oxygen to glucose consumption obtained from on-line OUR and off-line glucose measurements (figure 2, Schwabe et al., 1999).

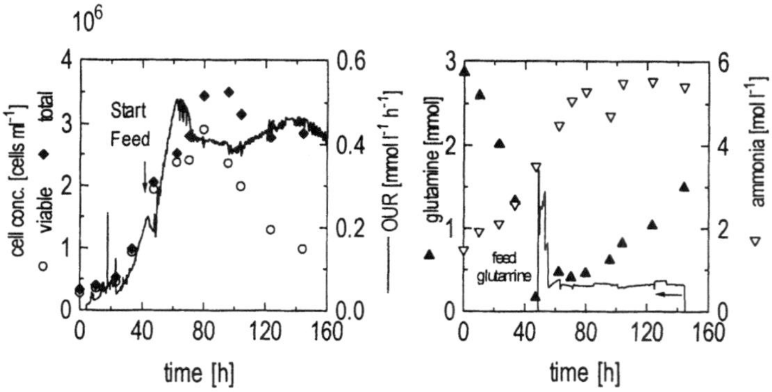

Figure 3: On-line characterisation: Fed-batch culture (Schwabe et al., 1999).

c) Linear Feed Trajectories

Determination of different feed trajectories for the simultaneous feed of two substrate solutions (1. nutrient concentrate with 100 mM glucose, slope: 0.167 ml h^{-2}, 2. 200 mM glutamine, slope: 0.022 ml h^{-2}) and application of these profiles (figure 3).

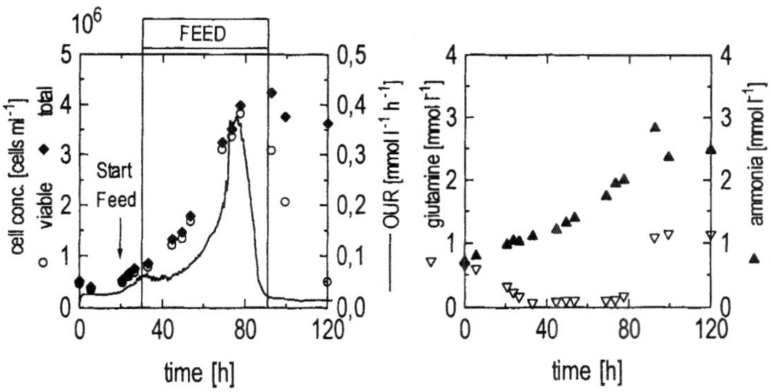

Figure 4: Linear feed trajectories.

4. Results

All approaches show comparable results with respect to cell concentration, whereas monoclonal antibody concentration varies among the strategies depending on culture time/feed period (a. 50 mg l^{-1} at 100 h, b. 40 mg l^{-1} at 144 h, c. 22 mg l^{-1} at 120 h). The choice of the method is dependent on the knowledge about the process and cell specific kinetics. Model based control is suitable for "a priori" determination of feed trajectories but does not consider actual changes or variations of the cell metabolism during the process. The approaches for modelling metabolism still insufficient or specific parameters for structured models are not available or difficult to obtain. The on-line feeding strategy by stoichiometric feeding needs balanced feed solutions. Coupling parameters between on-line and off-line measurement have to be updated in regular intervals and with high reliability, on-line analytic would significantly improve the strategy. The feed trajectory is a simple method for evaluating uptake and production rates which is able to realise high cell concentration at low substrate concentrations.

[1] Pörtner, R., Schilling, A., Lüdemann, I., Märkl, H.: High density fed-batch cultures for hybridoma cells performed with the aid of a kinetic model. Bioproc. Eng. 15 (1996) 117-124

[2] Schwabe, J. O., Pörtner, R., Märkl, H.: Improving an on-line feeding strategy for fed-batch cultures of hybridoma cells by dialysis and 'Nutrient-Split' feeding. Bioproc. Eng. (1999)

NON-INVASIVE ADAPTIVE CONTROL OF THE FEED RATE FOR HIGH CELL DENSITY CHO CULTURES ACCORDING TO THE pO2 PROFILE

DETHARDT MÜLLER, OTTO DOBLHOFF-DIER AND HERMANN KATINGER
Institute of Applied Microbiology (IAM), University of Agricultural Sciences, Muthgasse 18, A-1190 Vienna, Austria;
http://www.boku.ac.at/iam/; email: dmueller@edv2.boku.ac.at

Abstract
For mass manufacture of animal cell culture derived biologicals continuous perfused high cell density or fed batch systems are favourable. We established a non-invasive adaptive control capable to optimize the feed rates of media in order to maximize the product concentration in the harvest without triggering nutrient limitations.
Current control systems based on flow injection analysis or online HPLC are well approved but invasive, the same applies to oxygen consumption measurements due to changes in the setup of process parameters. As a consequence a big expenditure of work has to be put on the validation when applying those systems to a scalable production process.
The novel control algorithm for the feed rate is based on the oxygen consumption calculated from the online pO_2 profile. It is non-invasive and adaptable to occurring changes in the utilization profile of nutrients and the growth behaviour of the cells. In addition the algorithm is generally applicable to fed batch systems albeit it was developed and challenged using a continuous perfused fluidized bed reactor.
A correlation between the pO_2 profile and the measured oxygen consumption was found for the whole period of the process, even when the parameters for the pO_2 controller were modified. The concentrations of glucose and amino acids were adjusted to the chosen steady-state values within an acceptable range. As a consequence of maintaining optimal steady-state conditions the product concentration in the harvest was maximized.

Introduction

The development of the control algorithm was carried out in three fundamental steps:

- correlation between oxygen consumption and online pO_2 data
- linkage of consumption rates for oxygen and glucose as the controlled variable
- calculating the feed rate from mass balances for steady-state and non-steady-state conditions having regard to the specifig growth rate

Results

The experiments were carried out with recombinant CHO cells in a Cytopilot Mini™ / Cytoline1™ fluidized bed reactor system.

A. Bernard et al. (eds.), Animal Cell Technology: Products from Cells, Cells as Products, 171–173.

The data resulted from the online pO_2-profile were calculated from an oscillation generated by an on/off control mode as shown in Fig.1.

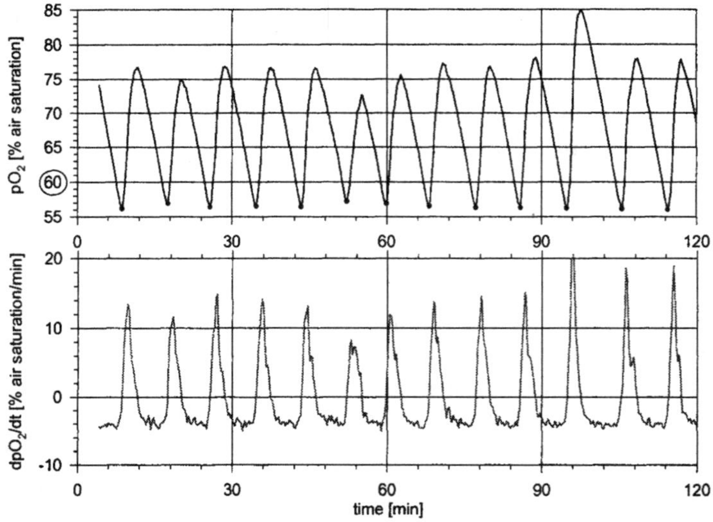

Fig.1: modified standard pO_2 output from a on/off controller mode and its first derivation

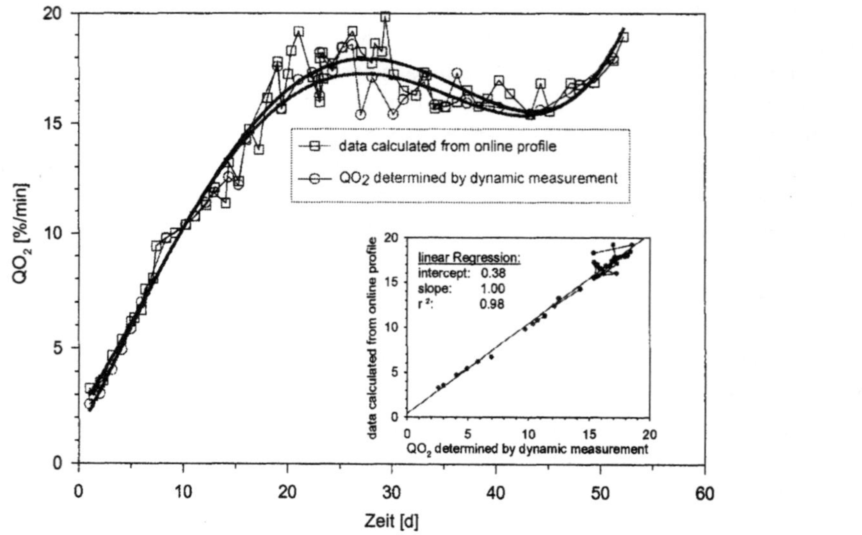

Fig.2: pO_2 data calculated from online profiles set into correlation with the oxygen consumption QO_2 determined by dynamic measurements

The results thus achieved were set into correlation with those coming from dynamic measurements carried out for the determination of the oxygen consumption rate. As

shown in Fig.2 we found a straight proportional function. In a next step the measured online QO_2 was linked to the volumetric (total) glucose consumption using constant specific rates coming from in-process control measurements.
With the following mass balance equation for a medium component i the actual feed rate F was calculated directly from the measured online QO_2 using steady-state and non-steady-state conditions respectively. In order to respect the growth behaviour the algorithm for the perfusion rate was upgraded with a term considering the specific growth rate.

$$\frac{d(V \cdot c_i)}{dt} = c_i \frac{dV}{dt} + V \cdot \frac{dc_i}{dt} = F \cdot c_i^M - F \cdot c_i - Q_i$$

As a consequence the glucose concentration was repeatedly controlled within an acceptable range as shown for one fermentation in Fig.3.

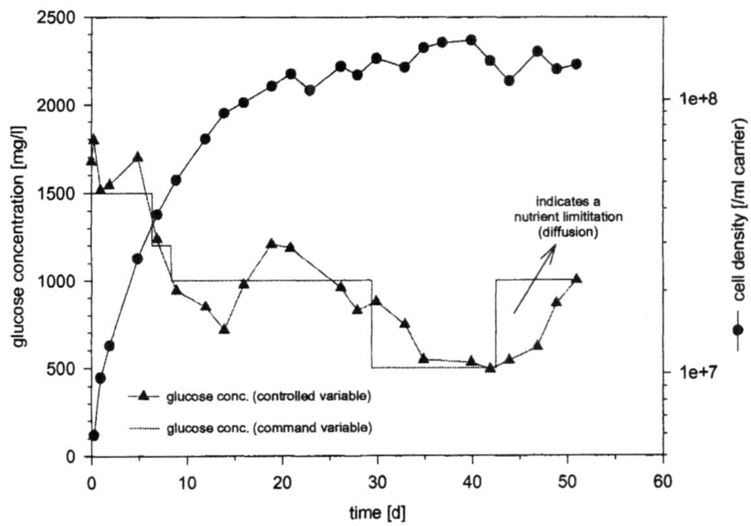

Fig.3: carrier cell density and glucose concentration profiles of a CHO fermentation using the developed control algorithm

Conclusion

This novel control algorithm complies with the following demands: it is

non-invasive - it does not interfere with the running process since the PID loops for pH, pO_2 and perfusion stay enabled during measurement and so it undergoes validation more easily

adaptive - the algorithm includes specific parameters characterizing different stages of the process and therefore it may be adapted to changing conditions

multifunctional - the control algorithm is applicable to different types of high cell density fermentations as continuous processes with or without cell retention/recycle or as fed batch processes

CELL GROWTH CONTROL BY THE IRF-1 SYSTEM IN PERFUSION CULTURE

A.V. CARVALHAL[1], J.L. MOREIRA[1] & M.J.T. CARRONDO[1,2]

1- IBET/ITQB, Apartado 12, 2780 Oeiras, Portugal,
2- Lab. Eng. Bioq., FCT/UNL, 2825 Monte da Caparica, Portugal

1. Abstract

Activation of the constitutively expressed interferon-regulatory-factor-1/estrogen receptor fusion protein (IRF-1-hER) in BHK cells was accomplished through the addition of estradiol to the culture medium, which enabled IRF-1 to gain its transcriptional activator function and inhibit cell growth. With the addition of 100 nM of estradiol at the beginning of the exponential phase, IRF-1 activation leads to a rapid cell growth inhibition but also to a significant decrease in cell viability. To apply this concept in industry, strategies to extend the stationary phase are required. Cycles of estradiol addition/removal were performed in 2 L stirred tank bioreactor operated under perfusion in order to reduce the time span of estradiol exposure by slow dilution after a step addition of 100 nM estradiol (perfusion rate between 0.7 and 1.4 day^{-1}). Cell growth inhibition of the regulated/non-producer clone was achieved for three consecutive times, showing that the cells were able to respond to estradiol addition independently of the cycle. Diluting the estradiol by perfusing medium without estradiol to concentrations lower than 10 nM lead to cell growth and viability recovery, independently of the perfusion rate used. These observations led to the definition of strategies to operate IRF-1 regulated cells by pulse estradiol addition followed by the longest possible period with estradiol and by the fast perfusion to low estradiol concentration. The response to IRF-1 activation and following estradiol removal by perfusion was also evaluated with a regulated/Factor VII producer clone, where the time of estradiol exposure and perfusion rate were varied. This clone presents a stronger response to IRF-1 activation without increase on factor VII specific productivity after cell growth inhibition; clearly indicating that the stationary phase obtained is clone dependent. The final conclusion is that it is possible to modulate the IRF-1 effect by the manipulation of cycles of addition/removal of estradiol.

2. Materials and Methods

Cell line and medium: BHK-21A cells (ATCC CCL10) and BHK-21A cells expressing blood coagulation Factor VII (FVII) (non-regulated/producer clone, kindly supplied by Dr Leif Kongerslev from Novo Nordisk, Gentofte, Denmark) were stably transfected with the IRF-1-hER construct using the calcium co-precipitation method by Dr Peter Muller (GBF, Braunshwig, Germany).

A. Bernard et al. (eds.), Animal Cell Technology: Products from Cells, Cells as Products, 175–177.

The IRF-1-hER regulated BHK clones were cultured in DMEM supplemented with 10% (v/v) FBS, 4.5 g l^{-1} of glucose, 5 mg l^{-1} of puromycin and 5 mg l^{-1} of vitamin K1 for the producer clone (all final concentrations). FBS was supplied by Sigma (St. Louis, MO); all others were supplied by Life Technologies (Glasgow, UK).

Culture system: Bioreaction studies were performed in BIOSTAT MD bioreactor (B.Braun, Melsungen, Germany) with a 2.0 liter working volume. The cultures were maintained at pH 7.2, oxygen tension of 20%, and agitation rate 80 rpm, and inoculated with a cell concentration of 3×10^5 cell ml^{-1}. Medium was supplemented with 0.01% of Pluronic F-68 (Sigma).

Perfusion system: The perfusion cultures were performed using the perfusion system BioPem (B.Braun) a magnetically stirred filtration cell with tangential flow. The culture broth was continuously recirculated through the BioPem, operated as an external loop of the bioreactor, with the help of a peristaltic pump (Watson Marlow, Falmouth, Cornwall, U.K.). Removal of estradiol from the culture medium was performed by the diluting it with estradiol-free fresh medium.

3. Results and Discussion

Previous experiments in static systems[1] showed that reducing the time span of estradiol exposure allowed to overcome the cell viability decrease. The reduction of time span of estradiol exposure could than be used as a strategy to extend the stationary phase. Estradiol removal is not very straightforward in static cultures, and in this case operation as perfusion was required.

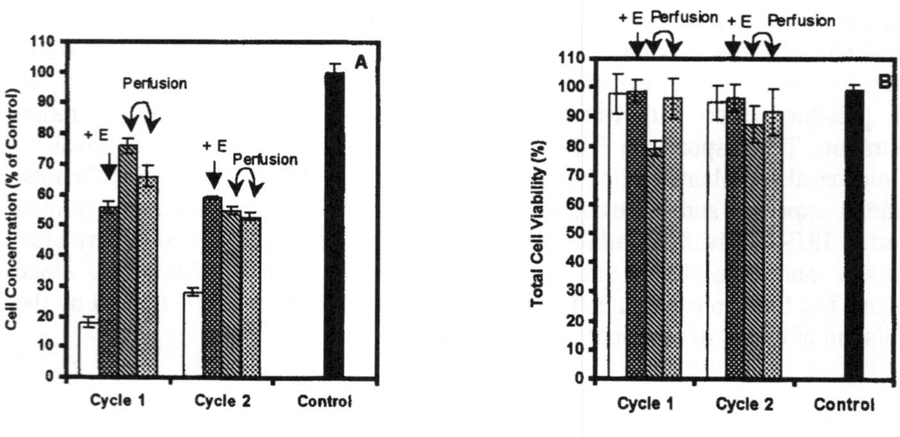

Figure 1: Effect of cycles of estradiol addition (100 nM, single step as batch culture) and removal (by perfusion at 1.4 and 0.7 day $^{-1}$, respectively, with estradiol-free fresh medium) upon cell concentration (A) and cell viability (B). For each cycle from left to right: □ Beginning of the cycle; estradiol addition (+E), start of perfusion, end of perfusion and ■ control experiment, with cells grown in the absence of estradiol.

In Figure 1 the results of two cycles of estradiol addition/removal, performed in a stirred tank bioreactor operated at a perfusion rate between 0.7 and 1.4 day^{-1} is presented. In these experiments 100 nM estradiol was added in a single step at the middle of cell exponential phase, still in the batch phase of the culture. Cell growth inhibition could be maintained for several consecutive times, showing that cells were able to respond to estradiol addition independently of the cycle (as observed in static cultures[1]).
The cell growth and viability recovery occurred at estradiol concentrations lower than 10 nM, independently of the perfusion rate used. The cell growth and cell viability of the regulated/Factor VII producer clone, during one cycle of addition/removal of estradiol and varying the time span of estradiol exposure was also evaluated. The regulated/FVII producer clone shows a stronger response to IRF-1 activation; these results being in agreement with previous observations in static and stirred flasks[1], thus the cell pattern response to estradiol addition/removal showed to be clone dependent.

4. Conclusions

The cell response to IRF-1 activation is reversible and cyclic, characterised by three distinct cell growth phases: slow exponential growth until two days after estradiol addition, followed by a death phase until the removal of estradiol to a critical concentration, after which there is a cell growth recovery.
Diluting the estradiol by perfusing medium without estradiol to concentrations lower than 10 nM lead to cell growth and viability recovery, independently of the perfusion rate used.
Cell growth inhibition of the regulated/non-producer clone could be maintained for three consecutive times, showing that cells were able to respond to estradiol addition independently of the cycle. Nevertheless, it was observed that the cell growth inhibition and cell viability decrease was less significant as the number of cycles increased.
This reversible process is dependent on the clone; the regulated/Factor VII producer clone presents a stronger response to IRF-1 activation not leading to the increase on factor VII specific productivity.
These observations led to the definition of strategies to operate IRF-1 regulated cells by pulsed estradiol addition (final concentration of 100 nM) followed by a period of estradiol exposure at 100 nM (48 to 72 hours in this case), and perfusion (1.4 day^{-1} in our experiment) to decrease estradiol concentration (lower than 10 nM) as fast as possible.

Acknowledgements
The authors are grateful to Dr. Leif Kongerslev and his team from Novo Nordisk and to Ms. Maria do Rosário Clemente from IBET/ITQB for technical support. The authors acknowledge and appreciate the financial support received from the European Commission (BIO4-CT95-0291) and from Fundação para a Ciência e Tecnologia-Portugal (FMRH/BIC/1788/95 and BIO1117/95).

References

[1]-Carvalhal, A.V., Moreira, J.L., Cruz, H., Mueller, P., Hauser, H. and Carrondo, M.J.T (1999) Manipulation of culture conditions for BHK cell growth inhibition by IRF-1 activation.– submitted.
[2]-Kirchhoff, S., Schaper, F. and Hauser, H. (1993). Interferon regulatory factor 1 (IRF-1) mediates cell growth inhibition by transactivation of downstream target genes. Nucleic Acids Res. 21:2881-2889 .
[3]- Kirchhoff S, Kröger A, Cruz H, Tümmler M, Schaper F, Köster M and Hauser H. (1996). Regulation of cell growth by IRF-1 in BHK-21 cells. Cytotechnology 22:147-156 .

MODULATION OF CELL CYCLE FOR OPTIMAL RECOMBINANT PROTEIN PRODUCTION

V. HENDRICK, O. VANDEPUTTE, A. RASCHELLA, T. MARIQUE, M. CHERLET, C. ABDELKAFI and J. WERENNE.
Université Libre de Bruxelles, Animal Cell Biotechnology (CP:160/17)
50, Av F.D. Roosevelt,1050 Bruxelles, Belgium

Abstract: Efficient t-PA production in recombinant CHO cells is of major interest for pharmaceutical industry . Contrary to the multigene metabolic engineering approach, our strategy allows investigations of recombinant cell lines already validated. Compared to 37°C, 32°C showed that t-PA productivity was significantly increased. The specific rate of t-PA secretion was enhanced at the lower temperature, in relation to the cell cycle modification. At this temperature, glycosylation is not significantly altered while serine protease activity is reduced. A similar study involving cytofluorimetric data and mathematical analysis was made for the other factors tested (PMA, TGF-β and butyrate). Our data not only emphasize the interest of a two step process for t-PA production (involving 1. a cell biomass production phase 2. a high protein productivity phase), but showed moreover that productivity can be further modulated by the extracellular environmental factors affecting cell cycle. On the basis of results obtained by our rapid screening method a multigen engineering strategy could be decided on a rational basis.

1. Introduction

Mammalian cell culture is becoming increasingly important for the production of human proteins especially in the pharmaceutical field. r-tPA (recombinant tissue plasminogen activator) produced by CHO cells is one example of such a glycoprotein of therapeutic value. The growing demand for such as product at low cost require the development of a long term and large scale production. Great deal of efforts has already been made in this field allowing us to further investigate the effect of the modification of the external parameters affecting cell cycle We have thus investigated the effect of temperature and other factors such as PMA, TGF-β and butyrate on the yield of t-PA produced in CHO cells in relation to the cell cycle.

2. Materials and methods

The Chinese Hamster Ovary (CHO) recombinant cells which produces the tissular activator of plasminogen (t-PA) are grown in suspension in a serum free medium within a 250 ml Techne spinner (speed: 50 rpm), placed in a 5% CO_2 and 100% humidity atmosphere. The cell concentration is determined manually with a hematocytometer and viability by trypan blue dye exclusion.

A. Bernard et al. (eds.), Animal Cell Technology: Products from Cells, Cells as Products, 179–181.

The cells inoculated at 1.5 x 10E5 cells $ml^{-1}$are treated with butyrate (1mM), PMA (10 mM) or TGF-β (5 ng/ml) . The shift of temperature (32°C) is also operated after 72 hours.
A sandwich ELISA test (Immulyse t-PA kit by Diagnostic International, Germany) is used for the quantitative determination of single-chain and two-chain t-PA antigen in cell culture supernatants.
The zymographic analysis allowed us not only to detect and evaluate the activity of the different active forms of t-PA but also the metalloproteinases present in the supernatents of the CHO-t-PA culture. After migration, the gel is treated by a nonionic detergent (Triton X-100) used to restore enzyme activity. Then, the gel is incubated in a substrate buffer over the night and next day the t-PA activity was estimated by the intensity of an unstained band visualized after coloration with coomassie blue. The system used for this electrophoresis is a Mini-Protean II cell (BIORAD).
Different phases of cell cycle (GOG1, S, G2M) are analysed by cytometry (FACSCalibur, Becton Dickinson). The cells are fixed in 75% PBS/25% methanol and placed at -20°C. Cells are centrifuged and the pellet is homogeneized in 2ml of phosphate-citrate buffer. After incubation during 30 min., 1ml of RNase (3mg/ml) is added to the solution. After incubation at 37°C during 20 min., pellets are resuspended in propidium iodide (50 μg/ml).

3. Results

The proportion of cells in G0G1 phase increased after the shift of temperature (37°C to 33°C) (Data not show). According to the equation of Slatter et al.$_{(2)}$, the length of the G1 phase appeared to be increased 4 times after the shift of temperature. By labeling t-PA for cytofluorimetry and using Collins-Richmond equation $_{(1)}$, accumulation of the recombinant protein during G1 phase was demonstrated (data not show).

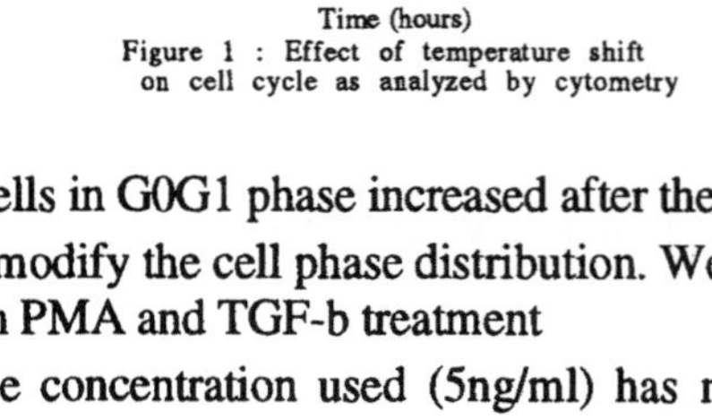

Figure 1 : Effect of temperature shift on cell cycle as analyzed by cytometry

The proportion of cells in G0G1 phase increased after the induction of butyrate but PMA and TGF-β did not modify the cell phase distribution. We have not observed an increase of GOG1% phase in PMA and TGF-b treatment
While TGF-β at the concentration used (5ng/ml) has no effect , butyrate, PMA and temperature shift increase t-PA secretion. TGF-β at the concentration used did no show effect on cell growth but the other inductors and temperature shift decrease the cell growth (Cf.table 1).
Moreover as the temperature shift leads to a reduction in proteases release (serine and metalloproteases)(data not show), these conditions appears for an improved procedure.

TABLE 1. Growth and productivity of CHOt-PA

	Control	32°C	Control	Butyrate	Control	PMA
μ (h^{-1})	0,01	0,005	0,09	0,006	0,02	0,01
q t-PA (pg/cell/hour)	0,04	0,12	0,02	0,08	0,025	0,05

Zymograms show that the different t-PA molecules produced in the different conditions are similar (no differences in glycosylation appear): Single chain t-PA (50-80 kDa) and Two chain t-PA (between 35-40 kDa) are observed. One extra band (32 kDa) appears at the end of culture (indicating probable presence of "degraded t-PA") (CF. *Figure 2*).

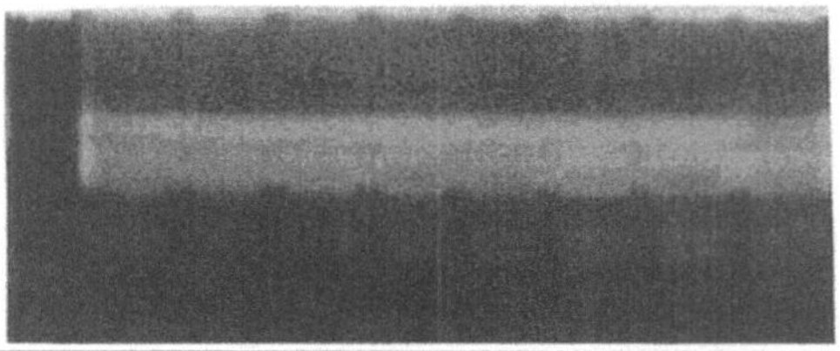

Figure 2. Zymogram of t-PA in different conditions of induction and temperature shift.

(1)37°C, (2)33°C; (3)37°C, (4)Butyrate; (5)37°C ,(6)PMA; (7)37°C, (8)TGF-β.

4. Conclusion

Optimal specific t-PA production is observed at 32°C and a shift of temperature process (biomass production at 37°C and t-PA production at 32°C) is therefore best suited. More efficient production is observed also in presence of butyrate or PMA despite some toxic effects. The specific t-PA production correlates well with the duration of the G1 phase as observed after temperatue shift or butyrate treatment. Further experimentation should clarify if the effect of PMA resides on a direct specific action on the promotor site.

5. References

(1) Fussenegger M.,Schlatter S., Datwler, Mazur X. et Bailey J.E., 1998, *Controlled proliferation by multigene metabolic engineering enhances the productivity of Chinese Hamster Ovary Cells.* Nature Biotechnoloy 16, 468-472.

(2) Slater M.L. , Sharrow S.O. et Gart J.J., 1997, *Cell cycle of Saccharomyces cerevisiae in population growing at different rates.* Proc.Natl.Acad.Sci.USA 74(9),3850-3854.

(3) Kronemaker S.J. et Srienc F., *Cell cycle kinetics of the acumulation of heavy and light chain immunoglobulin proteins in a mouse hybridoma cell line.* Cytotechnology 14, 205-218, 1994.

APPLICATION OF BIOASSAYS FOR SUSPENDED 293 CELL ENUMERATIONS IN CHEMICALLY DEFINED MEDIUM (CDM)

M.C. TSAO; B.A. JACKO; E. CURNOW; R.N. BERZOFSKY
BioWhittaker, Inc., Walkersville, Maryland 21793-0127, USA

1. Introduction

Suspended 293 cells grown in chemically defined medium (CDM) are difficult to accurately enumerate using the hemacytometer. Since 293 cells tend to form aggregated clumps of various sizes as they grow, inaccurate hemacytometer cell counts are likely due to errors in attempting to count aggregated cell clusters as well as uneven distribution of the aggregates in the capillary loaded counting chambers. For our work, it was imperative that we develop two different fluorescence based bioassays to 1) ascertain an accurate 293 cell enumeration and to 2) determine the cell vitality. We applied these bioassays in the development of CDM for single-cell suspension culture of 293 cells. Moreover, these fluorescence bioassays can be used as tools in monitoring the cells during viral and transient protein production. One of the bioassays measures the metabolic activity by Alamar Blue staining and the other measures the cell numbers by Sytox nucleic acid staining. Alamar Blue (AB) is a water soluble fluorometric/colormetric growth indicator based on detection of metabolic activity (REDOX). Sytox is a green nucleic acid stain that easily penetrates cells with compromised plasma membranes. These bioassays can be formatted from 96 well to 6 well plates. Both the Alamar Blue and Sytox assays have proved to be powerful tools in new media development and in the monitoring of spinner suspension cell growth kinetic experiments.

2. Material and Methods

2.1. CELL CULTURE

The human embryonic kidney cell line, 293, was obtained from ATCC and routinely passaged and maintained by BioWhittaker in the Cell Culture Department. These cells were expanded and quickly adapted to chemically defined 293 medium supplemented with 5 mg/L r-Insulin. After a few expanding passages, these cells were removed from t-flasks with trypsin to inoculate spinners with an initial cell density of 3-4x10^5 cells/ml in either 250 ml or 500 ml spinners with one-third filled medium volume.

2.2 ALAMAR BLUE AND SYTOX BIOASSAYS

Cell suspension dilutions were prepared based on hemacytometer cell counts. Diluted cell suspension (200 µl, Alamar Blue, and 100µl, Sytox) was added to each well of the 96 well plate (CoStar). The background formulation fluorescence values were subtracted from each case.

Plates were prepared for Days 0, and days ranging from 2 through 6, depending on the particular test. Sytox plates were incubated until the specified experiment day, then frozen (-20°C). Day 0 Alamar Blue plates were incubated at 37°C for 1 hour prior to the addition of 50 ul of Alamar Blue (Accu-med International) to each well including the background control well. Alamar Blue plates were run each day of the experiment. All Sytox plates were thawed at the end of the experiment and 120 µl of Sytox/Triton was added per well. Sytox-Gr (Molecular Probes) was made by preparing a 8uM Sytox solution in 1:40 Triton-X:UltraSaline A. Plates were then wrapped in aluminum foil, incubated at 37°C for 90 minutes (Alamar Blue) or 60 minutes (Sytox), and read on a CytoFluor II, fluorometric reader (PerSeptive BioSystems). Plates were scanned at 50 gain, 530/25 Excitation, 590/35 Emission (Alamar Blue), and 55 gain, 485/20 Excitation, 530/25 Emission (Sytox), 10 reads per well. Remaining plates, to be read at later days, were wrapped in polyvinyl wrap and incubated at 37°C with 5% CO_2 until the specified day.

A. Bernard et al. (eds.), Animal Cell Technology: Products from Cells, Cells as Products, 183–186.

2.3 TWO SPINNER EXPERIMENTS

The first experiment used serum-free 293 cells grown attached in t-flasks that had undergone 4 passages in a commercially supplied medium. Two suspension media were used in this experiment: Pro293s-CDM and the commercial medium 2. Cells were removed from flasks by incubating with Trypsin-Versene (BioWhittaker) for 20 minutes incubated at 37°C and inactivating with soybean trypsin inhibitor (1 mg/ml). The cell suspension volume was increased with UltraSaline (BioWhittaker) before centrifugation at 1000 rpm for 5 minutes and resuspending the cell pellet in UltraSaline. Spinners were seeded at approximately 400,000 cells/ml, in duplicate, with a starting volume of 200 ml. Each day, samples were taken, and counts done, as described above. Spinner cultures were fed 50 ml of the appropriate medium on Days 3 and 5.

The second experiment used spinner adapted serum-free 293 cells grown in Pro293s-CDM. The experiment compared four formulations: Pro293s-CDM and three commercially supplied 293 formulations, which were completed according to manufacturer instructions. Each media was tested in triplicate; 200 ml was added to each spinner, which was then placed on a spinner platform at 80 rpm, and incubated at 37°C with 5% CO_2. Each spinner was seeded at approximately 400,000 cells/ml, which was verified by hemacytometer count.

Daily Sytox samples: a 5 ml sample was frozen, while a second 5 ml sample was spun down (1000 rpm, 5 minutes), and cell pellet was frozen to be used for Sytox plates.

Daily hemacytometer count: A 1 ml aliquot was taken, and 25 µl of 2.5% trypsin solution added. Samples were incubated for 20 minutes at 37°C, and then counted by hemacytometer. Cell morphology was inspected each day. All spinners were fed 50 ml of the appropriate medium on Days 3, 4 and 5.

2.4 SYTOX SAMPLES FROM SPINNERS

Daily cell suspension and cell pellet samples were taken from each spinner. Analysis of cell suspension sample of cells resulted in the relative total cell numbers and analysis of the cell pellet resulted in the relative viable cell numbers respectively. The relative cell numbers were calculated from the standard curve, which was run concurrently with each day's samples. The viability for each spinner culture was determined by the ratio of viable to total cell numbers. Each pellet sample was resuspended to 5 ml using the appropriate medium. When cell number was below 800,000 cells/ml the samples were added directly to the 96 well plate (100 µl/well). As cell count (or pellet size, for cells too clumped to count) increased, dilutions were made prior to adding the sample to the plate. Resuspended pellets as well as complete suspension samples were compared in order to estimate viability within the spinners. To each well of a 96 well plate, 100 µl of cell sample from the spinners (or background medium without cells) was added and Sytox run as described above.

293 Cell Density Titrations by Alamar Blue Bioassay

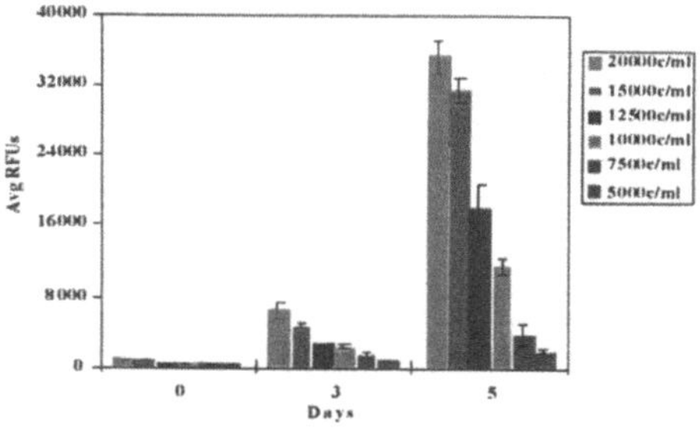

Figure 1

293 Cell Density Titrations by Sytox Bioassay

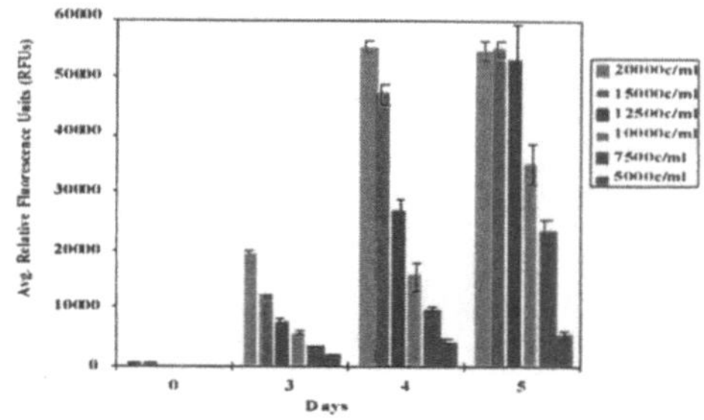

Figure 2

The data from both the Alamar Blue (Figure 1) and Sytox (Figure 2) bioassays support a direct relationship between cell concentration and average Relative Fluorescence Units (RFUs).

Suspension Growth in Pro293s-CDM

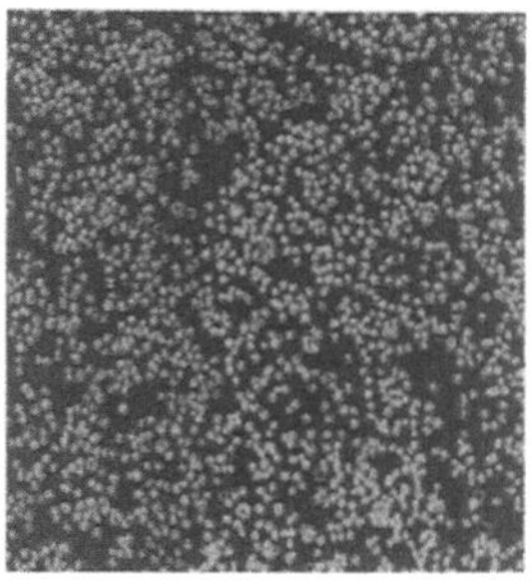

Figure 3

in Other Commercial Medium

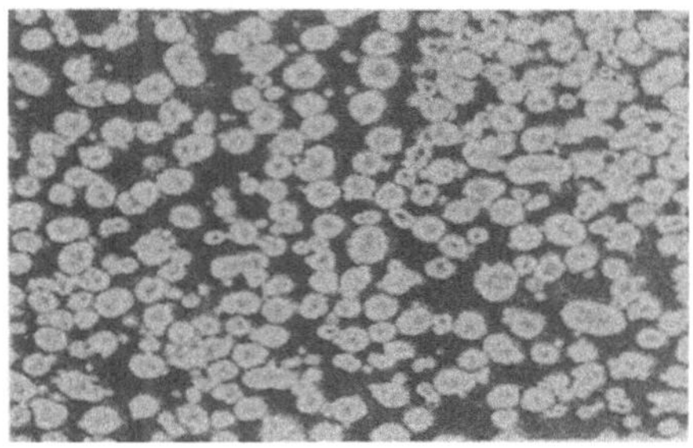

Figure 4

SF 293 Cells Grown Attached in T-flasks Cultured in Commercial Supplier 2 Medium Subcultured into Spinner Suspension Growth

Hemacytometer Trypan Blue Bioassay

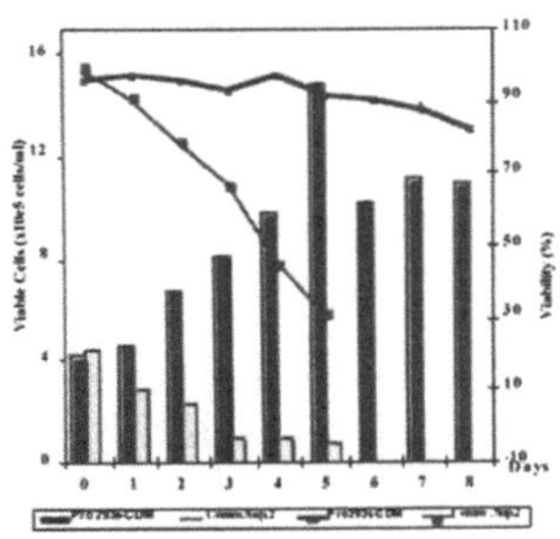

Figure 5

Sytox Bioassay

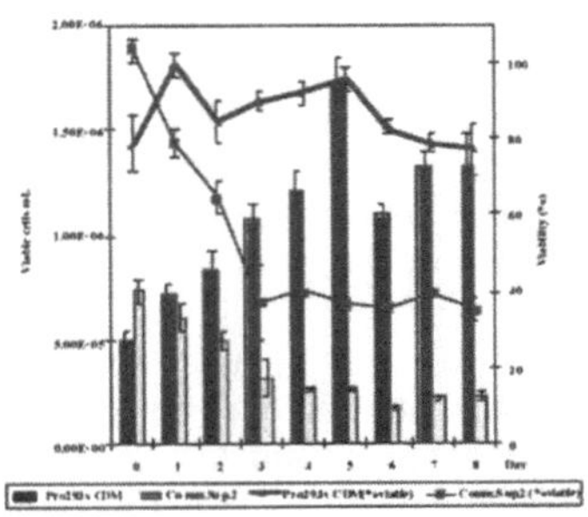

Figure 6

Cell numerations by hemacytometer (Figure 5) and Sytox (Figure 6) verified that Pro293s-CDM supported the one-step transition from attached to suspension culture while maintaining high density and high viability.

Sytox Bioassay
SF 293 Cells: Four Defined Commercial Media Compared

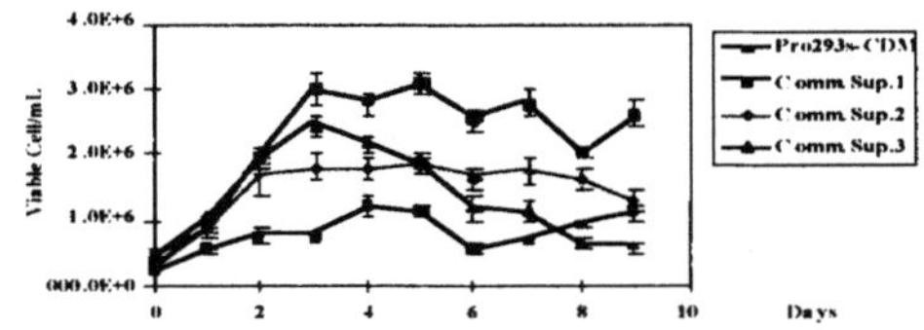

Fig. 7 – Viable Cells

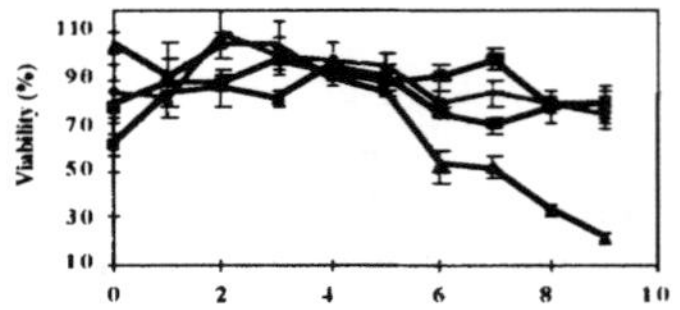

Fig. 8 - Total Cells/ml

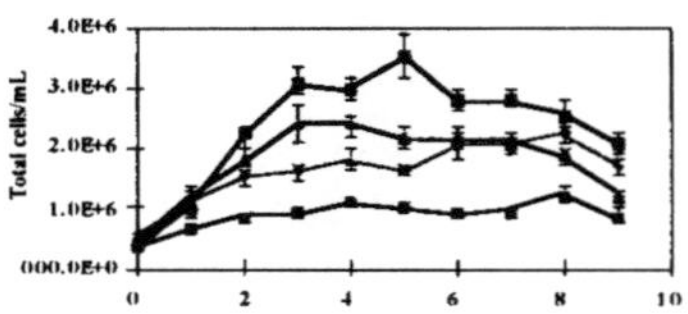

Fig. 9 - Viability (%)

The data is the average of 3 - 500ml suspension spinner cultures for each defined commercial media. On days 3, 4 and 5 each spinner was fed with 50 ml of the appropriate fresh defined medium

3. Results and Discussion

Because the 293 cells have tendency to clump with increasing cell density, the development of bioassays useful for the enumeration of cells was critical. Both Alamar blue and Sytox bioassays were utilized to monitor progress in the development of two distinctive chemically defined formulations: Pro293a-CDM to be used for quick serum-free adaptation and high density growth of adherent 293 cells, and Pro293s-CDM to be used for one-step adaptation to serum-free high-density single cell suspension.

The Sytox bioassay was used to monitor the growth and viability of cells in our Pro293s-CDM, compared with three commercially supplied 293 media. While cells growing in Pro293s-CDM exhibited single-cell morphology throughout the experiment, cells growing in the other formulations could not be accurately counted as early as Day 1 due to clumping. Sytox offered a method for comparing relative cell growth within and between formulations, and correlated well with hemacytometer counts. Adherent cells conditioned in a commercially supplied medium adapted easily to single cell suspension in Pro293s-CDM, but were unable to transition from attached to suspension growth in the same commercially supplied medium.

Full text with references provided upon request.

EFFICIENT CELL GROWTH BY AN OPTIMISED PROTEIN FREE MEDIUM DESIGN AND A CONTROLLED FEEDING STRATEGY

A. LOA[1,2], P. LEFEBVRE[1], G. KRETZMER[2]
[1]*BIO WHITTAKER, Parc Industriel de Petit- Rechain, B- 4800 Verviers*
[2]*Institut für Technische Chemie, Universität Hannover, Callinstr. 3, D- 30167 Hannover*

INTRODUCTION

Cultivating mammalian cells is a pretenious business. The basis of a cell culture is the medium. Regarding the industrial demands for a biotechnological process one can say that a medium for production should be cheap, easy to produce, induce a proper cell growth and be safe regarding the risk of contamination by products of animal origin. Another aspect is a facilitated down stream process by a well defined medium with low protein content.
The replacement of foetal calf serum (FCS) was the most remarkable advancement in this development. But in most media are still some proteins of animal origin present. Also the acceptability of the product will be increased using media without products of animal origin, especially those used for therapeutic purpose.

Most commercial mediums are available as batch media. They have to contain all the nutrients the cells will need during the duration of the cultivation. A change in the metabolism of cells is described in literature [1, 2, 3], when the culture conditions were changed. A high concentration of glucose for example can induce a faster consumption and thus a high production of lactate. Hence it can be advantageous to use two media for a cultivation of the cells. First a start medium and second a feeding medium. To avoid a waste of nutrients the concentrations should be very low but one has to take care that no limitation will be induced by an insufficient feeding. To prepare a feeding medium the stoichiometric model of L. Xie et al. [2, 3] was used. The medium was modified to adapt it to the start medium Bio Pro 1 P.F. (BIO WHITTAKER, Verviers).

A. Bernard et al. (eds.), Animal Cell Technology: Products from Cells, Cells as Products, 187–189.

EXPERIMENTAL WORK

The calculations are based on the composition of the cells. In literature one can find different cell compositions. For this we made calculations with three different compositions to prove the importance of a right composition. The calculations for medium I assume a high lipid content of the cells. Medium III assumes a low protein content and the lipid content for the calculations of medium II is in between.

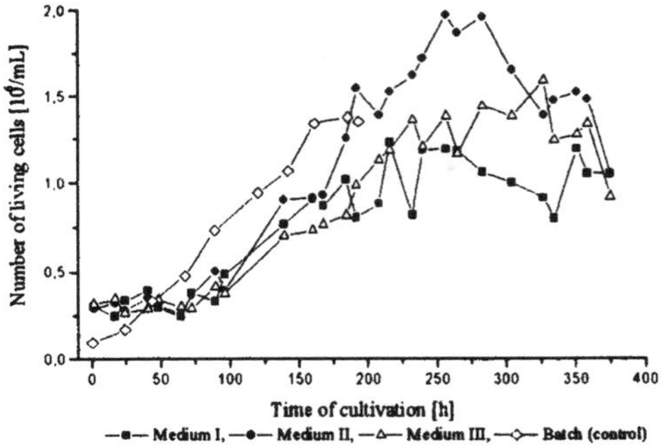

Fig. 1 Cell growth with different feeding mediums

Regarding the results (Fig. 1) one can see that the composition of the medium is important for the cell growth. Both media I and III with extremes in the lipid content concerning the cell composition grow worse. Only the culture with medium II growth better than the batch. The feeding was started after 80 h. In the beginning all cultures behave in the same way but after 160 h the behaviour of the cultures differs. It is the same point of time when the batch culture stops growing. Only the culture with feeding medium II seems to be advantageous in comparison to a batch culture. After 260 h the maximum of viable cells is reached. This is the point of time when a component is missing or inhibiting cell growth. Because of a lack of lipids in the feeding medium II it could be that the lipids were limiting the cell growth. It was decided to repeat the experiment with addition of lipids in the feeding medium. Regarding the amino acid results (not shown) one can see that the amino acids valine and threonine were depleted after 250 h. This can be a reason for reduced cell growth too. Feeding of these amino acids after 340 h (Fig. 1) does not lead to further cell growth. To avoid depletion of these amino acids we enriched the feeding medium II with these amino acids and take the change into account while re-calculating the medium II. The second experiment (Fig. 2) shows that lipid containing feeding is better than a feeding without lipids. One can see additional amino acids have no effect on cell growth. But it is remarkable that the maximum of cells is reached after 260 h at the same time as in the culture without additional lipids. The cells started to die at the same time of cultivation. Hence it is probably that something else is the growth inhibiting factor. The feeding was controlled by the concentration of glucose which was kept constant at 2.5 mM.

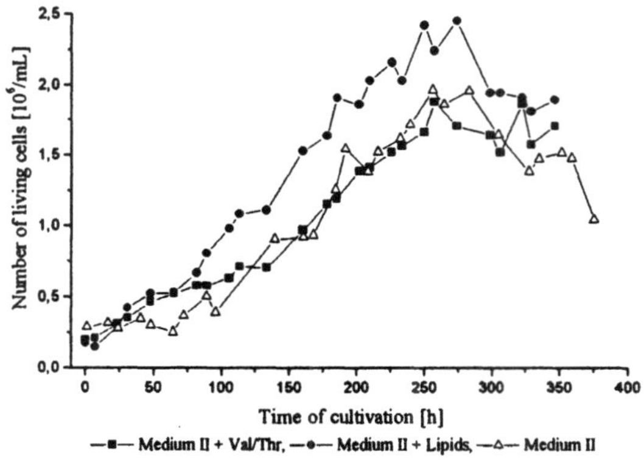

Fig. 2 Cell growth with an improved medium

CONCLUSIONS

The experiments show the importance of a feeding medium with a balanced composition (Fig. 1). Each change in the culture conditions (e.g. nutrients, concentrations, temperature, ...) changes the metabolism of the cells, hence it is quite difficult to determine the needs of the culture. To determine the consumption rate of the cells by chemostat experiments will have the same problems like the calculation of the need of nutrients by the stoichiometric model. The difficulties to determine the right cell composition are caused by changing cell composition during the life of a cell. Which way one choose one will only be able to assimilate the medium to the requirements of the cells and not to determine by a single experiment the optimal feeding medium. The advantage of the protein free medium is to have a well defined medium with a constant quality. So the improvement of the feeding medium is much more easier than with a medium less defined, because every change in the conditions can cause a change in the behaviour of the cells.

LITERATURE

[1] H.P.J. Bonarius, V. Hatzimanikatsis, K. P. H. Meesters, C. D. De Gooijer, G. Schmid, J. Tramper; Biotechnol. Bioeng.; 50: 299-318; (1996)

[2] L. Xie, D. C. Wang; Biotechnol. Bioeng.; 49: 1164-1174; (1994)

[3] L. Xie, D. C. Wang; Biotechnol. Bioeng.; 49: 1175-1189; (1994)

DEVELOPMENT OF A SERUM FREE MEDIUM FOR A HUMAN DIPLOID FIBROBLAST CELL LINE

S. PIROTTON, J. DEWELLE, F. ELIAERS, D. DELFORGE, M. RAES.
Laboratory of Cellular Biochemistry, University of Namur (FUNDP), 61 Rue de Bruxelles, 5000 Namur, Belgium.

The aim of this work was to formulate a serum-free medium (or a medium containing a reduced serum concentration) for MRC-5 cells (human diploid fibroblasts from foetal lung. The screening was performed in multi-well plates using the propidium iodide assay to follow cell proliferation. Among all the media tested, the prototype medium B (containing 0.5% foetal bovine serum (FBS) and several growth factors and hormones) stimulated the proliferation of MRC-5 cells in a similar way to our reference medium (Ultra-MEM + 2% FBS). The quality of medium B was confirmed by the incorporation of 5-bromo-2'-deoxyuridine (BrdU) in newly synthesised DNA. This medium also supported MRC-5 cell growth over 8 population doublings. Finally, medium B can be slightly improved by the addition of the synthetic peptide *cyclo*(GRGDSPA) acting through the interaction with integrins, confirming the hypothesis of a crosstalk between the integrin and the growth factor receptor activation pathways.

1. Introduction

Serum is essential for cell proliferation (it is a source of nutrients, growth and attachment factors), for trypsin inactivation in subculture protocols and for cryopreservation [1]. However, its use in the industrial field is questionned for several reasons : the cost, the variability between batches, its undefined composition, its animal origin and last but not least, the presence of pathogenic contaminants (viruses, bovine spongiform encephalopathy agents ...) [2]. The current tendency is to recommend the elimination of serum from culture media used in the manufacture of pharmaceuticals and its replacement by complete defined additives. Here we described the results obtained with one experimental medium containing only 0.5% of FBS + various growth factors and hormones on the proliferation of MRC-5 cells.

2. Methods

2.1 PROPIDIUM IODIDE ASSAY

MRC-5 cells were seeded in 96-well plates (4000 cells/well) and incubated in the different media at 37°C for 1, 2, 3, 4, or 7 days. The cells were then washed twice with

A. Bernard et al. (eds.), Animal Cell Technology: Products from Cells, Cells as Products, 191–193.

Phosphate Buffered Saline (PBS), permeabilised in ethanol and incubated with propidium iodide (10 μg/ml PBS) for 30 minutes in the dark. DNA-associated fluorescence was quantified with a fluorescence plate reader (excitation at 515 nm and emission at 612 nm).

3. Results and discussion

Several basal media supplemented with various growth factors or commercial serum replacements were tested on MRC-5 cell proliferation in multi-well plates and compared to our reference medium (Ultra-MEM + 2% FBS). For this screening, the propidium iodide assay appeared to be the most rapid, sensitive and reproducible method. Among all the tested media, the following one (called medium B and containing 0.5% FBS) has given the best results. Figure 1 illustrates the time course of proliferation of MRC-5 cells in both media. After 7 days of culture, the DNA content obtained in medium B represented 103% ± 12% of the one obtained in the reference medium (mean ± SD of 16 independent experiments).

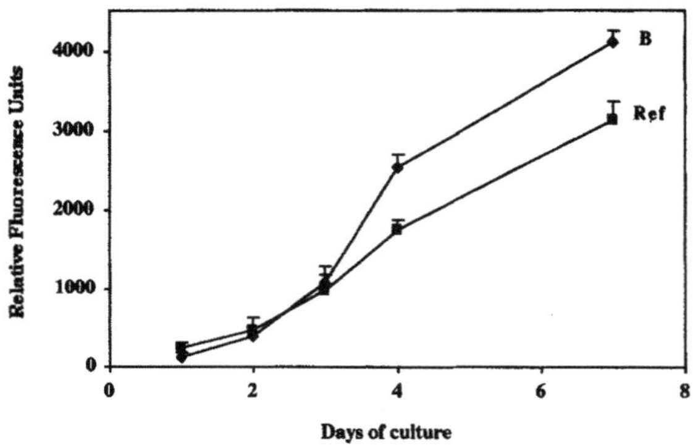

Figure 1. Time course of MRC-5 cell proliferation in medium B and in the reference medium (Ref). Results are expressed as mean ± SD of quadruplicate determinations in one representative experiment.

Incorporation of BrdU was performed, using the the Cell Proliferation ELISA BrdU Kit of Boerhinger Mannheim (colorimetric measurement), in order to further evaluate the efficiency of the medium B. BrdU incorporation into cells cultured in this medium represented 132% ± 6% of the incorporation in the control cells (mean of 2 experiments).

MRC-5 cells were also seeded in medium B in T-80 flasks (80 cm^2 of surface) for long-term cultures. In these conditions, the cells grew up, at a constant rate, during 8 population doublings.

In order to further improve the medium B, additional additives had to be defined. In a preliminary experiment, cyclic RGD peptides were tested. As shown in figure 2, the proliferation of MRC-5 cells in medium B was increased by 20-25% in the presence of 0.02 -2 μM c(GRGDSPA) peptides.

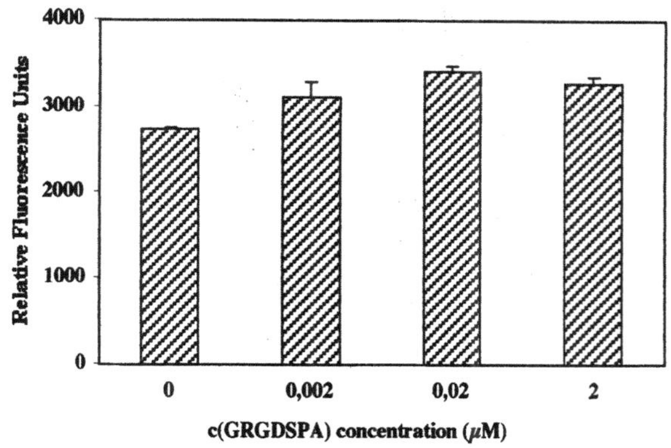

Figure 2. Effect of c(GRGDSPA) peptides on the proliferation of MRC-5 cells. Cells were cultured 7 days in medium B + different concentrations of the peptides before DNA quantification. Results are expressed as mean ± SD of quadruplicate determinations in one representative experiment.

4. Conclusions

The use of a basal medium enriched with defined growth factors and hormones allowed us to reduce serum concentration from 2% to 0.5% for the culture of MRC-5 cells.
The Arg-Gly-Asp (RGD) sequence is one of the major motifs of extracellular matrix proteins (fibronectin, collagens...) involved in the activation of their receptors (the integrins). Moreover, a positive cooperation has been reported between the integrins and the growth factor receptor signalling pathways [3]. This hypothesis was confirmed in our system, since the addition of the c(GRGDSPA) peptide slightly improved the quality of medium B.

5. Acknowledgements

This work was supported by the Wallon Region.

6. References

1. Barnes, D. and Sato, G. (1980) Serum-free cell culture : a unifying approach, *Cell* **22**, 649-655.
2. Merten, O.-W. (1999) Safety issues of animal products used in serum-free media in F. Brown, T. Cartwright, F. Horaud and J.M. Spieser (eds) *Dev. Biol. Stand.*, Karger, Basel, **99**, 167-180.
3. Schwartz, M.A. and Baron, V. (1999) Interactions between mitogenic stimuli, or, a thousand and one connections, *Curr. Opin. Cell Biol.* **11**, 197-202.

The Use of Peptones as Medium Additives for High-Density Perfusion Cultures of Animal Cells

R. Heidemann, C. Zhang, H. Qi, J. Rule, C. Rozales, S. Park, S. Chuppa, M. Ray, J. Michaels, K. Konstantinov and D. Naveh
Bayer Corporation, 800 Dwight Way, Berkeley, CA 94710, USA

Abstract
This paper describes the test of several new vegetarian hydrolysates (peptones of soy, rice, wheat gluten etc.) as protein-free medium supplements for the production of a recombinant therapeutic protein. Multiple peptone-supplemented, continuous perfusion bioreactor experiments were conducted. Cell specific rates and product quality studies were obtained for the various peptones and compared with peptone-free medium.
It was found that peptones confer a nutritional benefit, especially at low dilution rates, for a recombinant BHK cell line used in this investigation. The specific productivity increased 20-30% compared to the peptone-free controls. However, this benefit was also fully delivered by using fortified medium in place of the peptone-enriched media.

Introduction
The optimization of culture medium is one of the key steps for the production of any products in addition to optimization of physical cultivation parameters such as temperature, pH or DO. The goals of medium optimization include the improvement of productivity, quality and cost savings. In the past two decades, serum has been successfully eliminated for the production of mammalian cell derived proteins. However, these serum-free or even protein-free media often resulted in a decrease of specific productivity and sometimes changes in product quality.
The use of protein hydrolysates as a substitute for serum has been attempted by many in cell culture. More recent reports on this subject (*Jan et al. 1994, Dyring et al. 1994, Zhang et al. 1994, Keen and Rapson 1995, Xie et al. 1997, Nyberg et al. 1999)* indicate that addition of peptones resulted in higher cell growth and increased productivity. Nevertheless, Gu et al. 1997 reported negative effects of Primatone, expressed in decreased terminal sialylation at each of the glycosylation sites of CHO-secreted rInterferon. Whereas early work often employed Primatone or peptones derived from bovine milk, currently, these are not options since the elimination of animal- and human-derived proteins is a primary goal for the production of protein therapeutics. More recently, vegetable hydrolysates (peptones of soy, rice, wheat gluten) have become commercially available and have been used to supplement basal medium in order to fortify the amino acid content in small peptide form for batch and fed-batch fermentations.

Materials and Methods
- Recombinant BHK-21 cell line producing a therapeutic glycoprotein
- Protein-free medium
- 2-L or 12-L stirred tank perfusion bioreactors (Applikon, Schiedam, The Netherlands) equipped with cell retention devices (see *Chuppa et al. 1997*)

A. Bernard et al. (eds.), Animal Cell Technology: Products from Cells, Cells as Products, 195–197.

- Peptones: From Quest International (Naarden, The Netherlands), except for NZ-Soy (hydrolyzed using porcine enzymes) all were complete vegetarian hydrolysates.

Table 1: Peptone candidates tested in this investigation

Name	Source	Features	Reactors
NZ-Soy	Soy	Porcine Enzyme, High solubility	2L, 15L
BL-4	Soy	Like NZ, Di/tri-peptides	2L, 15L
HySoy	Soy	High carbohydrates	15L
HyPep 5603	Rice	Di/tri-peptides, low free amino acids	2L
HyPep 4601	Wheat	Di/tri-peptides, low free amino acids	2L

Results and Discussion

The first goal was to test the glycosilation and biological activity of the product. To this end, the target protein was purified and fully characterized during distinct phases in which the fermentor was alternately fed with NZ soy medium at 5 g/L or standard medium. SDS PAGE and western blot analyses showed that the molecule remained structurally intact during all periods of the cultures. Oligosaccharide mapping of the product from the peptone-containing medium and peptone-free phases showed no significant difference.

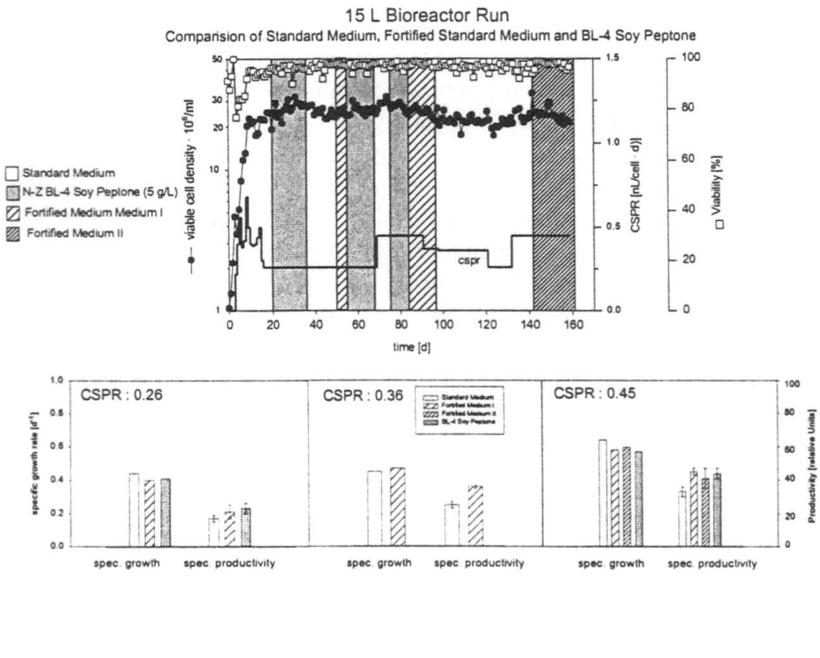

Figure 1: Time course and cell specific rates during a 15L high density perfusion culture. At low cspr specific productivity increased using fortified or peptone containing media, the growth rate was not affected. At higher cspr's productivity and growth rate showed overall higher values. Within this group no difference between the two fortified media or peptone medium was observed.

The time course of a 15-L perfusion culture together with cell specific rates is shown in Figure 1. The reactor was inoculated into standard medium at a cspr (cell specific perfusion rate) of 0.5 nl/(cell·d). The fermentor performance was characterized using NZ

BL-4 Soy peptone medium (5 g/L), two fortified media with additional amino acid content (I, 2x Gln, Asn, Ser and II, like I but 4x Ser) and the standard medium at three different cspr ranging between 0.26-0.45 nl/(cell·d). The fermentor was first run at a cspr of 0.26 nl/(cell·d). Supplementation of NZ BL-4 soy peptone (5 g/L) resulted in a 30% higher specific productivity compared to standard medium. Fortified medium I also gave similar results as the NZ BL-4 medium. Similarly, a titer decrease was also observed when the fermentor was switched from NZ BL-4 medium to standard medium. The fortified medium gave similar results to the NZ BL-4 medium, suggesting that the effect of BL-4 was simply nutritional. At a less-limiting cspr of 0.45 once again the NZ BL-4 and medium I outperformed the standard medium by 30% in terms of specific productivity. Additionally, the specific productivities at a cspr of 0.45 were almost 100% higher than those at cspr of 0.26 for all three media tested. The fermentor was then run at a cspr of 0.36 to confirm that NZ BL-4 and fortified medium gave higher specific productivity at various cspr. All three media (fortified medium II was only tested at high cspr) resulted in similar trends when comparing the relationship of cspr with specific productivity and growth rate.

Summary

Comparable product quality and glycosylation patterns were obtained using peptone-supplemented media compared to standard medium. While at a given cspr the specific rates for the main energy sources were similar regardless of the medium used, the addition of peptones had a nutritional benefit at lower cspr. Under these conditions, the specific productivity increased 20-30%. During a 15-L fermentor run peptone even showed positive results at higher cspr (>0.45 nl/(cell·day)) compared to standard medium. However, this benefit could also be delivered using media fortified with additional free amino acids

Consequently, this investigation shows that systematic medium development and, in particular, amino acid concentration optimization can successfully replace peptones. However, peptones can be a useful medium additive when time is limiting, and systematic media is not feasible.

References

Chuppa, S., Tsai, Y-S., Yoon, S., Shackleford, S., Rozales, C., Bhat, R., Tsay, G., Matanguihan, C., Konstantinov, K. and Naveh, D. (1997) Fermentor temperature as a tool for control of high-density perfusion cultures of mammalian cells. *Biotechnol. Bioeng.* **55:** 328-338

Dyring, C., Hansen, H. and Emborg, C. (1994) Observations on the influence of glutamine, asparagine and peptone on growth and t-PA production of Chinese hamster ovary (CHO) cells. *Cytotechnology* **16:** 37-42

Gu, X., Xie, L., Harmon, B. and Wang, D. (1997) Influence of Primatone RL supplementation on sialylation of recombinant human interferon-γ produced by Chinese hamster ovary cell culture using serum-free media. *Biotechnol. Bioeng.* **56:** 353-360

Jan, D., Jones, S. Emery, A. and Al-Rubeai, M. (1994) Peptone, a low cost growth-promoting nutrient for intensive animal cell culture. *Cytotechnology* **16:** 17-26

Nyberg, G., Balcarcel, R., Follstad, B., Stephanopoulos, G., Wang, D. (1999) Metabolism of peptide amino acids by Chines hamster ovary cells grown in a complex medium. *Biontechnol. Bioeng.* **62:** 324-335

Keen, M. and Rapson, N. (1995) Development of a serum-free culture medium for the large scale production of recombinant protein from Chinese hamster ovary cell line. *Cytotechnology* **17:** 153-163

Xie, L., Nyberg, G., Gu, X., Li, H., Möllborn, F and Wang, D. (1997) Gamma-Interferon production and quality in stoichiometric fed-batch cultures of chinese hamster ovary (CHO) cells under serum-free conditions. *Biotechnol. Bioeng.* **56:** 577-582

Zhang, Y., Zhou, Y. and Yu, J. (1994) Effects of peptone on hybridoma growth and monoclonal antibody formation. *Cytotechnology* **16:** 147-150

DETERMINATION OF ANIMAL CELL DENSITIES AND VIABILITIES BY THE NEW ANALYSER CEDEX2

T. WEHN, TH. LORENZ *, U. BEHRENDT *, C. WALLERIUS *, H. BÜNTEMEYER AND J. LEHMANN
Institute of Cell Culture Technology, University of Bielefeld
P.O. Box 10 01 31, 33501 Bielefeld, Germany
** Roche Diagnostics GmbH*
Nonnenwald 2, 82377 Penzberg, Germany

1 Introduction

For the determination of cell densities and viabilities of animal cell suspension cultures the trypan blue dye exclusion method is commonly used. Manually performed this is a time consuming method and the precision of the results depends on the handling of the sample (storage, dilution and staining of cells) and on the accuracy of the operator by counting the cells. To eliminate these disadvantages the "Cell Density Examination"-System CeDeX has been developed to perform the diluting, staining and counting process automatically.

Two years ago the concept of CeDeX was already presented at the 14th ESACT Meeting Conference [1]. Since then this system has been further improved and its second generation (CeDeX2) has now been tested in co-operation between the Institute of Cell Culture Technology, University of Bielefeld, Germany, and Roche Diagnostics GmbH, Penzberg, Germany. Samples from cultivation processes of different cell lines (hybridoma, CHO, HeLa, myelom) have been automatically analysed. In this paper data of dialysis cultivation [2] of hybridoma cells are discussed. The cell densities varied from 1 x 10^5 to 1 x 10^7 cells /ml. The data obtained were compared with the data generated by the manual trypan blue dye exclusion method.

2 Material and Methods

2.1 MANUAL TRYPAN BLUE DYE EXCLUSION METHOD

The cell densities were manually determined in duplicate with the trypan blue dye exclusion method: one determination was always performed by the same operator (trypan blue operator 1), while the other one was performed by different operators (trypan blue operator 2).

A. Bernard et al. (eds.), Animal Cell Technology: Products from Cells, Cells as Products, 199–201.

2.2 AUTOMATED TRYPAN BLUE DYE EXCLUSION METHOD - THE "CELL DENSITY EXAMINATION"-SYSTEM 2 (CEDEX2)

The CeDeX2-System was developed by Innovatis GmbH, Bielefeld, Germany. It combines a dilution system, a microscope with CCD-camera and a data processing computer which uses image processing routines to detect cells.

The trypan blue dye exclusion method has been completely automated in the CeDeX2-System. Defined sample preparation, standardized generation of sample images, data analysis and graphical presentation of the results are performed automatically with the objective to generate user independent data.

The two dimensional image information is used to detect and identify cells precisely. Both, local as well as global pattern recognition strategies contribute to a stable and reliable cell identification. Images as well as results are saved for further investigations.

3 Results

Figure 1 shows the variation between two different cell density determinations both performed manually, while figure 2 compares the results of the manual cell density determination with the results obtained by the CeDeX2 analyser. The figures show that there is hardly a difference to be seen between both comparisons.

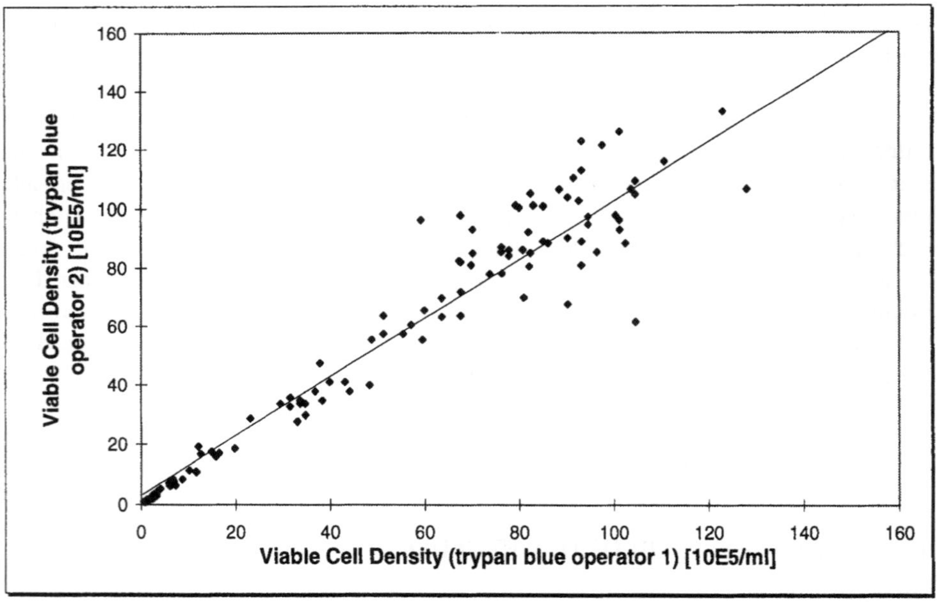

Figure 1: Comparison of two determinations of cell densities of viable hybridoma cells, both performed manually by the trypan blue dye exclusion method but by different operators

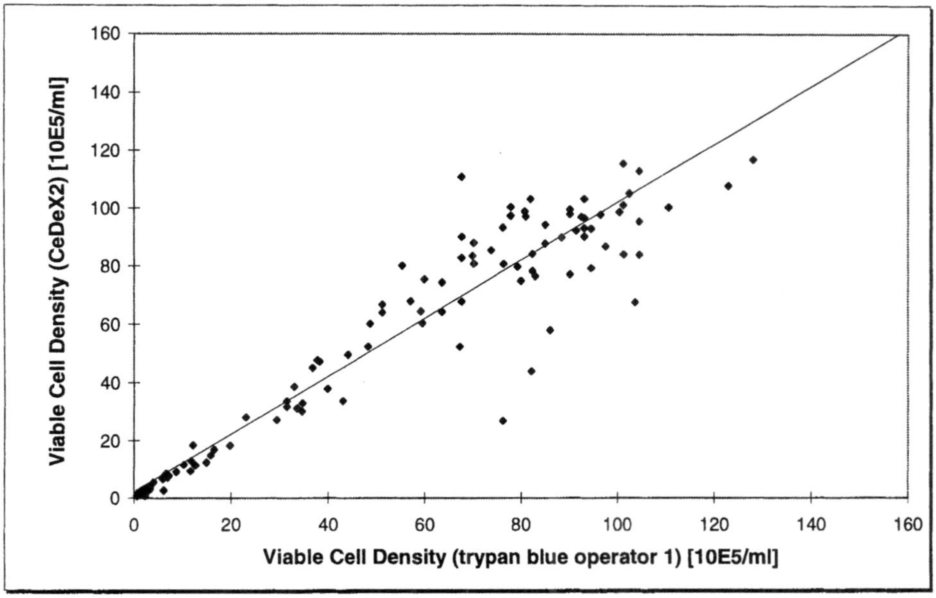

Figure 2: Comparison of two determinations of cell densities of viable hybridoma cells, obtained manually by the trypan blue dye exclusion method and by CeDeX2

For further details about the results and a comparison between the results obtained by the manual trypan blue dye exclusion method and the results of another cell counter (CASY 1, Schärfe, Reutlingen, Germany) see [3]. A comparison of further methods has been presented by A. Falkenhain [4] previously.

4 Summary

The results shown above allow the conclusion that the manual trypan blue dye exclusion method can be replaced by the CeDeX2-System due to its high precision, high reproducibility, high reliability and its user independency. A single measurement procedure with CeDeX2 takes less than four minutes.

5 References

1. Gudermann, F., Ziemeck, P., Lehmann, J.: CeDeX: Automated Cell Density Determination, Animal Cell Technology - From Vaccines to Genetic Medicine (1997), 301-305
2. Comer, M. J., Kearns, M. J., Wahl, J.; Munster, M., Lorenz, T., Szperalski, B., Koch, S., Behrendt, U., Brunner, H.: Industrial Production of Monoclonal Antibodies and Therapeutic Proteins by Dialysis Fermentation, Cytotechnology 3 (1990), 295-299
3. Wehn, T.: Untersuchungen zur Automatisierung der Zelldichte- und Viabilitätsbestimmung in eukaryontischen Suspensionszellkulturen (1999)
4. Falkenhain, A., Lorenz, Th., Behrendt, U., Lehmann, J.: Dead Cell Estimation - A Comparison of Different Methods, New Developments and New Applications in Animal Cell Technology (1998), 333-336

DETERMINATION OF CELL NUMBERS AND VIABILITY; COMPARISON OF THE MICROCYTE FLOW CYTOMETER WITH TRYPAN BLUE COUNTS AND COULTER COUNTS

NIENKE VRIEZEN
Leiden Pilot Facility, Centocor BV
P.O.Box 251, 2300 AG Leiden, the Netherlands

I. Introduction

Traditionally in mammalian cell culture the amount of biomass in a culture is expressed as a cell density. Cell densities are commonly measured by manually counting dilutions of a culture in the presence of a staining agent like trypan blue. For larger numbers of samples this method is not only time consuming but also prone to operator dependent variations. Nielsen et al. have shown that variability between individuals and counting of relatively small numbers of cells both contribute to a large margin of error in manual trypan blue counts [1]. Instrument mediated ways of cell counting can lead to a higher standardization of counting results and if a large number of cells is processed within one measurement the margins of error will also improve. To yield the same information as a manual count of a trypan blue stained cell sample the instrument mediated method should be able to distinguish between viable and non-viable cells. Recently a compact portable flow cytometer was marketed which can be used for cell enumeration. The use of the fluorescent nucleic acid stain TOPRO-3 allows simultaneous determination of total cell number and the number of non-viable cells. The measuring principle of the Microcyte flow cytometer is shown in Fig. 1.

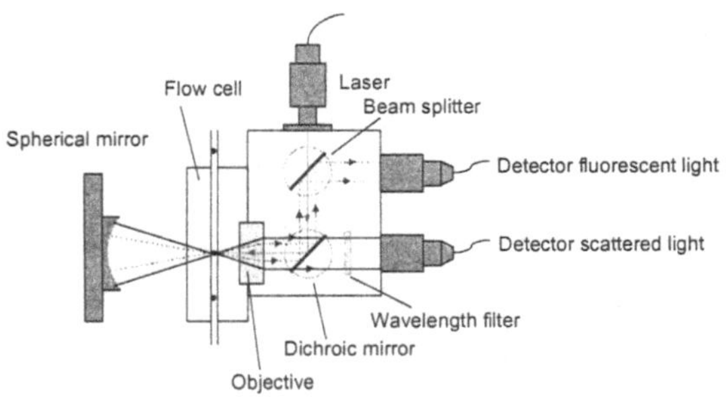

Figure 1. Measurement principle of the Microcyte flow cytometer

II. Materials and methods

Cell samples were taken from perfusion cultures and T-flask cultures of two types of hybridoma cells. Manual counts of trypan blue stained samples were performed using a Fuch-Rosenthal type heamocytometer. For Coulter counting samples were diluted to between 1 and 7 x 10^4 cells per ml. A Coulter multisizer II (Coulter, UK) equipped with a 100 µm aperture tube and Accucomp software version 2.01 were used. The Microcyte flow cytometer (Optoflow, Norway) was used according to manufacturers' instructions. The fluorescent stain SYTO-62 and TOPRO-3 were obtained from Molecular Probes (the Netherlands). Samples for the Microcyte were diluted to between 10^5 and 10^6 particles per ml. The

A. Bernard et al. (eds.), Animal Cell Technology: Products from Cells, Cells as Products, 203–205.

fluorescent stains were added to a final concentration of 5 nM for SYTO-62 and 3 nM for TOPRO-3. SYTO-62 stained samples were incubated in the dark for 10 minutes, TOPRO-3 stained cells were measured within 30 minutes after adding the stain.

III. Results and discussion.

The Microcyte flow cytometer can be used to quantify total cell number and cell viability for mammalian cells when fluorescent dyes are used. In this research two stains were used; SYTO-62 which binds to nucleic acid irrespective of a cell membrane being present and TOPRO-3 which binds to nucleic acid only when the cell membrane is permeable, thus indicating non-viable cells. SYTO-62 stained cell samples were measured on the Microcyte flow cytometer to indicate over what detection range cells were present. Based on these measurements the threshold between the regions of interest was set, separating the debris peak from the cell peak. Subsequent measurements were performed with TOPRO-3 stained cell samples. A dilution series of a cell culture sample showed that a wide range of cell concentrations could be measured reliably on the Microcyte flow cytometer (Fig. 2). To compare the performance of the flow cytometer to manual cell counts a series of samples was taken from cell cultures of two different cell lines at low and high cell concentrations and at low and high viability (Fig. 3). The results of these measurements show good agreement between the flow cytometer and manual counts for cell densities. The viabilities determined show good agreement for cell line A (Fig. 3, top). Viabilities for cell line B (Fig. 3, bottom) were somewhat lower when determined by the flow cytometer than found for the manual counts. This deviation is thought to be caused by the size distribution of the particles present in the samples, as can be seen in the Coulter size profiles (Fig. 3, right). Cell line A shows a distinct cell peak while for cell line B the cell and debris peaks are nearly merged. This size distribution will make it more difficult to set a correct threshold between the regions of interest on the Microcyte flow cytometer. Previous work published on the Microcyte flow cytometer did not show this kind of deviation in cell viability [2]. In that work however T-flask culture was the only culture method used. In the work described here most samples were taken from perfusion cultures. These culture will accumulate not only cells but also cell debris in the course of time.

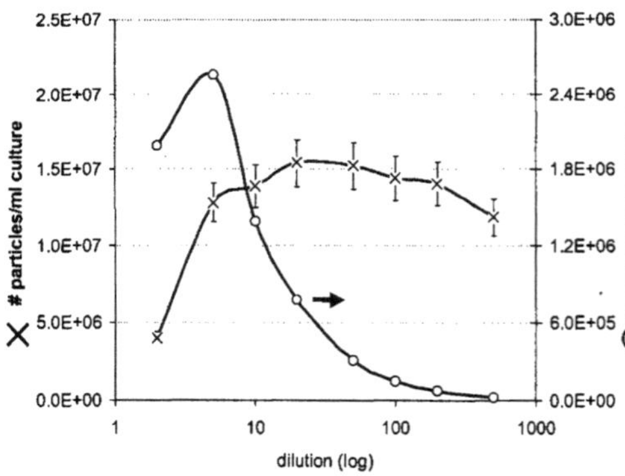

Fig 2. Results of a dilution series of cell culture sample measured on the Microcyte flow cytometer.

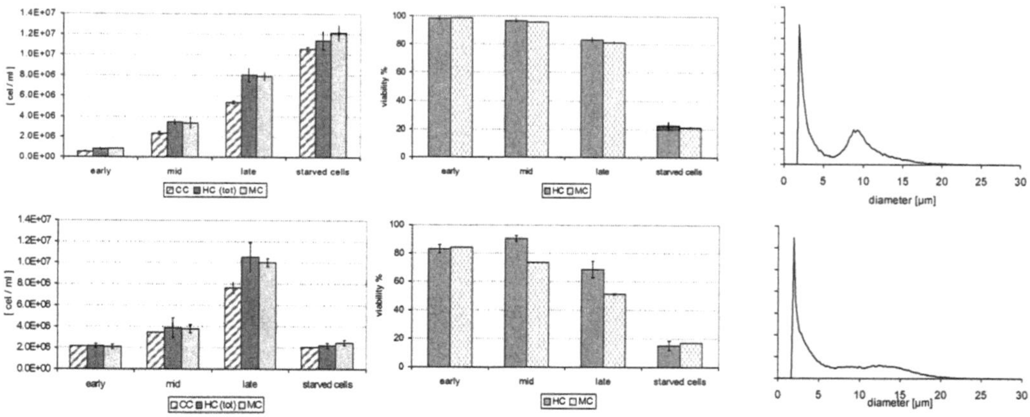

Fig 3. For cell lines A (top) and B (bottom) the results are shown of cell counts (left) and viabilities (middle) as measured by Coulter counting (CC), manually counted trypan blue measurements(HC) and the Microcyte flow-cytometer (MC). Particles size profiles from the Coulter method are shown on the right.

This will contribute to a larger amount of smaller particles being present in a culture which can lead to an underestimation of culture viability as measured on the Microcyte flow cytometer compared to manual counting using a haemocytometer.

The accuracy obtained by the flow cytometer was comparable to the accuracy obtained by a single operator performing manual counts. Because the flow cytometer offers an operator independent method for cell enumeration, the accuracy for the flow cytometer counts will be better than that for manual counts performed by different operators on different haemocytometers.

The time a single measurement on the Microcyte flow cytometer takes is shorter than a traditional manual haemocytometer count. The dilution and staining of the sample is comparable for both methods, but the actual measuring is much quicker on the flow cytometer. To increase the accuracy of a determination multiple measurements of a sample can easily be performed on the flow cytometer.

IV. Conclusion

The Microcyte flow cytometer can be used as an operator independent and quick method for measuring cell density and culture viability in mammalian cell cultures. In cultures that contain a large amount of debris particles, that are not significantly smaller than the cells in the culture, an underestimation of cell culture viability may be found.

V. References

1 Nielsen, N.K., Smyth, G.K., Greenfield, P.F. 1991. Haemocytometer cell count distributions: implications of non-Poisson behaviour. Biotechnol. Progress 7: 560-563

2 Harding, C.L., McFarlance, C., Al-Rubeai, M. 1998 Use of Microcyte flow cytometer for determination of animal cell number and viability. ESACT UK.

MID-INFRARED SPECTROSCOPIC MONITORING OF ANIMAL CELL CULTURE BIOREACTOR PROCESSES

M. Rhiel[1], T. Ziegler[2], P. Ducommun[1], U. von Stockar[1], I. W. Marison[1]
[1]*Institute of Chemical Engineering, Swiss Federal Institute of Technology (EPFL), 1015 Lausanne, Switzerland;*
[2]*Department of Process Development, Laboratoires Serono SA, 1804 Corsier-sur-Vevey, Switzerland*

ABSTRACT

Mid-Infrared (MIR) spectroscopy in combination with partial least-squares (PLS) regression analysis was used to monitor the concentrations of glucose, lactate, ammonia, and asparagine *in situ* during bioreactor cultivation of CHO/SSF3 cells. Simple PLS calibration models were established using referenced *in situ* collected single-beam spectra of one immobilzed culture. The models were applied to another immobilized culture and one suspension culture. In general, glucose predictions were unbiased for both culture types. Application of the lactate model resulted in a biased prediction during the first part of both cultures. Prediction results of the ammonia and asparagine models had both superior performance for the same type culture used in calibration. A negative bias of approximately 1 mM was observed for all the predictions of ammonia in the suspension culture.

INTRODUCTION

Bioprocess monitoring and control of key analytes, e.g., glucose, lactate, ammonia, and amino acids, is a prerequisite for the optimum and consistent production of medically important proteins. Desirable sensors in a cell culture production environment should be able to measure the concentrations of analytes of interest *in situ* with minimal sensor maintenance during long-term operation.

Spectroscopic sensors offer the advantage to measure all above listed analytes simultaneously, *in situ*, and without any wet chemical reagents. Among the various spectral regions, mid- infrared (MIR) spectroscopy offers enhanced sensitivity and selectivity due to the information content of the "fingerprint" region. This study was undertaken to screen the MIR absorbance features of key analytes and quantitatively analyze *in situ* collected spectra.

MATERIALS AND METHODS

A. Bernard et al. (eds.), Animal Cell Technology: Products from Cells, Cells as Products, 207–209.

Cell Line, Culture Medium, Reaction Vessels, and Culture Conditions. CHO/SSF3 cells propagated in ChoMaster HP1 medium (Ferrucio Messi Cell Culture Systems, Zürich, Switzerland) were used for all cell culture experiments. Cells were cultivated in a stirred tank bioreactor (BioLafitte, St-Germain-en-Laye, France). Experiments were performed with both free suspended cells and cells immobilized on macro-porous microcarrier Cytopore 2 (Pharmacia Biotech, Uppsala, Sweden).

Reference Analysis. Samples taken at specific times during the experiments were subjected to off-line reference analysis. Glucose and lactate concentrations were determined on an HPLC system with refractive index detector (Hewlett-Packard, Waldbronn, Germany) with relative standard errors of $\pm$ 2%. Amino acid concentrations were determined on an HPLC system after derivatization with OPA (Kontron Instruments, Zürich, Switzerland) usually with relative standard errors of $\pm$ 2%. Ammonia concentrations were measured with an enzymatic assay kit (Boehringer Mannheim, Mannheim, Germany) with a relative standard error of $\pm$ 2%.

Spectra Collection and Analysis. Single-beam spectra at 4 cm^{-1} resolution and 128 co-added scans were collected *in situ* with a ReactIR™ 1000 system (ASI Applied Systems, Millersville, MD) equipped with a DiComp™ diamond ATR probe. Partial least-squares (PLS) calibration models were generated with selected spectral ranges of single-beam spectra using one immobilized culture and applied to another immobilized culture and one suspension culture.

RESULTS AND DISCUSSION

Analyte specific calibration models were based on 20 referenced *in situ* collected single-beam spectra of one immobilized culture. Selected spectral ranges were based on characteristic absorbance features of the respective analyte. For example, glucose has a very characteristic absorbance band centered around 1080 cm^{-1}. Limiting PLS analysis for glucose to 1108 to 1069 cm^{-1} yielded an optimal calibration model requiring only 2 PLS factors. Application of this model to *in situ* collected spectra of other independent cultures is described in Figures 1a and 1e for immobilized and suspension cultures, respectively. In both cultures the glucose concentration was accurately monitored, except for a small bias towards the end of the suspension culture (Fig. 1e).

The lactate calibration model required 5 PLS factors when utilizing two spectral regions, 1592 to 1400 cm^{-1} and 1400 to 1127 cm^{-1}, in which lactate absorbances are significant. A negative bias is observed in the beginning of the cultures when the model is applied to both immobilized culture #2 (Fig. 1b) and suspension culture #2 (Fig. 1f). The negative bias diminishes in the second half of the culture. The reason of the bias may be an interference of other medium components at low lactate concentration and/or probe mis-alignment between calibration and application cultures.

Establishment of the models for both ammonia and asparagine was more challenging due to their naturally low concentration values during the bioreactor runs. An optimum ammonia calibration model could be established using 3 PLS factors and the spectral range 1482 to 1426 cm^{-1}. Application of this model was better for the second suspension culture (Fig. 1c) than for the second suspension culture (Fig. 1g), in which a negative prediction bias of approximately 1 mM was present throughout the culture.

The calibration model for asparagine required 3 PLS factors based on 5 spectral ranges, 1611 to 1607 cm^{-1}, 1502 to 1498 cm^{-1}, 1405 to 1401 cm^{-1}, 1360 to 1356 cm^{-1}, and 1316 to 1312 cm^{-1}. Prediction results for asparagine are overall better for the immobilized culture (Fig. 1d) compared to the suspension culture (Fig. 1h) in terms of following the trend of the concentration profile accurately.

CONCLUSIONS

MIR spectroscopy in combination with PLS analysis is suitable to monitor important animal cell culture analytes in the lower mM concentration range. Simple PLS calibration models based on one culture can be established for each of glucose, lactate, ammonia, and asparagine by selecting characteristic spectral features for each analyte of interest. These models seem to be at least valid for cultures of the same type and could be used to monitor if a process is within defined specifications. Models insensitive to batch-to-batch variations may be established by combining multiple processes. A comparison of both models and spectra may yield insights into the sources of potential process variability.

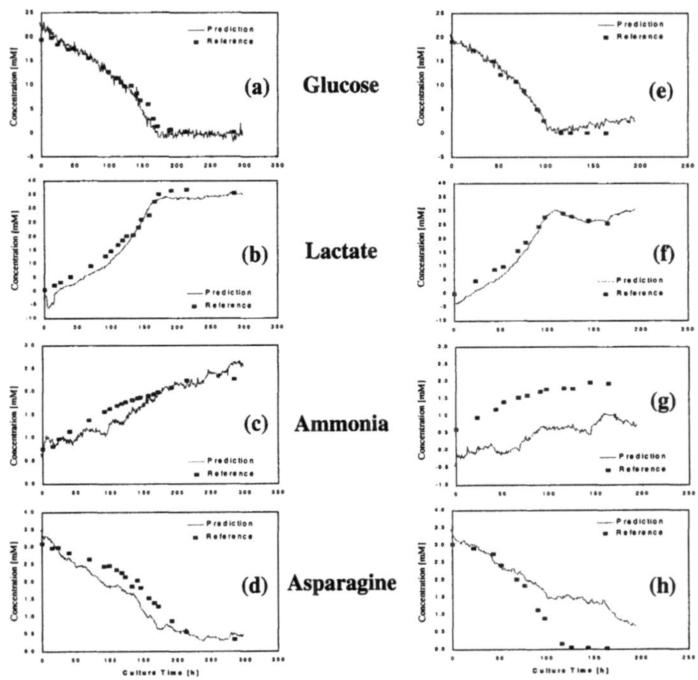

Figure 1: Metabolite profiles during cultivation of CHO/SSF3 cells in a 2L stirred bioreactor for both immobilized culture #2 (a-d) and free suspended cells (e-h). Symbols represent concentration values determined by off-line reference analysis. Continuous line represents concentration values predicted from *in situ* collected MIR spectra.

MAMMALIAN CELL MONITORING USING THE MICROCYTE FLOW CYTOMETER

C.L.HARDING[1], D.R.LLOYD[2], C.M. MCFARLANE[2] &
M. AL-RUBEAI[2]
[1]*Aber Instruments Ltd, Science Park, Aberystwyth, SY23 3AH, UK*
[2]*Centre for Bioprocess Engineering, School of Chemical Engineering, University of Birmingham, Edgbaston, Birmingham, B15 2TT, UK*

1. Introduction

Flow cytometry is a valuable research tool but is generally considered unsuitable for routine monitoring of mammalian cell cultures, because until now, instrumentation was large, expensive and required considerable operator training. The introduction of the Microcyte, a portable and robust flow cytometer, means that these restrictions can now be overcome. The Microcyte uses a diode laser as the light source, greatly reducing the size and weight of the instrument compared to conventional flow cytometers. The Microcyte also possesses a novel optical arrangement, housed within a solid aluminium block for stability, negating the need for laser alignment before use. Excitation maximum is 635 nm, and the instrument operates at a constant controlled sample uptake rate, so that absolute counts per volume of sample are obtained. This work aims to demonstrate the utility of the Microcyte for cell counting and viability determinations and show that the Microcyte can reliably detect apoptotic cells.

2. Materials and Methods

Stirred batch cultures of mouse:mouse hybridoma cells producing antibody against human IgG (TB/C3.pEF, Simpson *et al*, 1997) were grown at 37°C in RPMI 1640 supplemented with 5% fœtal calf serum.

Cells were counted in triplicate using an Improved Neubar counting chamber. Cells excluding trypan blue (final concentration 0.25%) were deemed viable. Triplicate samples were analysed on the Microcyte flow cytometer (Aber Instruments Ltd). Cell counts were determined from the forward scatter light signal after gating as appropriate for the TB/C3 cell line. Viability was assessed by addition of the cell membrane impermeable, fluorescent, nucleic acid stain TO-PRO-3 iodide (Molecular Probes, final concentration 0.1 μM). Viability was calculated form the total and dead cell counts.
Apoptosis was induced by the addition of camptothecin (3μM). The Annexin V binding assay (Vermes *et al*, 1995; Ishaque and Al-Rubeai, 1998; reviewed by van Engeland *et al*, 1998) was adapted for use with the Microcyte. A TACS™ Annexin V-biotin

A. Bernard et al. (eds.), Animal Cell Technology: Products from Cells, Cells as Products, 211–213.

apoptosis detection kit (Genzyme) was used in accordance with the manufacturer's instructions, followed by staining with streptavidin-allophycocyanin (Molecular Probes), see *figure 1*.

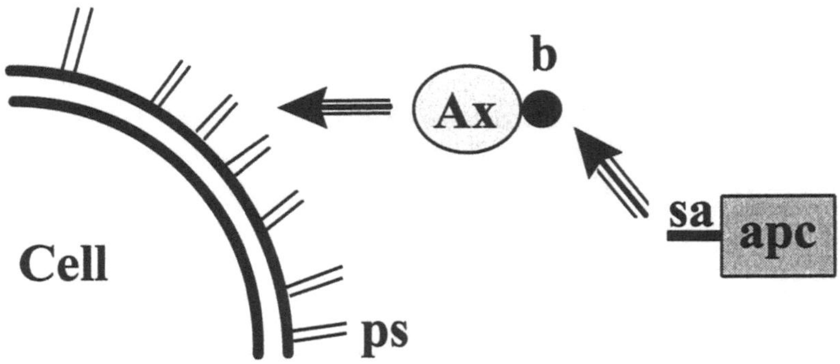

Figure 1. Annexin-V (Ax) conjugated to biotin (b) binds to externalised phosphatidylserine (ps). In turn, streptavidin (sa) conjugated to allophycocyanin (apc) binds to the biotin.

3. Results

4.1 CELL COUNTING AND VIABILITY

The growth curves of a batch culture produced by Microcyte and the traditional manual method were essentially identical. However, the point to point curve generated from the Microcyte data was smoother than that generated by the manual method, possibly indicating greater error in the manual method. From the same batch culture, the two methods gave viability curves with similar profiles. However, when viability was below approximately 80% the results obtained by Microcyte were consistently higher than those obtained by microscopy. This apparent discrepancy between the two methods may be simply explained by the use of different stains, since it is well recognised that different exclusion stains exhibit different properties; a dying cell membrane may be permeable to one stain but not to another. Nevertheless, since this difference only occurs in lower viability samples, it is unlikely to be of great practical importance.

Key benefits of the Microcyte for cell counting and viability determination are that results are obtained quickly, are less subjective and thus less likely to be subject to operator error or interobserver variation.

4.2 DETECTING APOPTOSIS

The Microcyte can also be used to monitor the onset of apoptosis. In the untreated culture, spontaneous apoptosis occurred at very low levels and barely changed during the experiment. However in the camptothecin treated culture, the number of apoptotic cells started to rise dramatically after 4 hours, with all cells apoptotic by 9 hours. The Microcyte's speed of analysis allows cultures to be sampled more frequently, thus pinpointing the onset of apoptosis (between 3.5 and 4 hours) much more accurately. At the same time the size (as forward scatter) of the apoptotic sub population was recorded, confirming that apoptotic cells are smaller than the general cell population.

4. Conclusions

The Microcyte is an excellent tool for the rapid and accurate enumeration of animal cells in suspension culture. It provides similar results to the established manual counting and staining techniques, with advantages of reduced intra- and inter-observer variation, increased speed of analysis and the ability to count large numbers of cells, leading to more statistically valid results. The Microcyte can also be used to detect apoptotic cells, using the annexin-V-affinity assay, not only at the high levels seen when apoptosis is induced but also when apoptosis is at low background levels in a "normal" culture. Furthermore the size of the apoptotic subpopulation can be compared with that of the general population. Having demonstrated the Microcyte's ability to count annexin-V stained cells allows the possibility of using the instrument to monitor cell surface markers or intracellular epitopes during culture processes and correlate these observations with cell size.

5. Acknowledgements

The DTI Teaching Company Scheme (CLH) & BBSRC (DRL) for financial support.

6. References

Ishaque, A. and Al-Rubeai, M. (1998) Use of intracellular pH and annexin-V flow cytometric assays to monitor apoptosis and its suppression by *bcl-2* over-expression in hybridoma cell culture *J Immunol Meth* **221**: 43-57

Simpson, N., Milner, A.N. and Al-Rubeai, M. (1997) Prevention of hybridoma cell death by bcl-2 during sub-optimal culture conditions *Biotech. Bioeng.* **54**: 1-16

van Engeland, M., Nieland, L.J.W., Ramaekers, F.C.S., Schutte, B. and Reutelingsperger, C.P.M. (1998) Annexin V-affinity assay: A review of an apoptosis detection system based on phosphatidylserine exposure *Cytometry* **31**: 1-9

Vermes, I., Haanen, C., Steffensnakken, H., and Reutelingsperger, C. (1995) A novel assay for apoptosis – flow cytometric detection of phosphatidylserine expression on early apoptotic cells using fluorescein-labelled Annexin V *J. Immunol. Meth.* **184**: 39-51

INDIRECT BIOMASS DETERMINATION IN CASE OF NON-CONSTANT METABOLIC RATES

P. DUCOMMUN[1], T. ZIEGLER[2], M. RHIEL[1], U. VON STOCKAR[1], I. W. MARISON[1]
[1]Institute of Chemical Engineering, Swiss Institute of Technology, 1015 Lausanne, Switzerland
[2]Department of Process Development, Laboratoires Serono SA, 1804 Corsier-sur-Vevey, Switzerland

1. Abstract

A novel approach for indirect biomass determination in case of non-constant specific metabolic rates is reported. Specific glucose consumption rate was determined during a first calibration culture for both free suspended and immobilized cells. It was expressed as a linear function of the limiting substrate, i. e. glucose for suspension and oxygen for immobilized cell cultures. For validation, these time-independent models were applied on a second culture presenting a different inoculum cell density. On-line monitoring of the limiting substrate enabled a continuous determination of the specific metabolic rate and thus on-line cell number prediction, which was verified with an off-line cell counting technique.

2. Introduction

The lack of accurate methods for on-line determination of animal cell concentration makes bioprocess monitoring and control difficult[1], particularly in the field of immobilized cell culture. Methods for the direct determination of cell concentration would be preferred but are generally not applicable for on-line measurement purposes. Indirect methods based on measurement of metabolic rates are therefore commonly used. However, these rely on the assumption that the specific metabolic rates are constant and known. In many cases these rates vary continuously during part of, or throughout, the culture process[2].
A new method was developed to determine biomass in cases where specific metabolic rates are not constant. This method is based on the continuous determination of specific metabolic rates for CHO cells as a function of the limiting substrate concentration during the whole culture process.

A. Bernard et al. (eds.), Animal Cell Technology: Products from Cells, Cells as Products, 215–217.

3. Materials and Methods

Cell Line and Growth Conditions. CHO SSF3 cells (Novartis, Basel, Switzerland) were grown at 37.0 °C and at a 7.3 controlled pH in the serum- and protein-free medium ChoMaster HP-1 (Dr. F. Messi, Cell Culture Technologies, Zürich, Switzerland). Batch cultures were performed with both free suspended cells and cells immobilized on macro-porous microcarriers Cytopore 2 (Pharmacia Biotech, Uppsala, Sweden) in a 2 liters stirred tank bioreactor (Biolafitte, St-Germain-en-Laye, France) with a 1550 ml working volume. Bubble-free aeration was achieved using a PTFE tubing (W.L. Gore and Associates GmbH, Putzbrunn, Germany) and the pO_2 maintained at 80% air saturation.

Analytical Methods. Culture samples were removed at intervals and cells counted using a haemocytometer. Total cell number was determined with the crystal violet staining method. Glucose was measured *in situ* with a ReactIR 1000 system (ASI Applied Systems, Millersville, MD) equipped with a DiComp diamond ATR probe, and consumed oxygen was determined through oxygen uptake rate on-line measurement[3].

4. Results and discussion

Specific glucose consumption rate was shown to be non-constant during batch cultures of CHO SSF3, even during the exponential growth phase. A first calibration culture was performed for both free suspended and immobilized cells. The consumed glucose was fitted as a function of time using an asymmetrical sigmoidal curve (Richards curve)[4]:

$$f(t) = a(1+(b-1)\exp[-c(t-d)])^{\frac{1}{1-b}} \tag{1}$$

The function was then differentiated and divided by the total cell number, determined off-line by crystal violet staining method, in order to calculate the specific glucose consumption rate:

$$q_{Glucose} = \frac{1}{X}\frac{dc_{Glucose}}{dt} \tag{2}$$

The specific glucose consumption rate could finally be reported as a linear function of the limiting substrate in the culture, which was identified as glucose itself for suspension and oxygen for immobilized cell cultures:

Suspension Cell Cultures

$$q_{Glucose} = 1.78 \cdot 10^{-14} + 8.11 \cdot 10^{-14} \cdot c_{Glucose} \tag{3}$$

Immobilized Cell Cultures

$$q_{Glucose} = 9.26 \cdot 10^{-13} + 3.71 \cdot 10^{-7} \cdot c_{Consumed\ oxygen} \tag{4}$$

($q_{Glucose}$ (*mol cell*$^{-1}$ *h*$^{-1}$), $c_{Glucose}$ (10^{-3} *mol l*$^{-1}$), $c_{Consumed\ oxygen}$ (10^{3} *mol l*$^{-1}$))

For validation, these time-independent models were applied on a second culture presenting a different inoculum cell density than the one used for model calibration. On-line monitoring of the limiting substrate enabled a continuous determination of $q_{Glucose}$. Equation (1) was applied to differentiate $c_{Glucose}$ with respect to time, and thus biomass could be predicted using Equation (2). Cell number prediction was verified off-line by crystal violet staining method. Results are shown in Figures 1 and 2:

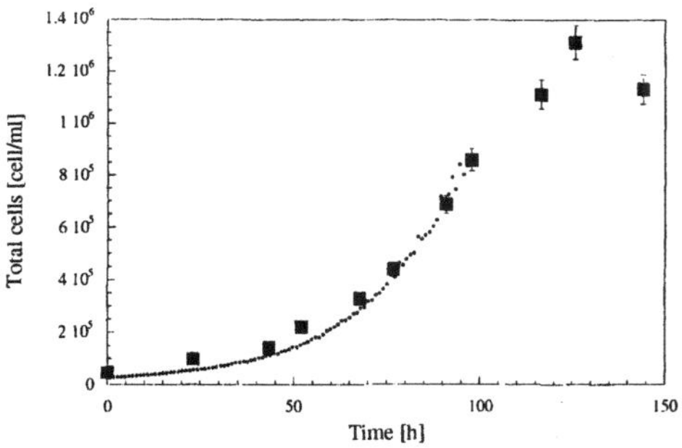

Figure 1. Total cells determined by crystal violet staining (squares) and by the model (curve) for suspension.

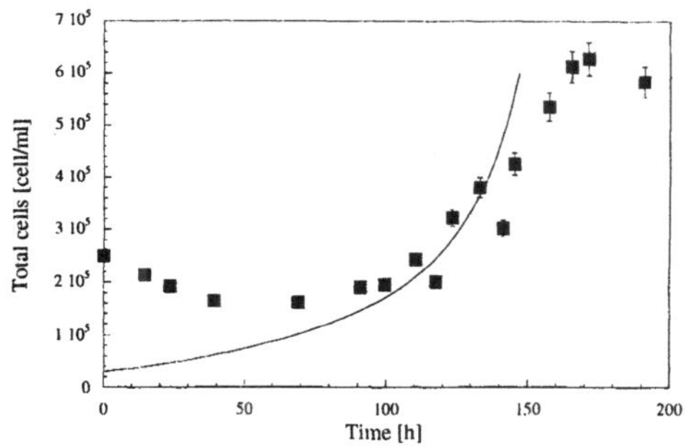

Figure 2. Total cells determined by crystal violet staining (squares) and by the model (curve) for immobilization.

5. Conclusions

Biomass was determined indirectly for both free suspended and immobilized CHO SSF3 cells in case of a non-constant specific glucose consumption rate. The time-independent models could be used in batch cultures presenting a different inoculum density than the calibration culture, and glucose as well as the limiting substrate on-line monitoring were sufficient to enable on-line biomass determination. Results were verified off-line by crystal violet staining. They were in good agreement during the whole exponential growth phase, until glucose was totally depleted, despite the non-constant specific glucose consumption rate.

6. References

1. Konstantinov, K., S. Chuppa, E. Sajan, Y. Tsai, S.J. Yoon, and F. Golini, *Real-Time Biomass-Concentration Monitoring in Animal-Cell Cultures.* Trends in Biotechnology, 1994. 12 (8): p. 324-333.
2. Pörtner, R., A. Bohmann, I. Ludemann, and H. Markl, *Estimation of specific glucose-uptake rates in cultures of hybridoma cells.* Journal of Biotechnology, 1994. 34 (3): p. 237-246.
3. Ruffieux, P.-A., *Détermination des flux métaboliques pour des cellules animales lors de cultures continues.* Chemistry Department, 1998, Lausanne: EPFL. 224 p.
4. Seber, G.A.F. and C.J. Wild, *Growth models,* in *Nonlinear regression.* 1989, John Wiley & Sons: New York. p. 325-365.

WHOLE CELL - SMART MICROSENSOR: A FIRST APPROACH TOWARDS INTERFACING LIVE CELLS AND ELECTRONICS

O. CHARLIER, M.-J. GOFFAUX, T. MARIQUE, V. HENDRICK, L. DE VOS* and J. WERENNE.
Université libre de Bruxelles (CP:160/17)
*Animal Cell Biotechnology and *Animal andCell Biology*
50, Av F.D. Roosevelt
1050 Bruxelles
Belgium

Abstract: In order to use whole eukaryotic cells as an active element in the detection and amplification of biological signals, for both *in vitro* and *in vivo* applications, we have undertaken a first approach to interface live cells and integrated circuit, and evaluate the possibility to develop a microbioreactor.

1. Introduction

The recent development of microelectronics allows emergence of smart microsensors, permitting detection, amplification and signals treatment on the same micro integrated circuit (1), as already generalized for physical parameters, examplified by numerous applications in the car industry. More direct strategies involving integration of biosignals processing, and design of microbioreactor (2) are under development.

1.1 THEORETICAL CONSIDERATIONS

The construction of a smart Whole-cell / integrated circuit device implies that:

° The cells can be seen as another component (analogous to a transistor, capacitor or resistor) fully integrated in the designed circuit.
° The cells, if used as a smart microsensor, could be engineered to reach the expected target.
° The cells can adhere to the integrated circuit without losing their full bioactivities. Appropriate immobilisation / or encapsulation treatment could be necessary for specific approaches.
Also, if multiple use is required, a gentle procedure for cells detachment from the microsensors should be developed; this may require appropriate coating and/or pretreatment of the microship, modifying appropriately its adhesive properties.
° The integrated circuit should be designed to perform the desired functions.

A. Bernard et al. (eds.), Animal Cell Technology: Products from Cells, Cells as Products, 219–221.

2. Materials and methods

The Chinese Hamster Ovary (CHO) and endothelial cells (either immortalized or not) were cultivated as reported elsewhere (3,4). The same medium conditions are used in presence of the microship.

The behaviour of the cells on the chips is observed either with a simple binocular or by Scanning Electron Microscopy.

For the t-PA experimentation recombinant CHO cells used are maintained as described in another section of this proceeding (4), and the same system is used for activity detection on the Beer Lambert relations.

3. Results and potentialities

We showed that CHO and endothelial cells attach, spread and behave happily on the silicium microcircuit device used (adhesion on oxide and nitride silicium is confirmed for both CHO and endothelial cells as we can see on the pictures)

Close-up of attached CHO cells as seen by SEM.

We showed that the design of the chip permits its use as a microbioreactor using eukaryotic cells as bioeffector.(see pictures and figure1)
The microreactor could be connected directly to an appropriate sensor for different kinds of measurements and/or control (the possible signals to measure could result for example from enzyme-substrate, receptor-ligand, antibody-antigen interactions,... for wich appropriate sensors should be designed).

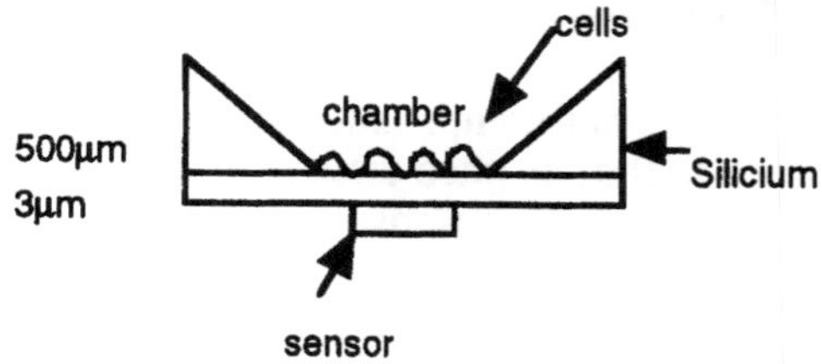

As an exemple, we describe here the possible use of this system to measure the productivity of t-PA in recombinant CHO cells, on the following chromophore based reaction:

$$CH_3\text{-}SO_2\text{-D-Phe-Gly-Arg-4-nitanilide} + HO$$
$$\rightarrow$$
$$CH_3\text{-SO2-D-Phe-Gly-Arg-OH} + \text{4-nitaniline}$$

using the Beer-Lambert law.

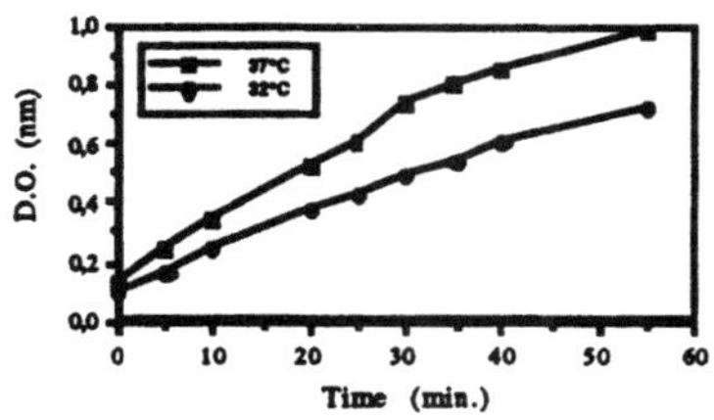

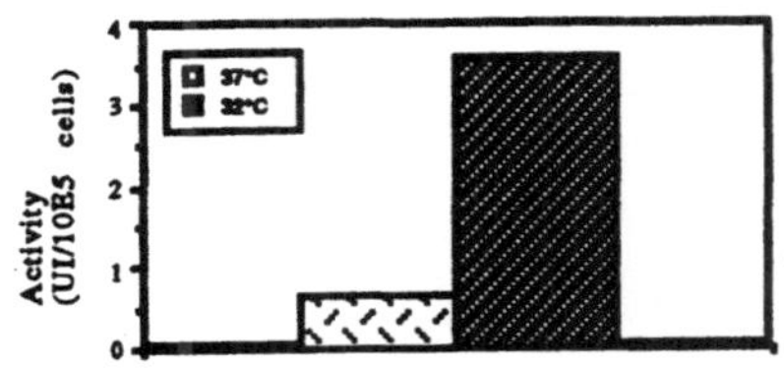

We confirmed that the productivity of t-PA is better at 32°c than 37°c (see figure 2).

4. Conclusion

Biocompatibility of microchips is confirmed for endothelial and CHO cells and microbioreactor could be developed, permitting on line analysis on the basis of the design of an appropriate sensor system.

5. References

(1) Bousse L., 1995,*Whole cell biosensors.*The 8th international Conference on solid state sensors and actuators, and Eurosensors IX, Stockholm, Sweden, 483-486.
(2) Son M., Peddie F., Yeow T; and Haskard M., 1995,*Whole cell biosensors.* The 8th international Conference on solid state sensors and actuators, and Eurosensors IX, Stockholm, Sweden, 894-897.
(3) Marique T., Blankaert, V. Hendrick, A. Raschella, B. Declerck, C. Alloin, I. Teixera-Guerra, D. Sandron, M. Cherlet, D. Parent, C. Kirkpatrick, J.P. Van Vooren and J.Wérenne,1997, *Biological response of endothelial cells and its modulation by cytokines: prospects for therapy and bioprocesses,* Cytotechnology **25**, 183-189.
(4) Hendrick V., Vandeputte O., Raschella A., Marique T., Cherlet M., Abdelkafi C.and Wérenne J.,1999, *Modulation of cell cycle for optimal recombinant protein production,* in this proceeding.

This is part of "CERBERE 2" project (Center of Engineering and Research in Biotechnology: Electronic Response and Eukaryotes) of wich O.C., M.-J.G.and J.W. are members.

COMPARISON OF IMMOBILISED GROUP SPECIFIC AFFINITY LIGANDS FOR THE BIOSEPARATION OF ANTIBODIES BY HIGH PERFORMANCE MEMBRANE AFFINITY CHROMATOGRAPHY

L. G. BERRUEX and R. FREITAG
Centre of Biotechnology, Institute of Chemical Engineering, Swiss Federal Institute of Technology, CH-1015 Lausanne, Switzerland.

1. Introduction

High Performance Membrane Affinity Chromatography (HPMAC) using Convective Interaction Media (CIM) disks has a high potential as fast multipurpose separation method in downstream processing and Quality Control of biopharmaceuticals [1, 2].

Protein A, protein G and protein L have been immobilised on epoxy groups of poly(glycidyl methacrylate-*co*-ethylene dimethacrylate) macroporous disks (BIA, Slovenia) and used as affinity chromatographic stationary phases for the separation of human and bovine polyclonal and recombinant monoclonal IgGs. The specificity of the affinity ligands for antibodies (protein A and protein G for Fc fragments [3, 4] and protein L for kappa light chains [5]) allows isolation of IgGs, for example, from cell culture supernatants.

2. Materials and Methods

Immobilisation. Protein A, G and L were immobilised on disks at 30°C, 16h, in 0.1M carbonate buffer, pH 9.3 [6].
Chromatography. Affinity chromatography was carried out on FPLC system (Pharmacia) by step elution from 100% binding buffer A to 100% eluent B, at a flow rate of 4ml/min, and with detection wavelength λ=280 nm. A = Phosphate buffer (PB) 50 mM pH 7.5 + 0.12 M NaCl, and B = glycine 0.1M pH 2.0 for protein A HPMAC; A = PB 50 mM pH 7.4 + 67 mM NaCl and B = HCl 0.01 M pH 2.0 for protein G HPMAC; A = Tris 20 mM pH 9.0 + 12 % Na_2SO_4 (anhydrous) and B = glycine 0.1M pH 2.0 for protein L HPMAC.
Preparation of samples. Samples of polyclonal human IgG and human IgG1-κ (Sigma), bovine IgG (Fluka) and recombinant monoclonal IgG1-κ (r-antiD Ab) and CHO serum-free DMEM/F12 cell culture supernatant were diluted in the binding buffer A corresponding to affinity ligand before injection.

A. Bernard et al. (eds.), Animal Cell Technology: Products from Cells, Cells as Products, 223–225.

3. Results and Discussion

3.1. BINDING OF DIFFERENT IgGs ON AFFINITY DISKS

All IgGs bind to the disks, albeit in different proportions (Figure 1). Bovine IgG binds the least well. Polyclonal human IgGs containing λ light chains do not bind to protein L. As expected, only the monoclonal r-antiD Ab showed high affinity binding to all three ligands.

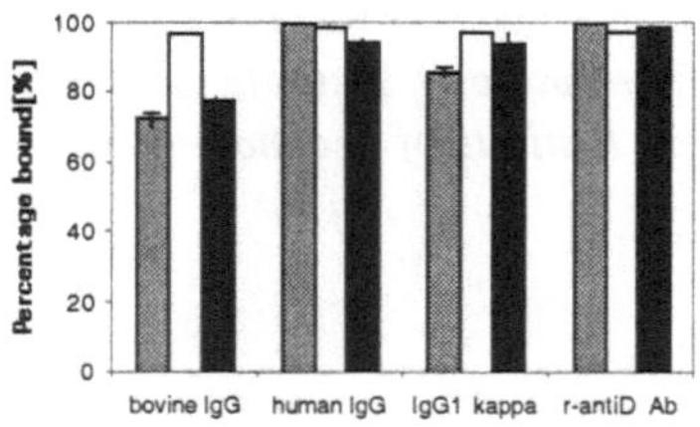

Figure 1: Percentage of binding of different IgGs on protein A (grey), protein G (white) and protein L (black) disks.

3.2. BINDING OF r-antiD Ab ON DISKS

A linear calibration curve was established between peak area and IgG amount for all three disk types. Identical amounts of antibody loaded onto disks in different volumes and concentrations give same peak areas. This means quantitation is possible by loading bigger samples in the case of low concentrations (data not shown). For practical reasons we limited the maximum sample volume to 2 ml, which corresponds to a LOQ of 0.5 mg/L IgG. Within this range, a linear correlation between the HPMAC quantitation of the product antibody (r-antiD Ab) [7] and that by ELISA [8] is established for all three affinity ligands (Table 1). HPMAC thus offers a fast alternative to ELISA and opens the way to quasi on-line monitoring of production of mAb in the bioreactor.

Table 1: Calibration of disks for the three affinity ligands: a) linear correlation between ELISA and HPMAC quantities of r-antiD Ab, LOQ = 1 µg for all three affinity ligands;
b) linear correlation between percentage of FCS in DMEM/F12 culture medium and HPMAC peak area.

	a) r-antiD		b) %FCS	
	r2	n	r2	n
Protein A	0,9983	15	0,9983	6
Protein G	0,9984	13	0,9997	6
Protein L	0,9855	13	0,9981	6

SDS-PAGE of IgGs isolated from CHO serum-free cell culture supernatant shows relatively pure IgGs (Figure 2). A further purification by cation exchange chromatography would be required to remove the remaining impurities.

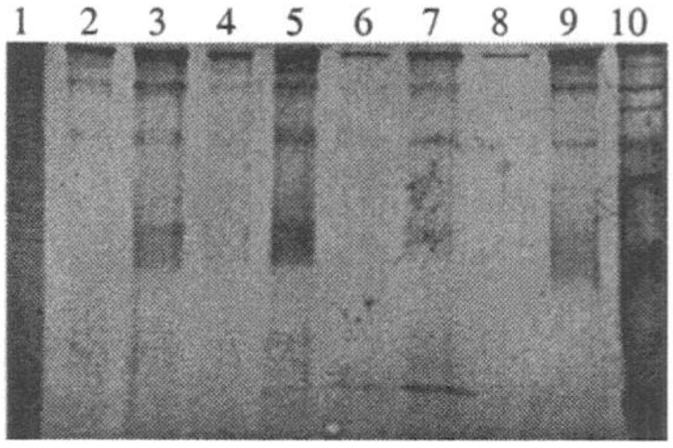

Figure 2: SDS-PAGE of IgGs isolated on three different affinity disks. 1 Supernatant, 2-3 on protein A, 4-5 on protein G, 6-7 on protein L, 8-9 standard purified on protein A column, 10 Molecular weights standards.

3.3. CORRELATION BETWEEN FCS CONTENT AND PEAK AREA

Bovine IgGs in culture medium containing foetal calf serum (FCS) also bind to all the disks; peak areas and FCS percentages are linearly correlated (Table 1). Isolation and/or quantitation of r-antiD Ab from supernatant containing bovine IgGs is therefore not possible in one step.

4. Conclusions and perspectives

High Performance Membrane Affinity Chromatography (HPMAC) is a fast way of isolating recombinant monoclonal antibody on a small scale. The method appears to be a fast alternative to ELISA (for concentrations above 0.5 mg/l), which could be used for quasi on-line monitoring of production of mAbs in the bioreactor. The use of HPMAC as sample preparation (concentration / enrichment) step prior to further analysis (hyphenated techniques) should also be investigated.

5. References

[1] Josic D, Schwinn H, Strancar A, Podgornik A, Barut M, Lim YP and Vodopicev M (1998). *Journal of Chromatography A* **803**, pp. 61-71.

[2] Strancar A, *Separation of biopolymers with different techniques of liquid chromatography*, Dissertation Thesis, University of Ljubljana (SLO), 1997.

[3] Forsgren A and Sjöqvist J (1966). *Journal of Immunology* **97**, p. 822.

[4] Björck L and Kronvall G (1984). *Journal of Immunology* **133**, p. 969.

[5] Björck L (1988). *Journal of Immunology* **140**, p. 1194.

[6] Kasper C, Meringova L, Freitag R and Tennikova T (1998). *Journal of Chromatography A* **798**, pp. 65-72.

[7] Amstutz HP et al. (1999) ESACT '99, Oral presentation No: 05.06, Poster No: 05.03

[8] Jordan M, Fraboulet D, Fourmestraux G, Wurm F, and Freitag R (1999) ESACT '99, Poster No: 05.09

Acknowledgements: This work was supported by the Swiss Priority Project (SPP) Biotechnology. We also wish to thank Prof. Tatiana Tennikova for all her help and advice.

POST-TRANSLATIONAL MODIFICATIONS

Chapter III

Posttranslational modifications limit high level expression of functionally active chimeric P-Selectin Glycoprotein Ligand-1 in rCHO cells.

Martin S. Sinacore, Troy Richards, Linda Francullo, Amy Woodard, Mark Hardy, Richard Cornell, Steve Koza, Monique Davies, Deb Ellis and Scott Harrison.
Genetics Institute, Andover, MA USA.

Introduction

P-Selectin Glycoprotein Ligand-1 (PSGL-1) is a dimeric mucin-like transmembrane glycoprotein identified as a functional ligand for P-Selectin (1). Functional activity of PSGL-1 is dependent upon at least two key posttranslational modifications made to the amino terminus. Core 2 O-linked oligosaccharide structures at Thr^{16} bearing the sialyl-Lewis^x (SLe^x) antigen and sulfation of one or more of the NH_2-terminal tyrosine residues have been shown to be essential for binding of PSGL-1 to P-Selectin. In addition, it has been shown that a polypeptide containing the NH_2-terminal 47 amino acid sequence of human PSGL-1 is sufficient for high-affinity binding to P-Selectin (2).

A chimeric construct containing the NH_2-terminal 47 amino acid sequence of human PSGL-1 fused to a human IgG Fc heavy chain sequence, called rPSGL-Ig, has been constructed and expressed in recombinant Chinese hamster ovary (rCHO) cells. The rCHO cell line construction strategy included coexpression of core 2 GlcNAc transferase (C2GnTase) and fucosyltransferase VII (FTase) in order to build core 2 oligosaccharides containing SLe^x. Evaluation of several rCHO cell lines for rPSGL-Ig expression indicated that an inverse relationship between total rPSGL-Ig antigen cellular productivity and P-Selectin binding activity existed. This report discusses studies designed to investigate the relationship between cellular productivity and posttranslational modification of rPSGL-Ig protein.

Results and Discussion

The rPSGL-Ig - expressing rCHO cell lines were constructed by sequential transfection of preadapted CHO cells (3) with expression vectors encoding rPSGL-Ig, C2GnTase and FTase. Amplified expression of rPSGL-Ig and FTase was achieved through selection of rCHO cells at increasing concentrations of methotrexate in the medium. Resultant rCHO cell lines were screened for rPSGL-Ig expression using an antigen specific ELISA and P-Selectin binding activity (expressed as relative binding units [RBU]) was determined using a competitive P-Selectin binding ELISA. A plot of the cellular productivity obtained for several clonal rPSGL-Ig – expressing cell lines versus RBU indicated that a negative relationship existed between cellular productivity and P-Selectin binding activity (Figure 1).

Investigations were carried out in an attempt to elucidate the underlying cause of the differences in P-selectin binding activity in these rPSGL-Ig – expressing rCHO cell lines. Firstly, the capacity of rCHO cells to synthesize core 2 O-linked oligosaccharide structures bearing SLe^x was investigated. To confirm that the rCHO cell line construction strategy used results in the expression of C2GnTase and FTase, cell extracts were prepared and directly assayed for both enzyme activities. In all cases similar ranges of C2GnTase and FTase enzyme activities were detected in rCHO cell extracts while no activity was detected in untransfected controls (data not shown). In order to confirm that the FTase and C2GnTase activities were capable of constructing core 2 O-linked oligosaccharide

A. Bernard et al. (eds.), Animal Cell Technology: Products from Cells, Cells as Products, 229–235.

structures bearing SLe^x, HPAEC-PED oligosaccharide fingerprint analysis was performed on rPSGL-Ig protein affinity - purified from conditioned medium. In this analysis oligosaccharide fingerprint of rPSGL-Ig proteins expressed by rCHO cells coexpressing FTase and C2GnTase was compared to control non-coexpressing rCHO cells (Figure 2). The results of the analyses indicated that coexpression of FTase and C2GnTase resulted in the appearance of core 2 type oligosaccharide structures and a peak corresponding to core 2 glycans bearing the SLe^x structure was evident in the chromatograms. In contrast, the majority of glycans detected in samples from rCHO cells not coexpressing C2GnTase and FTase were of the core 1 type. With this information we concluded that the rCHO cell lines coexpressing C2GnTase and FTase were capable of building the appropriate glycan structures on the rPSGL-Ig molecule. The HPAEC-PED analysis could not provide positional information and we were unable to determine if the glycan structure at Thr^{16} was of the core 2 – SLe^x type. However, rPSGL-Ig proteins expressed in rCHO cells lacking heterologously expressed FTase and C2GnTase were inactive in the P-Selectin binding assay suggesting that the appropriate core 2 – SLe^x glycan was present at position Thr^{16} (data not shown).

Since sulfation of N-terminal tyrosine residues is also required for P-selectin binding activity we examined the sulfation status of rPSGL-Ig proteins expressed by several rCHO lines. The sulfation status of rPSGL-Ig was assessed using an analytical anion exchange HPLC procedure in which the rPSGL-Ig dimeric species containing 6, 5, 4, 3, 2, 1 or 0 sulfated tyrosine residues were resolved. Figure 3 shows the anion exchange HPLC chromatograms comparing the sulfation profile of affinity – purified rPSGL-Ig protein expressed by rCHO lines with differing cellular productivity phenotypes. The data indicated that an increase in the proportion of hyposulfated rPSGL - Ig protein species correlated with an increase in cellular productivity and a decrease in RBU. We interpreted these data to mean that the inverse relationship between cellular productivity and P-selectin binding activity observed was driven by an increased proportion of hyposulfated rPSGL-Ig.

The impact of rPSGL-Ig hyposulfation on P-Selectin binding activity was directly assessed. Peak fractions isolated by preparative anion exchange HPLC containing 6, 5, 4 or 3 moles of sulfated tyrosine per mole of rPSGL-Ig dimer were assayed using the P-Selectin binding ELISA. The results shown in Figure 4 indicated that the absence of a single tyrosine – SO_4 residue/dimer resulted in a ~50% decrease in binding activity. rPSGL-Ig species lacking 3 or more tyrosine – SO_4 residues/dimer scored at or below the limit of detection for the assay. These results directly support the notion that incremental hyposulfation of rPSGL-Ig attenuates the P-Selectin binding activity of the molecule.

The negative relationship between cellular productivity, rPSGL-Ig hyposulfation and P-selectin binding activity was investigated further using an individual rCHO cell line displaying rPSGL-Ig protein expression instability. The rationale for these experiments comes from the formal possibility that previous results obtained using several individual rCHO clonal isolates could potentially be a reflection of other phenotypic differences specific to those clones. Accordingly, a model system involving a single rCHO cell clone would provide a means to experimentally isolate rPSGL-Ig cellular productivity as a key variable influencing the sulfation status of the rPSGL-Ig molecule. The rPSGL-Ig expression phenotype of a particular rCHO line was shown to be unstable after 155 days of continuous culture in methotrexate-containing culture medium. Coincident with the observed decline in cellular productivity was an increase in P-selectin binding activity of rPSGL-Ig (Table 1). Examination of the sulfation status of rPSGL-Ig species expressed at timepoints early and late in the cell line development process indicated that the increase in RBU correlated with the disappearance of hyposulfated rPSGL-Ig protein species in the anion exchange HPLC chromatograms (Figure 5). HPAEC-PED oligosaccharide fingerprint analysis of these same samples indicated that there was no detectable change in the glycan profiles (data not shown). Therefore, these results further support the hypothesis that the capacity of rCHO cells to carry out tyrosine sulfation limits high level expression of active rPSGL-Ig.

Presumably, tyrosine sulfation of rPSGL-Ig was carried out by a tyrosylprotein sulfotransferase (TPST) activity endogenous to rCHO cells. The levels of endogenous TPST activity in several rCHO cell lines was evaluated by direct assay of cell lysates using a protocol described by Ouyang, et al (4). The results shown in Figure 6 indicated that TPST activity was detected in all rCHO cell lines examined. In addition, the relative levels of TPST in the rCHO cell extracts appeared to vary suggesting that population heterogeneity may exist with respect to this phenotype.

Table 1: Mean cellular productivity of a unstable rCHO cell line early and late in development

rCHO Lineage	Cellular Productivity (ug/10^6 cells/day)	P-Selectin binding (RBU)
Early	8.1	0.33
Late	2.5	1.50

Conclusions

Using a rCHO cell model system in which amplified expression of rPSGL-Ig protein was achieved we have observed that expression of functionally active rPSGL-Ig protein was inversely proportional to cellular productivity. This fundamental observation led us to investigate the underlying cause of the decline in P – selectin binding activity. HPAEC-PED HPLC analysis of O-linked glycans released from rPSGL-Ig proteins indicated that the appropriate SLe^x – bearing core 2 O - linked oligosaccharides were present in the molecule. Examination of the sulfation status of rPSGL-Ig proteins by anion exchange HPLC analysis indicated that the fraction of hyposulfated rPSGL-Ig protein species increased in proportion to cellular productivity. In addition, the P-selectin binding activity correlated with the appearance of hyposulfated rPSGL-Ig protein species. These results indicated that the observed decrease in P-selectin binding activity was caused by hyposulfation of rPSGL-Ig protein and supported the hypothesis that high level expression of active rPSGL-Ig was limited by the sulfation capacity of these rCHO cell lines. Direct assay of rCHO cell lysates confirmed that detectable TPST enzyme activity was present in all rCHO cell lines examined and evidence for population heterogeneity was obtained. We are currently investigating if a direct relationship exists between endogenous TPST activity and the capacity of rCHO cells to carry out rPSGL-Ig N-terminal tyrosine sulfation.

References

1. Sako, D., et al, 1993. Cell.75: 1179.
2. Sako, D., et al, 1995. Cell. 83: 323.
3. Sinacore, MS., et al 1996 Biotechnol. Bioeng. 52: 518 – 528.
4. Ouyang, et al. 1998. J. Biol. Chem. 273: 24770 –24774.

Figure 1: Negative Relationship Between Cellular Productivity and P-Selectin Binding Activity

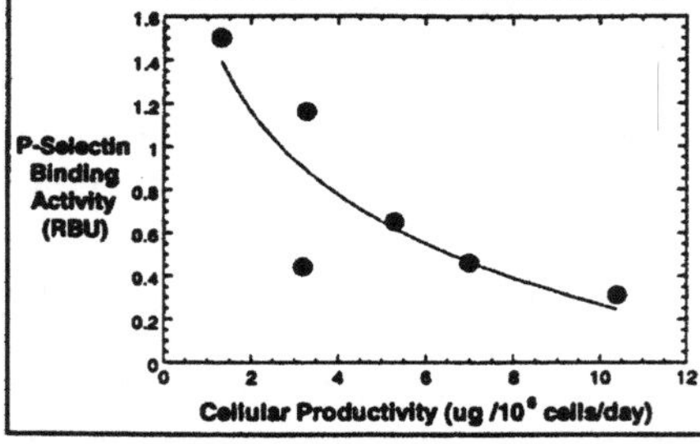

Figure 2: Effect of C2GNTase & FTase coexpression on HPAEC-PED Profiles of rPSGL-Ig O-Linked Alditols

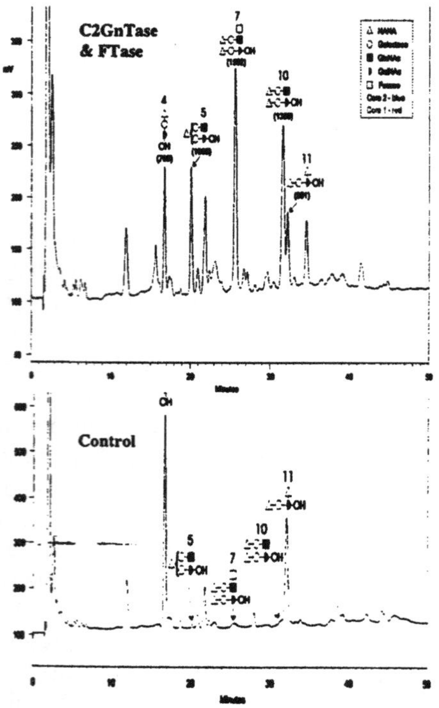

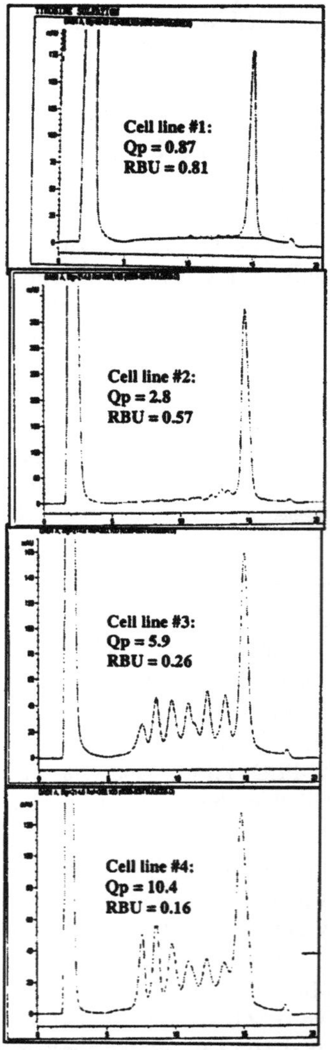

Figure 3: Sulfation status of rPSGL-Ig proteins expressed by rCHO cell lines with differing cellular productivity phenotypes.

Figure 4: Effect of rPSGL-Ig Sulfation Status on P-Selectin Binding Activity

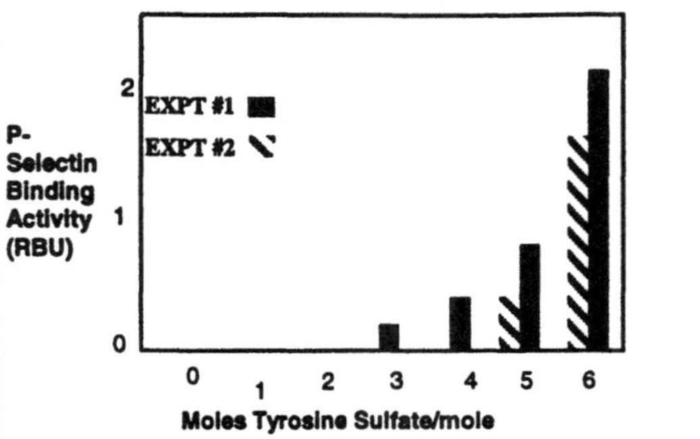

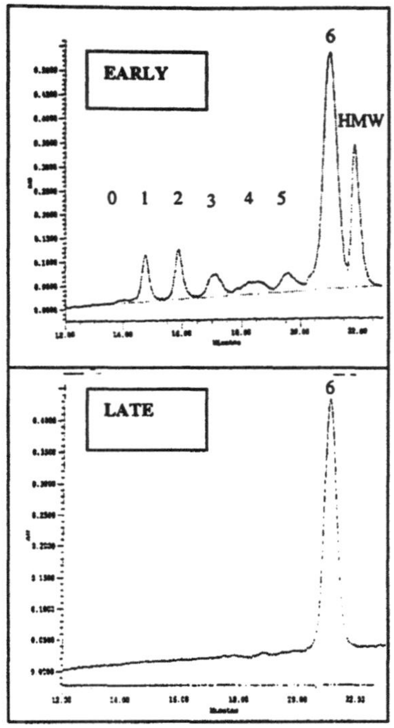

Figure 5: Sulfation analysis of rPSGL-Ig expressed by an unstable rCHO cell line

Figure 6: TPST Activity Assay of rCHO Cell Lysates

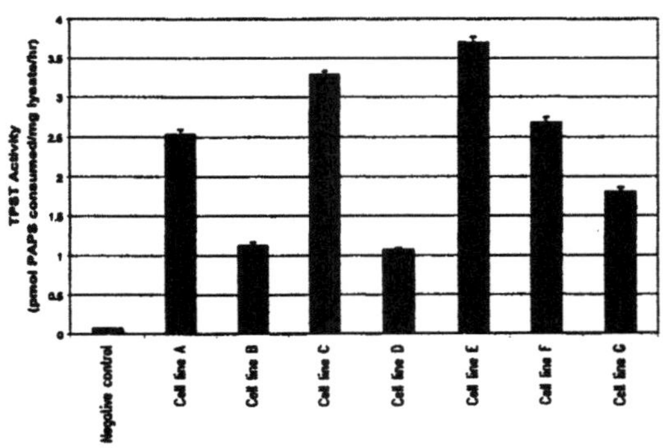

Discussion (Sinacore)

Naveh: Do you think you can sulphate this molecule by chemistry rather than by selecting clones?

Sinacore: I can only speculate that chemical sulphation of protein would be destructive to the molecule - particularly one where we have very critical post-translational modifications. Our approach is to screen for clones which have high endogenous activity, or to clone out the tsp activity from CHO cells and to overexpress this enzyme in CHO cells.

Piret: We have seen changes in the glycosylation when you engineer a membrane protein for secretion. I wonder if the sulphation of your protein is influenced by the fact that you have expressed it in the secreted form?

Sinacore: I have no experimental work to answer that question, but it is possible that a non-secretary pathway may have a different set of modifications.

Bailey: Some groups have had difficulty maintaining core 2 in a transfected CHO cell line. Have you observed any problems along these lines?

Sinacore: No, we have had good success with the CMV promotor - gives good stable robust expression. We have had problems with maintaining stable expression of glucosyl tranferase with this system but we have, by careful screening, obtained clones that manage that enzyme OK.

GLYCOSYLATION ANALYSIS OF NANOMOLAR AMOUNTS OF GLYCOPROTEIN COMBINING IN-GEL ENZYMATIC DIGESTION AND FACE™

C. KLOTH[1], H. LEIBIGER[2], U. VALLEY[1], E. YALCIN[1], R. BUCHHOLZ[3], F. EMMRICH[1], U. MARX[1]
[1]*University of Leipzig, Institute for Immunology and Transfusion Medicine, Delitzscher Str. 141, D-04129 Leipzig, FRG*
[2]*Humboldt-University Charité, Clinic for Dermatology, Schumannstr.20/21, D-10098 Berlin, FRG*
[3]*Technical University Berlin, Institute of Biotechnology, Ackerstr 71-76, D-13355 Berlin; FRG*

1. Introduction

The carbohydrates on glycoproteins, such as monoclonal antibodies, may result in an enormous loss of effectiveness, e.g. increased clearance from bloodstream, or induce an immune response if used for therapy [1, 2, 3]. Therefore, a sensitive glycosylation analysis for product control of glycoproteins is needed. Furthermore, the glycosylation control during the production process may favour specification of harvest batches [4, 5]. To meet these requirements, we have chosen an simple to perform yet powerful glycosylation analysis technique. This method combines in-gel enzymatic digestion of the glycoprotein and the Fluorophore Assisted Carbohydrate Electrophoresis (FACE™).

Purification and concentration of the target protein is achieved by an immunoprecipitation step, followed by SDS PAGE or IEF. Bands of interest were excised from the gel and the glycans were cleaved off from the protein with endoglycosidases. The oligosaccharides were analysed using the FACE™ [6]. The advantage of the FACE™ technique is simplicity as well as sensitivity when compared to HPAEC-PAD, HPLC, MS and NMR [7]. The profile analysis of N-linked oligosaccharides in combination with a monosaccharide composition analysis provides information about oligosaccharide structures and their heterogeneity. Afterwards single oligosaccharides can be sequenced using highly specific exoglycosidases. In addition, multiple samples can be run simultaneously while the expenses of the equipment are low. Therefore, this technique is a suitable method for the in-process control, batch-to-batch consistency analysis and quality control of final bulk products.

A. Bernard et al. (eds.), Animal Cell Technology: Products from Cells, Cells as Products, 237–239.

2. Methods and Materials

In-gel digestion:

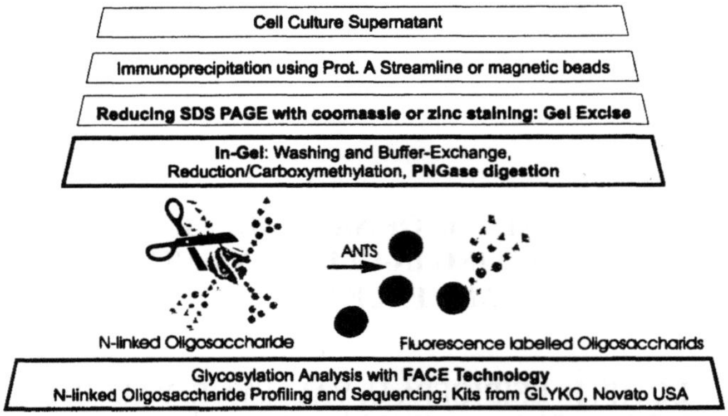

Cell culture:
A monoclonal anti-human CD4 antibody (IgG1,κ) producing murine hybridoma cell line was cultured for 31 days in a hollow fibre bioreactors (HFBR) with either serum-free medium or medium supplemented with 5% FCS. The antibody was harvested daily, 11g antibody were produced in total. The harvests from days 15-20 and 25-30 were pooled and 100µg unpurified antibody of each pool were used for the oligosaccharide analysis.

3. Results

The released and labelled oligosaccharides of each antibody pool were analysed simultaneously on an electrophoresis gel. The electrophoretic mobilities of the single oligosaccharide lines in each lane and their fluorescence intensity were compared to a glucose polymer standard G_n (n=1..20). The predominant oligosaccharides are complex biantennary, corefucosylated structures. The ratios of the oligosaccharide structures to each other changed during cultivation time in both the serum containing as well as in the serum-free culture.

HFBR culture	with 5%	serum	without	serum
pooled harvests from day	15-20	25-30	15-20	25-30
glycan structures with terminal mannose [%]	20 ± 3	55 ± 2,5	44 ± 3	59 ± 1

While the proportion of sialylated and galactosylated structures decreased, the proportion of oligosaccharides containing terminal mannose residues dramatically increased within the cultivation process time. In the serum containing culture 2.7 times more oligosaccharides containing terminal mannose were determined in the harvest of the days 25-30 than in the harvests of the days 15-20. A similar result was obtained from serum-free culture. In the pooled harvests of the days 15-20 the proportion of terminal mannose containing oligosaccharides was already twice as high as in serum-containing culture. Exposed mannose residues may lead to increased clearance rate from the bloodstream by the mannose-binding protein (MBP) and mannose receptors on hepatocytes [8]. That might reduce the biological effectiveness of this antibody in therapy.

4. Conclusion

Using the described methods of immunoprecipitation, in-gel digestion and the FACE™, 90% of all oligosaccharides of the antibody could be analysed. During a 31 day cultivation period of an anti-human CD4 antibody in hollow fibre biorecactors a significant increase of terminal mannose containing structures in late harvests (day 25-30) was determined. Thus, expression of certain oligosaccharides like terminal mannose residues may serve as an criterion for optimisation of an production process.

5. Acknowledgement

The ESACT is acknowledged for granting the ESACT bursary.This project was partly granted by the BMBF (No. 031131313). Cellular Products GmbH, Leipzig / FRG is acknowledged for providing the FACE™ system, so is the Boehringer Ingelheim Foundation for granting a one month working visit at the CCRC in Athens, Georgia / USA.

6. References

[1] Frodin, J.E.; Lefvert, A.K.; Mellstedt, H.H.: Pharmacokinetics of the mouse monoclonal antibody 17-1A in cancer patients receiving various treatment schedules, *Cancer Res.* 50 (16) (1990), 4866-71.

[2] Hamadeh, R.M.; Jarvis, G.A.; Galili, U.; Mandrell, R.E.; Zhou, P.; Griffiss, J.M.: Human natural anti-Gal IgG regulates alternative complement pathway activation on bacterial surfaces, *J. Clin. Invest.* 89 (1992), 1223-1235.

[3] Noguchi, A.; Mikuria, C.J.; Suzuki, E.; Naiki, M.: Immunogenicity of N-glycolylneuraminic acid-containing carbohydrate chains of recombinant human erythropoietin expressed in chinese hamster ovary cells, *J. Biochem.* 117 (1995), 59-62.

[4] Jenkins, N.; Parekh, R. B.; James, D. C.: Getting the glycosylation right: Implications for the biotechnology industry - Review article-, *Nature Biotechnol.* (1996), 975-981.

[5] Geisow, M.J.: Glycoprotein glycans-roles and controls, *Trends in Biotechnol.* 10 (1992), 333-335.

[6] Hu, G.F.: Fluorophore-assisted carbohydrate electrophoresis Technology and application-Review, *J. of Chromatogr. A* 705 (1995), 89-103.

[7] Jackson, P.: The analysis of fluorophore-labelled carbohydrates by Polyacrylamid Gel Electrophoresis, *Mol. Biotechnol.* 5 (1996), 101-123.

[8] Emmrich, F.; Schulze-Koops, H.; Burmester, G.: Anti CD4 and other antibodies to cell surface antigens for theraphy, *Immunopharmacology of joints and connective tissue*. Academic press (1994), 87-117.

[9] Cumming, D.A.: Glycosylation of recombinant protein therapeutics: Control and functional implication, *Glycobiology* 1 (1991), 115-130.

GLYCOSYLATION PATTERNS OF A Rec-FUSION PROTEIN EXPRESSED IN BHK CELLS AT DIFFERENT METABOLIC STATES

H.J. CRUZ[1], H.S. CONRADT[2], C.M. PEIXOTO[1], P.M. ALVES[1], M. NIMTZ[2], E.M. DIAS[1], H. SANTOS[1], J.L. MOREIRA[1], M.J.T. CARRONDO[1,3]
1 - IBET/ITQB, Ap. 12, 2780 Oeiras, Portugal
2- Protein Glycosylation, GBF, Braunschweig, Germany
3 - Lab. Eng. Bioq., FCT/UNL, 2825 Monte da Caparica, Portugal

1. Abstract

BHK-21 cells expressing a rhIgG-IL2 fusion protein were grown under different nutrient conditions in a continuous system. At very low glucose (<0.5 mM) or glutamine (<0.2 mM) concentrations, a shift towards an energetically more efficient metabolism was observed. Cell specific productivity was maintained under metabolically shifted growth conditions and at the same time a constant energy state was observed. No significant differences in the oligosaccharide structures were observed from the rhIgG-IL2 obtained under the different metabolic states. Only neutral diantennary oligosaccharides with or without core α1-6-linked fucose were detected that carried no, one or two Galβ1-4 linked galactose. The data obtained point to the presence of the classical NeuAcα2-3Galβ1-3GalNAc structure for O-linked oligosaccharides that are present in the IL-2 moiety of the protein.

2. Introduction

It is possible to increase the final titre of a recombinant product by confining the cells to more efficient metabolic states (minimising the production of toxic metabolites) without any loss in cell productivity[1]. Thus, it is important to investigate the influence of different cell metabolic states on the carbohydrate structure of polypeptides with biotechnological application, such as clinical treatment.

3. Materials and Methods

BHK 21A cells producing a rhIgG-IL2 fusion protein were obtained from Merck KGaA, Darmstadt, Germany, and grown in continuous cultures (D=0.3 day^{-1}) in a 2.0 lt. (1.2 lt. working volume) bioreactor (Biostat MDC, B. Braun, Melsungen, Germany) in DMEM supplemented with different glucose and glutamine concentrations, as presented in Table 1. The carbohydrate structural analysis were performed, after oligosaccharide release by automated hydrazinolysis in a GlycoPrep 2000 instrument (Oxford Glycosystems, UK), by several techniques, such as HPAEC-PAD mapping with a BioLC System (Dionex, Sunnyvale, CA) equipped with a CarboPac PA1 column, with and without removal of proximal fucose performed by using 10 μg of beef-kidney α-fucosidase (Boehringer Mannheim, Mannheim, Germany), and MALDI-TOF/MS.

A. Bernard et al. (eds.), Animal Cell Technology: Products from Cells, Cells as Products, 241–243.

TABLE 1. Summary of the culture conditions, i.e., nutrient and metabolite concentrations, and also metabolic quotients, under which the samples (A-E) of the fusion protein were obtained.

Sample	Glc feed (mM)	Gln feed (mM)	Glc (mM)	Lac (mM)	$Y_{Lac/Glc}$	Gln (mM)	Amm (mM)	$Y_{amm/Gln}$
A, B	16.7	4.0	1.23	20.83	1.35	1.22	2.70	0.97
C	1.1	4.0	0.14	~0	~0	0.34	4.46	1.22
D	5.6	1.0	1.26	4.93	1.15	0.14	1.29	1.56
E	5.6	2.5	0.28	2.49	0.47	0.21	2.61	1.12

Legend: Glc- glucose; Gln- glutamine; Lac- lactate; Amm- ammonia.

4. Results and Discussion

The rhIgG-IL2 contains two potential N-glycosylation (at Asn297 of each C_H2 moieties of the IgG heavy chains[2]) and O-glycosylation sites (at the Thr03 of the IL2 moiety[3,4]).

4.1. MALDI/TOF MASS SPECTRA

Several signals were observed for the reduced and permethylated glycan mixture from all preparations. A major signal corresponding to the sodium adducts ($M+ Na^+$) of a diantennary structure lacking two galactose as well as proximal fucose (m/z= 1679) was detected as well as signals corresponding to a proximally fucosylated diantennary glycans with (m/z= 2261) or without one (m/z= 2057) and without two galactose residues (m/z= 1853), respectively.

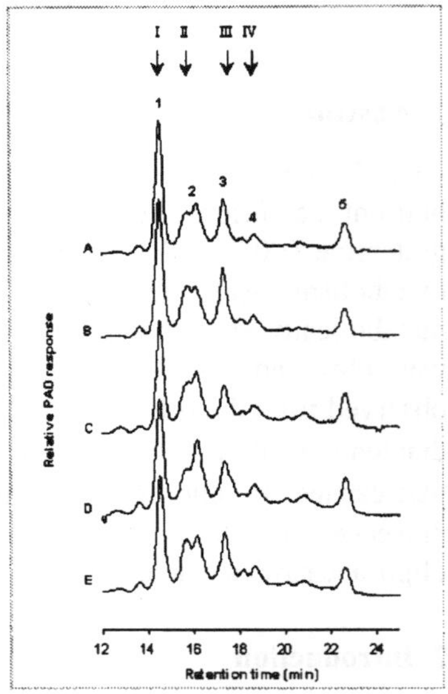

Fig. 1. HPAEC-PAD mapping of N- and O-glycans of the fusion protein after glycan release by hidrazinolysis. Samples identified from A to E are as described in Table 1. Standard oligosaccharides retention times are indicated by arrows: complex type a-galactosylated diantennary struture with (I) and without (II) proximal fucose and fully galactosylated diantennary struture with (III) and without (IV) proximal fucose.

4.2. HPAEC-PAD MAPPING

HPAEC-PAD mapping of oligosaccharides yielded elution profiles shown in Fig. 1. Peak 1 corresponds to the major structure ($GlcNAc_2Man_3$ GlcNAcFucGlcNAc), peaks 2a,b comprise the nonfucosylated structure thereof as well as the oligosaccharide from peak 1 containing an additional galactose. Peak 3 contains the monogalactosylated biantennary oligosaccharide form lacking the proximal fucose; the accompanying shoulder visible in profiles C-E of Fig. 1 indicates the presence of the fully galactosylated Gal_2 $GlcNAc_2Man_3GlcNAc$[Fuc]GlcNAc structure. The minor peak 4 was identified as the di-galactosylated glycan minus proximal fucose. These interpretations are consistent with the results obtained after enzymatic defucosydation of the native oligosaccharides, in which removal of proximal fucose by α-fucosidase resulted in the detection of only three peaks (G0, G1 and G2) representing the agalacto-, mono- and di-galactosylated biantennary structures lacking proximal fucose. In summary, from both the MALDI/TOF analysis as well as HPAEC-PAD mapping results, it is evident that the N-glycan structures of rhIgG-IL2 fusion protein consist of a mixture of biantennary structures lacking one or two galactose residues and their variants without or with proximal fucose. The presence of at least the hexasaccharide $GlcNAc_2Man_3GlcNAc_2$ in the C_H2 has been reported to be required for IgG binding to Fcγ receptors whereas galactosylated branches are not essential for an efficient recognition[5]. All carbohydrate structures detected in the IgG-IL2 in this study are in agreement with these requirements. The partial presence of unfucosylated glycans observed here should not interfere with *in vivo* clearance mechanisms since such variant structures (25 % of the total) are also observed in similar percentages in human serum IgG preparations[6]. Although no efforts were made to study also in detail the O-linked oligosaccharide structures that are present in the IL-2 moiety of the protein, the data obtained point to the presence of the classical NeuAcα2-3Galβ1-3GalNAc structure, that has also been described for the recombinant wild-type huIL2 expressed from BHK-21 cells[4].

5. Conclusions

It is of interest to learn how the carbohydrate structures of a polypeptide respond to variations in cell metabolic state, especially in potential therapeutics to be used in humans. From the present work, it can be concluded that by confining the metabolism of the cells to an energetically more efficient state, the overall product quality with respect to posttranslational modification with carbohydrates is maintained. This is an important consideration in the design and development of optimised biotechnological cell culture processes where cell metabolism is to be kept at a defined and controlled state, in order to yield larger amounts of product and reproducible final product preparations including reproducible posttranslational modification of carbohydrates.

Acknowledgements

European Commission (Biotech ERBIO4-CT960721), Merck KGaA, Darmstadt, Germany (Drs. E. Rieke, C. Burger and R. Dunker), Fundação para a Ciência e a Tecnologia (PBICT/BIO/20333/95, PRAXIS XXI/BD/2764/94 and PRAXIS XXI/BD/2721/95) and FEBS (short-term fellowship atributed to HJ Cruz).

References

1. Cruz, H.J., Ferreira, A.S., Freitas, C.M., Moreira, J.L., and Carrondo, M.J.T. (1999) Metabolic responses to different glucose and glutamine levels in BHK cell culture. Appl. Microbiol. Biotechnol., *in press.*
2. Youings, A., Chang, S.-C., Dwek, R.A., Scragg, I.G. (1996) Site-specific glycosylation of human immunoglobulin G is altered in four rheumatoid arthritis patients. Biochem. J. 314: 621-630.
3. Conradt, H.S., Geyer, R., Hoppe, J., Grotjahn, L., Plessing, A., Mohr, H. (1985) Structures of the major carbohydrates of natural human interleukin-2. Eur. J. Biochem. 153: 255-261.
4. Conradt, H.S., Nimtz, M., Dittmar, K.E.J., Lindenmaier, W., Hoppe, J., Hauser, H. (1989) Expression of human interleukin-2 in recombinant baby hamster kidney, Ltk^-, and chinese hamster ovary cells. J. Biol. Chem. 264: 17368-17373.
5. Lund, J., Takahashi, N., Pound, J.D., Tyler, R., Goodall, M., Nakagawa, H., Jefferis, R. (1995) Oligosaccharide-protein interactions in IgG can modulate recognition by Fcγ receptors. FASEB J. 9: 115-119.
6. Parekh, R.B., Dwek, R.A., Sutton, B.J., Fernandes, D.L., Leung, A., Stanworth, D., Rademacher, T.W., Mizuochi,T., Taniguchi, T., Matsuta, K., Takeuchi, F., Nagano, Y., Miyamoto, T., Kobata, A. (1985) Association of rheumatoid arthritis and primary osteoarthritis with changes in the glycosylation pattern of total serum IgG. Nature 316: 452-457.

ELIMINATION OF N-GLYCOLYLNEURAMINIC ACID RESIDUES IN RECOMBINANT GLYCOPROTEINS: A GENE KNOCK-OUT APPROACH

OMBRETTA FONTANA, NEVIE COVINI, MONICA PIGHINI, DANIELA CARPANI, PASQUALINA BUONO, MARCO R. SORIA AND LUCIA MONACO
Dibit, Department of Biological and Technological Research
San Raffaele Scientific Institute, via Olgettina 58, Milano, Italy

The enzyme CMP-N-acetylneuraminic acid hydroxylase (CNAH) is responsible for the conversion of N-acetylneuraminic acid (NeuAc) into N-glycolylneuraminic acid (NeuGc). Incorporation of NeuGc residues into glycan chains of recombinant glycoproteins of therapeutic interest is not desirable, since NeuGc is potentially antigenic in man.

To the aim of knocking-out the *CNAH* gene in CHO cells, for the production of recombinant glycoproteins devoid of NeuGc residues, a genomic library from CHO DUX-B11 cells was screened with the whole murine *CNAH* cDNA [1]. Four independent clones were identified. Southern blot analysis of these clones revealed that they carried progressively overlapping genomic inserts, spanning a region of approximately 30 kb. A second screening of the library with an intronic probe at the 5'-end of the 30 kb region yielded one additional clone, extending the cloned region by approximately 14 kb at the 5'-end.

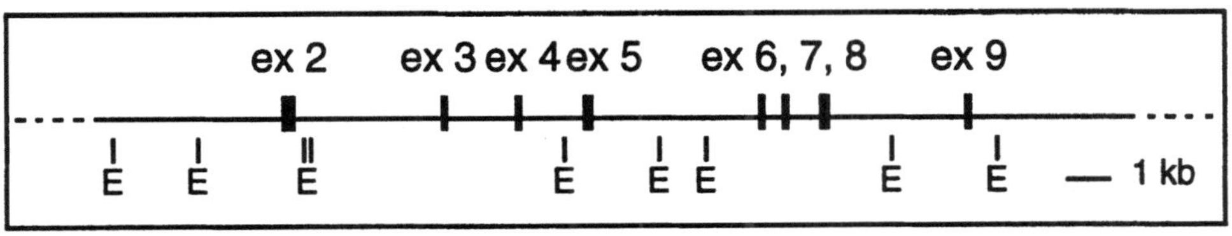

Figure 1. Partial genomic structure of the hamster *CNAH* gene. Vertical bars represent exons. E: *Eco*RI sites.

Open reading frames corresponding to exonic regions homologous to the murine *CNAH* cDNA were identified by sequencing (Figure 1). Canonical exon-intron boundaries limit all exonic regions. The deduced protein sequence spans positions 5-370 of the 577 aminoacids of murine *CNAH*. Identity between the murine and hamster deduced protein sequences is 96.2%. Alignment of the deduced hamster CNAH protein to the mouse [1], human [2] and pig [3] proteins in shown in Figure 2.

A. Bernard et al. (eds.), Animal Cell Technology: Products from Cells, Cells as Products, 245–249.

```
                ↓2                                                     =======
Ha     1 ..........KQTAETLLSLSPAETANLKEGINFFRNKTTGKEYILYKEKNHLKACKNLCKHQGGL
Mo     1 MMDRKQTAETLLTLSPAEVANLKEGINFFRNKTTGKEYILYKEKDHLKACKNLCKHQGGL
Hu         ............................................................
Pi         ............................................................

         ==2Fe2S====↓3====================           ↓4
Ha    57 FIKDIEDLDGRSVKCTKHNWKLDVSTMKYINPPGSFCQDELVVEMDGNDGLFLIELNPPN
Mo    61 FMKDIEDLDGRSVKCTKHNWKLDVSTMKYINPPGSFCQDELVIEMDENNGLSLVELNPPN
Hu     1 ...........................................MDENNGLLLLELNPPN
Pi         ............................................................

                       ↓5
Ha   117 PWDSDPRTPEELAFGEVQITYLTHACMDLKLGDKRMVFDPWLIGPAFARGWWLLHEPPSD
Mo   121 PWDSDPRSPEELAFGEVQITYLTHACMDLKLGDKRMVFDPWLIGPAFARGWWLLHEPPSD
Hu    17 PWDLQPRSPEELAFGEVQITYLTHACMDLKLGDKRMVFDPWLIGPAFARGWWLLHEPPSD
Pi     1 ..........................MDLKLGDKRMVFDPWLIGPAFARGWWLLHEPPSD

                   ==Fe== ↓6
Ha   177 WLERLCKADLIYISHMHSDHLRYPTLKQLSQRRPDIPIYVGDTERPVFWNLDQSGVQLTN
Mo   181 WLERLCKADLIYISHMHSDHLSYPTLKQLSQRRPDIPIYVGDTERPVFWNLDQSGVGLTN
Hu    77 WLERLCQADLVYISHLHSDHLSYPTLKKLAGRRPDIPIYVGNTERPVFWNLNQSGVQLTN
Pi    35 WLERLSRADLIYISHMHSDHLSYPTLKKLAERRPDVPIYVGNTERPVFWNLNQSGVQLTN

                ↓7                              ↓8
Ha   237 INVVPFGVWQQVDKNLRFMILMDGVHPEMDTCIIVEYKGHKILNTVDCTRPNGGRLPEKA
Mo   241 INVVPFGIWQQVDKSLRFMILMDGVHPEMDTCIIVEYKGHKILNTVDCTRPNGGRLPEKV
Hu   137 INVVPFGIWQQVDKNLRFMILMDGVHPEMDTCIIVEYKGHKILNTVDCTRPNGGRLPMKV
Pi    95 INVVPFGIWQQVDKNLRFMILMDGVHPEMDTCIIVEYKGHKILNTVDCTRPNGGRLPMKV

               ===CMP-NeuAc===↓9  =======b5=======
Ha   297 ALMMSDFAGGASGFPMTFSGGKFTEEWKAQFIKAERRKLLNYKAQLVKDLQPRIYCPFAG
Mo   301 ALMMSDFAGGASGFPMTFSGGKFTEEWKAQFIKAERRKLLNYKAQLVKDLQPRIYCPFAG
Hu   197 ALMMSDFAGGASGFPMTFSGGKFTEEWKAQFIKTERKKLLNYKAWLVKNLQPRIYCPFAG
Pi   155 ALMMSDFAGGASGFPMTFSGGKFTEEWKAQFIKTERKKLLNYKARLVKDLQPRIYCPFPG

Ha   357 YFVESHPSDK..................................................
Mo   361 YFVESHPSDKYIKETNTKNDPNQLNNLIRKNSDVVTWTPRPGAVLDLGRMLKDPTDSKGI
Hu   257 YFVESHPSDKYIKETNTKNDPNELNNLIKKNSDVITWTPRPGATLDLGRMLKDPTDSKGI
Pi   215 YFVESHPADKYIKETNIKNDPNELNNLIKKNSEVVTWTPRPGATLDLGRMLKDPTDSKGI

Ha         ............................................................
Mo   421 VEPPEGTKIYKDSWDFGPYLEILNSAVRDEIFCHSSWIKEYFTWAGFKNYNLVVRMIETD
Hu   317 IEPPEGTKIYKDSWDFEPYLEILNAALGDEIFLHSSWIKEYFTWAGFKDYNLVVRMIETD
Pi   275 VEPPEGTKIYKDSWDFGPYLNILNAAIGDEIFRHSSWIKEYFTWAGFKDYNLVVRMIETD

Ha         ............................................................
Mo   481 EDFSPFPGGYDYLVDFLDLSFPKERPSREHPYEEIHSRVDVIRYVVKNGLLWDDLYIGFQ
Hu   377 EDFNPFPGGYDYLVDFLDLSFPKERPQREHPYEEIHSRVDVIRHVVKNGLLWDELYIGFQ
Pi   335 EDFSPLPGGYDYLVDFLDLSFPKERPSREHPYEEIRSRVDVIRHVVKNGLLWDDLYIGFQ

                 ==Fe===
Ha         ............................................................
Mo   541 TRLLRDPDIYHHLFWNHFQIKLPLTPPNWKSFLMHCD........................
Hu   437 TRLQRDPDIYHHLFWNHFQIKLPLTPPNWKSFLMCCEQNGPAILQECKTT...........
Pi   395 TRLQRDPDIYHHLFWNHFQIKLPLTPPDWKSFLMCSG........................
```

Figure 2. Deduced protein sequence for hamster (Ha), mouse (Mo), human (Hu) and pig (Pi) CNAH. Mismatched residues are underlined. Arrows indicate exon boundaries in the hamster sequence. Double dashes indicate functional domains. 2Fe2S: Rieske center; Fe: iron binding site; CMP-NeuAc: substrate binding site; b5: cytochrome b5 binding site.

The coding region of hamster *CNAH* identified so far spans putative exons 2-9. The putative exon 1, corresponding to the first 4 aminoacids in the murine sequence, has not been identified yet. DNA sequences coding for biologically active domains postulated on the basis of the mouse and pig cDNAs [3] are all conserved in the hamster sequence and fall within discrete exons. In particular: exon 2 and exon 3 encode one box each of the putative binding site for an iron/sulfur Rieske centre; exon 5 codes for one of the putative mononuclear iron-binding sites; exon 8 contains the sequence for the CMP-NeuAc binding site and exon 9 the sequence for the cytochrome b_5 binding site (Figure 2). Interestingly, a deletion mutant of murine *CNAH* described to code for an inactive enzyme [4] lacks a sequence corresponding to exon 8 in our hamster sequence, encoding the binding site to the substrate.

A plasmid for *CNAH* gene homologous recombination has been prepared, including: a region from the hamster *CNAH* gene, corresponding to 2.3 kb of sequence 5' to the putative second exon of the *CNAH* gene; a replacement cassette flanked by *loxP* sites, carrying the zeocin resistance gene and the thymidine kinase gene for positive and negative selection, respectively; a region from the hamster *CNAH* gene, corresponding to 5.0 kb of sequence 3' to the putative fifth exon. As a result of the homologous recombination event, a genomic region of the hamster *CNAH* gene should be deleted, spanning the functional domains for the Rieske center and for the first mononuclear iron binding site. Recombinant human CNAH, which lacks the Rieske center, was shown to lack hydroxylase activity [5].

The homologous recombination plasmid will be transfected into both CHO DUK X-B11 cells which had been used for the construction of the CHO library and CHO cells which we had engineered to express the rat α2,6-sialyltransferase (*α2,6-ST*) cDNA [6], since obtaining *α2,6-ST*(+) and*CNAH*(-) CHO cells would combine very useful properties.

Acknowledgments

This work was supported by the European grant BIO4-CT96-0767.

References

1. Kawano, T., Koyama, S., Takematsu, H., Kozutsumi, Y., Kawasaki, H., Kawashima, S., Kawasaki, T. and Suzuki, A.: Molecular cloning of cytidine monophospho-N-acetylneuraminic acid hydroxylase. Regulation of species- and tissue-specific expression of N-glycolylneuraminic acid, *J. Biol. Chem.* **270** (1995), 16458-16463.
2. Chou, H.H., Takematzu, H., Diaz, S., Iber, J., Nickerson, E., Wright, K.L., Muchmore, E.A., Nelson, D.L., Warren, S.T. and Varki, A.: A mutuation in human CMP-sialic acid hydroxylase occurred after the *Homo-Pan* divergence, *Proc. Natl. Acad. Sci. USA* **95** (1998), 11751-11756.
3. Schlenzka, W., Shaw, L., Kelm, S., Schmidt, C.L., Bill, E., Trautwein, A.X., Lottspeich, F. and Schauer, R.: CMP-N-acetylneuraminic acid hydroxylase: the first cytosolic Rieske iron-sulphur protein to be described in Eukarya, *FEBS Lett.* **385** (1996), 197-200.
4. Koyama, S., Yamaji, T., Takematsu, H., Kawano, T., Kozutsumi, Y., Suzuki, A. and Kawasaki, T.: A naturally occurring 46-amino acid deletion of cytidine monophospho-N-acetylneuraminic acid hydroxylase leads to a change in the intracellular distribution of the protein, *Glycoconj. J.* **13** (1996), 353-358.
5. Irie, A., Koyama, S., Kozutsumi, Y., Kawasaki, T. and Suzuki, A.: The molecular basis for the absence of N-glycolylneuraminic acid in humans, *J. Biol. Chem.* **273** (1998), 15866-15871.
6. Bragonzi, A., Distefano, G., Buckberry, L.D., Acerbis, G., Foglieni, C., Lamotte, D., Marc, A., Soria, M.R., Jenkins, N. and Monaco, L.: A new Chinese hamster ovary cell line expressing α-2,6-sialyltransferase used as universal host for the production of human-like sialylated recombinant glycoproteins, *submitted.*

Discussion (Monaco)

Baetge: Do you know about the way in which this DNA inserts into the genome - is it one site or multiple sites?

Monaco: We have done Southern Blotting of a number of stable clones and we observe a number of integration sites. It mainly depends upon the type of transfection vector we use, so in some cases there are only 1 or 2 integration sites - in other cases more. In one of these sites there is the long polymer. So if you cut the genome with the site which fragments the polymer into monomers, you can see intense bands which can be quantified and you can measure the number of copies that are integrated.

FUCOSYLTRANSFERASE III PRODUCTION BY MAMMALIAN AND INSECT CELLS FOR THE SYNTHESIS OF THERAPEUTIC OLIGOSACCHARIDES

C.Benslimane[1], S.Chenu[1], H.Tahrat[1], V.Deparis[1], C.Augé[2], M.Cerutti[3], P.Delannoy[4], J.L.Goergen[1], A.Marc[1].

1 LSGC-CNRS BP 172, F-54505 Vandoeuvre - 2 LCOM-Univ. Paris XI, F-91405 Orsay - 3 S. R. de Pathologie Comparée, INRA-CNRS, F-30380 St-Christol-lez-Alès - 4 LCB, USTL F-59655 Villeneuve d'Ascq.

Abstract : A soluble form of the α(1,3/4)-fucosyltransferase (Fuc-TIII) was produced using suspension cultures of CHO and *Sf9* cells. CHO Fuc-TIII was tested for its storage stability. Fucose transfer on different oligosaccharide acceptors was assayed with both recombinant enzymes.

1. INTRODUCTION

Oligosaccharides such as Lewis a (Le^a) or Lewis x (Le^x) types are involved in complications surrounding organ transplants, cellular adhesion or inflammation [1]. Molecules bearing such structures could be used as antagonists for treatment of various diseases. Chemical synthesis of these therapeutic oligosaccharides is both difficult and time consuming. An alternative method is a chemical-enzymatic approach that requires the α(1,3/4) - fucosyltransferase activity of the Lewis blood group Fuc-TIII. A method to provide enough enzyme is the production in recombinant animal cells. The cDNA of this enzyme was engineered to get a soluble and active form of the protein in stably transfected CHO cells and baculovirus/*Sf9* transient expression system.

2. MATERIALS AND METHODS

2.1 VECTORS CONSTRUCTION & CELL TRANSFECTION

- The Fuc-TIII cDNA was depleted from the transmembrane and cytoplasmic regions, amplified by PCR, inserted in a vector and transfected in CHO cells. [2,3]. Positive clones were selected and S7 clone adapted to suspension culture.
- For baculovirus construction, the truncated *fut3* gene was inserted in the p119 transfer vector [4] after the EGT signal peptide, a 6-His tag and 4 *fut3* codons upstream of the *avr*II restriction site. The vector was co-transfected in *Sf9* cells with DNA of the wild type AcNPV baculovirus. A recombinant baculovirus expressing Fuc-TIII was purified.

A. Bernard et al. (eds.), Animal Cell Technology: Products from Cells, Cells as Products, 251–253.

2.3 CELL LINES AND CULTURE MEDIA

CHO cells were grown at 37°C in αMEM medium with ribo- and deoxyribonucleosides (Life technologies) supplemented with 10% fetal calf serum and 4mM glutamine (ATGC, France). The batch culture was performed in a 250 ml spinner flask. *Sf9* cells were grown at 27°C in serum-free medium.

2.4 ENZYME ASSAY

Recombinant fucosyltransferaseactivity in supernatant towards type 1 or 2 acceptors was tested at 37°C in 50 μl of the reaction mixture [2].

3. RESULTS AND DISCUSSION

3.1 MASS PRODUCTION OF THE Fuc-TIII USING SUSPENSION CELLS

CHO S7 clone was cultivated in suspension in batch mode for 8 days. The enzyme activity in the supernatant increased with the cell density until a maximal value of 12 $mU.ml^{-1}$. The maximal yield of enzyme produced per cell was equivalent to 10^{-5} $mU.cells^{-1}$. *Sf9* cells were cultivated in spinner flasks and infected with the recombinant baculovirus at a density of 5.10^5 cells mL^{-1} with a MOI of 5 pfu mL^{-1}. Fuc-TIII production started four days post-infection and reached a maximum after 12 days.

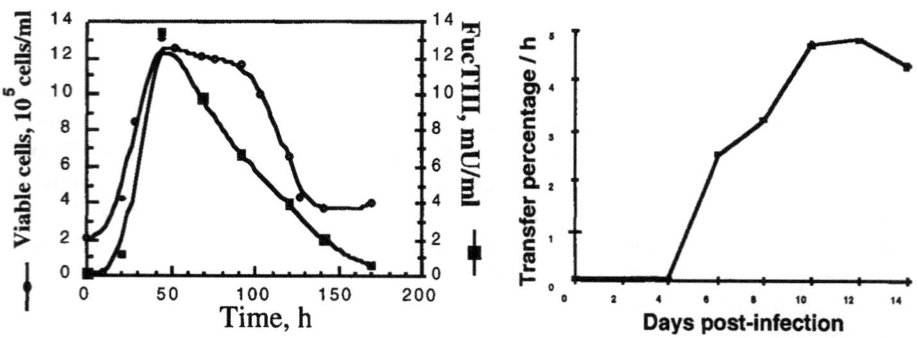

***Figure** : Kinetics of cell growth and enzyme production with suspension cells (left : CHO, right : Sf9) cultivated in spinner flask*

3.2 STABILITY OF THE CHO-Fuc-TIII ENZYME

Stability of the FucTIII activity was tested for different storage durations and conditions of supernatants harvested after 75 h or 120 h of culture. The enzyme activity was stable during 100 days at -20°C+glycerol (30%). A rapid decrease of enzyme activity was observed at 37°C, whereas at -80°C the enzyme activity remained maximal for at least 20 days. A cocktail of protease inhibitors did not prevent enzyme degradation.

Storage duration (Days)	Fuc-TIII activity (%)							
	37°C pH 7,3	4°C pH 7,3	-80°C pH 7,3	-20°C pH 7,3	-20°C+glycerol pH 7,3	-20°C pH 6	-20°C+protease inhibitors	-20°c w/o protease inhibitors
0	100	100	100	100	100	100	100	100
10	2	85	104	54	98	51	22	51
20	0	90	100	48	114	51	13	43
30	0	66		40	112	37	13	36
100	0	19		10	110	10	12	14

3.3 ENZYMATIC FUCOSE TRANSFER ON OLIGOSACCHARIDES.

The soluble FucTIII was able to transfer fucose residues on either type 1 acceptors with α (1,4) linkage or type 2 acceptors with α (1,3) linkage. Enzymatic activity was higher for type 1 acceptors and increased with the type 2 acceptor concentration. Type 1 acceptor sialylation strongly reduced α (1,4) activity.

	Acceptor (concentration)	CHO	*Sf9*
Type 1	Gal-β-(1-3)-GlcNAc-β-O-octyl (0.6 mM)	100	100
	Gal-α-(1-3)-Gal-β-(1-3)-GlcNAc-β-O-octyl (0.6 mM)	125	112
	NeuAc-α-(2-3)-Gal-β-(1-3)-GlcNAc-β-O-*pm*benzyl (0.6 mM)	42	53
Type 2	Gal-β-(1-4)-GlcNAc-β-O-benzyl (6 mM)	9	4
	Gal-β-(1-4)-GlcNAc (10 mM)	32	18
	Gal-β-(1-4)-GlcNAc (20 mM)	44	46
	Gal-β-(1-4)-Glc (10 mM)	43	31
	Gal-β-(1-4)-Glc (20 mM)	59	55

4. CONCLUSION

Both r-CHO and *Sf9*/baculovirus expression systems are able to produce an active, secreted and soluble form of the FucTIII during suspension cultures. The produced enzymes show a large specificity which is of interest for their future use in oligosaccharides synthesis.

REFERENCES

[1] Walz G, Aruffo A, Kolanus W, Bevilacqua M, Seed B (1990) *Science*. **250**, 1132-1135.
[2] Chenu S, Tahrat H, Augé C, Delannoy P, Mollicone R, Marc A, Goergen JL (1998) Proceeding Forum For Applied Biotech., Brugges September 1998, pp. 1123-1126.
[3] Kukowska-Latallo JF, Larsen RD, Nair RP, Lowe JB (1990) *Genes Dev*. **4**, 1288-1303.
[4] Missé D, Cérutti M, Schmidt I, Jansen A, Devauchelle G, Jansen F, Veas F (1998) *J.Virol*. **72**, 7280-8.

This work has been achieved in the frame of the French network "GT-rec" supported by MENRT and of the CNRS PIR-GP.

CONTROL OF THERAPEUTIC MONOCLONAL ANTIBODY GLYCOSYLATION

A.E. HILLS[1], A.K. PATEL[2], P.N. BOYD[2], D.C. JAMES[1]

[1] *Research School of Biosciences, University of Kent at Canterbury, Canterbury CT2 7NJ, UK.* [2] *Biotechnology Analytical Labs, Glaxo Wellcome, Langley Court, Beckenham, Kent BR3 3BS, UK.*

Results NSO CELL SPECIFIC PRODUCTIVITY

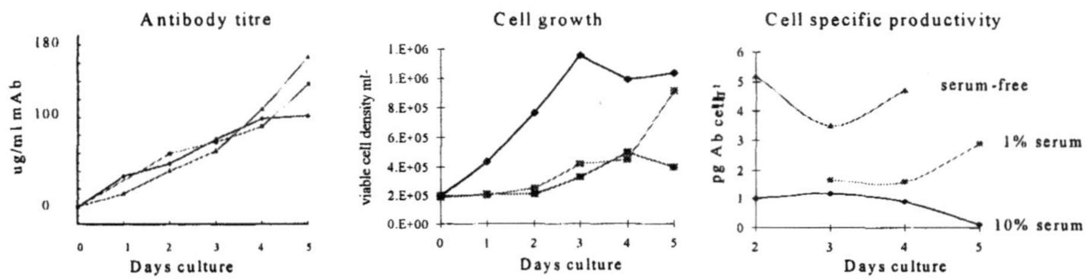

Fig. 1. Antibody titre and cell specific producitvity of NSO cells during weaning into serum-free medium NSO cells grown in serum-free medium have a higher cell specific productivity. In this medium it is likely that cells are under higher levels of stress compared to medium that contains even low levels of serum. We speculate that adaptation to serum-free growth may induce stress proteins implicated in mAb folding and oligomerisation.

INDUCTION OF ENDOPLASMIC RETICULUM STRESS PROTEIN EXPRESSION IN RESPONSE TO SERUM DEPRIVATION IN NSO CELLS

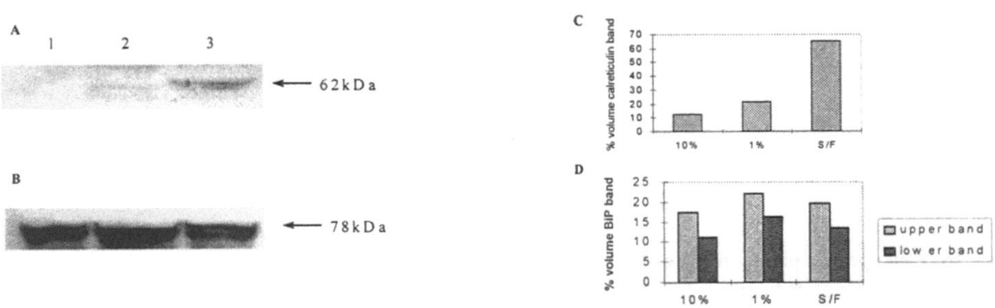

Fig 2. Western blot of calreticulin expression (A) and BiP/GRP78 (B) expression in NSO cell lysates. Lane 1 NSO cells grown in 10% serum; lane 2 NSO cells grown in 1% serum; lane 3 NSO cells grown in serum-free medium. Mr 62kDa = calreticulin. Mr 78kDa = BiP/GRP78. **C & D.** Densitometry of ECL

A. Bernard et al. (eds.), Animal Cell Technology: Products from Cells, Cells as Products, 255–257.

exposed Western blots. Calreticulin acts in a lectin-like manner to aid protein folding in the endoplasmic reticulum[1]. Expression is increased in NS0 cells grown in serum-free medium, whereas expression of BiP stays the same. BiP ensures proper assembly of antibody subunits and also prevents intracellular aggregation of polypeptides, therefore it is likely to play a quality control function in cells that are engineered to produce high levels of mAb.

ANALYSIS OF Fc N-GLYCANS DURING WEANING INTO SERUM-FREE MEDIUM

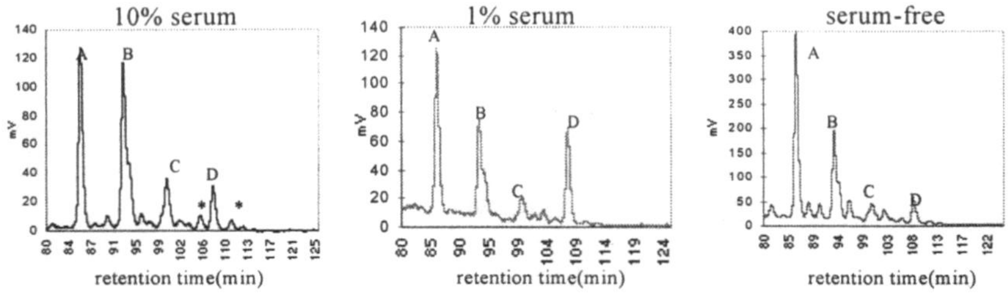

Fig. 4. Normal phase chromatography of 2-AB labelled glycans from Fc fragments of IgG1
HPLC method adapted from Guile *et al.* [10] Exoglycosidase digestions were carried out on samples to confirm structures. A= A2G0F; B=A2G1F; C=A2G2F; D=Man 8. * denotes Galα1,3Gal epitopes.

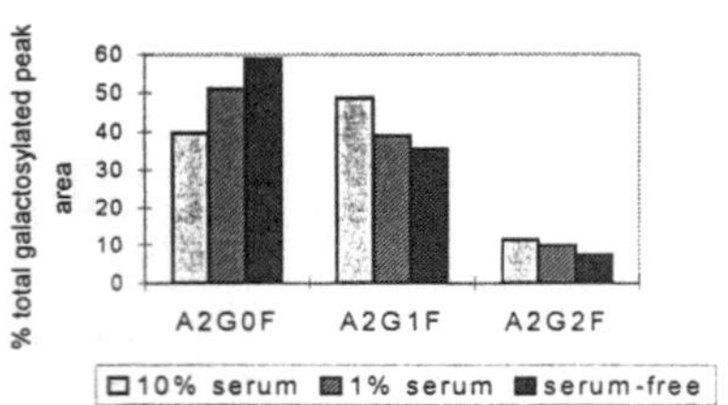

Fig.4. Peak area analysis of b1-4 galactosylated biantennary glycans produced on Fc fragments in high, low and serum-free conditions
An increase in cell specific productivity during growth of NS0 cells in serum-free medium results in a decrease in Fc N-glycan galactosylation. This may be due to nucleotide sugar availability or an increase in the rate of N-glycan processing

INDUCTION OF ER PROTEINS FOR IgG FOLDING AND SECRETION

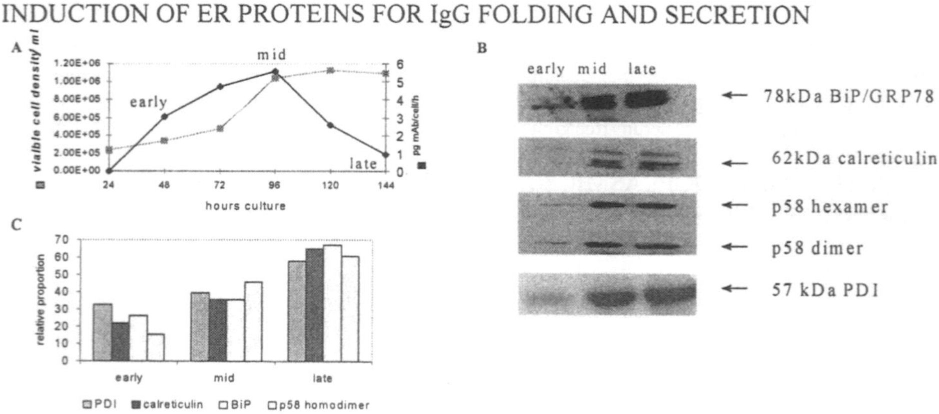

Fig. 5A; Cell specific productivity of NS0 cells in serum-free batch spinner culture. 5B; Western blot of NS0 cell lysates (non-reducing 4-20% SDS-PAGE). **BiP** = immunoglobulin binding protein; **p58** = recycling intermediate compartment protein; **PDI** = protein disulphide isomerase

ROLE OF NUCLEOTIDE SUGAR SUBSTRATES ON CELLULAR GLYCAN PROCESSING

Protein N-glycosylation is regulated by the availability of nucleotide sugar substrates. Pels Rijcken *et al.*[3] have proposed that elevated cytosolic levels of UDP-N-acetylhexosamine impaired the transport of CMP-NeuAc into the Golgi, which lead to decreased sialylation. Addition of glucosamine to cell culture medium has resulted in an increase in antennarity of N-glycans [4] . By manipulating the levels of nucleotide sugar precursors available to NS0 cells (by additions to the cell culture medium) a strategy for controlling product glycosylation could be devised.

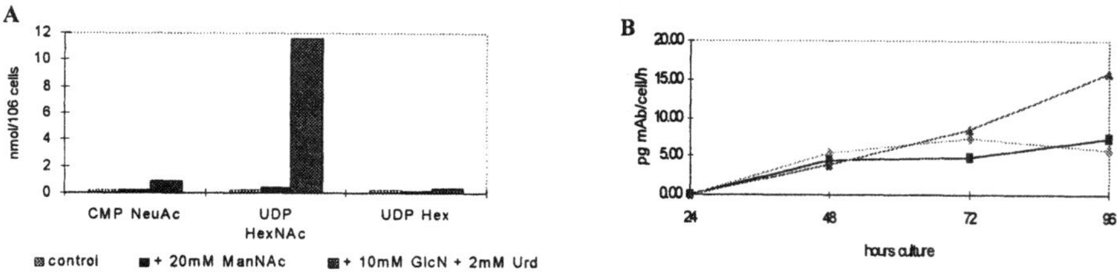

Fig.6A. Assay of nucleotide sugars in NS0 cells by HPLC and **B.Cell specific productivity of NS0 cells**; Additons of 20mM N-acetylmannosamine, and 10mM glucosamine + 2mM uridine were made to serum-free medium after 48 hours of culture. **CMP NeuAc** CMP-*N*-acetylneuraminic acid; **UDP HexNAc** UDP-*N*-acetylgalactosamine or UDP-*N*-acetylglucosamine; **UDP Hex** UDP-glucose or UDP-galactose. N.B NS0 cells are likely to have levels of CMP-NeuGc rather than CMP-NeuAc.

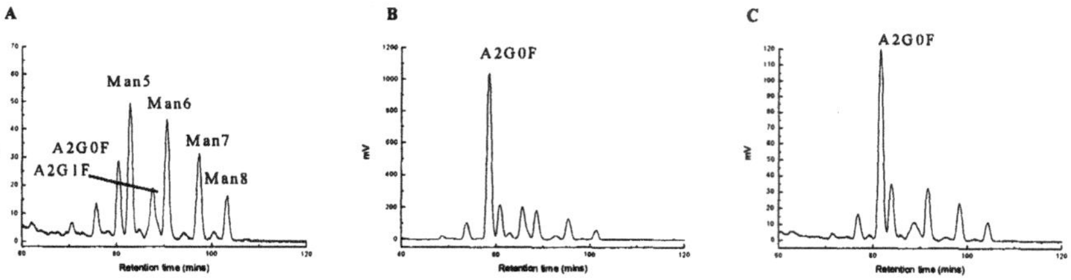

Fig 7. Normal phase chromatography of 2-AB labelled Fc glycans from IgG1 produced in serum-free batch spinner culture A; Control, B; +20mM N-acetylmannosamine, C; +10mM glucosamine + 2mM uridine. Compared to the control culture, addition of nucleotide sugar precursors did not affect antennarity or sialylation of the Fc N-glycans. Steric hinderence in the Fc region could prevent branching or additional processing of the glycans, although the nucleotide sugar additions have caused the proportions of ungalactosylated glycans (A2G0F) to be increased.

References and Acknowledgements

1. Heal, R., McGivan, J. (1998) Induction of calreticulin expression in response to amino acid deprivation in Chinese Hamster Ovary cells. Biochem.J.329 389-394.

2. Guile, G.R, Rudd, P.M., Wing, D.R., Prime, S.B., Dwek, R.A. (1996) A Rapid high-resolution high performance liquid chromatographic methodfor separating glycan mixtures and analyzing oligosaccharide profiles. Anal. Biochem, 240 (2) 210-26.

3.Pels Rijcken, W.R., Ferwerda, W., Van den Eijden, D.H., Overdijk, B. (1995) Influence of D-galactosamine on the synthesis of sugar nucleotides and glycoconjugates in rat hepatocytes. Glycobiology 5, 495-502.

4. Grammatikos, S.I., Valley, U., Nimtz, M.,Conradt, H.S., Wagner, R. (1998) Intracellular UDP-N-Acetyl hexosamine Pool Affects N-Glycan Complexity: A Mechanism of Ammonium Action on Protein Glycosylation. Biotechnology Progress 14, 410-419.

I would like to thank Glaxo Wellcome for sponsoring this Industrial Research Studentship, and ESACT for the bursary to attend this conference.

THE EFFECT OF CELL LINE, TRANSFECTION PROCEDURE AND REACTOR CONDITIONS ON THE GLYCOSYLATION OF RECOMBINANT HUMAN ANTI-RHESUS D IGG1

S. NAHRGANG, E. KRAGTEN, M. DE JESUS[2], M. BOURGEOIS[2], S. DÉJARDIN[3], U. VON STOCKAR, I.W. MARISON
EPFL, DC-LGCB, DC-LBTC[2], CH-1015 Lausanne, Switzerland
ZLB, Bern, Switzerland[3]

ABSTRACT

An anti-Rhesus D antibody was produced in various cell-lines and cell-types which were either stable or transient transfected and grown under varying culture conditions. Glycosylation analysis of the IgG1 was carried out mainly by HPAEC methods. The antibody was N-glycosylated only at the conserved glycosylation site at Asn-297 carrying complex type N-glycans as usually present on IgG. The core structure was entirely fucosylated and did not carry intersecting GlcNAc for all antibodies analyzed. Production in SP2/0, CHO and HEK-293 cells resulted in significantly different oligosaccharide patterns. The influence of the cultivation conditions on the glycosylation was usually less than the influence of cell-type. The glycosylation pattern of IgG produced in transient transfected CHO cells was different than in stable transfected CHO cells but the differences observed might be also due to different culture conditions. Further factors influencing the glycosylation were medium composition and the type of bioreactor used for the cultivation.

INTRODUCTION

IgG1 comprises one conserved N-glycosylation site in the hinge region at Asn-297. N-glycans present at this site usually are of the complex type and may differ with regard to fucosylation, galactosylation and sialylation. Glycosylation is a not a template directed process but is influenced by cell type, cell line, reactor type and other production parameters [1-3].

MATERIALS AND METHODS

PURIFICATION. Purification was performed on recombinant protein A using 0.1 M citrate buffer, pH 3.0 for elution. Some samples were further purified by cation exchange chromatography on SP-Sepharose which did not affect the glycosylation profile obtained.

HPAEC. The oligosaccharides were released from the protein by PNGase-F addition in 50 mM PBS, pH 7.5 and 50 mM Tris/HCl pH 8.0. Protein was precipitated by addition

A. Bernard et al. (eds.), Animal Cell Technology: Products from Cells, Cells as Products, 259–261.

of MeOH 80%, followed by a sample cleanup ot the supernatant on a SepPak C18 column. After evaporation of the solvent the glycans were resuspended in water. Carbohydrate analysis was carried out on a High Performance Anion Exchange Chromatography (HPAEC) system equipped with a pulsed amperometric detector (PAD) using a gradient starting with 150 mM NaOH and 7.5 mM NaOAc to 150 mM NaOH and 150 mM NaOAc.

EXOGLYCOSIDASE TREATMENT. Exoglycosidases (α-galactosidase, β-galactosidase, fucosidase and sialydase) obtained from Oxford Glcosciences (Oxford, UK) were used according to the protocol provided by the manufacturer to assign HPAEC signals to carbohydrate structures by observed peak shifts upon enzymatic digestion.

RESULTS

Independently of cell type and cultivation conditions all N-glycans were entirely fucosylated and carried no intersecting GlcNAc. The overall structure is displayed in figure 1. Terminal oligosaccharides residues were influenced by various parameters including the cell type, culture conditions and possibly the transfection method.

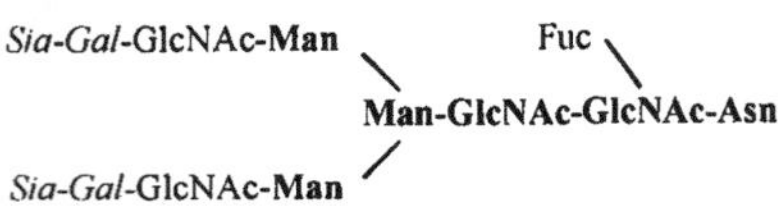

Figure 1. Carbohydrate structures found on the anti-Rhesus D antibody. In bold the pentasaccharide core structure. Monosaccharide residues present on all samples are written in regular, those in italics represent the variable part of the N-glycan structure and were not present on all samples.

Transfection Method. CHO cells were transfected either stable or transient. The pattern obtained varies significantly but stable CHO cells were cultivated in serum-free in a STR whereas the medium for transient transfected CHO cells contained 10 % IgG-free serum. Different stable transfections of CHO or SP2/0 cells and transient transfections of HEK-293 cells normally led to similar glycosylation patterns per cell type.

Cell type. Cultivation of CHO was carried out in serum-free medium whereas SP2/0 required serum for growth. The influence of serum on the glycosylation still has to be determined. SP2/0 contained minor amounts of sialylated and possibly α-galactose structures not present in the other samples.

Parameter	Transfection Method			Cell Type		
cell type or transfection	CHO MDJ 8S	CHO MDJ-1	CHO transient	CHO MDJ 8S	HEK-293	SP2/0-3
0-galactose	63	62	22	63	24	24
1-galactose	33	34	40	33	53	56
2-galactose	4	4	38	4	23	20

Table 1. Comparison of transfection method and cell type. CHO MDJ 8S and CHO MDJ-1 are two different stable clones. All productions mentioned in the cell type section were carried out in a STR, only the CHO cells are cultivated serum-free, HEK-293 cells are transient transfected. All numbers are relative amounts in percent.

Reactor type. CHO cells have been cultivated as adherent cells in roller bottles and in suspension in an STR. No major differences in glycosylation were observed. SP2/0 galactosylated the IgG to a larger extent if cultivated in a STR compared to a hollow fibre reactor.

Culture medium. Galactosylation could be slightly improved by addition of glucose to the culture medium. The use of different media for the production also resulted in minor variations in the ratio of the galactosylated structures.

Parameter	Reactor Type				Culture Medium	
reactor type or medium composition	SP2/0, Hollow Fibre Reactor	SP2/0 in STR	CHO, Roller Bottles	CHO in STR	CHO MDJ8S, standard	CHO MDJ 8S, Glc-addition
0-galactose	65	24	63	63	63	53
1-galactose	31	56	33	32	33	41
2-galactose	4	20	4	5	4	6

Table 2. Comparison of glycosylation in different reactor types. SP2/0 has been cultivated in serum containing medium. CHO cells in serum free medium. Glucose addition to CHO medium improved slightly the galactosylation. All numbers are relative amounts in percent.

DISCUSSION

As already shown before, the glycosylation of a protein is mainly influenced by the type of host cell employed and to a lesser extent by the culture conditions. For the CHO cells studied here the glycosylation pattern only changed little except when transient and stable transfected cells were compared. This small variation might be due to the overall poor glycosylation of the protein in CHO cells thus leaving little possibilities for variation. The small amount of galactosylation might influence the therapeutic performance of the protein since antibody dependent cellular cytotoxicity (ADCC) mediated by K-cells is correlates inversely to the level of galactosylation [4]. The influence of certain parameters still has to be clarified further. Transient transfections of CHO cells have only been performed in serum containing medium but the stable transfections shown here grow serum independent. The serum may also play a role in the glycosylation of proteins [5] though it was found that the antibody was less galactosylated in serum containing medium. This glycosylation did not vary per transfection contrary to the observations made in [6].

REFERENCES

1. Lund, J., *et al.*, *Control of IgG/Fc glycosylation: a comparison of oligosaccharides from chimeric human/mouse and mouse subclass immunoglobulin Gs.* Mol Immunol, 1993. **30**(8): p. 741-8.
2. Lifely, M.R., *et al.*, *Glycosylation and biological activity of CAMPATH-1H expressed in different cell lines and grown under different culture conditions.* Glycobiology, 1995. **5**(8): p. 813-22.
3. Kumpel, B.M., *et al.*, *Galactosylation of human IgG monoclonal anti-D produced by EBV-transformed B-lymphoblastoid cell lines is dependent on culture method and affects Fc receptor-mediated functional activity.* Hum Antibodies Hybridomas, 1994. **5**(3-4): p. 143-51.
4. Hadley, A.G., *et al.*, *The glycosylation of red cell autoantibodies affects their functional activity in vitro.* Br J Haematol, 1995. **91**(3): p. 587-94.
5. Patel, T.P., *et al.*, *Different culture methods lead to differences in glycosylation of a murine IgG monoclonal antibody.* Biochem J, 1992. **285**(Pt 3): p. 839-45.
6. Cant, D., *et al.*, *Glycosylation and functional activity of anti-D secreted by two human lymphoblastoid cell lines.* Cytotechnology, 1994. **15**(1-3): p. 223-8.

METABOLIC ENGINEERING OF RECOMBINANT PLASMINOGEN SIALYLATION IN ANIMAL CELLS BY TETRACYCLINE-REGULATED EXPRESSION OF MURINE CMP-NeuAc SYNTHETASE AND HAMSTER CMP-NeuAc TRANSPORTER GENES.

K. N. BAKER[1], I. D. JOHNSON[2], G. ROBERTS[2], A. COOK[2], A. BAINES[1], R. GERARDY- SCHAHN[3], AND D. C. JAMES[1]

[1]Research School of Biosciences, University of Kent, Canterbury, Kent, CT2 7NJ, U.K. [2]British Biotech Plc, Watlington Rd, Oxford, OX4 5LY, Oxon, U.K. [3]Institute of Medical Microbiology, University of Hannover, Carl-Neuberg-Strasse 1, 30625 Hannover, Germany.

Introduction

It has been well documented that the degree of sialylation can affect the pharmacokinetics of a potential therapeutic product[1]. To date, the majority of efforts to augment sialylation have concentrated on expression of recombinant sialyltransferases in CHO and BHK cell lines, demonstrating variable results[2-4]. As there is now strong evidence to suggest that the availability of nucleotide-sugar substrate (CMP-NeuAc) in the Golgi lumen is an important factor limiting the rate of *N*-glycan sialylation[5-6], we have targeted this metabolic control of sialylation. Our aim is to increase the rate of nucleotide sugar synthesis and transport into the Golgi lumen by co-expression of the recently cloned CMP-NeuAc synthetase[7] and CMP-NeuAc transporter[8] genes. Although high levels of gene expression are desirable to increase sialylation, these proteins may be lethal to the cell if over-expressed. To overcome this problem, a CHO cell line expressing a recombinant human plasminogen variant[9] with well characterised N- and O-linked glycosylation sites[10-13], was transfected with 5' FLAG-tagged murine CMP-NeuAc synthetase or 3' HA-tagged CMP-NeuAc transporter genes. Stable cell lines were produced with these genes under the control of a tetracycline-repressed promotor using the Tet-Off expression system (Clontech). The tetracycline (Tet)-regulated expression of a glycosyltransferase has previously been demonstrated as an effective means of altering the glycosylation of an engineered humanised IgG molecule[14]. Mutants of both genes have also been transfected into the tetracycline-regulated CHO cell line expressing the recombinant plasminogen variant to provide "inactive" negative controls. Thus, the level of expression of either or both genes can be varied to assess the degree of recombinant plasminogen sialylation in relation to the level of synthetase and/or transporter gene expression. Furthermore, expression can be tailored to control sialylation while minimising toxicity of the gene products to the cell.

Materials and Methods

CMP-NeuAc Synthetase and Transporter Vector Construction. The FLAG-tagged CMP-NeuAc synthetase gene and its deletion mutant were released from the vectors pAM16F and pBg7 and these genes cloned into the pTRE vector (Clontech). The vectors were used to transform JM109 cells and positive clones (pKB1 and its mutant pKB2) were screened and sequenced. The HA-tagged CMP-NeuAc transporter genes from the pME10HA vector and its mutant pLec2HA were amplified by PCR, cloned into pTRE and used to transform JM109 cells. Positive clones were screened and sequenced (pKB3 and its mutant pKB4).

Generation of Tetracycline-sensitive Stable CHO Cell Lines Expressing Plasminogen. A stable CHO cell line expressing a recombinant human plasminogen variant was made tetracycline-dependent by stable transfection with Tet-Off DNA. Cells were grown under G418 selection and positive clones screened by transient infection with a luciferase-containing gene (pTRE-Luc) using Superfect (QIAgen) as the transfection agent. Transiently-transfected clones were then grown in the presence or absence of Dodoxycycline (Dox), a tetracycline derivative, at 1 μg mL^{-1} and their resultant luciferase activity recorded. Clones which demonstrated significant differences in luciferase activity $\pm$ Dox were screened further for their plasminogen concentrations by ELISA.

A. Bernard et al. (eds.), Animal Cell Technology: Products from Cells, Cells as Products, 263–265.

Generation of Double Stable Cell Lines. The stable cell line which demonstrated high sensitivity ±Dox and maintained high plasminogen concentrations by ELISA underwent stable transfection with the CMP-NeuAc synthetase or transporter vector constructs, together with the hygromycin–resistant selection vector pTK-Hyg. Each of the four cell lines were grown in the presence of hygromycin and G418, to allow selection of clones expressing both Tet-Off and the gene of interest, and 1μg mL^{-1} Dox to inhibit toxicity of the expressed product to the cell. 72 clones from each transfection were selected, of which 16-30 were grown in the presence or absence of Dox (1μg mL^{-1}). These were screened using western analysis by probing cell lysates obtained by detergent extraction for their FLAG- or HA-tag using relevant antibodies. Clones with high protein expression were then re-screened by ELISA for continuing plasminogen expression.

Results and Discussion

Vector Construction and Generation of Tet-Off Stable Cell Lines

Vectors encoding the genes for CMP-NeuAc synthetase (pKB1) and its mutant (pKB2), CMP-NeuAc transporter (pKB3) and its mutant (pKB4), were constructed as described above and used to generate stable cell lines under the control of Tet-Off repressor. Twenty-four clones positive for Tet-Off under G418 selection were transiently transfected with pTRE-Luciferase. These clones were screened, using a luminescent LucLite® luciferase reporter gene assay (Packard), for high induction of luciferase activity and low background following removal of Dox (a tetracycline derivative) from the culture medium. Six of these clones demonstrated high induction of luciferase activity with low background activity (8-10:1; Figure 1). These clones were then screened for their plasminogen expression using ELISA (Figure 2). The highest plasminogen-producing clone with high luciferase activity (clone 1D1) was chosen for future stable transfections.

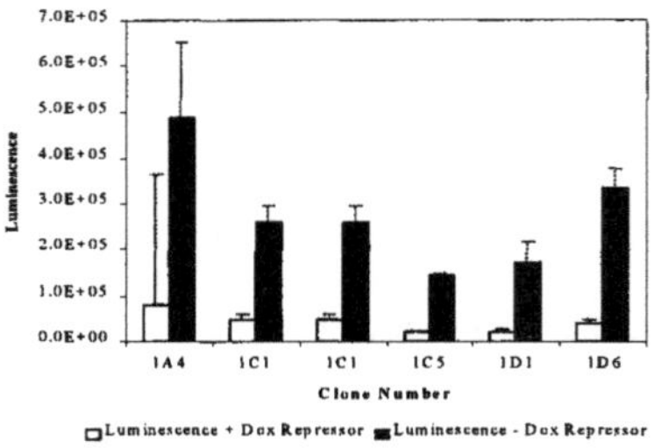

Figure 1: Luminescence of recombinant plasminogen clones following transient transfection with pTRE-Luc ± Repressor (Dox, 1 μg mL^{-1}).

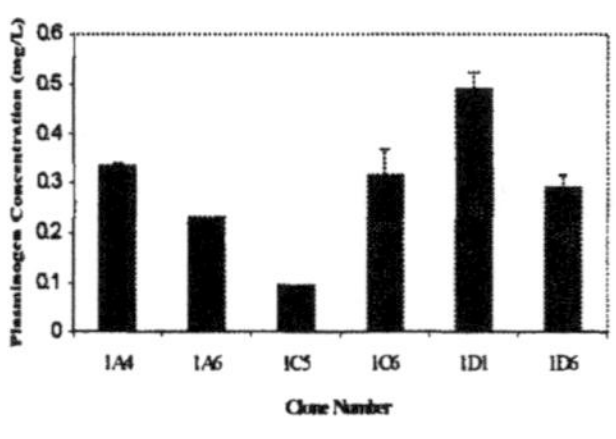

Figure 2: Recombinant plasminogen concentration (mg L^{-1}) of the best performing tetracycline-regulatable clones determined by ELISA.

Generation of Double Stable Cell-Lines

The stable Tet-Off cell line 1D1 underwent stable transfection with pKB1-4 (40 μg) and a hygromycin-resistance selection marker pTK-Hyg (2μg). To prevent problems associated with overexpression of the potentially lethal transporter gene, the repressor Dox (1 μg mL^{-1}) was also included in the cell culture medium. 72 individual clones resistant to both hygromycin and G418 (thus containing the gene of interest and Tet-Off) were selected. Of these clones, 16-24 were grown in the presence or absence of Dox (1 μg mL^{-1}) for 72 hours to allow transient expression of the genes of interest. The levels of CMP-NeuAc synthetase (pKB1 and pKB2) proteins were detected by probing cell lysates with an anti-FLAG antibody using slot blot analysis. The levels of CMP-NeuAc transporter (pKB3 and pKB4) proteins were detected by probing cell lysates with an anti-HA antibody using slot blot analysis. Clones demonstrating high protein expression were re-screened for their plasminogen concentration by ELISA and one clone from each chosen for use in future experiments.

Future experiments

We plan to express varying amounts of the synthetase and transporter proteins by varying the Dox concentration and purify recombinant plasminogen from each fermentation. The recombinant plasminogen will then be analysed to investigate how changes in the CMP-NeuAc synthetase or CMP-NeuAc transporter expression levels affect plasminogen sialylation (and glycosylation). Then, any changes in the total sialylation (and other glycosylation) or changes at a site-specific level will be determined, and their effect on the biological activity of recombinant plasminogen examined by biological assay. Construction of a dual vector (pKB6) in which the expression of both genes will be comparable, using a bi-directional Tet-Off vector (Clontech) is underway.

Conclusions

We have constructed a series of CHO cell lines expressing recombinant human plasminogen, which have been engineered to allow manipulation of plasminogen sialylation and/or glycosylation via expression of either CMP-NeuAc synthetase or CMP-NeuAc transporter genes. In addition, these genes are under the control of a tetracycline-repressor promoter, allowing manipulation of the level of gene expression and thus the downstream sialylation of the product.

References

1. Marzowski, J *et al*; (1995); "Fluorophore-labeled carbohydrate analysis of immunoglobulin fusion proteins – Correlation of oligosaccharide content with *in vivo* clearance profile" ; *Biotech. Bioeng.* **46(5)**:399-407.
2. Minch, S. L *et al* .; 1995; "Tissue plasminogen activator co-expressed in Chinese hamster ovary cells with α(2,6)-sialyltransferase contains NeuAcα(2,6)Galβ(1,4)Glc-N-AcR linkages"; *Biotech. Prog.* **11**: 348-351.
3. Grabenhorst, E. *et al.*; "Construction of stable BHK-21 cells co-expressing human secretory glycoproteins and human Gal(β1-4)GlcNAc-R α2,6-sialyltransferase: α2,6-linked NeuAc is preferentially attached to the Gal(β1-4)GlcNAc(β1-2)Man(α1-3)-branch of diantennary oligosaccharides from secreted recombinant β-trace protein"; *Eur. J. Biochem.* **232**: 718-725.
4. Monaco, L *et al* (1996); "Genetic engineering of α2,6-sialyltransferase in recombinant CHO cells and its effect on the sialylation of recombinant interferon-γ"; *Cytotechnology* **22**: 197-203.
5. Pels Rijcken, W.R. *et al* (1995); "The effect of increasing nucleotide-sugar concentrations on the incorporation of sugars into the glycoconjugates in rat hepatocytes"; *Biochem. J.* **305**: 865-870.
6. Hooker, A. D *et al*, (1999, *in press*); "Constraints on the transport and glycosylation of recombinant IFN-γ in Chinese hamster ovary and insect cells". *Biotech. Bioeng.*
7. Münster, A. K *et al*, (1998); "Mammalian cytidine 5'-monophosphate N-acetylneuraminic acid synthetase: A nuclear protein with evolutionarily conserved structural motifs"; *Proc. Natl. Acad. Sci. U.S.A.* **95(16)**:9140-9145
8. Eckhardt, M. *et al*, (1996); "Expression cloning of the Golgi CMP-sialic acid transporter"; *Proc. Natl. Acad. Sci, U.S.A.*; **93**:7572-7576.
9. Dawson, K.M *et al*, (1994); "Plasminogen mutants activated by thrombin: Potential thrombus-selective thrombolytic agents"; *J. Biol. Chem.* **269(23)**: 15989-15992.
10. Hayes, M.L. *et al*, (1979);"Carbohydrate of the human plasminogen variants I. Carbohydrate composition, glycopeptide isolation and characterization"; *J. Biol. Chem.* **254(18)**: 8768-8771.
11. Hayes, M. L *et al*, (1979); "Carbohydrate of the human plasminogen variants II. Structure of the asparagine-linked oligosaccharide unit"; *J. Biol. Chem.* **254(18)**: 8772-8776.
12. Hayes, M. L *et al*, (1979); "Carbohydrate of the human plasminogen variants III. Structure of the O-glycosidically-linked oligosaccharide unit"; *J. Biol. Chem.* **254(18)**: 8777-8780.
13. Pirie-Shepherd, S. R *et al*, (1997); "Evidence for a novel O-linked sialylated trisaccharide on Ser-248 of human plasminogen 2"; J. Biol. Chem. **272(11)**:7408-7411.
14. Umaña, P *et al.*; (1999); "Engineered glycoforms of an anti-neuroblastoma IgG1 with optimized antibody-dependent cellular cytotoxic activity"; *Nature Biotech.* **17**: 176-180.

REACTOR DESIGN AND OPERATION

Chapter IV

AIR-TRAP-TECHNOLOGY: BUBBLE-FREE GAS TRANSFER IN LARGE SCALE

Membrane-Free Gas Transfer Based on Surface Aeration Under the Liquids Level

M. B. DAESCHER[1], B. SONNLEITNER[1], A. GEORG[2]
[1] *Universitiy of Applied Sciences (ZHW), Winterthur, Switzerland*
[2] *Fluitec AG, Winterthur, Switzerland*

The Air-Trap-Technology realises a completely bubble-free aeration without the need of membranes. It is based on the principle of surface aeration under the liquids level. So called Air-Traps inhibit the free rising of gas and allow a total exploitation of the gas provided. Metabolic gas can be stripped out in the same way. k_La values of 15 h^{-1} were achieved. This novel device is tested in industry and capable for scale-up.

1. Introduction

There is an increasing interest in the production of important biologicals, such as vaccines, hormones and therapeutics, by cell cultures. To meet the demands, more efficient technology in quantity, purity and safety for these products is required, preferably at reduced cost.
Scale-up of the traditional batch-culture has in practice several technical and economic limitations resulting from the low product titre and poor volumetric productivity. Because of that, high cell density cultures are propagated for an intensified process.
Meeting the oxygen demand in large-scale or in high-cell-density cultures becomes a serious problem. In particular, high energy input and its local intensity may be too high for the integrity of fragile cells. The growth-rate can be negatively influenced, already at relative low turbulence.
The Oxygen Transfer Rate (OTR) is defined as:

$$\mathrm{OTR} = k_L a\,(c^*_L - c_L) \qquad (1)$$

Thereby is c^*_L the maximum possible oxygen concentration in the liquid and c_L the current concentration in the liquid. The term k_La is a combination of two factors: the mass transfer coefficient k_L and the specific interfacial area a.

A. Bernard et al. (eds.), Animal Cell Technology: Products from Cells, Cells as Products, 269–275.

2. Comparison of Standard Methods for Aeration of Cell Cultures

2.1. SURFACE AERATION

Aeration at small scale, for example in T-flasks or culture dishes, is normally no problem. With an increasing working volume the surface-area a decreases and the diffusion of oxygen becomes a limiting factor to the space-time yield. Stirring becomes necessary to increase k_L. Unfortuntaly, energy dissipation by stirring is mainly located close to the stirrer, causing high local shear forces. Working at increased pressure requires very expensive equipment. It also leads to higher risks and problems - for example with sealings around the stirrers shaft.

2.2. AERATION BY SPARGING

Nowadays, a standard cell-culture-bioreactor includes a sparger system. The formation of bubbles increases the specific interfacial area a, and the difference of the velocities of the two phases k_L. This method seems to be a proper solution for all the problems mentioned above, but it is generating several new, serious disadvantages.

Several studies have indicated that sparging cell damage is mainly a result of cell interaction with bursting bubbles at the air-liquid interface. The high-speed liquid motion of a bursting bubble generates intense hydrodynamic stresses and causes severe cell damage in the liquid layer surrounding the bursting bubble.

The discussion about the pros and cons of aeration with small or large bubbles is still very controversial. Small bubbles possess a larger interfacial area and have a longer residence time in the liquid, but the increased cell - bubble contact results in a higher shear stress. Large bubbles, on the other hand, have a lower efficiency in gas transfer and a higher gas flow is required. Investigations proved, that the specific cell-death-rate increases almost linearly with the gas flow rate.

Viscosity increasing agents may reduce the shear-forces, but they also clearly restrict mass transport in the liquid phase and cause serious problems in down-stream purification. Cell stabilising agents like calf serum, are very expensive, a serious hazard for infections, and make validation often very problematic. Other disadvantages of aeration by bubbles need to be mentioned: the formation of foam which can make the addition of anti-foam agents necessary, aerosols which can block the exhaust filter, poor exploitation of the air provided due to the very short residence time of the gas and released intracellular compounds, such as DNA, proteases and lipases, from damaged cells

2.3. AERATION BY MEMBRANES

Bubble-free aeration is mostly realised by the use of internal or external membranes such as silicon tubes. However, up to 3'000 meters of silicon tube per cubic meter liquid are necessary to provide sufficient amounts of oxygen. Problems are caused by the space required, integrity and sanitation of the equipment and the often limited long term stability. Another clear disadvantage is the price of the silicon tubes.

3. The Air-Trap-Technology

Still the preferred method is the gentle diffusion over the liquids surface, which avoids all the problems described. By inserting so called gas traps under the liquids level, the Air-Trap-Technology inhibits the free rising of the gas and realises a surface aeration under the liquids surface (*Figure 1*). This device is capable of increasing the ratio of area to volume by a factor of 50.

Figure 1: The Air-Trap-Aeration unit as it is used for the aeration of hybridoma cells in a standard 25 litre bioreactor

The mechanical traps are of the shape of the upper half of a torus. Each trap is equipped with a floating valve. Air is provided to the lowest trap, filling it completely up with gas. The gas is accumulated because the floating valve still closes the only possible outlet for the gas. As soon as the gas film reaches a certain thickness, the floating valve swims on the liquids surface and falls down. It thereby opens the outlet of the chimney. The gas released will not simply rise up to the surface in form of bubbles, it is trapped in the gas phase of the next air traps placed above where the same procedure will take place. At any gas flow rate, the continuous gas film is guided trap by trap up to the surface, where it is released bubble-free into the head-space. Experiments proofed that metabolic gas, such as CO_2, can be stripped out in the same way.

If the gas supply is shut down, then the valves will rise up and close the outlets. The gas is caught and remains in contact to the liquid (*Figure 2.*). It is then possible to transfer the gas completely into the liquid phase, allowing a 100% exploitation.

4. Experiments

Due to the constant interfacial area a, the OTR is only a function of the k_L. A specially developed stirrer generates a strongly defined fluid dynamic in radial and axial direction, which was characterised by the use of tracers and different stirrer types. High surface exchange rates are achieved that allow a significantly improved gas transfer rate. Investigations proofed that the dependence of the OTR on the revolution rate is in almost direct proportion. The achieved k_La of 15 h^{-1} is remarkable high - it is comparable to sparged systems. Mass transfer values were evaluated by the dynamic method using water/air or water/oxygen and in biological systems with strictly aerobic micro-organisms and hybridoma cells leading to comparable results. The influence of the gas flow rate was also investigated.

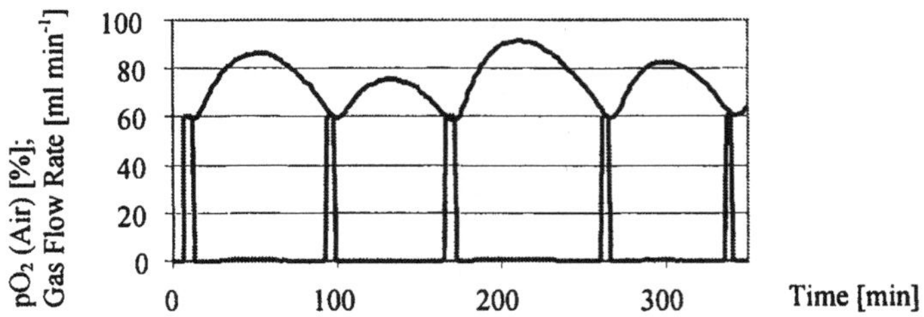

Figure 2: In a cultivation of hybridoma cells, a minimum pO_2 of 60% needed to be held (upper curve). The fresh added gas (lower curve) remained under the traps until it was transferred completely. Due to the long contact time, the amount of gas required could be reduced to less than 5 % compared to spargers. Using a sparger, bubbles would have to be released almost continuously, causing high shear stress.

The gentle aeration improves the chance of a culture surviving critical conditions. In an experiment, we split hybridoma cells in two identical bioreactors, but one was equipped with a sparger system, the other one with an Air-Trap-System. During the fed-batch cultivation, we diluted the culture too much with fresh media. In both reactors, the viability decreased significantly. In the sparged one, the cells even died after 200 hours, probably because of the additional stress - while in the Air-Trap-Reactor 250% more antibodies per litre were produced and a cultivation time of 600 hours was reached. Such critical conditions are especially likely to happen in research and development projects.

5. Applications

In contrast to membranes, the Air-Trap-Technology has a stable long term condition. The device is not only suitable for continuous cultures, it also suitable for fed-batch cultivations. In contrast to most of the other systems, performance in large scale even becomes better due to a higher mixing efficiency and a better use of the space available in the reactor. The number of traps is only limited by the height of the reactor. The Air-Trap-unit can be easily inserted into almost any kind of reactor, usually without any alterations necessary (*Figure 3*). In the same simple way it is removed. It is available in the form of a basket or fixed to the cover lid of the reactor (for glass bottom reactors). The aeration unit can be in-situ autoclaved and is easily maintained. Cleaning is realised by heating up with alkaline solution or surfactants, replacing all the gas by liquid. A special design for large scale allows CIP-cleaning.

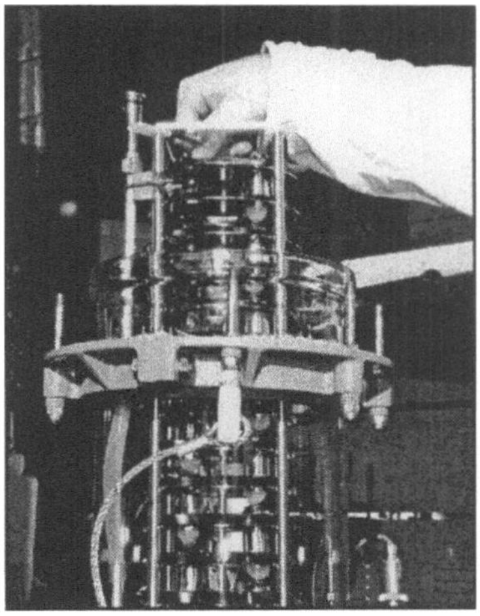

Figure 3: The Air-Trap-unit is easily inserted into almost any kind of reactor, usually without any alterations necessary.

Finally, there is the possibility of a continuous, emulsion-free in-situ product removal. For this the gaseous phase needs to be substituted by a lighter, inmiscible liquid (organic) phase which climbs continuously up, from the bottom to the top of the reactor. For the use of Perfluorocarbons, the unit is inserted in upside down position.

6. Conclusions

Experiments proofed that the Air-Trap-Technology may not only solve problems with aeration of shear force sensitive cells, it also removes metabolic gas and avoids the generation of foam and aerosols completely. It works without membranes and stable over long term. It saves costs for gas and is suitable for scale-up.

Discussion (Däscher)

Gray: I was wondering whether you had validated the cleaning procedure for your device?

Däscher: Yes, if you heat it up with sodium hydroxide, for example, so the gas will be replaced by liquid and thereby you do not have to remove it from the reactor.

Carrondo: I have a comment and some advice. The comment is that if the problems related to surface aeration are half as bad as you mentioned, then this room would be far emptier as industry would not have been able to afford to send people to meetings like this one - the fact is that industry has been able to cope with these problems throughout the years. Nevertheless, there is always room for improvement and my advice is that if this system cannot be simplified to permit cleaning in place and sterilisation in place, then it will not become an industrial application.

The Improvement of Aeration in 8000l Animal Cell Culture Vessels.

T.M. Clayton, I. Jenkins, P. Steward

GlaxoWellcome R&D,
Langley Court, South Eden Park Road, Beckenham, Kent. BR3 3BS UK.

1. Abstract

The mixing and oxygen transfer in 8000 L cell culture vessels was significantly improved to allow high cell density and high productivity fed batch fermentations to be performed without the aid of top pressure or oxygen enriched air. For the last 20 years GlaxoWellcome have successfully used 8000 l working volume cell culture vessels for the culture of CHO, NS0 and Namalwa cells. We can achieve significant densities of cells in our serum and protein free medium and increase them further using fed batch techniques. The magnetic couplings used in our agitation system limited power input to the fermenters and limited the oxygen transfer capability of the system. Consequently oxygen enriched air was required to maintain dissolved oxygen tension in the latter stages of the fed batch culture. We have fully characterised the oxygen transfer capabilities of the vessel and evaluated different agitator and sparger designs and sizes. A modified agitation system has been installed and has increased oxygen transfer sufficiently to eliminate the need for oxygen enriched air in fed batch culture of NSO cells producing high levels of monoclonal antibodies for therapeutic use.

2. Introduction

Oxygen transfer is one of the limiting factors in the culture of animal cells. Fed batch processes are now used to increase product titre and cell densities in our culture systems. Increases in the density of cell cultures used at GlaxoWellcome have resulted in the need for oxygen enriched air in the final stages of cultures. Our existing systems are low power input stirred tank reactors and there is relatively little written about our systems. We collaborated with the University of Birmingham to establish some baseline data (Refs). The original configuration and design of our vessels precluded the use of multiple impellers and high power input. Our aim was to identify the main design features that control oxygen transfer in our vessels and the effect of scale on these features. We then modified the agitation system to confirm our findings.

A. Bernard et al. (eds.), Animal Cell Technology: Products from Cells, Cells as Products, 277–283.

3. Methods

All the oxygen transfer work was carried out using our serum free medium or a solution that mimicked the properties of the medium. Two sizes of vessel were used, 80 L SGi vessel working at 40 L volume and an 8000 L working volume vessel operating at 2000L to 8000L. At the small scale three impellers were used there were a scaled down rushton turbine (9 cm diameter), Sgi impeller (14.5 cm diameter) and lightnin' a310 impeller (16 cm diameter). At the larger scale three impellers were used, a 45 cm diameter Rushton turbine, a 63 cm diameter Rushton turbine and a 76 cm diameter two blade low shear impeller (Mixertech).
Two ring spargers were made using plastic water pipe and drilled with 100 1mm holes. The spargers were 45 and 76 cm in diameter. Gas flow rates of 0 to 100 L/ min were used in the 8000 l vessels and flow rates of up to 10 L/ min were used in the 80 l vessels. The original fermenter operating conditions and configuration are shown below.

- Operate at 2000 to 8000 l working volumes.
- Impeller position fixed by minimum operating volume.
- Speed and size of impeller limited by the magnetic coupling and agitation system design.
- Single 2.5cm internal diameter point sparger.
- Diameter 2 metres.
- Maximum operating height 2.6 metres.
- Agitator diameter 0.45 m.
- Maximum agitation rate in normal process conditions 90 rpm.
- Gas flow rate up to 150 l/min.
- Magnetically coupled agitation system.
- Agitator positioned low in the vessel to allow operation at low volumes.

4. Results and discussion

Impeller type and agitation speed had little effect on oxygen transfer in the 80 L vessel (figure 1) with the greatest impact being seen when the gas flow rate increases (figure 2). The cell culture system could achieve kla values that were similar to the values achieved in the 8000 L vessel. Most of the gas transfer appears to be mediated by a realtively high volumetric gas flow rate. The vessel was running at values of up to 0.25 vvm and superficial gas velocities of 1.33×10^{-3} m/s.
At the 8000 litre scale agitation rate and gas input had a significant effect on oxygen transfer (figure 3). The controlling factors appeared to be gas flow rate and agitator tip speed. Low shear impellers did not perform at all well as oxygen transfer devices. Ring spargers showed no advantage over the point sparger and the role of the impeller in breaking up gas bubbles was confirmed by this work. The agitator was never at the point where it flooded and the gas flow rates used for the cultures were remarkably small compared to the 80 L vessel with a maximum of 0.0125 vvm at 8000L operating volume and 0.05 vvm at 2000 L operating volume with maximum superficial gas velocities of 5.3×10^{-4} m/s.
The operating volume of the 8000 L vessel had little effect on oxygen transfer (figure 4).

Figure 1

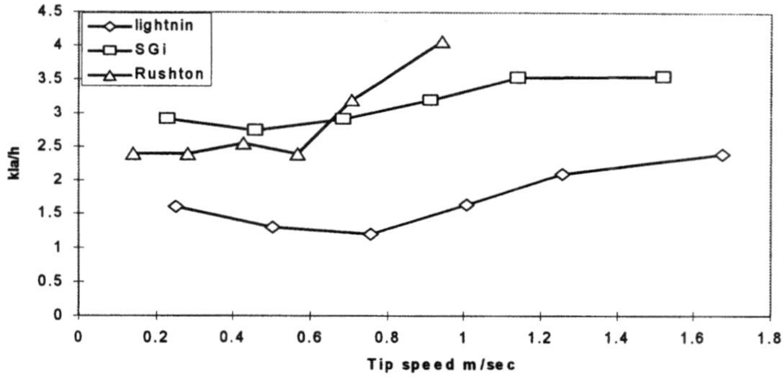

Effect of tip speed on kla at constant gas flow rates in SGi vessel at 300 l/h gas flow rate and 40 l operating volume

Figure 2

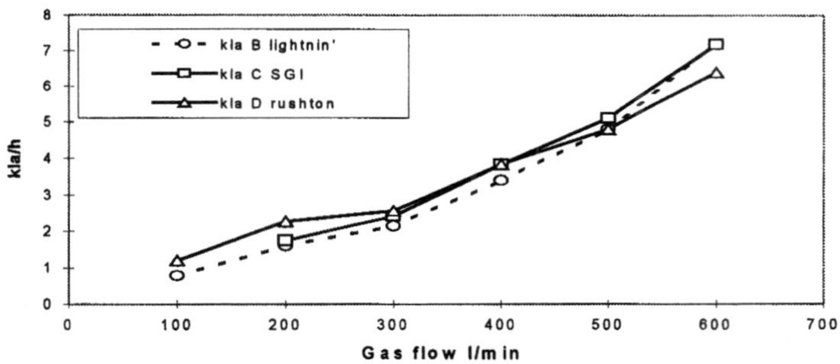

The effect of gas flow rate on oxygen transfer in a 40l working volume with the vessel being agitated at 60rpm

Figure 3

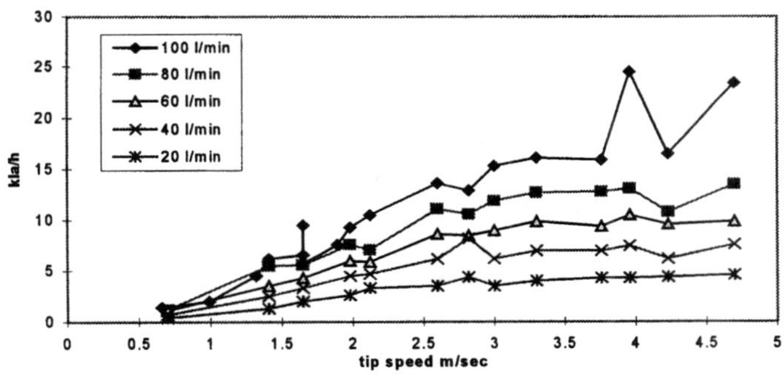

Relationship between kla and impeller tip speed combining the data for the 0.45m and 0.63 m diameter impellers at different gas flow rates

Figure 4

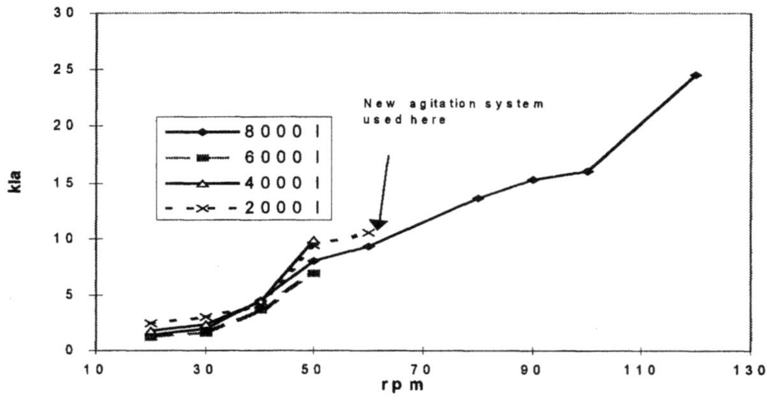

Kla in 8000l vessel operated at 100 l/m in air flow rate, different operating volumes and agitation rates using a 63cm diameter impeller

4.1. MODIFICATIONS

Our results suggested that using a faster agitation speed or a larger Rushton turbine would be the most appropriate method of increasing gas transfer whilst limiting the gas flow into the culture system. The agitation system used in the fermenter was extensively modified to allow larger agitators to be used. A stronger magnetic coupling was made and the position and type of thrust bearing was modified to allow for the greater strength of the magnets used. This arrangement allowed us to improve the flexibility of our system and to run fed batch processes without the use of oxygen enriched air.

4.2. COMPARISON OF THE VESSELS

The kla achieved in cell culture systems for normal cultures is very low (3 to 6/h) and this is sufficient to allow for growth of cells in most situations. Modifications to cell culture vessels will allow the improvement of the kla but appropriate methods must be used. Our results suggest that in small tanks agitator design is largely irrelevant from an oxygen transfer perspective. The main concern should be choosing an impeller design and growth regime that gives good mixing and keeps the cells in suspension without killing them. Scale down models based on geometric design or power input will not accurately reflect scale effects because of the slow tip speeds of the impellers used. The impeller design and tip speed is much more important in larger vessels where the tip speeds are high enough to shear bubbles and increase the interfacial area. Gas flow rates in the larger cultures are much lower that are seen in the small scale cultures because of the change in balance between kla improvements resulting from gas flow and agitation rate. Ion the 8000 L vessels the improvement in kla caused by increased gas flow rate is linear within the ranges investigated and a 5 fold increase in flow rate gives a five fold increase in kla. Therefore, the low power input into the system is still not limiting the oxygen transfer capability by allowing the impeller to become flooded and there are further benefits to be gained from increasing the gas flow rate. Operating the culture vessel at differing volumes gave similar kla values and this indicates that the hydrostatic head and surface aeration have a limited effect on oxygen transfer in these systems. Sparger design was irrelevant to oxygen transfer in our non coalescing system with the larger agitators used in the 8000 L vessel dispersing gas as smaller bubbles than the sparger system could achieve. As would be expected the main priority is to ensure that the gas flow hits the impeller. The mixing times in the 8000 L vessels are slow and this indicates the low average velocity of the culture medium in the vessels. A result of the low circulation rate is that the gas holdup in the vessel is very low and the air bubbles rise directly to the surface where they disengage.

4.3. LIMITATIONS

Our work was limited by the access ports on the production vessels we used for the experimental work and the by the capacity of the existing equipment. The couplings used for the agitation system were replaced to increase the torque transmission into the fermentation system and allowed us to increase the maximum agitation rate of our existing impeller to at least 200 rpm and the new impeller (63 cm diameter) to ~120 rpm.

5. Conclusions

- The requirements of an animal cell culture vessel are that it should be able to provide a kla of 3 to 6/h. This will be achieved by air sparging in small scale vessels and a combination of sparging and agitation in larger vessels.
- Larger vessels use much lower quantities of air than lab scale vessels.
- The main function of the impeller in small scale vessels is mixing.
- Rushton turbines are good for gas dispersion and have been used successfully for 8000L animal cell culture for many years.
- To date our oxygen transfer limitations are a result of caution and mechanical limitations and the next challenge is to see how far the system can be pushed before the cell culture shows signs of damage.

References

Nienow, A.W., Langhenrich, C., Stevenson, N.C., Emery A.N., Clayton, T.M. and Slater, N.K.H.(1996) Homogenisation and oxygen transfer rates in large agitated and sparged animal cell bioreactors: some implications for growth and production, *Cytotechnology* 22,87-94.

Langhenrich, C., Nienow, A.W., Eddleston, T., Stevenson, N.C., Emery A.N., Clayton, T.M. and Slater, N.K.H.(1998) Liquid homogenisation studies in animal cell bioreactors of up to $8m^3$in volume, *Trans IChemE* 76, Part C, 107-116.

Discussion (Clayton)

Zhou: I was surprised to see that the Kla was not much affected by agitation or gas flow rate. This is in contrast with published results. Have you any explanation for your results of why Kla is independent of agitation rate?

Clayton: In our system the agitators are not turning enough to shear the gas bubbles into smaller bubbles. There is no evidence which we have seen that bubble size distribution would change. In some collaborative work with the University of Birmingham we looked at the effects of scaling the impeller on gas transfer mixing. When we tried to scale in terms of power input, we saw no effect on bubble size at all.

Zhou: Have you looked at the gas flow rate? What would be the saturation level at which the Kla would be independent of gas flow rate?

Clayton: We have never gone that far. We are working within the parameters which we think that we need to run our existing systems. I think that we could go far higher with gas flow rate.

Spier: There is, presumably, a good reason why you just stuck with the open tube sparger of a given diameter. Is there any reason why a sintered glass/metal sparger cannot be used as they actually break up the bubbles before you have the agitator break up the bubbles? Certainly at the small scale you really do not have to agitate at all, once you have this type of sparger in the system, except of course to keep the cells in suspension which is a different exercise.

Clayton: From our point of view we wish to keep things simple, and we already have a system which works, and an open tube is very easy to clean compared to sinters. Also, we do not get blockage during long, high cell density cultures which you may do using sinters.

Spier: I do not think you will get a blockage problem with sinters as the air is cleaning it as it blows out. It is easy enough to take the tube out of the vessel to clean it.

Clayton: It is not that easy with an 8000L vessel!

A NOVEL CONTINUOUS SUSPENSION CULTURE SYSTEM FOR HEMATOPOIETIC CELLS

S. Schmidt, N. Jelinek, U. Hilbert, S. Thoma, C. Wandrey, M. Biselli
Institute of Biotechnology 2, Forschungszentrum Jülich GmbH
D-52425 Jülich, Germany

ABSTRACT

The novel reactor is capable of extensive cell expansion, up to clinically relevant cell numbers. The system can be inoculated with a single donor derived sample, while full process control at any time of cultivation is possible. A cell density of more than 10^7 c/mL and nearly a 1500fold cell expansion can be realized. The system was verified by successful expanding a hematopoietic cell line, primary human lymphocytes and primary human hematopoietic progenitor cells.

1. INTRODUCTION

The ex vivo cultivation of primary hematopoietic cells is of growing clinical interest. In applications like cancer treatment and gene therapy large amounts of cells are needed. But the number of cells in a donor derived sample are usually very small. Hematopoietic cells do have enormous proliferation capability, but they need very defined conditions to grow in a controlled way without unwanted differentiation.

Therefore a culture system is needed, that is able to handle the very small cell number of the donor sample and the finally high cell output on the one hand. On the other hand, the environment of the cells like the medium components, the temperature, the pH, and the pO2-value has to be kept in a very small range at every time.

Commercially available systems do meet only one of these two requirements. Either they offer the capability of working with small cell numbers (e.g. well-plates, culture bags or spinner vessels), or they provide controlled physical and chemical conditions (like common stirred fermenter).

2. CONTINUOUS CULTURE SYSTEM

We developed a novel culture system trying to meet the requirements of small cell numbers at the beginning of cultivation, a sufficient expansion rate for getting clinical relevant cell numbers and of complete process control at any time of cultivation.

The novel reactor is a continuously operated stirred vessel with retention of the cells. For the cell retention the outlet of the reactor contains a metal filter (pore size < 9mm) and is placed directly above the magnetic stirrbar (Figure 1).

A. Bernard et al. (eds.), Animal Cell Technology: Products from Cells, Cells as Products, 285–287.

Other main features of the reactor are:

- The magnetic stirrbar is especially designed to minimize the shear force.
- Working volume between 40 and 550mL.
- Electrodes for measuring pH, pO_2 and temperature.
- Optional space for other electrodes.
- Surface aeration ($k_l a$ value up to 5.5 1/h).
- Large port for taking very small samples at any time of culture.

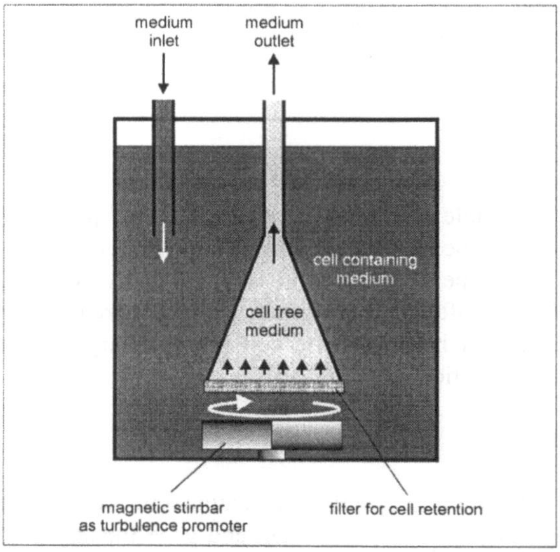

Figure 1. The cell retention by filtration

3. CONCLUSIONS

The results of the different applications (Table 1) show, that the novel suspension culture system is suitable for producing clinical relevant numbers of hematopoietic cells. The cultivation process has been proved to run reliable for over one month. The cell retention by filtration worked very effective and reliable. High retention rates (> 90%) and no blocking of the filter could be observed.

The reactor is capable to cultivate hematopoietic cells up to a density over $1x10^7$ c/mL. Therefore, it is comparable to other high density culture systems, like those with a cell retention by spin filtration or by continuous centrifugation. Figure 2 shows a comparison of the maximal expansion capability of different reactor types calculated for a cell density of 10^5 c/mL in the inoculum at minimal working volume. Due to the much smaller minimal volume, and the comparable high cell density the novel reactor can exceed the expansion capability even of the highly efficient system of continuous centrifugation by 72%.

TABLE 1. Different applications of the novel culture system

Application	Cells	Medium	Result
Human hematopoietic cell line	$CD34^+$ cell line KG-1 [1]	IMDM + 10% FCS	Starting with a cell density of $2.5x10^5$ cells/mL in 40mL, the final cell concentration exceeded $1x10^7$ cells/mL in 550mL.
Primary T-cells	$CD3^+$ cells from peripheral blood	RPMI 1640 + 10% FCS + IL-2	After a 4 day stimulation the cells were expanded successfully in continuous culture to obtain enough cells to inoculate a 20L stirred fermenter (fed-batch).
Hematopoietic progenitor cells	$CD34^+$ cells from umbilical cord blood	CellGro SCGM + IL-3 + IL-6 + SCF + GM-CSF + flt-3	First results show similar growth of the progenitor cells as observed in static small scale cultures over a period of 8 days.

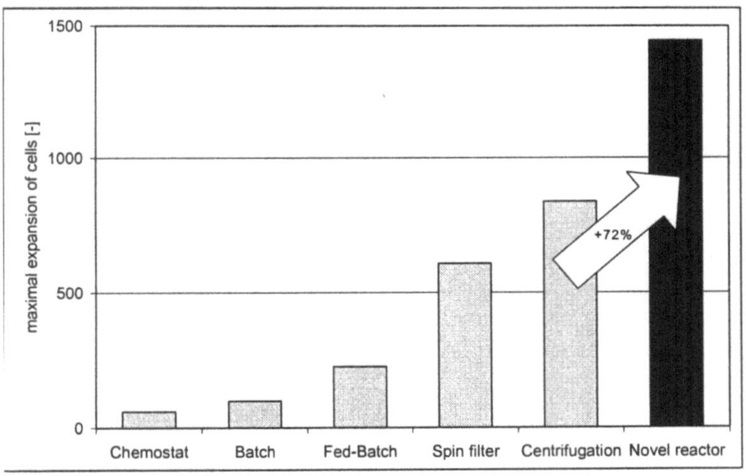

Figure 2. Comparison of the maximal expansion of cells in different types of reactors

4. ACKNOWLEDGEMENTS

We like to thank Cornelia Herfurth and Thomas Noll for their help and MainGen Biotechnologie GmbH, Frankfurt/M., Germany for supporting our work.

5. REFERENCE

[1] Koeffler, H. P. and D. W. Golde (1978) Acute Myelogenous Leukemia: A Human Cell Line Responsive to Colony-Stimulating Activity, *Science* **200**, 1153-1154.

LARGE SCALE PRODUCTION OF HIV-1 gp120 USING RECOMBINANT VACCINIA VIRUS IN PACKED-BED BIOREACTOR

JEANNE KAUFMAN[2], MICHAEL W. CHO[1], MYUNG K. LEE[1], JOSEPH SHILOACH[2*].

[1] *Laboratory of molecular Microbiology, NIAID, NIH, Bethesda, MD, USA*
[2] *Biotechnology Unit LCDB, NIDDK, NIH, Bethesda, MD, USA*
**Corresponding author (phone: 301 496 9719, fax: 301 496 5239)*

The HeLa cell-vaccinia virus system is an attractive method for producing recombinant mammalian proteins with proper post-translation modifications. This system is especially applicable to the production of the HIV-1 envelope glycoprotein, gp120, because more than half of its total mass is carbohydrates. A recombinant vaccinia virus/T7 RNA polymerase expression system to produce large amounts of gp120 was developed. The packed-bed bioreactor configuration was examined for pilot-scale production and was found to be an efficient system for both cell growth and virus infection resulting in high levels of gp120.

1. Introduction

Understanding the structural, biochemical, functional, and immunological properties of HIV-1 gp120 would be important for the design of an effective vaccine against HIV-1. Because such studies require large amounts of the protein, it is necessary to develop a high-expression system that allows easy purification of the protein. HIV-1 gp120 is a large protein, heavily glycosylated with 23 to 24 potential N-linked glycosylation sites, and has some O-linked modifications throughout the length of the protein. Expressing gp120 extracellularily by infecting HeLa cells with the recombinant vaccinia virus/T7 RNA polymerase system was found to be an efficient method to express the properly modified protein. Several options are available for growing anchorage dependent mammalian cells for the production of extracellular proteins. The packed-bed bioreactor approach was found to be suitable for cell growth, expression and protein production.

2. Materials and Methods

HeLa cells, grown in tissue culture flasks (in DMEM with 4.5 g/l glucose, supplemented with 10% fetal bovine serum, 20 mM glutamine, 0.1% pluronic acid and non-essential

A. Bernard et al. (eds.), Animal Cell Technology: Products from Cells, Cells as Products, 289–291.

amino acids), were transferred to a 2.2 L (1.6 L working volume) packed-bed bioreactor equipped with an internal retention device (basket) and a vertical mixing system (Celligen Plus™, New Brunswick Scientific, Edison, NJ). Cells were trapped onto Fibra-cel™ disks (inside the retention assembly) made of polyester non-woven fabric laminated to a polypropylene screen. Dissolved oxygen was kept at 30% saturation and the pH at 7.0 using a four-gas mixing system. Agitation was 80 rpm and the temperature was 37°C. Perfusion was conducted by pumping fresh medium into the bottom of the vessel and removing spent medium from the top. Perfusion flow rate was adjusted to keep the glucose level at 1.0 g/l.

When the cells reached maximum density, the medium was replaced with serum-free medium. The perfusion was stopped and the virus was added at an MOI of 3. During the virus infection phase, more glucose, glutamine and non-essential amino acids were added.

At the end of the virus infection phase the medium was collected and clarified by centrifugation; Na_2HPO_4 was added to a final concentration of 20mM and the solution sat overnight at 4°C. The solution was then centrifuged and the supernatant was mixed slowly for 4 hours with Ni NTA (Qiagen Hilden Germany) at a ratio of 1 ml resin per 100 ml of the protein solution. This was packed into a column, washed with sodium phosphate buffer containing 0.5 M NaCl, and then eluted with the same buffer containing 200mM imidazole.

3. Results

The overall process (cell growth, medium replacement and protein production) is shown in Fig 1. In the growth phase, lasting approximately 110 hours, the cell mass increased from $8x10^8$ to 10^{10} (based on glucose consumption rate of 1 ng/cell/day). The medium replacement phase (replacing the medium with a serum-free one) lasted approximately 40 hours with the perfusion rate being increased to its maximum value of 200 ml/min towards the end of this phase. In the production phase, the perfusion was stopped, the two viruses were added, and the process was continued for additional 100 hours. During this phase, a concentrated glucose solution containing glutamine and nonessential amino acid was added twice. Production of gp120 was detected (7 mg/l) 10 hours after infection with the virus. Sixty hours into the infection, the concentration of the protein increased exponentially to 30 mg/l. At this point, due to high concentration of lactate, the medium (with gp120) was replaced. Afterwards, during the next 15 hours, gp120 concentration increased from 5 mg/l to 16 mg/l when the production process was terminated. After clarification, adsorption and elution, 8.3 mg of pure protein were recovered from one liter of medium, approximately 0.5% of total protein present in the medium.

4. Discussion and conclusion

Extracellular, native and recombinant, proteins can be produced from anchorage

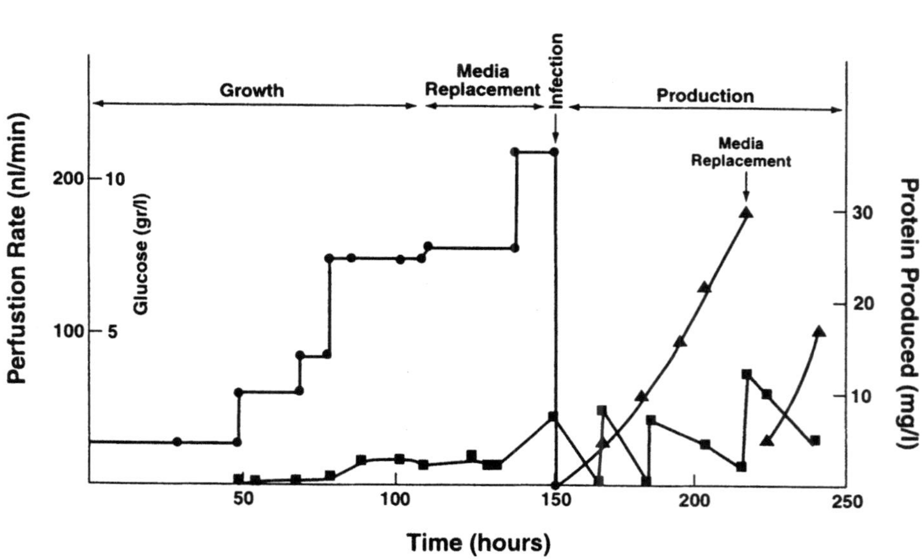

Fig.1: Production process of gp120 using the HeLa / vaccinia system
● perfusion rate, ■ glucose concentration, ▲ protein concentration

dependent mammalian cells in several configurations. Production of secreted proteins from immobilized cells on a packed-bed bioreactor offer the following advantages: obtaining large concentration of cells thanks to continuously perfused medium through the culture; relatively easy medium replacement to accommodate different growth and production conditions; and continuous collection of secreted products (1). This work introduces the ability to infect immobilized cells in a packed bed with recombinant viruses for the expression of protein. The gp120 expression was found to be more efficient when cells grew on a surface rather than in suspension (unpublished results), suggesting the importance of cell-to-cell spread of vaccinia viruses.

Because a culture cannot be perfused with medium during the infection phase, we added to the medium a batch of a concentrated nutrient solution, prolonging the production process and increasing the amount of the protein produced. The production process of gp120 from immobilized HeLa cells on a packed-bed bioreactor by infection with recombinant vaccinia virus was shown to be an efficient process, most likely due to effective cell-to-cell spread of the vaccinia viruses.

References

1. Shiloach, J., Kaufman, J., Trinh, L. and Kemp, C. (1997) Continuous production of the extracellular domain of recombinant human CA++ receptor from HEK 293 cells using novel serum-free medium, in M.J.T. Carrondo, B. Griffiths and J.L.P. Poreira (eds.), *Animal Cell Technology From Vaccines to Genetic Medicine*, Kluwer Academic Publishers, Boston, pp. 535-540.

The "Shear Tester".

A universal tool for characterization of the shear sensitivity of animal cell lines under various process conditions.

A. FLAGMEYER, K. KONSTANTINOV, H.-J. HENZLER*
Bayer Corp., Berkeley, CA 94701
*Bayer AG, 42096 Wuppertal, Germany

Introduction

Before a new type of mammalian cells or medium is introduced to large-scale production, it is important to know how the cells will respond to the new shear conditions. Accurate quantification of the optimum process conditions or parameter ranges at this early stage is highly desirable, resulting in a significant reduction of the large-scale optimization work. Since variables related to shear are of paramount importance at production scale, a simple and efficient method for shear testing at small scale is required. Such a test would be used for:

- new cell line characterizations
- evaluation the protective effect of new medium formulations
- prediction the effect of agitation, temperature, and other fermentation conditions on the shear damage of cells in the production fermentor.

To address this issue, a 1.5 L model of the large-scale production fermentors used at Bayer Corporation was built.

Materials and Methods

A shear sensitive particle system with well established performance [1] has been used to evaluate the shear forces in a large-scale fermentor as well as in the model and to compare them. The particles are mineral-polymer-flocs made of 5 g/L clay powder and 0.01 g/L Praestol™ BC 650, a cationic polymer, as flocculation agent. The density of these flocs is similar to this of animal cells.
The diameter of the flocs exposed to shear stress exponentially approaches a power-specific equilibrium diameter $d_{eq.}$ which is directly proportional to the shear stress in the reactor and can be used to characterize different reactors regarding shear. The relationship between the equilibrium diameter and the volume-specific power input (P/V) or the agitation speed (n) can be described as:

$$d_{eq.} = A \cdot \left(\frac{P}{V}\right)^{-\frac{1}{3}} = A \cdot n^{-1} \tag{1}$$

A. Bernard et al. (eds.), Animal Cell Technology: Products from Cells, Cells as Products, 293–297.

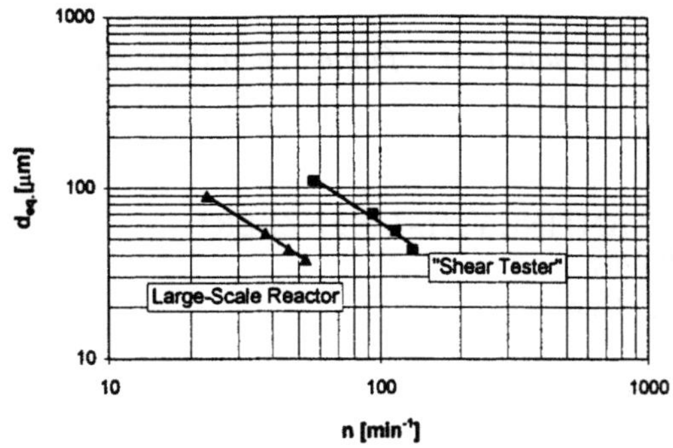

Figure 1: Comparison of the equilibrium diameters measured in a large-scale reactor and the "shear tester".

By plotting the equilibrium diameters versus the impeller rpm, the proportionality constant A can be determined and used to calculate shear equivalent agitation speeds which are necessary to create the same shear forces in the model as in the production fermentors (see Figure 1).

For shear tests, 1.5 L of BHK cell culture were exposed to shear stress for 3 hours under non-sterile conditions. The shear tests were performed at pH 6.8- 7.1, DO around 50 % saturation, 34°C-37°C, and volume-specific power inputs (P/V) of 12, 55, 97 W/m^3 corresponding to 20, 33, 40 rpm in the production fermentors.
Every experiment was carried out three times to prove reliability and reproducibility. The repetitions of the experiments showed that the trends indicated by the shear tests are well reproducible even if the absolute values sometimes were not exactly the same.
At time 0, 30, 60, 120, and 180 minutes, samples were taken and cell counts and viability microscopic examinations using Trypan Blue staining were carried out. At time 0, 60, and 180 minutes the content of LDH, representing the degree of cellular damage, was determined.

Results

Shear tests at three different power inputs were performed to estimate shear damage in the production fermentors and to compare an HSA containing medium with a protein-free (PF) medium enriched with Pluronic F68. Figure 2 summarizes the results of these agitation experiments. The profiles reveal the difference between the HSA and the PF medium.
Using the HSA medium the viability was stable at 12 and 55 W/m^3. At 97 W/m^3 the viability lessened by about 15%. LDH analysis provided similar results: At 12 W/m^3 the LDH concentration was constant, at 55 W/m^3 after 3 hours it was about twice as high as in the beginning of the test, and at 97 W/m^3 it was 6.5 times higher. From the viability results, it can be concluded that the shear forces in the production fermentors at agitation speeds in the 30 rpm range may not be lethal for the BHK cells growing in the HSA medium, but the doubling of the LDH content within a 3 hour test period points to a significant cell damage. This is confirmed by the results of a production fermentor study in which the effect of agitation speed and temperature on harvest quality and BHK cell damage at large scale has been investigated. This study used the same HSA medium. Figure 3 shows one important result of this investigation: the dependence of the specific LDH concentration on temperature and agitation speed. An up to 3.5 times higher

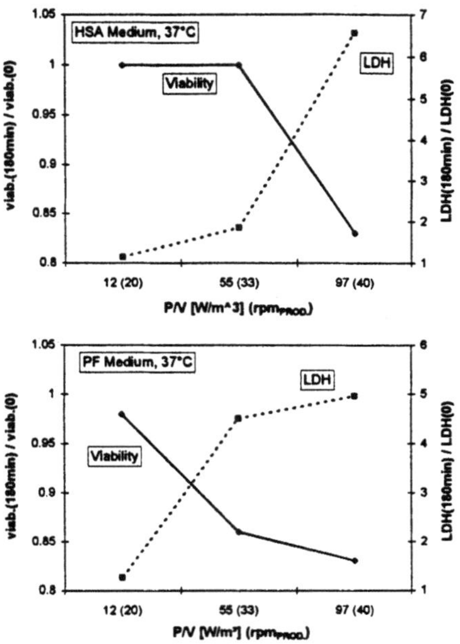

Figure 2: Shear damage at different agitation speeds using HSA medium (upper chart) and PF medium (lower chart).

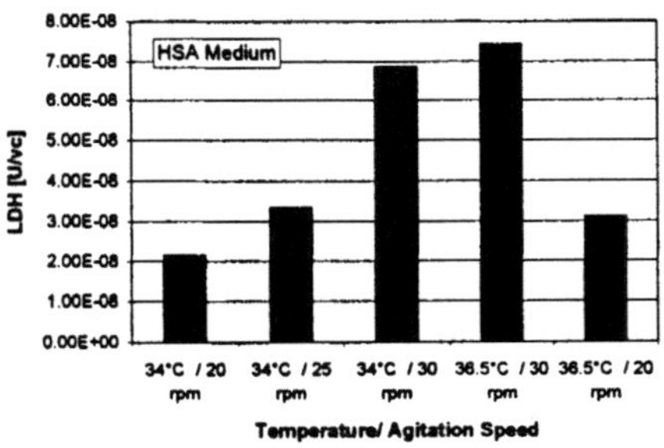

Figure 3: Specific LDH content at different temperatures and agitation speeds.

harvest LDH concentration and a worse harvest filterability at 30 rpm compared to 20 rpm indicated a negative impact of the increased shear.

Using the PF medium the viability decreased and LDH rose sharply at 55 W/m³. Both at 55 W/m³ and 97 W/m³ the viability lessened by about 15% while LDH at the end of the test was up to five times higher than in the beginning.
In our case Pluronic F 68 obviously did not provide the same protection as HSA which led to the conclusion that in case of PF medium, large-scale fermentors should be agitate very gently at <20 rpm.

During the repetitions of some experiments at 34°C cells suffered significantly less shear damage than at 37°C.

Conclusions

The data generated by the shear tests enabled selection of the optimum fermentation conditions for BHK cells in respect to shear. Results from large-scale production runs showed that the shear tester represents adequately the shear damage in the large-scale system. Although the 1.5 L model was not designed to replace the large scale optimization runs under real production conditions, it has proved to be a useful tool for the characterization of new cell lines, media development, and significant reduction of the optimization space defined by the key process variables.

Reference

1. Henzler, H.-J., Biedermann, A.: Modelluntersuchungen zur Partikelbeanspruchung in Reaktoren. *Chem.- Ing.- Tech.* **68** (1996), 1546-1561.

Discussion (Konstantinov)

Rhiel: I have 2 questions on the experiment where you compared optimal versus a non-optimal perfusion run. What was the most significant parameter which had an effect on the optimum run?

Konstantinov: We used many variables, but in this case we used a sub-optimal inoculum.

Rhiel: Did you design an experiment to look at the different parameters, or did you just try them one by one?

Konstantinov: You are asking how we optimise the various variables? We do this one by one, the hard way!

Brown: You talk about a residence time in hours, and a perfusion rate in the range of 8-9 vessel volumes per day. Is that rate also consistent with efficient utilisation of media under your conditions?

Konstantinov: That is a good question because in many cases you just do not have options. It is a relatively large volume of medium which you need to perfuse in these cases. If the residence time which you can afford is significantly higher, then the medium volume will go down. So you have to think about storage space, preparation of medium, etc. The medium does not need to be very concentrated, the glucose concentration may be lower but you do not benefit much from this. Basic issues related to high perfusion rates are manufacture of medium and storage space.

MEMBRANE OXYGENATION OF MAMMALIAN CELL CULTURE FERMENTERS USING DUPONT TEFLON AF-2400 TUBING

Jonathan Blackie, Paul Wu and David Naveh
Bayer Biotechnology
800 Dwight Way, Berkeley Ca 94701

1. INTRODUCTION

For animal cell bioreactors, gas exchange to and from the culture is accomplished by a combination of the following methods: Surface or headspace aeration, gas permeable membrane aeration or direct sparging. Bubble free membrane aeration in various configurations has been used industrially for oxygenation of animal cells. Membrane types used for oxygenation are mainly open pore polypropylene or PTFE membranes and silicone diffusion membranes. In the microporous membranes, a gas-liquid interface is held stationary in the pores of a hydrophobic membrane. For the silicone solution diffusion membranes, the gas dissolves in the hydrophobic membrane material itself and diffuses through (Henzler and Kauling, 1993)

DuPont Teflon AF-2400 (AF means "amorphous fluoropolymer") is a copolymer of terafluoroethelyne and perfluro-2,2-dimethyl-1,3 dioxole. It is reported by its manufacturer that the oxygen permeability is amongst the highest ever reported for a nonporous polymer. By way of comparison, the permeability of silicone tubing, or PDMS, is reported to be about 367×10^{-13} cm^3 (273.15K; 1.013×10^5 Pa) cm / (cm^2 x s x Pa) while that of AF-2400 is reported to be about 742×10^{-13} cm^3 (273.15K; 1.013×10^5 Pa) cm / (cm^2 x s x Pa). This means that for a given dimension and set of conditions, tubing of AF-2400 should deliver about twice as much oxygen per unit time as silicone tubing. In addition, because of the higher tensile strength and lower elongation, it should be possible to use thinner walled tubing, even at elevated pressures. This would result in even higher levels of oxygen delivered.

It is the goal of this project to determine the mass transfer properties of this novel material with respect to oxygen delivery and to perform a relatively short mammalian cell fermentation to determine that maximum cell density that can be supported by a fermenter.

2. MATERIALS AND METHODS

2.1 *Teflon AF-2400 Tubing*

The tubing material used for these studies was fabricated by Random Technologies (San Francisco, Ca). Approximately 100 feet of tubing was produced with an outside diameter of approximately 0.0125" and a wall thickness of 0.006. The tubing was wrapped horizontally around the fermenter tubing cage apparatus in a single layer. The tubing ends were again, heat shrink sealed to the headplate connections to allow high pressure operation.

A. Bernard et al. (eds.), Animal Cell Technology: Products from Cells, Cells as Products, 299–301.

2.2 *Measurement of Oxygen Mass Transfer Coefficient*

For k_La measurements of tubing alone, the reactor was filled with PBS solution and heated to 36 deg C. k_La for Teflon tubing and silicon tubing were measured using the dynamic method. The system performs automatic measurement of k_La by both integral and differential calculations. Measurements were performed at different agitation rates and tubing pressure conditions.

2.3 *Fermentation System*

Experiments were performed in a 20 L stirred tank reactor (Applikon, Holland). Agitation was provided by a 4 vertical/straight blade impeller. The agitation was maintained at 70 rpm. Cell densities were estimated from the on-line determination of oxygen consumption rates. The process control system displays the current acquired values for the fermenter variables, such as temperature, pH, DO, flow rates (gas, harvest, base, etc.). The feed pump was triggered by the control panel of the fermenter scale so that the weight of the fermenter was kept constant. The temperature and DO was kept constant at 36.5^{o}C and 50% air saturation, respectively . Temperature was controlled by recirculating water bath through an internal heat exchanger (tubing cage). The pH was maintained at 6.7 - 6.8 by the use of 0.3 N NaOH fed to the medium line.

3. RESULTS AND DISCUSSION

3.1 *Comparison of Oxygen Transfer Coefficient With Silicon Tubing*

Initial series of measurements were performed to look at measured k_La of both AF-2400 and standard silicon tubing in a true "side-by-side" comparisons. At an agitation rate of 75 rpm and a "typical" pressure drop from 12 $psi_{(in)}$ to 7 $psi_{(out)}$, k_La for silicone tubing was measured to be an average of 0.99 (1/hr) and AF-2400 resulted in 1.68 (1/hr).This represents an improvement of approximately 70% under these conditions. At a higher agitation rate of 150 rpm the improvement was much higher. An average k_La of 1.20 (1/hr) for silicone tubing compared to 2.80 (1/hr) for AF-2400 was recorded during the test. This represents a 135% improvement under these conditions.

3.2 *Effect of Pressure and Agitation of k_La for AF-2400 Tubing*

The next series of trials was performed following the assembly of the tubing sections into the full length (95 feet) single tubing and integration to the fermenter for high pressure operation. The runs were performed as before with data collected at both 75 and 150 rpm while also varying the inlet and back pressure of the tubing. As expected, the k_La increased as pressure increases. Values ranged from 3.6 (1/min) at pressure of $15_{in}/0_{out}$ to 7.2 (1/min) at $60_{in}/30_{out.}$ at 75 rpm. At 150 rpm the results ranged from 5.4 to 9.6 under the same pressure conditions.

3.3 *Perfusion Fermentation Evaluation*

The 20L fermenter with AF-2400 tubing was inoculated on day 0 with 15L of cells. The high inoculum allowed immediate perfusion at a CSPR of 0.35 nl/c/d. The fermenter quickly reached 31 million on day 4. Cell density remained in the 30 - 35 million range at CSPR of 0.3 - 0.4 nl/c/d. On day 15 perfusion was reduced to the 0.25 - 0.35 CSPR. Cell densities in the 35 - 40 million range were maintained for the remaining 5 days. During this time gas pressures were reached a maximum of 52 psi inlet and 45 psi back pressure. On the final day the cell density was measured at 40 million cells and the tubing pressure was increased to the physical limit for the tank/regulator system installed of 52 psi. At this condition the system was not able to maintain a DO concentration of 50% indicating that a limit for oxygen transfer had been reached. Additional increases in tubing pressure would not have resulted in significant oxygen support for, higher cell densities. This run was terminated on day 19. The successful result from this run were obtained using a tubing of approximately 95 feet in length. This was a single layer wrapped horizontally around the cage. This length of material compares to approximately 240 feet of silicone tubing that is typically used with the same reactor. If oxygen limitation was the only consideration, then 100 million cells/ml cell density can be reached in a 20L fermenter with AF-2400 tubing, based on this study.

3.4 *Discussion: Material and Design*

While the performance of the material with respect to gas transfer were positive, there were several problems that would make a tubing configuration in this current form impractical. The cost of the material is considerably higher than current material cost. The rigid nature of the material limits the bend or flex of the tubing. The thin walled material was extremely fragile and developed several cracks and breaks that were difficult to patch.

4. CONCLUSIONS

It is generally accepted that in order to maximize production of a secreted product from mammalian cell fermentation, highest possible cell densities are targeted, provided an optimal environment for the cell can be maintained. Using silicone tube aeration, cell densities in the range of 20 - 30 x 10^6 Vc/ml are achievable. The oxygen limit can be overcome by supplementing with direct sparging of oxygen. This is often done, however, care is taken to minimize foaming and cell damage in the absence of pluronic or anti-foam agents. In this study we have evaluated a prototype tubing that allows bubble free oxygenation at significantly higher levels than can be achieved with the current silicone material. Forty million c/ml was reached in a 20L fermenter, at tubing pressure of 52 psi. This with 35% of its usual tubing length. The improvements in mass transfer coefficient and the high cell density fermentation demonstrate a viable alternative to current standards. Based on the fermentation results, we could postulate that an appropriately constructed aeration cage made from this material could support cell concentrations up to 100 million/ml.

The authors would like to acknowlege the following people for contributions to this project: Amos Gottlieb (Random Technology, San Francisco CA), K. Konstantinov, S Shackleford, J.L. Rule, P. Kramer and H. Qi (Bayer Corp.).

CONFIGURATION OF A SIMPLE DIALYSIS MEMBRANE BIOREACTOR SUITABLE FOR LARGE-SCALE APPLICATION

B. KLEUSER, H.P. KOCHER, K. MEMMERT, M. STENZ,
B. STRAUBHAAR, M. ZURINI
Novartis Pharma Inc., Research-CTA-BMP, Basel, Switzerland

Abstract

We have developed a prototype dialysis reactor of 10L working volume, which combines the advantages of batch and of continuous fermentation processes and allows further scalability. Batch fermentations with and without an integrated dialysis membrane were compared. We could demonstrate considerable improvements regarding cell densities, cell viability and product titer by the integration of a dialysis membrane. Evaluation of the transport of lactate and glucose across the dialysis membrane indicated an efficient removal of waste products and a sufficient supply of nutrients. The system has a high potential for industrial application.

1. Introduction

For large-scale production of biologically important molecules using mammalian cells mainly four different fermentation processes are employed: batch-, fed-batch-, continuous chemostat- and continuous perfusion cultures. Each system has its advantages and disadvantages (Mizrahi A., 1989). The integration of a dialysis membrane into the bioreactor represents a possibility to combine the advantages of a continuous process and of a batch process and has already been described for small scale application (Hagedorn J. et al., 1990, Poertner R. et al., 1997).

2. Methods

2.1 CELL CULTURE

Recombinant CHO cells were adapted to serum-free suspension growth in an in-house developed medium (Plumin) according to the following procedure: Cells were seeded at a density of 20000/cm^2 into T-flasks in MEM alpha- containing 10% FCS. Two days after seeding, 90% of the culture medium was replaced by Plumin. After additional two days, floating cells were harvested and seeded in a mixture of 20% conditioned medium and 80% Plumin. After additional 2-3 days again the floating cells were harvested and seeded into Plumin. The selective harvest of floating cells and seed into serum-free Plumin was performed for additional 2-3 passages until adherent cells disappeared in

A. Bernard et al. (eds.), Animal Cell Technology: Products from Cells, Cells as Products, 303–305.

the culture. This adaptation scheme allows the selection for suspension growth under serum-free conditions. Plumin: Ham's F12 5.31 g/L, MEM alpha- 5.04 g/L, $NaHCO_3$ 1.688 g/L, D-(+)-glucose 6 g/L, human serum albumin 1 g/L, ferric citrate 50 µM, Pluronic F68 1 g/L, beef or recombinant insulin 5 mg/L, lecithin 50 mg/L putrescine 0.1 mg/L hydrocortison 3.6 µg/L, sodium selenite 5 µg/L, L-glutamin 2 mM, ethanolamine 150 µM, pH 7.2-7.4. Dialysis Plumin: Plumin without human serum albumin and without lecithin but with 2g/L Pluronic F68 and 4.4g/L glucose.

2.2 BIOREACTOR DESIGN AND ANALYSIS OF PROCESS PARAMETERS

10L Glass fermenters equipped with a marine-type impeller for agitation and a sintermetal sparger for aeration were used. For temperature control the fermenter was placed into a waterbath (37° C). Dissolved oxygen concentration was maintained between 40-60% air saturation by controlled addition of pure oxygen to the air stream. The pH was adjusted to 7.0-7.4 by manual addition of 2M NaOH. For dialysis fermentation 10m of a dialysis membrane (12-14 kD, Spectra/Por^{R}2, Socochim Lausanne) was placed into the fermenter as described below and the system was sterilized by autoclaving. To evaluate potential improvements by the integration of a dialysis membrane, a recombinant CHO cell which show a low performance in conventional batch cultivation was used as a test organism. In batch as well as in dialysis fermentations cells were seeded at a density of about $1x10^5$/ml in serum-free Plumin into the 10L reactor. Cell counts and viability were determined by trypanblue exclusion using a hemacytometer. Lactate and Glucose was measured by a dual-channel biochemistry analyzer (YSI 2700D Select, IG Instrumenten Gesellschaft AG, Zuerich).

3. Results and conclusions

To stabilize the dialysis membrane a robust dialysis tube, about 10m long, was constructed (in collaboration with Heraeus) by stretching the dialysis membrane onto a steel spring. The tube was then coiled on a cylindrical device consisting of 8 supporting rods fixed to the fermenter cover-plate. Because of the limited surface area (0.184 cm^2) and volume (294 ml) available for diffusion, the tube was connected to a supply and a harvest tank allowing a high flow through rate (20-40 L/day) of the dialysis Plumin in order to maintain high concentration gradients across the dialysis membrane. The integrity of the tube was periodically monitored by analyzing the dialysate for the absence of HSA and/or recombinant protein (SDS-PAGE, Novex, 4-20 % gradient gel). The result is a two-compartment fermentation; a cell compartment (the 10L biorector) and a dialysis compartment (the tube). In dialysis fermentation a 3-4 fold increase in maximum cell density and product titer was obtained as compared to the batch fermentation. Cell viability could be maintained at high levels ($\geq$ 90 %) in dialysis fermentation whereas cell viability decreased dramatically beyond day 4 in batch fermentation (Fig. 1). Analysis of purified product (SDS-PAGE) showed that product quality is comparable to that obtained from the conventional batch and/or the continuous fermentation process (data not shown). At flow rates $\geq$ 20 L/day, the glucose concentration in the dialysis stream could be maintained constantly at about 4 g/L preventing glucose limitation in the reactor. The lactate concentration in the dialysis stream could be kept at low levels of 200-300 mg/L, thus avoiding excessive accumulation of lactate within the reactor (Fig. 2a). The linear correlation of the glucose consumption and the lactate production with the transport rates of glucose and lactate indicates that under the given conditions the surface area of the dialysis membrane is not a rate limiting factor for the transport of low molecular weight substances (Fig. 2b).

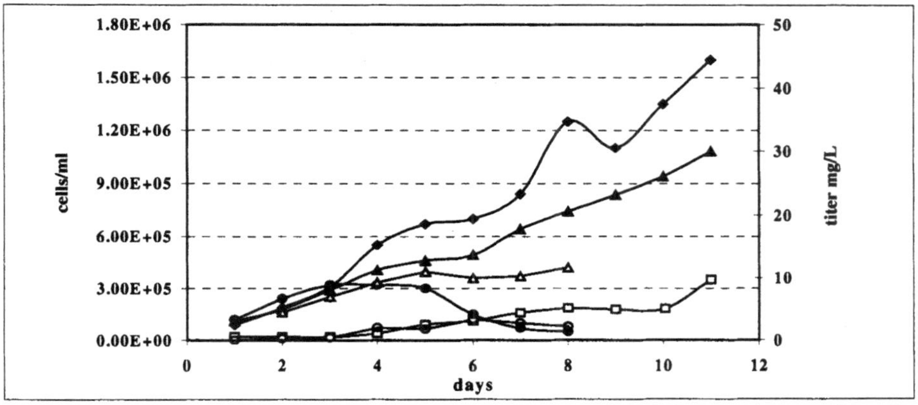

Fig. 1: Comparison of growth, viability and productivity in batch and dialysis fermentation: ♦ viable cells dialysis, □ dead cells dialysis, • viable cells batch, o dead cells batch, ▲ titer dialysis, Δ titer batch

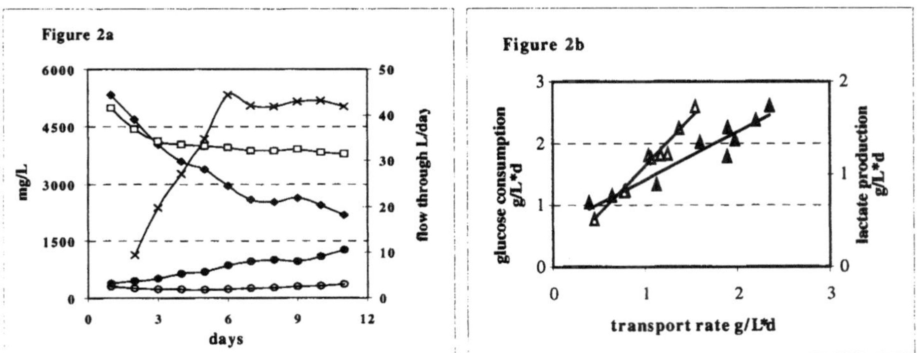

Fig. 2a and 2b: Nutrients and waste products in dialysis fermentation. 2a): Glucose and lactate concentrations in dialysis fermentation: ♦ glucose reactor, □ glucose dialysis stream, • lactate reactor, o lactate dialysis stream, x flow through. 2b): Correlation of transport rates for glucose and lactate with ▲ glucose consumption, Δ lactate production.

References

Mizrahi A. (1989) Techniques and equipment for animal cell cultivation, in R.E. Spier, J.B. Griffiths, J. Stephenne and P.J. Crooy (eds), *Advances in animal cell biology and technology for bioprocesses*, Kluwer Academic Publishers, Dordrecht, pp. 314-321.

Hagedorn J. and Kargi F. (1990) Coiled tube membrane bioreactor for cultivation of hybridoma cells producing monoclonal antibodies, *Enzyme Microbiol Technology* **12,** 824-829

Poertner R., Luedemann I. and Maerki H. (1997) Dialysis cultures with immobilized hybridoma cells for effective production of monoclonal antibodies, *Cytotechnology* **23,** 39-45.

ERYTHROPOIETIN PRODUCTION FROM CHO CELLS GROWN IN A FLUIDIZED-BED BIOREACTOR WITH MACROPOROUS BEADS

M-D. WANG, M. YANG, N. HUZEL AND M.BUTLER
Department of Microbiology, University of Manitoba, Winnipeg, Canada R3T 2N2

Abstract

A stable CHO cell line that expresses human erythropoietin (huEPO) was grown in a Cytopilot fluidized-bed bioreactor for 48 days with a variable perfusion rate. The cells were entrapped in porous microcarriers (400 ml) within the main column of the bioreactor (2 litre). EPO accumulated to a total production of 28,000 kUnits over the culture period. The specific EPO production increased during the later part of the culture probably in response to a higher glucose concentration and an addition of sodium butyrate. The cell density increased to 23 x 10^6 per ml beads. The culture profile in the fluidized-bed bioreactor was compared with three modes of batch culture. The high volumetric yield of EPO attained in the Cytopilot bioreactor indicates its potential as a system for large-scale production.

1. Introduction

The Cytopilot is a fluidized-bed bioreactor designed for the large-scale production of mammalian cells (Reiter et al, 1991). The principle of operation is that the cells are entrapped in macroporous beads which are fluidized by an upward liquid flow within the main column of the bioreactor (2 litre). The porous microcarriers (Cytoline 1) are made of polyethylene and weighted by silica. The pores (10-400 μm) of the microcarriers (20 mm diameter) allow the cells to populate the inner space where the cells are anchored or entrapped. The cells inside the carriers can grow to high densities and are protected from any shear or environment stresses which could cause cell damage. The system can be run in a perfusion mode in which the liquid medium is easily separated from the beads. The continuous supply of nutrients, control of O_2 and pH allows the growth of cells to high densities in the beads. In this report the characteristics of cell growth and productivity of a CHO cell line transfected with the human erythropoietin gene are analysed in the bioreactor and compared to simple stirred and stationary batch cultures.

2. Results

2.1: Cell growth in the Cytopilot culture A cloned stable transfectant (EPO-81) which expresses human erythropoietin was derived from a CHO-K1 cell line transfected with a plasmid containing the gene for huEPO. The transfected CHO cells were maintained in a proprietary serum-free formulation designated CHO-SFM2.1. The Cytopilot culture was established with 400 ml Cytoline macroporous microcarriers in 2L serum-free medium containing 42.5 mM glucose and with an inoculation of 1x10^5 cell/ml. The culture was maintained with the dissolved oxygen at 50% air saturation and pH at 7.1. The stirrer speed was 230 rpm and this maintained a bed expansion of 20-25%.

A. Bernard et al. (eds.), Animal Cell Technology: Products from Cells, Cells as Products, 307–309.

The cell concentration increased to a maximum of 23 x 10^6 cells per ml beads at day 26 (Fig. 1). The maximum growth rate occurred between day 20 and 26. The initial lag period of low growth up to day 9 is unexplained but did not occur in all cultures. Perfusion was started on day 9 and was adjusted daily to maintain the glucose concentration in the culture between 10-20 mM. By day 23 this necessitated a perfusion rate of 4 litre/ day (= 2 vol/ day). At day 24 the feed concentration of glucose was increased to 75 mM to prevent depletion of the substrate without having to increase the perfusion rate further. From this point the glucose concentration of the culture was allowed to increase to 40 mM. EPO synthesis was boosted by sodium butyrate (0.5 mM) for 3 days from day 27. This was achieved by an initial addition of 50 ml stock butyrate (18 mM) followed by perfusion at 120 ml/day for 3 days. The effect of butyrate was to cause a decline in the cell concentration but an increase in EPO concentration. At day 36 the growth medium was substituted for a protein-free medium.

Fig 1: Continuous culture of EPO-81 CHO cells in the Cytopilot

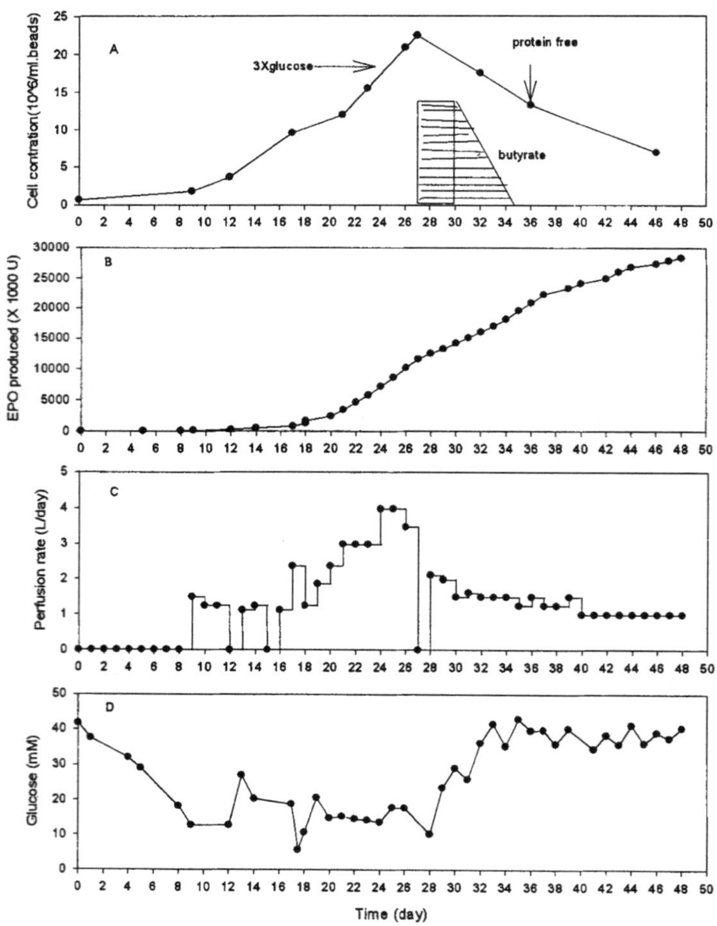

2.2: EPO production EPO was analysed from culture samples using an ELISA previously developed in our laboratory. A total of 28,000 kUnits of EPO was produced over the 48 day period of culture. The EPO concentration increased to a maximum of 950 U/ml on day 35 and this was maintained over the subsequent 9 days. The overall specific production rate of EPO was calculated by integration of the growth curve from time zero and indicated a value of 125 U/10^6 cell-days. This represents an overall mean determined by regression analysis. However, analysis between selected time points indicates that the specific EPO production increased significantly during the later part of the culture (>23 days). This was likely to be due to the combination of the increased glucose concentration of the medium and the added bolus of sodium butyrate. Previous work in our lab (unpublished) has shown that butyrate has a significant effect on protein expression in this cell line. The EPO was shown to be glycosylated from analysis performed by capillary electrophoresis (data not shown).

2.3: Comparison of production from different cultures Table 1 shows summary data in which the growth and productivity of the EPO-81 cells in the Cytopilot was compared with 3 modes of batch culture. Batch cultures were established in a stirred tank bioreactor (Braun Cell Optimizer; 2.3 litre), a spinner flask (100 ml) and a 25 cm^2 T-flask (7 ml) over a 7 day period. The Cytopilot was operated as a perfusion system resulting in gradual increases in cell density with increased perfusion rates. The data shows that the growth rate was low in the Cytopilot and the specific EPO production rates were highest for the low volume cultures in the spinner and T-flasks. However, the Cytopilot culture attained a significantly higher cell density and maximum EPO concentration than any of the other cultures. This shows the potential of the Cytopilot bioreactor for use in the large-scale production of EPO.

Table 1: Summary of results from 4 cultures in different bioreactors

Culture	Maximum cell density (x10^6/ ml)	Maximum EPO (U/ ml)	Specific growth rate (h^{-1})	Specific productivity (U/10^6 cell-d)
Cytopilot	5.5	950	0.0044	125
Batch STR	1.9	400	0.0184	69
Spinner	0.7	200	0.0187	178
T-flask	0.8	210	0.0180	180

Reference Reiter, M., Bluml, G., Gaida, T., Zach, N., Unterluggauer, F., Dobihoffdier, O., Noe, M., Plail, R., Huss, S. and Katinger, H.. (1991) Modular integrated fluidized bed bioreactor technology Bio/Technology **9**, 1100-1102.

Acknowledgements This work was supported by a grant from NSERC and Cangene Corp. The Cytopilot was provided by Amersham/ Pharmacia. A graduate studentship from the Manitoba Health Research Council to M.Yang is gratefully acknowledged.

SCALE-UP OF FIXED-BED REACTORS FOR THE CULTIVATION OF ANIMAL CELLS

DIETER FASSNACHT, INGO REIMANN, RALF PÖRTNER
Technische Universität Hamburg-Harburg, Bioprozeß- und Bioverfahrenstechnik, Denicke Str. 15, D-21071 Hamburg, Germany

1. Abstract
A successful scale-up of fixed-bed reactors is presented in this study. This was achieved by changing the medium flow inside the bed from axial to radial.

The scale-up was achieved in three steps. First and for preliminary scale-up studies, a 13 litre radial-flow fixed-bed was constructed to investigate and describe the hydrodynamic flow and mass-transfer characteristics of large-scale systems. Then, an autoclavable 1.5 litre radial-flow fixed-bed reactor was built and used to cultivate a hybridoma cell line over a duration of 3 weeks. In the last step and based on the previous results a 5 litre fixed-bed was designed which was then integrated into a standard 19 litre stirred-tank fermentor.

The optimal dimensions of the fixed-beds (inner and outer radius, height) were modelled 'a priori' by means of a reaction-diffusion model for oxygen. This model took into account the reaction-diffusion of oxygen within the carrier material, the diffusive flux over the boundary layer of the carrier and the intraparticle convection.

2. Fixed-bed reactors for animal cells
Cultures of animal cells in laboratory-scale fixed-bed reactors are applied for the production of proteins or antibodies from adherent (rCHO, r293) or non-adherent (hybridoma, transfectoma) cell lines, the classical 'products from cells'. Furthermore, they proved to be especially suitable for 'cells as products', such as the cultivation of immortalised hepatocytes (Fassnacht et al., 1998) for the future development of artificial liver support systems, or the expansion of hematopoietic cells for bone marrow transplantation or gene therapy.

The main advantages of fixed-bed reactors are: low shear-stress, due to the immobilisation and the separation of the aeration vessel from the fixed-bed, high cell densities, simple scale-up, and cell retention.

Until now, only small-scale laboratory systems are established with fixed-bed volumes up to one litre (Fig. 1 and 2). For a successful scale-up and reactor design detailed knowledge of the fluid flow, mass-transfer and reaction within the packed-bed is essential. The main problem is a sufficient oxygen supply. This problem can be solved by pumping the fluid via a distributing tube located in the centre of the fixed-bed radial through the fixed-bed to the outside.

A. Bernard et al. (eds.), Animal Cell Technology: Products from Cells, Cells as Products, 311–313.

3. Oxygen Limitation Model

The model used in this approach is a reaction-diffusion model which considers the oxygen gradient within the carrier that results from the substrate consumption (reaction) and the transport of substrate from the bulk phase to the cells into the carrier (diffusion). Three major effects influence the performance of a single porous carrier with immobilised cells (Fig. 3):

I: Reaction-diffusion limitation within the carrier: Substrates have to diffuse into the carrier to supply the immobilised cells. A limitation can occur in the centre of the carrier, if the diffusion is relatively slow compared to the consumption rate of the cells. Furthermore, inhibition can occur from products which accumulate inside the carrier to inhibiting concentrations. The diffusion into the carrier depends on the shape of the pores (tortuosity) but also on the presence or absence of cells.

II: Boundary layer: The substances also have to pass the external boundary layer surrounding the carrier. This effect depends mainly on the diffusion coefficients of the substances in the medium and on the thickness of the boundary layer which is influenced by the Reynolds number (or the flow velocity) of the surrounding fluid.

III: Intraparticle convection: Previous experiments showed that the increase of the carrier performance at increasing Reynolds numbers is larger than the boundary layer effect alone explains (De Backer and Baron, 1994). This was taken into account by assuming a convective flux through parts of the carrier which then improved the supply of substrates. The effect can be significant when using large pore carriers. However, this should not imply the existence of a convective flow through the whole carrier.

Please refer to Fassnacht and Pörtner, 1999 for details on the model.

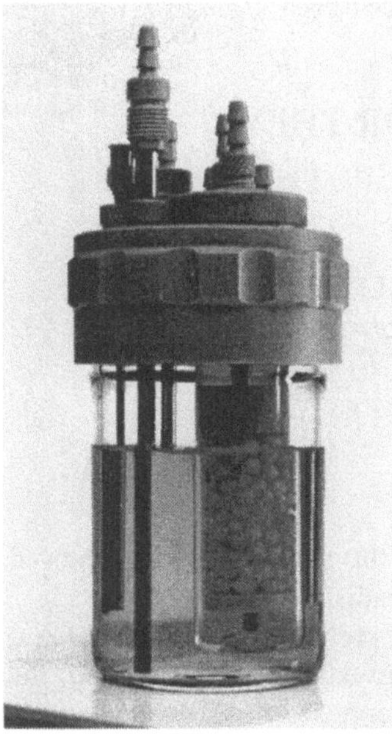

Figure 1. Small-scale reactor with integrated 5 ml fixed-bed for basic studies (carrier material, media).

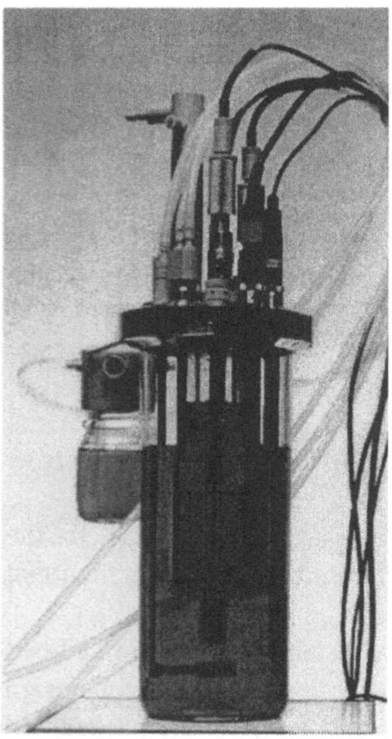

Figure 2. Laboratory-scale 2 litre reactor with 100 ml fixed-bed situated within the conditioning vessel. On the left the bottle used for inoculation.

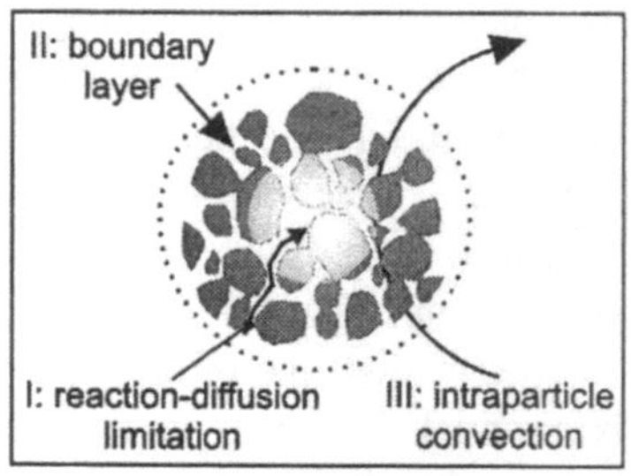

Figure 3. Mass transfer limitation of a single porous carrier containing cells.

4. Scale-up

Small-scale fixed-bed systems (Fig. 1 and 2) are operated by pumping the medium axial through the fixed-bed. These reactors were successfully used for the continuous cultivation of hybridoma, transfectoma, rCHO, r293 and immortalised hepatocytes. However, because of oxygen limitation the maximal length is limited to approx. 10 - 15 cm. The limitation is caused by the low oxygen solubility and the slow flow velocity of the medium within the fixed-bed. This problem can be solved by pumping the fluid via a distribution tube located in the centre of the fixed-bed radial through the fixed-bed to the outside.

Figure 4. Pilot large-scale reactor with integrated 1.5 litre radial-flow fixed-bed.

For preliminary scale-up studies a non-autoclavable 13 litre (height: 90 cm; diameter 25 cm) radial-flow fixed-bed was constructed to investigate the fluid flow and mass-transfer characteristics of large-scale systems. The residence time distribution of tracer experiments showed that the flow can be modelled well by assuming 'radial plug-flow' with radial dispersion. It was found that the Péclet number was constant and that the fluid distribution over the height and the circumference of the fixed-bed was uniform. These positive findings proved that a simple scale-up to the technical scale is possible. The oxygen limitation model was then used to optimise the performance and to find the optimal reactor geometry (radius, height) for the next two scale-up steps 'a-priori':

In a first scale-up a reactor with integrated 1.5 litre fixed-bed was constructed (Fig. 4) and was successfully used to cultivate a hybridoma cell line over a period of 3 weeks at a dilution rate of 4.5 litres medium per day.

Based on the data a 5 litre fixed-bed was designed (Fig. 5) which was integrated into a standard 19 litre fermentor. This reactor is capable of utilising up to 35 litres medium per day which corresponds to a 100 litre stirred tank reactor.

Figure 5. 5 litre radial-flow fixed-bed to be integrated into a standard 19 litre stirred-tank reactor.

De Backer, L. D., and Baron, G. (1994). Residence time distribution in a packed bed bioreactor containing porous glass particles: Influence of the presence of immobilised cells. Journal of Chemical Technology and Biotechnology 59: 297-302

Fassnacht, D., Rössing, S., Stange, J., Pörtner, R. (1998). Long-term cultivation of immortalised mouse hepatocytes in a high density, fixed-bed reactor. Biotechnology Techniques 12: 25-30

Fassnacht, D., and Pörtner, R. (1999). Experimental and theoretical considerations on oxygen supply for animal cell growth in fixed-bed reactors. Journal of Biotechnology (in press)

A NOVEL CONICAL SHAPED BIOREACTOR FOR SMALL-SCALE INVESTIGATIONS UNDER DEFINED CONDITIONS

J.O. SCHWABE, H. MATSUOKA+, H. JUNGKEIT*, R. PÖRTNER
Technische Universität Hamburg-Harburg, Bioprozeß- und Bioverfahrenstechnik, D-21071 Hamburg, Germany
+Teikyo University of Science and Technology, Dept. of Biosciences, Uenohara 2525, Yamanashi 409-0193, Japan
**Meredos GmbH, Alte Dorfstr. 37, D-37120 Bovenden, Germany*

1. Introduction

The reliability of metabolic or kinetic data determined in small-scale culture systems is often insufficient, as these systems cannot be run continuously and parameters such as dissolved oxygen and pH are not controlled. Here a new conical shaped bioreactor (Meredos GmbH, D) for the cultivation of animal cells is presented. The reactor (figure 1) can be operated at a volume range of about 50 ml to a maximum volume of 400 ml. The conical shape allows the integration of probes and various tubes even at small volumes (figure 2). The reactor has all features of a conventional glass reactor with up to 9 openings for inlet, outlet, level control and sampling, temperature sensor, an exhaust air cooler and ports for dissolved oxygen and pH probes. Low shear stress mixing is achieved by a 3-blade propeller stirrer with a diameter of 36 mm. Temperature can be controlled over a water jacket with a conventional water bath via the internal sensor. The main applications are:

- studies of batch and fed-batch cultures at small scale
- expanding cell cultures in fed-batch under defined conditions (pH, DO)
- long-term continuous cultures at high dilution rates for kinetic studies with low medium consumption.

The reactor can be modified to a fixed-bed reactor (figure 3) of 20 ml packed bed volume for the cultivation of adherent cell lines and testing of different carrier material.

A. Bernard et al. (eds.), Animal Cell Technology: Products from Cells, Cells as Products, 315–317.

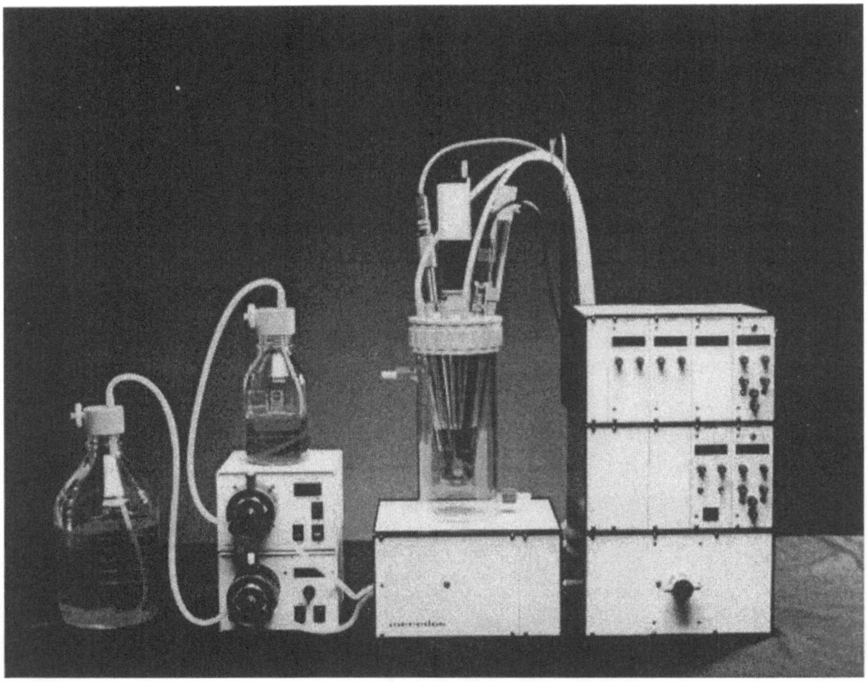

Figure 1. Conical Shaped Bioreactor: Reactor setup for chemostat culture with control unit and pumps.

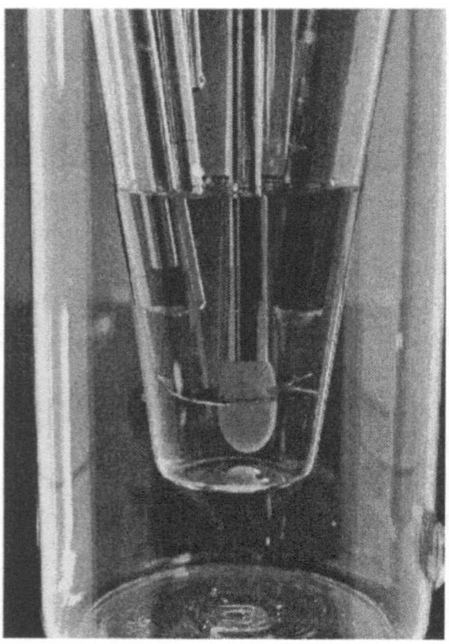

Figure 2. Conical Shaped Bioreactor: Close up with probes and stirrer.

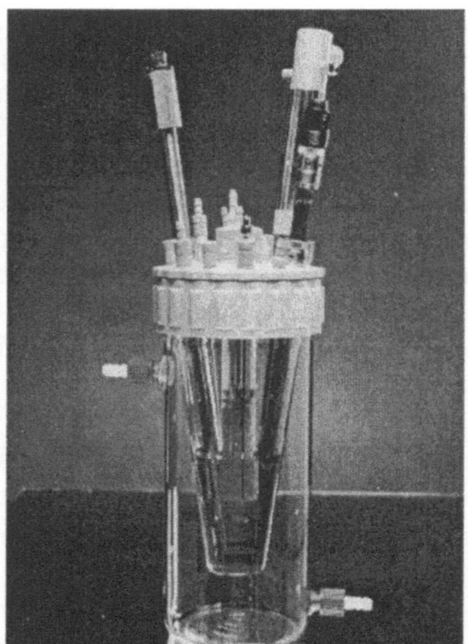

Figure 3. Conical Shaped Bioreactor with integrated fixed bed.

2. Cultivation Examples

Continuous Culture

A hybridoma cell line (IV F 19.23) was cultivated for 2 months in a chemostat for kinetic studies with different media formulations. Figure 4 shows a comparison of cell concentrations with a 2L-reactor (viable cell concentration of $x_v = 2.5\cdot10^6$ cells ml^{-1}). In the conical reactor a cell concentration of $4\cdot10^6$ cells ml^{-1} (monoclonal antibodies $c_{MAb} = 35$ mg l^{-1}) was reached for standard medium (1:1 mixture IMDM/ Ham's F12, 3 % horse serum) at a dilution rate of 0.5 d^{-1}.

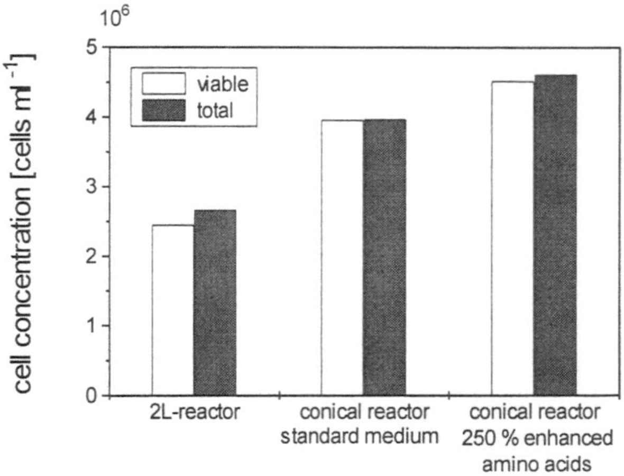

Figure 4. Chemostat culture (Hybridoma IV F 19.23): comparison of viable/total cell concentration of a 2L-reactor and the conical reactor (steady states).

The amino acids were enhanced to 250 % by addition of a nutrient concentrate (10-fold) and resulted in cell concentration of $4.5\cdot10^6$ cells ml^{-1}. The chemostat was run at a constant volume of 200 ml which allowed a steady-state within 6 days with a media consumption of only 100 ml d^{-1}.

Fed-batch Culture

In a fed-batch culture (figure 5) standard media was supplied using a linear feed profile. The initial volume of 80 ml was expanded to 330 ml. A final cell concentration of $2.2\cdot10^6$ cells ml^{-1} was reached and the total viable cell number was 20-fold increased. During feeding the cell concentration was initially reduced because feed profile and composition were not yet optimised in this experiment.

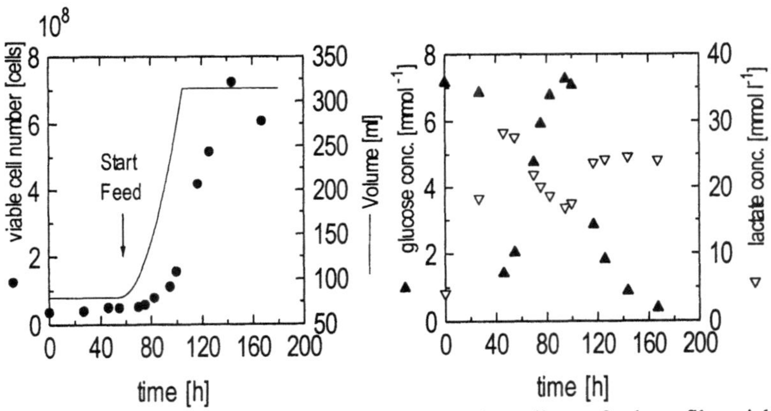

Figure 5. Fed-batch culture (Hybridoma IV F 19.23) using linear feed profile: viable cell number, volume, glucose and lactate concentration (medium/feed IMDM/Ham's F12 medium, 3 % horse serum, concentration of monoclonal antibody $c_{MAb} = 14$ mg l^{-1} at 144 h).

CONTINUOUS PERFUSED FLUIDIZED BED TECHNOLOGY
Increased productivities and product concentrations

D.MÜLLER, F.UNTERLUGGAUER, G.KREISMAYR, C.SCHMATZ, S.WIEDERKUM, S.PREIS, K.VORAUER-UHL, A.ASSADIAN, O.DOBLHOFF-DIER, AND HERMANN KATINGER
Institute of Applied Microbiology (IAM), University of Agricultural Sciences, Muthgasse 18, A-1190 Vienna, Austria;
http://www.boku.ac.at/iam/; email: dmueller@edv2.boku.ac.at

Introduction

The following items emerged to be critical when optimizing this proteinfree high cell density perfusion process

- the initial cell density, cell viability and the stage of cell growth
- the procedures for the cell attachment phase
- a well balanced medium for proteinfree perfusion
- the control of the perfusion rate to ensure optimal conditions

Materials & Methods

The experiments were carried out with recombinant CHO cells producing a monoclonal antibody using a Cytopilot MiniTM / Cytoline1TM fluidized bed reactor system as shown in Fig.1.

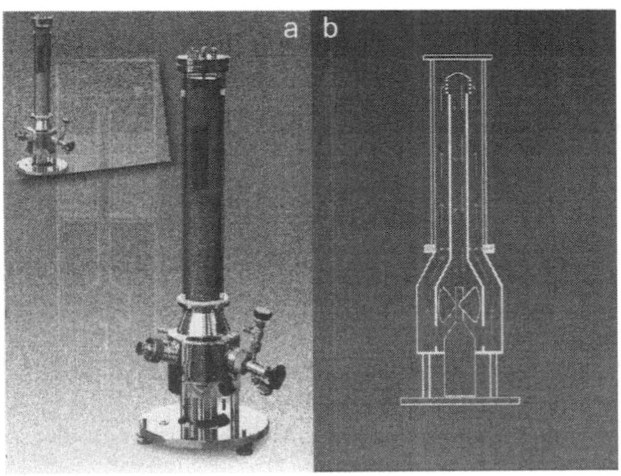

Fig.1: Cytopilot Mini^{TM} (a), functionality of the reactor (b)

A. Bernard et al. (eds.), Animal Cell Technology: Products from Cells, Cells as Products, 319–321.

Carrier cell densities were determined measuring the cell nuclei with a Coulter Counter after cell disintegration. Parts of the carrier were stained with acridine orange and examined under a confocal laser microscope (Fig.2).

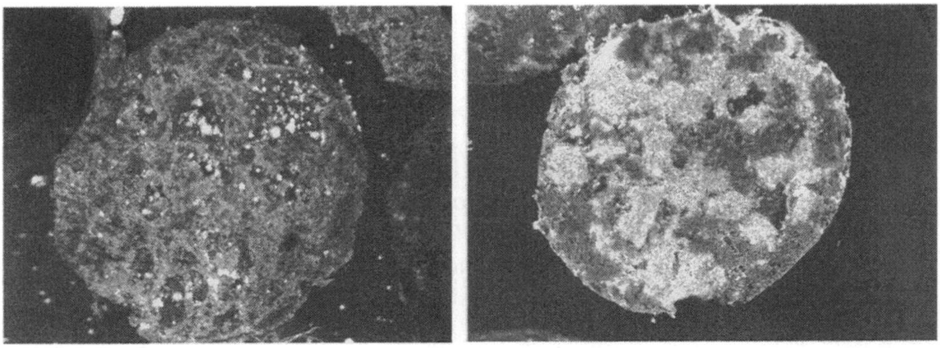

Fig.2: Cytoline1TM 24h after inoculation (left) and at high cell densities (right)

Concentrations of carbohydrates and amino acids in the medium and supernatant were determined by means of HPLC, the product concentrations using an in-house developed ELISA. The perfusion rate was adjusted by a non-invasive control algorithm according to the pO_2 profile.

Results

The cells started to grow without any lag phase due to their good condition when entering the reactor. They settled to the carrier with a recovery of more than 60% within six hours and of more than 95% after 24 hours. The maximum cell density of $1.5*10^8$ cells/ml carrier was reached after three weeks (Fig.3). From the moment the cells entered the stage of "maintenance metabolism", their requirement of nutrients was reduced which was reflected in decreasing specific consumption rates (data not shown). Thus the (specific) perfusion rate was reduced gradually as described in Fig.3.

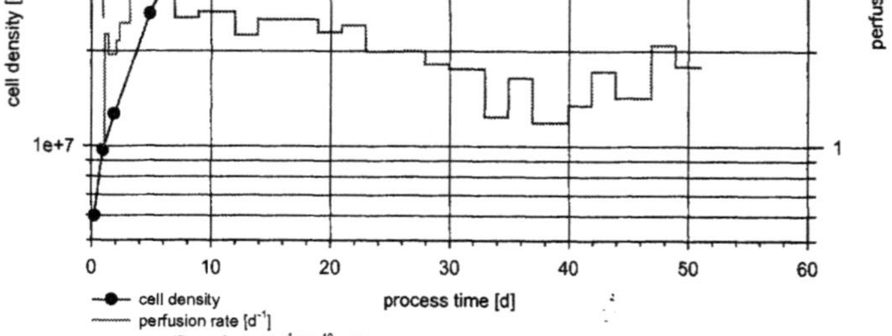

Fig.3: cell density and (specific) perfusion

At the same time the product concentration which remained constant during exponential growth was elevated twofold to more than 100 mg/l. With a specific production rate between 5 and 8 µg mAB/(d*10^6 cells) the productivity reached almost 1 g mAB/(d*l carrier) (Fig.4).

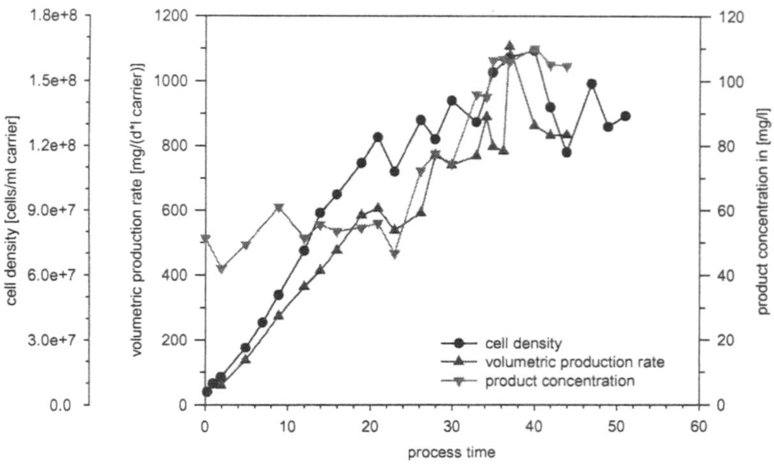

Fig.4: cell density, productivity and product concentration

Conclusions

Since this carrier technology is scalable and the control algorithm for the perfusion rate is in the state of automation we have optimized a system for stable long-term production under proteinfree conditions at high prductivities and high product concentrations.

A fluidized bed reactor has been scaled up to 100l and a carrier volume of 20l to 25l at the IAM/POLYMUN (GMP-) pilotplant and a CHO fermentation has been started recently. A scale-up to 500l carrier should be within the bounds of possibility.

SMALL SCALE BIOREACTOR SYSTEM FOR PROCESS DEVELOPMENT AND OPTIMIZATION

P. GIRARD°, P. MEISSNER°, M. JORDAN°, M. TSAO# and F. M. WURM°
° Laboratory of Cellular Biotechnology, Swiss Federal Institute of Technology, Lausanne, Switzerland. # BioWhittaker Inc., Walkersville, MD, USA.

Keywords: High throughput, process optimization, suspension, milliliter scale

1 Abstract

An agitated 12-well microtiter plate system with a working volume of 2 ml was investigated. Agitation enhances mass transfer and assures homogeneity in wells, thus improving pH stability. The pH can be adjusted by altering the carbon dioxide content of the gas phase. To monitor pH phenol red is used either visually or, for high throughputs, in combination with a spectrophotometric plate reader. Cell growth is assessed non-invasively using stable GFP expressing cells and a fluorescence plate reader. The basic setup is simple and inexpensive, it can be automated and allows several hundred reactors to be run in parallel.

2 Introduction

Process optimization for mammalian cells in suspension is impractical in full-scale bioreactors or spinner flasks due to labor and material costs. Static microtiter plates do not allow suspension culture or high cell density experiments. Screenings with high throughput milliliter-scale systems are only useful if the results correlate with the larger scale.

3 Materials and Methods

Cell Culture. CHO and HEK 293 cell lines were cultured with or without FCS (SeraTech, Germany) in DMEM/F12 (GIBCO, Scotland) or in chemically defined media (BioWhittaker, USA). Prior to the experiments the cells were cultured in spinners. 3 l bioreactors were used (ADI1030, Applikon, Netherlands).

Small-Scale Microtiter Plate System. A rotational shaker (KS250 basic, IKA, USA) was installed in a CO_2-incubator (BB16, Heraeus, Germany). 12-well microtiter plates (TPP, Switzerland) were stacked on the insulated agitator plate allowing to run several hundred 'bioreactors' in parallel.

Growth Assessment. The cells were analysed with a CASY1 Counter (Schärfe System, Germany) and with trypan blue (Sigma, USA) counting method. The

A. Bernard et al. (eds.), Animal Cell Technology: Products from Cells, Cells as Products, 323–327.

fluorescence of cells stably expressing the green fluorescent protein (GFP) was measured in a plate reader (CytoFluor®Series 4000, PerSeptive Biosystems, USA) [2].

pH Assessment. Phenol red (Sigma, USA) was used as pH indicator (15 mg/l). Its absorption at 588 nm was measured with a spectrophotometer (SpectraMAX 340, Molecular Devices, USA).

4 Results and Discussion

4.1. GROWTH ASSESSMENT

Cell density can be quantified using the fluorescence signal of stable GFP expressing cell lines. The method is fast, non-invasive and there is no need for sampling.

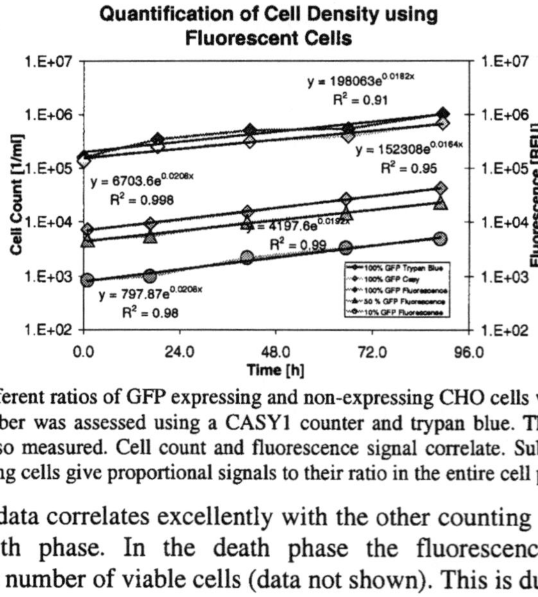

Figure 1: Different ratios of GFP expressing and non-expressing CHO cells were cultivated. The cell number was assessed using a CASY1 counter and trypan blue. The fluorescence signal was also measured. Cell count and fluorescence signal correlate. Subpopulations of GFP expressing cells give proportional signals to their ratio in the entire cell population.

The fluorescence data correlates excellently with the other counting methods during the exponential growth phase. In the death phase the fluorescence declines not as pronounced as the number of viable cells (data not shown). This is due to the stability of GFP. Correlation might be improved using cells expressing destabilized, faster degrading GFP mutants.

4.2. pH

At this scale, culture pH is estimated visually. The use of a spectrophotometric plate reader enhances the throughput and generates numeric data that is easier to evaluate. The pH and absorption at 588 nm were measured and modelled using:

$$pH = pK_a + A - B * \log \frac{C - Abs_{588}}{Abs_{588} - D} \qquad (1)$$

with A and B correcting for optical parameters and the electronics, C and D corresponding to the absorptions when all the indicator shifted to either of its forms. For

HEK 293 cultures no significant absorption at 588 nm due to the presence of cells was observed. Other cell lines have not been tested. Mass transfer is enhanced by agitation, convection instead of diffusion, in- and outgazing are improved by at least one order of magnitude. pH regulation with the CO_2 content in the gas phase is rather slow, hours compared to minutes in a bioreactor. This excludes the use of pH shifts in processes, but it is possible to regulate the overall pH.

4.3. CORRELATIONS

The transfection efficiency, the part of the expressing cells, and the day of highest transient protein expression are the same for reactor, spinner, shaken microtiter plate. Cell growth as aggregates that are induced by high calcium concentration [3] ($CaPO_4$ transfection) is also similar. In plates cells express up to twice more protein than in the reactor. Protocols and methods can be up- and downscaled linearly. The correlation to the larger scale is sufficient for the development of a transient transfection process.

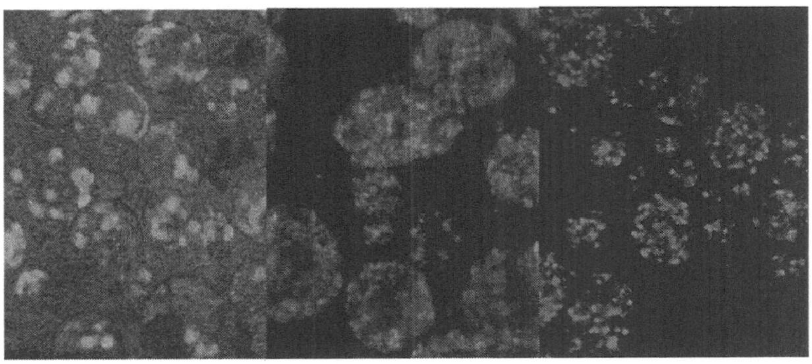

Figure 2: Transiently transfected[1] HEK 293 cell growing as aggregates. Growing in a 3 l bioreactor (left), in a 100 ml spinner (center) or in a 2 ml shaken microtiter plate (right). Aggregate size and transfection efficiency are identical.

5 Conclusion

The agitated 12-well system correlates to larger scale spinners and reactors. Combined with fluorescence based growth assessment, the system is a very powerful tool for optimization and development of processes and media.

6 References

[1] Jordan, Köhne and Wurm (1998) Calcium-phosphate mediated DNA transfer into HEK-293 cells in suspension: control of physicochemical parameters allows transfection in stirred media
Transfection and protein expression in mammalian cells, *Cytotechnology*,**26**,p. 39-47.

[2] Hunt, Jordan, DeJesus, Wurm (Submitted) GFP expressing mammalian cells for fast, accurate, non-invasive and kinetic-mode-type cell growth assessment. *Nature Biotechnology,*

[3] Peshwa, Kyung, McClure, Hu (1993) Cultivation of Mammalian Cells as Aggregates in Bioreactors: Effect of Calcium Concentration on Spatial Distribution of Viability, *Biotechnol Bioeng*,**41**,p. 179-187.

Discussion (Meissner)

Grammatikos: Do you have any data, or can you speculate, as to the reproducibility of this method in terms of the quality of the recombinant proteins?

Meissner: We tested the antibody with ELISA and it was always of the same quality. For the other products, I cannot say at the moment.

Cho: While you were transfecting, your cells were changed to DMEM medium. DMEM has a high calcium concentration and aggregated the cells. Was this to help transfection efficiency?

Meissner: We used this medium because it is optimal for calcium phosphate transfection. We transfected the cells in suspension before the large aggregates were formed.

Bernard: Could you give us a personal impression on how far we can go with scale in this? The reason I am asking is that my personal prejudice is that we are going to have problems of pH control at larger scale. You are talking about controlling pH very accurately from 7.4 to 7%. Within a large bioreactor you are going to have these variations anyway.

Meissner: If you are working on medium optimisation, maybe we do not have this problem because we can remove the precipitate, without reducing the pH, by altering the feeding strategy. This is one solution. If we can reduce the amount of DNA, then there is no limitation to scale-up to, let us say, 100L.

Bergemann: Can you please tell us about the quality of the plasmid DNA which you need to get such high transfection efficiency?

Meissner: The quality is not the main problem. We are not really testing if it is LPS-free, or for freedom from toxicity.

Bergemann: So you can use standard kit-form preparation of DNA and not use caesium chloride centrifugation?

Meissner: Yes.

TRANSIENT GENE EXPRESSION

Chapter V

MAMMALIAN TRANSIENT GENE EXPRESSION-POTENTIAL, PRACTICAL APPLICATION AND PERSPECTIVES

H.D. Blasey, R. Hovius[+], L. Rey, H. Vogel[+] and A.R. Bernard

Serono Pharmaceutical Research Institute, 14 Chemin des Aulx, CH 1228 Plan les Ouates, Switzerland, [+] EPFL, Lausanne, Switzerland

Abstract

The Semliki Forest Virus Expression system, the most rapid and also very powerful viral transient tool had been scaled up to 12 liters yielding 3-8·10^6 5-HT3 receptors per cell. Most recently, with HEK293EBNA and pCEP4 we could express up to 10^7 5-HT3 receptors per cell in a stirred bioreactor. We conclude that both systems allow to produce functional recombinant protein within days to weeks at very high levels.

Introduction

This last decade transient mammalian gene expression was successfully scaled up to reactor scale, rapidly yielding milligram quantities of protein for functional and structural studies (Bernard & Blasey, in press; Wurm & Bernard, 1999). Transient expression is defined as time limited, vector mediated protein expression in non-clonal biological systems. Despite the short time of productivity, transient expression is of increasing interest due to the short time span for yielding product: weeks or in some cases only days. Furthermore, it is the preferred tool for gene co-expression or expression of toxic products.

Transient expression is either performed by infection or transfection of the appropriate vector (Fig. 1). Within the group of viral expression vectors Vaccinia, Semliki Forest virus and Adenovirus have very successfully been scaled up at least to tens of liters. Adenovirus was run in a reactor yielding 90mg/l protein tyrosine phosphatase (Garnier et al., 1994). The largest scale described for a mammalian viral system however has been known for Vaccinia. In the biopharmaceutical drug production it is being used up to the 40 litre scale (microcarrier perfusion, Barrett et al., 1989) and run under P3 safety conditions to produce a recombinant HIV vaccine.

The most recent viral expression system which was successfully scaled up (Blasey et al., 1998) is based on the Semliki Forest Virus (SFV) (Liljeström and Garoff, 1991). The SFV infection with the recombinant virus leads to rapid high level expression. The only

A. Bernard et al. (eds.), Animal Cell Technology: Products from Cells, Cells as Products, 331–337.

drawback lies in the viral production protocol, because the SFV expression system is not based on a replicating virus.

DNA vector based systems are widely established for transient expression. Among those, the alphavirus derived DNA expression vectors (Berglund et al., 1998) are very useful for vaccination. However, of particular interest for large scale transient expression are the plasmids which are maintained or replicated intracellulary to yield high level expression: the EBV derived and the SV40 derived expression system.

Fig. 1 Mammalian Transient Expression Vectors

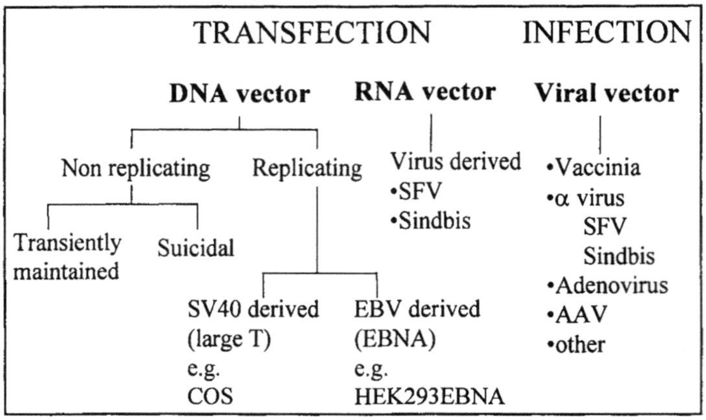

COS cells, which belong to the SV40 derived systems, were successfully scaled up to the one litre volume and in one published example, produced 30 mg of the fusion protein CD40Fc purified from 10 litres of the serum free supernatant (Blasey et al., 1996). However, this approach suffers limitations such as requirements for large amounts of DNA (~ 1 mg/litre culture) and the demand for a large scale transfection method (we face a similar problem when transfecting cells with RNA, generated in-vitro from the recombinant pSFV plasmid). The bottle neck of large scale transfection was recently overcome by using continuous electroporation system (Parham et al., 1999) or by applying Calcium phosphate transfection protocols at reactor scale (Jordan et al., 1998). On the other hand, with the HEK293EBNA system, expression can be initiated by performing transfections at small scale which requires little DNA and only a small amount of cells. Expanding then the transfected cell population under selection pressure allows to reach larger volumes of culture.

For the work presented here, the objective was to compare the SFV and HEK293EBNA expression system. As an example for the protein production we expressed the mouse 5-HT3 receptor with the SFV in a 11.5 litres reactor and with HEK293EBNA in various systems such as Cellcube and stirred tank reactors.

Literature reports on many proteins which already have been expressed with the SFV (Bernard & Blasey, in press) and also on such which had been expressed with the EBV based HEK293/EBNA system: soluble tumor necrosis factor receptor sTNFR, neurotrophins, prolactin, hepatocyte growth factor HGF, vascular endothelial growth factor VEGF, tissue plasminogen activator tPA, secreted alkaline phosphatase SEAP (Cachianes et al., 1993), ß Gal (Tomiyasu et al., 1998), soluble Interleukin receptors, 5-

HT3 receptor, TNF signaling kinasis, blue fluorescent protein (Blasey, pers. communication), hu-LIF, complement factor hu-DAF (Geisse et al., 1999), cDNA expression libraries, phosphoribosyltransferase, CD8 (Margolskee et al., 1988).

Materials and methods

5-HT3 production by SFV

Recombinant 5-HT3 SFV virus generation and the receptor production were described earlier (Blasey et al., 1997 and Blasey et al., 1998).

5-HT3 production by HEK293EBNA

Fig. 2: Transient Expression with the HEK293EBNA System

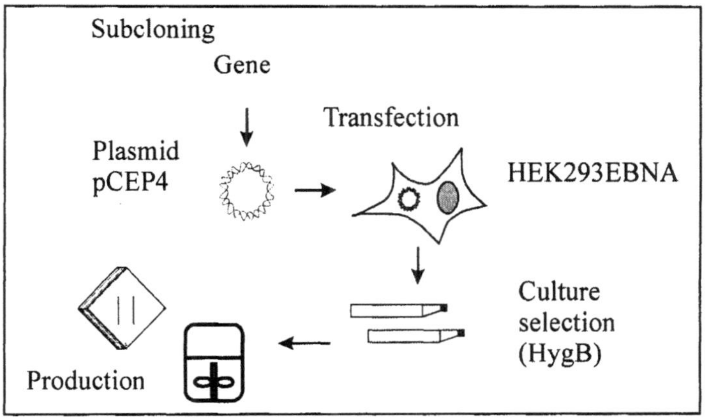

For expression of the 5-HT3 receptor by HEK293EBNA (Fig 2), the 5-HT3 gene was cut out from the pSFV1 by BamH1 and inserted into the corresponding site of pCEP4 (Invitrogen). HEK293EBNA were electroporated with 40 μg DNA (0.4 mm cuvettes, 290V, 960 μF, Gene Pulser, Biorad) and the transfected cells were selected with 300 μg/ml HygromycineB (Boehringer Ma) starting from 3 days post transfection. A cell stock was frozen at passage 3 post transfection and used as a production bank. For large scale production we used the Cellcube system in a configuration described earlier (Blasey, 1995). Stirred tank reactors from Chemap (Switzerland, equipped with a marine impeller) and MBR (Switzerland, equipped with a VIBROMIX) were used for growing HEK293EBNA in suspension. We used medium composed of: Iscove's/Ham's F12 1+1, 10% FCS, 2 mmol Gln.

Results

Up to this date a large number of different types of proteins have been expressed using the SFV. After having demonstrated the feasibility of using the SFV for large scale protein production we produced in 5 different batches cumulated 60 litre of culture expressing the 5-HT3R at $3 \cdot 10^6$ receptors per cell, corresponding to 15 mg purified receptor protein per 12 litre batch. Optimisation (expression at pH 7.3) allowed to further increase expression levels to $8 \cdot 10^6$ 5-HT3 receptors per cell. The receptor was characterised and found to be fully functional.

Two DNA-vectors which both offer the advantage of replication are used for transient expression: the SV40- and EBV derived system respectively. We had recently established the EBV based expression system using the HEK293EBNA cells and the plasmid pCEP4 (non-clonal culture selection).

HEK293EBNA cells were found to change their adhesion kinetics in flask cultures when transfected with pCEP4-5HT3. In T-flasks, 3 hours after seeding, 85% of the HEK293EBNA were attached while the same cell transfected with pCEP4 containing the 5HT3 gene showed only 70% of attached cells. In consequnce we allowed cells to attach for 4 hours when inoculating a Cellcube to start an adherent culture.

HEK293EBNA cells expressing 5-HT3 receptor were also grown in stirred tank reactors. In a STR with a marine impeller the formation of aggregates at 30 to 50 cells was observed. Growth in single cell suspension culture was made possible only by using the Vibromix as an agitation device. However, probably as a consequence of the shear forces induced, the doubling time increased to 75 hours and the final cell yield was $4 \cdot 10^5$ cells per ml. Despite this poor growth performance, the culture expressed 10 million receptors per cell (Tab. 1).

Table 1: Expression of the 5-HT3 Receptor in Different Rector Systems

System	Cell Yield	Cell growth	T_d [h]	Receptor Yield: $*10^6$ R/c
T-flask (225 cm^2)	$3 \cdot 10^7$/flask	attached	20	0.5-1.2
Cellcube (4.4 m^2)	$5.2 \cdot 10^9$	attached	-	1.4-3.7
STR: Marine Impeller	$6.3 \cdot 10^8$ per litre	suspension, aggregates up to 30/50 cells	58	1.5
STR: VIBROMIX	$4.4 \cdot 10^8$ per litre	single cell suspension	75	9.7

Conclusion

We compared the 5-HT3 receptor expression with the viral SFV system and in HEK293EBNA cells, a DNA vector mediated transient expression system. Both systems are characterised by a very short time interval between subcloning and expression and by very high level of recombinant protein expression. In the case of the 5-HT3 receptor the expression level reached 8 and 10 million receptors per cell for the SFV and the HEK293EBNA respectively. Furthermore, the HEK293 EBNA system with the pCEP4 plasmid allows selection of the transfected population with the antibiotic HygB. The selected culture can be scaled-up for production like 'conventional' cell lines.

Future perspective: we anticipate that transient expression systems in general will further grow in importance and that in particular SFV and HEK293EBNA will be scaled up to allow routine runs at between 10 and 100 litres. The objective is certainly the production of hundreds of mg recombinant protein per batch. The field of application can be expected to extend to pre-clinical and clinical grade protein production.

References:

Barrett N, Mitterer A, Mundt W, Eibl J, Eibl M, Gallo RC, Moss B and Dorner F (1998) Large-scale production and purification of a vaccinia recombinant-derived HIV-1 gp160 and analysis of its immunogenecity. Aids Research and Human Retrovirus 2: 159-171.

Berglund P, Smerdou C, Fleeton MN, Tubulekas I & Liljeström P (1998) Enhancing immune response using suicidal DNA vaccines. Nature Biotechnology 16: 562-565.

Bernard A.R. and Blasey H.D. (in press) Transient expression systems, in M.C. Flickinger and S.W. Drews (eds.), The Encyclopedia of Bioprocess Technology: Fermentation, Biocatalysis & Bioseparation, John Wiley & Sons.

Blasey HD, Isch C. & Bernard AR (1995) Cellcube: A new system for large scale growth of adherent cells. Biotechnology Techniques 9 (19): 725-728.

Blasey HD, Aubry JP, Mazzei GJ & Bernard AR (1996) Large Scale Transient Expression with COS Cells, Cytotechnology 18: 183-192.

Blasey HD, Brethon B, Hovius R, Lundström K, Rey L, Vogel H, Tairi AP, & Bernard AR (1998) Large Scale Application of the Semliki Forest Virus System: 5-HT3 Receptor Production. In: New Developments and New Applications in Animal Cell Technology; Kluywer Academic Publishers; eds: O.W. Merten, P. Perrin and B. Griffiths, pp. 449-454.

Cachianes G, Ho C, Weber RF, Williams SR, Goeddel DV & Leung DW (1993) Epstein-Barr Virus derived vectors for transient and stable expression of recombinant proteins. BioTechniques 15 (2): 255-259.

Garnier A, Côté J, Nadeau I, Kamen A & Massie B (1994) Scale-up of the adenovirus expression system for the production of recombinant6 protein in human 293S cells. Cytotechnology 15: 145-155.

Geisse S. (in press) Float! - and express more: converting the HEK.EBNA/oriP System to suspension culture. InProducts from Cells - Cells as products, eds.: AR Bernard & F. Wurm.

Jordan M, Köhne C & Wurm FM (1998) Calcium-phosphate mediated transfer into HEK.293 in suspension: control of physicochemical parameters allows transfection in stirred media. Cytotechnology 26: 39-47.

Ki-ichiro T, Satoh E, Oda Yohei, Nishizaki K, Kondo M, Imanishi J & Mazda Osam (1998) Genne transfer in vitro and in vivo with Epstein-Barr Virus-Based Episomal Vector Results in markedly high transient expression in rodent cells. Biochem. and Biophys. Res. Comm. 253: 733-738.

Liljeström P & Garoff H. (1991) A new generation of animal cell expression vectors based on the Semliki Forest virus system Bio/Technology 9: 1356-1361.

Margolskee RF, Kavathas P & Berg P (1988) Epstein-Barr Virus Shuttle Vector for Stable Episomal Replication of cDNA Expression Libraries in Human Cells. Meolecular and Cellular Biology 8, 7: 2837-2847.

Parham JH, Marie A, Overton LK & Hutchins JT (1998) Optimization of transient gene expression in mammalian cells and potential for scale-up using flow electroporation. Cytotechnology 28: 147-155.

Wurm F & Bernard AR (1999) Large-scale transient expression in mammalian cells for recombinant protein production. Current Opinion in Biotechnology 10: 156-159.

Discussion (Blasey)

Pollitt: The 293 adeno system was originally touted as a semi-stable production system. Can you comment on the stability of expression? Is it transient or stable?

Blasey: It is reported in the literature that ERV expression systems show a certain integration rate of between 70-30% of cells which integrate the gene. So there is a tendency towards stable expression. What we have done is to allow the cells to grow a bit after transfection for a few days, then apply a selection pressure and scale-up the cells so that we can freeze down some ampoules. We can then expand further. So we did experiments where we grew cells over 50 days and the expression did not change. In the literature it is stated that stable expression exists for over 6 months with the EBV system.

Cho: Are your selection pressures throughout the system?

Blasey: We left out hydromycin after we had frozen down our first ampoules of cells. For the experiments shown here we had kept hydromycin in the culture, but at that time we were unsure whether this was necessary or not. On the large-scale we showed that leaving out hydromycin and continuing to grow the culture without selection pressure for 50 days did not make a difference to expression .

Cho: So productivity was not modified or decreased?

Blasey: Not during the 50 day period.

FLOAT! - AND EXPRESS MORE:

Converting the HEK.EBNA/oriP System to Suspension Culture

S.GEISSE, O.ORAKÇI, A.PABLER, A.PATOUX, D.RINALDI,
J.WATKINS*, R.SCHMITZ, H.GRAM
Novartis Pharma Inc.
Core Technology/Biomolecules Production
Bldg. S-506.3.04
CH-4002 Basel/Switzerland
** Imutran Ltd., Cambridge CB2 2EH, U.K.*

1. Abstract

Serum-free growth of recombinant mammalian cells in suspension offers several advantages over adherent cultivation in fetal calf serum containing medium. We report here the development of a serum-free medium for HEK.EBNA cells supporting growth in suspension as single cells or small aggregates, thus facilitating the scale-up of the culture to large scale. In addition, we describe a suitable transfection protocol for these suspension-adapted cells. When applied in combination, the HEK.EBNA/oriP expression system becomes highly efficient, giving rise to production of 50-100 mg of recombinant protein within approx. 3 months.

2. Introduction

An efficient mammalian expression system for the rapid production of recombinant proteins is based on the episomally driven expression of foreign transgenes in HEK 293 cells by means of genetic elements derived from Epstein-Barr-virus (EBNA-1/oriP). High level transient expression plus the option to select for stable transfectants via antibiotic resistance, as well as the human origin of the host cell line (in view of secondary modifications of the transgene product) are the main advantages of this system [1,2].

However, the natural adherent growth of the HEK 293 cell line precludes easy adaptation/scaling to larger culture vessels. Few serum-free media have been described to date for HEK 293 cells allowing continuous growth of fully adapted cells. Furthermore, the HEK 293 cells exhibit a strong tendency towards aggregate formation in suspension culture [3].

By using the commercially available HEK.EBNA cell line we tried to develop a

A. Bernard et al. (eds.), Animal Cell Technology: Products from Cells, Cells as Products, 339–345.

medium composition, which allows serum-free growth of HEK.EBNA cells as single cells or small aggregates in suspension. In a subsequent step, we wanted to establish a suitable transfection protocol for HEK.EBNA cells grown in suspension.

3. Materials and Methods

Cell Line: HEK 293 cells carrying a stably integrated copy of the EBNA-1 gene were bought from Invitrogen [293-EBNA cells, HEK.EBNA cells]. This cell line is routinely maintained in Dulbeccos' Modified Eagle medium (DMEM) plus 10 % fetal calf serum at 37°C/5% CO_2 in humidified atmosphere.

Media for serum-free suspension growth:

TABLE 1: Composition of two different media for serum-free suspension growth of HEK.EBNA cells

Components	M2α-Medium	LowCa-Medium	Suppliers
Ham's F12 nutrient mixture (powder)	5.31 g/l	5.31 g /l	Gibco/Life Technologies
MEM Alpha medium (powder)	5.04 g/l	-	Gibco/Life Technologies
S-MEM medium (powder)	-	5.29 g/l	Gibco/Life Technologies
$NaHCO_3$	1.69 g/l	1.69 g/l	Merck
D(+)-Glucose	1.0 g/l	1.0 g/l	Merck
Ferric citrate soln.	50 μM	50 μM	Merck
Insulin soln.(beef)	5 mg/l	5 mg/l	Eli Lilly
Human serum albumin	2 g/l	2 g/l	Swiss Serum Inst. Bern/CH
Hydrocortisone	10 nM	10 nM	Fluka
Putrescine	0.62 μM	0.62 μM	Fluka
Selenous acid, Na-salt	5 μg/ml	5 μg/ml	Fluka
Ethanolamine	150 μM	150 μM	Fluka
Lecithin soln.	50 mg/l	50 mg/l	Central Soya Overseas BV
L-Glutamine	2 mM	2 mM	Gibco/Life Technologies
HEPES soln.	10 mM	10 mM	Gibco/Life Technologies
Biotin soln.	-	0.8 μM	Fluka
Vitamin B12 soln.	-	0.074 μM	Fluka
Non-essential Amino Acids soln. (100 x)	-	20 ml	Gibco/Life Technologies
Sodium Pyruvate	-	0.5 mM	Merck

All components were dissolved in 1 litre of distilled water.

In addition, the following media additives were evaluated: Dextran sulfate (Sigma, 100 µg/ml), Polyvinylsulfate (Sigma, 100 µg/ml), Heparin (Sigma, 6 mg/ml).
The adherently growing HEK.EBNA cells were gradually adapted to either M2α-medium or LowCa-medium by stepwise reduction of the fetal calf serum containing DMEM medium. Upon full detachment of the cells from the plastic surface, the culture was transferred to roller bottle (Costar), and the cells were maintained by regular dilution and aeration of the bottles (94% air/6% CO_2/20 seconds) in roller culture on a roller device (turning speed:6.5 rpm).

Transfection of HEK.EBNA cells in serum-free suspension:
For all transfection optimisation experiments shown HEK.EBNA cells adapted to M2α-medium were used, but cells transfected in LowCa-medium yielded comparable results. As transgenes of choice expression plasmids harbouring either the cDNA coding for hu-LIF (Leukemia Inhibitory Factor) or hu-DAF (Decay Accelerating Factor) were selected. These plasmids feature in addition a Zeocin resistance marker.
For electroporation the cells were washed twice in phosphate-buffered saline and adjusted to a cell density of $2x10^7$ cells/ml. 0.8 ml of cell suspension was transferred to an electroporation cuvette (0.4 cm electrode gap) and mixed with 10 or 15 µg of plasmid. Following incubation on ice for 10 minutes, electroporation using varying conditions (160-300 V/960 µF, BioRad gene pulser) was carried out. Subsequently, the cells were seeded onto T75 flasks in 20 ml medium.
The lipofection experiments were performed using Superfect (Qiagen), DAC-30 (Eurogentec) and DMRIE-C (Gibco/Life Technologies) reagents in comparison. Initially the experiments followed the recommendations of the vendors for cell concentration, DNA and reagent concentrations, and were then carefully analyzed for optimal ratio of DNA vs. reagent concentration.
As read-out for all transfections experiments ELISAs determining product titers of hu-LIF and hu-DAF were established.

Establishment of stable transfectants by selection with Zeocin:
Two days after transfection selection for antibiotic resistance was started by transfer of the cells from small scale (6-well-plate/35 mm petri dish) to T75 tissue culture flasks (in case of the lipofection experiments) and addition of 100 µg/ml of Zeocin (Invitrogen) to the medium. The selection process with occasional feeding and/or complete exchange of medium, in dependence on the condition of the cells, required approx. 4 weeks, until a logarithmically growing cell population emerged. These cell pools were then transferred back to roller culture for maintenance and production analysis.

4. Results

The M2α-medium described in Table 1 was originally developed for serum-free suspension growth of CHO cells (see abstract B. Kleuser et al., this meeting). It proved to be suitable as well for growth of HEK.EBNA cells in suspension; however, large cell clumps with necrotic areas inside formed readily, which could not be disintegrated without compromising viability. Addition of anionic polymers (such as dextran sulfate,

polyvinylsulfate or heparin) to the medium, which has been shown to successfully prevent aggregation of insect cells [4], had no significant effect on reduction of aggregate formation of HEK.EBNA cells.
A sevenfold reduction of the calcium content of the medium by exchange of the basal powder medium from MEM Alpha to S-MEM (LowCa-medium, Table 1) led to a marked reduction in aggregate formation. The HEK.EBNA cells grew as single cells or, upon prolonged cultivation, as small to medium-sized lose aggregates without any visible sign of necrosis. Thus, correct cell counting as well as viability determinations were greatly facilitated. Adherently growing transfection pools could be readily adapted to this medium for scale-up of the production process; however, the adaptation to suspension culture required approx. 3-4 weeks. In order to circumvent this delay, we evaluated different transfection techniques with the aim of direct transfection in suspension.
By comparing the individually optimized protocols for electroporation and three different lipofection reagents, clear differences in efficiency with respect to product titers became visible (Fig. 1). By using DMRIE-C as the most effective reagent, a simple, rapid and reliable protocol for transfection of cells in suspension was developed.

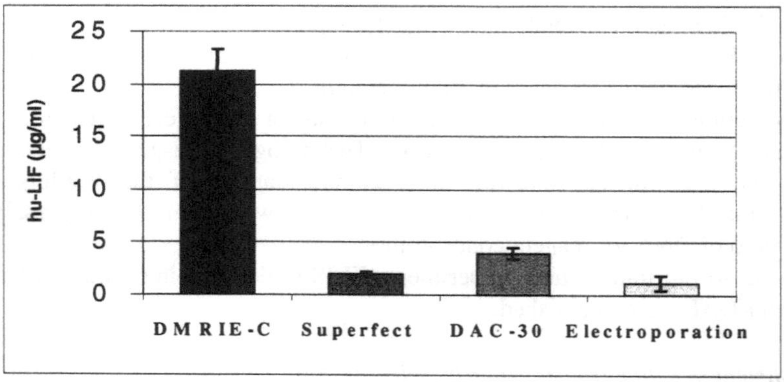

Figure 1: Comparative production of hu-LIF using different transient transfection methods

For establishment of stable transfectants,1 x 106 suspension cells were transfected in 6-well-plates by means of DMRIE-C/lipofection. Subsequently, the cultures were gradually scaled up during the selection process using Zeocin as antibiotic. After approx. four weeks of selection on flasks, logarithmically growing transfection pools were re-transferred to roller culture. Production titers obtained from transfected suspension cells on roller ranged from 7.5 mg/l (hu-DAF) to 10 mg/l (hu-LIF) (Fig.2) Furthermore, these cell pools retained stable recombinant protein expression over several weeks to months.

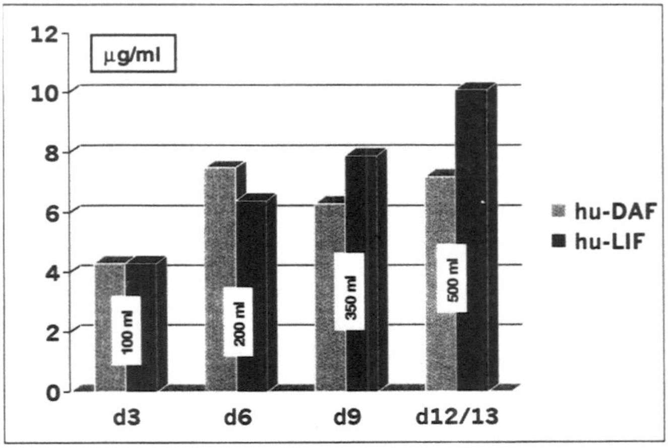

Fig. 2: Production of hu-LIF and hu-DAF in Fed-batch Roller Culture

Thus, for production of >10-100 mg of recombinant protein, expansion of a well-producing pool to the multi-liter scale in roller culture represents a simple and rapid possibility. Alternatively, the recombinant HEK.EBNA cells can also be grown in high-density membrane modules (e.g. Integra CELLINE). Yields are severalfold higher when compared to roller batch production; however, harvest volumes are small (10-30 ml per harvest). Inoculation of a bioreactor with suspension-adapted HEK.EBNA cells is an option, if larger quantities of protein are needed. Initial experiments performed with HEK.EBNA cells growing in LowCa-medium in a 5-l-bioreactor in continuous perfusion culture and aeration via tubing demonstrated the feasibility of that approach.

5. Conclusion

The modification of HEK.EBNA/oriP expression system to complete suspension culture in serum-free medium offers several advantages over the conventional cultivation in fetal calf serum containing medium: a) it promotes scale-up of the production process, b) facilitates downstream processing and c) transfection in suspension eliminates the need for subsequent adaptation to suspension associated with time delays. The HEK.EBNA suspension system thus allows production of 50-100 mg of recombinant protein from transfection to product within approx. three months.

References:

1. Cachianes, G., Ho, C., Weber, R.F., Williams, S.R., Goeddel, D.V.,and Leung, D.W.: Epstein-Barr Virus-derived Vectors for Transient and Stable Expression of Recombinant Proteins, Biotechniques **15** (1993), 255-259.
2. Sclimenti, C.R. and Calos, M.P.:Epstein-Barr Virus Vectors for Gene Expression and Transfer, Curr. Opinion Biotechnology **9** (1998), 476-479.
3. Côté, J., Garnier, A., Massie, B., and Kamen, A.: Serum-Free Production of Recombinant Proteins and Adenoviral Vectors by 293SF-3F6 Cells, Biotechnol.Bioeng. **59** (1998), 567-575.
4. Dee, U.D., Shuler, M.L., and Wood, H.A.: Inducing Single Cell Suspension of BTI-TN5B1 Insect Cells: The Use of Sulfated Polyanions to Prevent Cell Aggregation and Enhance Recombinant Protein Production, Biotechnol.Bioeng. **54** (1997), 191-205.

Discussion (Geisse)

Wurm: What is the cost of your transfection cocktail (DMRIE) at the 100L scale?

Geisse: We do not transfect on the 100L scale but on a 6 well plate - for 8 SWF. So it is a reasonable cost and why we use a commercial mixture. At a 10L scale we would use calcium phosphate transfection.

Cho: Your expression vector only contains *orp*?

Geisse: It has only the *orp*.

Cho: Did you check whether the *orp* vector was maintained for a long term?

Geisse: No, we have not looked for stable integration versus epiposomal replication. We expect that some of the copies will eventually integrate and become stable expressions.

Massie: Do you have to maintain selective pressure in roller bottle culture as you scale-up, or just for the inoculum?

Geisse: We maintain the selective pressure as we scale-up but for the final expansion step you may as well leave it out.

Massie: What cell density do you reach with your media formulation?

Geisse: In roller culture around $1x10^6$/ml and in fermenters up to $5x10^6$/ml. I think there is room for improvement.

Massie: Did you ever try batch replacement where you just put more medium in every day and just keep harvesting?

Geisse: No.

Massie: It is very simple and works very well, even in roller cultures.

MULTIPLICITY AND TIME OF INFECTION AS TOOLS TO MAXIMIZE VIRUS-LIKE PARTICLE PRODUCTION BY INSECT CELLS

L.A. PALOMARES, S. LOPEZ, AND O.T. RAMIREZ*
Instituto de Biotecnología
Universidad Nacional Autónoma de México
A.P. 510-3
Cuernavaca, Morelos
62250 México

Abstract

Commercial use of vaccines based on virus-like particles (VLPs) demand effective production strategies. However, the production of VLPs is very complex, as several recombinant proteins have to be simultaneously produced, and efficiently assembled. In this work, rational strategies to obtain recombinant double shelled rotavirus-like particles (dRLPs), using the insect cell-baculovirus expression system, are presented.

1. Introduction

One of the advantages of the widely used insect cell-baculovirus protein expression system is the possibility of simultaneous expression of different proteins which assemble successfully in multimers (1). Such simultaneous expression can be very complex, as proteins, very different in nature, may be required at different times in different stoichiometrical relations. One of the most complex cases is the production of virus-like particles (VLPs), particularly rotavirus-like particles (RLP), which are structurally identical to native virions but lack the genomic material. If economically produced, VLPs can become a viable novel immunization alternative for many viruses. Rational production strategies, that maximize yields and productivity, can make this possible. RLP are formed by three concentric layers of four viral proteins (Figure 1).

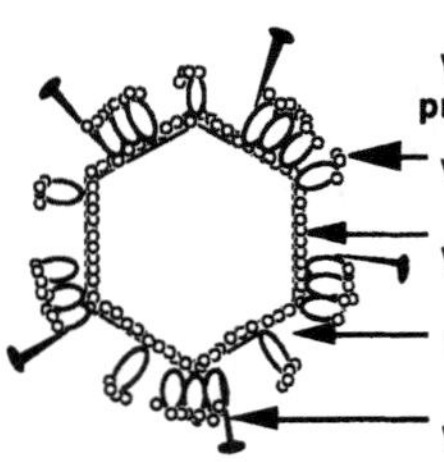

Viral protein	Molecules/ virion	Molecular weight	% Total protein
VP7	780	33 kDa	30
VP2	200	102 kDa	15
VP6	780	44 kDa	51
VP4	120	86 kDa	1.5

Figure 1. Structure of RLP (adapted from 2).

A. Bernard et al. (eds.), Animal Cell Technology: Products from Cells, Cells as Products, 347–349.

The successful production of RLPs in this expression system implies that the four viral proteins encoded by four different recombinant baculoviruses have to be simultaneously produced by the infected insect cell culture. As an initial approach to explore the optimal conditions for the production of RLPs, single and double shelled RLP (sRLP and dRLP, respectively) were produced by simultaneous expression of VP2 and VP6.

2. Materials and Methods.

The insect cell lineSf9 (ATCC 1711) was used in all cultures. Cells were grown in Sf900 II SFM (GIBCO, 10902-088) in 200 mL spinner flasks (working volume 60 mL). Cell concentration was determined with a Coulter counter II and viability by trypan blue. Cultures were infected at $1x10^6$ cell/mL with a multiplicity of infection (MOI) of 5pfu/cell, unless otherwise indicated. Recombinant baculoviruses producing VP2 (strain RF, bac2) and VP6 (strain SA11, bac6) under the polyhedrin promoter were obtained from Dr. Estes (3). Recombinant proteins were detected by SDS-PAGE and Western blot analysis using a rabbit antiserum to rotavirus and then quantified by densitometry against a standard curve of a purified GST-VP8 fusion protein of rotavirus. This curve allowed a relative quantification of VP6 and VP2. To separate sRLP and dRLP, cell pellets from 1mL samples were treated with deoxycholate lysis buffer and sonicated. Supernatants were run in 0.6% Tris-agarose gels. Standards of viral particles (single or double shelled) were obtained from CsCl gradients. Multimers separated by agarose gels were transferred to a nitrocellulose membrane and immunodetected. Multimer concentrations were then compared by densitometry. Unassambled proteins were quantified by 12% polyacrylamide gels under non-denaturing conditions.

3. Results and Discussion.

Twelve cultures were individually infected with bac2 or bac6 at MOIs of 1, 5 or 10 pfu/cell. It was found that the protein production rate was dependent on the MOI employed in each culture. Also, the specific production rate and maximum concentration of both proteins produced in cultures infected at a MOI of 5 pfu/cell were dependent on their molecular weight. Recombinant VP2 formed sRLP, as has been observed before (4). sRLP formation kinetics closely followed VP2 concentration. All VP2 detected was assembled in sRLP.

To analyze kinetics of simultaneous expression of VP2 and VP6, cells were simultaneously infected with bac2 and bac6. Expression using the same MOI of bac2 and bac6 yielded a VP6/VP2 concentration ratio of 16 after 40 hours postinfection. Manipulation of the MOI of bac2 and bac6 yielded different VP6/VP2 ratios, as shown in Figure 2. Another strategy used to manipulate protein ratio was through time of infection (TOI). Cultures were initially infected with bac2, 3 or 6 hours later bac6 was added. As shown in Figure 3, this strategy was also effective to regulate the stoichiometry of production of proteins. Multimer formation was affected by VP6/VP2 ratio. Namely, lower ratios yielded higher concentrations of dRLP but lower of sRLP.

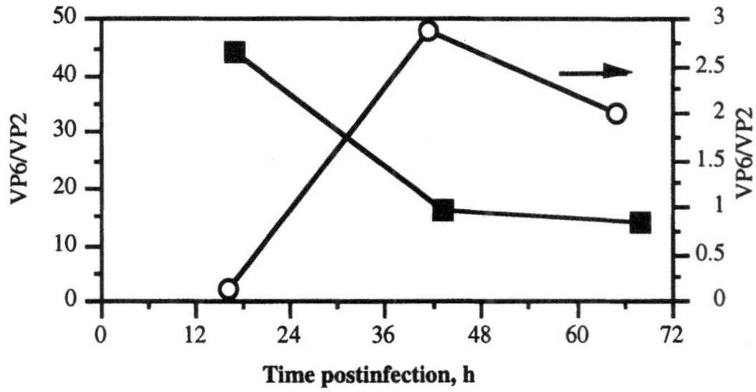

Figure 2. MOI as a tool to manipulate VP6/VP2 ratio. ■, culture simultaneously infected with 5 pfu/cell of bac2 and bac6. ○, culture simultaneously infected with 5pfu/cell of bac6 and 17 pfu/cell of bac 2.

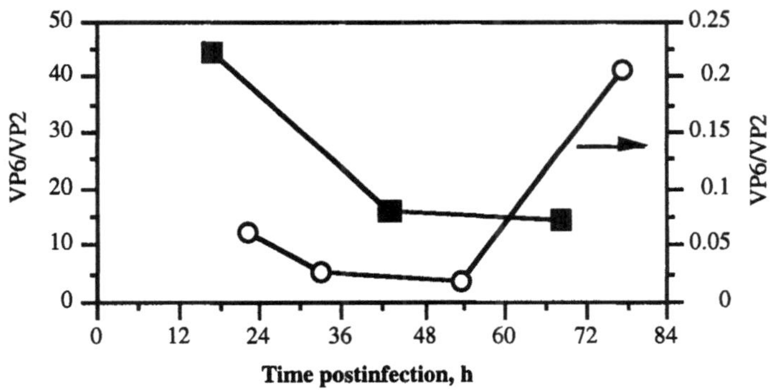

Figure 3. Non simultaneous infection with bac2 and bac6 allowed the manipulation of protein concentration. ratio. ■, culture simultaneously infected by bac2 and bac6. ○, infected with bac2 at time 0 and with bac6 6h later.

4. Conclusions

We have explored the manipulation of MOI and TOI to acquire a correct stoichiometrical production of individual protein subunits. Using this strategy, we were able to manipulate the production rate and concentration of each recombinant protein. The influence of such parameters on sRLP and dRLP assembly was determined.

Acknowledgements

Technical support N. Osorio, A. Gómez and S. Coria. Financial support by CONACyT 25164-B and PAEP 202355. L.A. Palomares acknowledges support by CONACyT 85797 and DGAPA-UNAM.

References

1. Palomares, L.A., Ramírez, O.T. Insect cell culture: Recent advances, bioengineering challenges and implications in protein production. In: Advances in Bioprocess Engineering II. Galindo, E., Ramírez, O.T. eds. pp. 25-52. Kluwer Academic Publishers. The Netherlands, 1998.
2. Mattion, N.M., Cohen, J., Estes, M.K. (1994). The rotavirus proteins. Pp. 160-247. In: A. Kapikian (ed.) Virus Infections of the gastrointestinal tract. 2a. ed. Marcel Dekker, New York.
3. Crawford, S.E., Labbé, M., Cohen, J., Burroughs, M.H., Zhou, Y.J., Estes, M.K. (1994). Characterization of VLPs produced by the expression of rotavirus capsid proteins in insect cells. J. Virol. 68, 5945-5952.
4. Labbé, M., Charpilienne, A., Crawford, S.E., Estes, M.K., Cohen, J. (1991). Expression of rotavirus VP2 produces empty corelike particles. J. Virol. 65, 2946-2952.

PROCESS DEVELOPMENT FOR TRANSIENT GENE EXPRESSION IN MAMMALIAN CELLS AT THE 3 L SCALE: 10-50 MG OF R-PROTEIN IN DAYS

P. MEISSNER, P. GIRARD, A. KULANGARA, M.C. TSAO°,
M. JORDAN, F.M. WURM
Centre of Biotechnology UNIL-EPFL, Swiss Federal Institute of Technology, 1015 Lausanne, Switzerland.
°BioWhittaker, Walkersville, MD, USA.

1. Abstract

Transient gene expression in mammalian cells at scales of 1-100 L may become a powerful tool of producing large quantities of proteins in a very short period of time.
In this study, we continued to optimize process parameters for transient transfection of HEK-293 cells with the calcium-phosphate method cultivated in a 3 Liter bioreactor. Two different HEK-293 cell lines (expressing EBNA or T-Antigen) were adapted to suspension culture using a newly developed medium, 293G (BioWhittaker). Process parameters, during and after transfection, were optimized using a vector that expresses green fluorescent protein (GFP) as a reporter. Cells were expanded in standard spinner culture. During the exponential growth phase, cells were recovered by centrifugation, resuspended in DMEM-F12 medium and transferred into a bioreactor at an initial cell density of $3x10^5$ cells/ml. The transfection cocktail with 0.5-1.0 mg of supercoiled plasmid DNA was transferred into the reactor by syringe injection, 2 hours after inoculation. The setpoint for pH, initially at pH 7.4, was shifted to pH 7.1 4-6 hours post-transfection. With transfection efficiencies ranging between 70 and 90% - as established with GFP vectors - anti-human RhD IgG1 antibody was secreted at concentrations of up to 20 mg/L within a time frame of less than one week.

2. Introduction

Production of recombinant proteins in mammalian cell culture has become equivalent in frequency of use to microbial production processes as reviewed for the production of antibodies by Verma and coworkers (1998). There are many different mammalian systems applied and stable expression from genome integrated plasmid DNA sequences are preferred. The development of these systems is costly and very time consuming. Contrary to this, the transient system promise to be a much faster and cheaper approach to produce high amounts of r-proteins, as already reported for COS cells (Blasey et al.,

A. Bernard et al. (eds.), Animal Cell Technology: Products from Cells, Cells as Products, 351–357.

1996), HEK-293 cells (Jordan et al., 1998) and compared also with BHK cells (Wurm and Blasey, 1999).
In this work we describe the development of transient gene expression (TGE) technology in stirred bioreactors at the scale of 1-3 liters. Suspension adapted HEK-293 cells (expressing EBNA or T-Antigen) were transfected using the calcium-phosphate method and maintained subsequently in a simple batch culture for up to one week. Improvement of a number of parameters in overall transfection and cell culture technique allowed us the production of more than 20 mg/L of secreted r-protein in 5-7 days.

3. Material and Methods

TGE in a bioreactor with 1 L working volume. Transient transfection of suspension adapted HEK-293 cells was carried out in DMEM/F12 medium supplemented with 29 mM sodium bicarbonate, 10.9 mM HEPES, 2.5 mg/L human transferrin, 2.5 mg/L insuline, 0.1 mM diethanolamine, 0.1 mM L-proline, and 2% FCS (hereinafter reactor medium). Prior to transfection, cells were expanded in 0.5-3 L spinner flasks in 293G medium (BioWhittaker) supplemented with 1% FCS. The bioreactor (Applicon) was inoculated with 3×10^5 cells/ml in a total volume of 800 ml reactor medium and the cells were cultivated at 37°C for 2 hours at pH 7.4. Some minutes before the transfer of the transfection cocktail, 2 ml 2.5 M $CaCl_2$ were injected through a septum. Transfection was carried out with 1 mg supercoiled plasmid DNA precipitated in a 40 ml transfection cocktail (20 ml 250 mM $CaCl_2$ + 20 ml 1.4 mM phosphate in 50 mM HEPES + 140 mM NaCl). After short and gently mixing the precipitation reaction was allowed to proceed for 1 minute. Within the same timeframe, the transfection cocktail was taken up into a 50 ml syringe and, at the 60 second time point, injected into the bioreactor through a septum.
4 hours after transfection, 200 ml fresh reactor medium were added and pH was reduced from 7.4 to 7.1. Temperature, pO_2 and pH were controlled and measured on-line. Cell number and product concentration were analyzed off-line by taking samples once or twice a day.

Anti-human RhD IgG1 ELISA. 96-well plates were coated with goat anti-human kappa light chain (Biosource) overnight at 4°C. Plates were washed 3 times with TBS-T (0.2‰ Tween 20 in TBS, pH 8.0). Then the plates were incubated with blocking solution (1% w/v casein hydrolysate + 0.5‰ Tween 20 in PBS, pH 7.1) at room temperature for 1 hour. At 2-fold serial dilution standard-IgG1 and samples were incubated for 30 min at 37°C, followed by 3 times washing with TBS-T. For the detection of the bound protein alkaline phosphatase-conjugated goat anti-human gamma chain (Biosource) was added and the plates were incubated for 30 min at 37°C, followed by 4 times washing. After addition of the substrate solution (NPP + diethanolamine + 0.5 mM $MgCl_2$, pH 9.8) plates were incubated in the dark at 37°C. Enzymatic reaction was stopped after 15 minutes with 3 N NaOH. Absorption was measured at 405 nm against 490 nm using an ELISA reader (Molecular Devices).

4. Results and Discussion

Adaptation of HEK-293 cells to suspension without change of transfectability. In this study we started with a transfection protocol for calcium-phosphate precipitation developed by Jordan and coworkers (1996) for adherent growing HEK-293 cells in DMEM/F12 medium + 2% FCS. With the use of a newly developed growth medium (293G from BioWhittaker) it took only 4 weeks to get the cells growing in single-cell suspension. Using DMEM/F12 medium supplemented with 2% FCS for the transfections, we found exactly the same transfection efficiencies of more than 80% with EGFP (enhanced green fluorescent protein) as a reporter, comparing cells growing before in adherent or suspension cultures (results not shown).

Transfection of suspension adapted HEK-293 cells in a 1-3 L bioreactor. As already described in material and methods HEK-293 cells were transfected at an initial cell density of $3x10^5$ cells/ml. For a final volume of 1 L we used 1 mg of plasmid DNA. In first experiments EGFP was transiently expressed, 1 day post-transfection the transfection efficiency (% of green fluorescent cells) could be estimated (Tab. 1). The most important modifications of the suspension transfection protocol compared to the adherent system are (1) use of suspension adapted cells, which were expanded in a special developed growth medium. (2) The growth medium (293G) was totally exchanged by DMEM/F12 + 2% FCS (transfection medium). (3) The time between cell seeding and transfection was dramatically reduced from 16 hours to 2 hours. (4) The volume of the transfection cocktail and DNA amount was reduced at 50%.

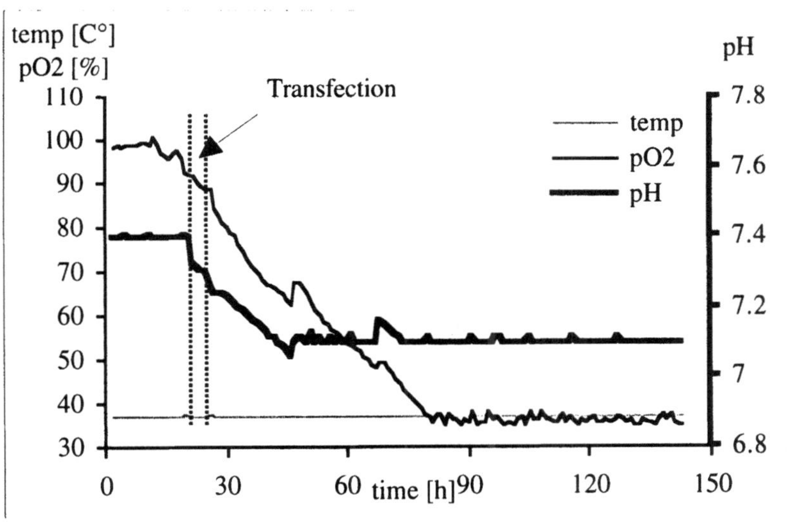

Figure 1. On-line measurement of temperature, pH and dO_2: HEK-293 EBNA cells were transiently transfected in a 3 L bioreactor in DMEM/F12 based medium.

Transient gene expression was carried out with plasmids encoding for different r–proteins. In Fig. 1 the on-line measurement of temperature, pH and dissolved oxygen of one representative production run is shown. Cells were inoculated at time point 19 and 2 hours later transfected. The setpoint of pH (7.4 before and during transfection) was changed to 7.15 (time point 25) after transfection, regulation of aeration was starting at 35% dO_2, and cultivation temperature was kept constant at 37°C during the whole cultivation time.

Production of r-proteins with TGE in HEK-293 cells using different plasmid constructs. After co-transfection of HEK-293 cells with two plasmids encoding for either heavy or light chain of anti-human RhD IgG1, we were able to produce more than 20 mg/L of secreted antibody (Fig. 2) in 6 days. During the whole cultivation time samples were taken once or twice a day, cells were counted and the product concentrations were analyzed by ELISA.

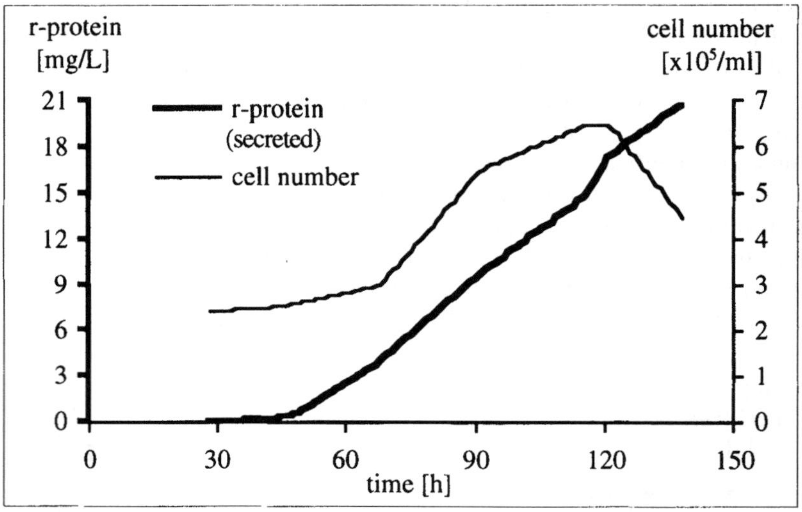

Figure 2. Off-line analysis of cell number and product concentration after transfection of HEK-293 EBNA cells with plasmid DNA encoding for a secreted r-protein (IgG1).

The developed process for TGE of GFP (intracellular protein) and secreted anti-human RhD IgG1 was reproducible for more than 3 times (data not shown).

TABLE 1. Transfection efficiencies (estimated by GFP expression in parallel experiments) and max. product concentrations (analyzed by ELISA) after transient transfection of HEK-293 EBNA cells in a 3 L bioreactor.

Product	Cells	Promoter	Transfection efficiency [%]	Production time [d]	Max. product concentration [mg/L]
GFP	HEK-293 EBNA	EF1a	80-95	3	
Anti-human RhD IgG1	HEK-293 EBNA	EF1a	80-90	5-7	>20
Soluble receptor	HEK-293 EBNA	CMV	80-85	5-7	>20

Additionally to that, other secreted proteins could be produced with equal product titers (e.g. secreted receptor molecule as shown in Tab. 1), which is demonstrating the generic aspect of our developed process.

5. References

Blasey, H.D., Aubry, J.-P., Mazzei, G.J. and Bernard, A.R. (1996) Large scale transient expression with COS cells, *Cytotechnology* **18**, 138-192.

Jordan, M., Schallhorn, A. and Wurm, F.M. (1996) Transfecting mammalian cells: optimization of critical parameters affecting calcium-phosphate precipitate formation, *Nucleic Acid Res* **24**, 596-601.

Jordan, M., Köhne, C. and Wurm, F.M. (1998) Calcium-phosphate mediated DNA transfer into HEK-293 cells in suspension: control of physicochemical parameters allows transfection in stirred media, *Cytotechnology* **26**, 39-47.

Verma, R., Boleti, E. and George, A.J.T. (1998) Antibody engineering: Comparison of bacterial, yeast, insect and mammalian expression systems, *J Immunol Methods* **216**, 165-181.

Wurm, F.M. and Bernard, A. (1999) Large-scale transient expression in mammalian cells for recombinant protein production, *Curr Opin Biotechnol* **10**, 156-159.

Discussion (Meissner)

Grammatikos: Do you have any data, or can you speculate, as to the reproducibility of this method in terms of the quality of the recombinant proteins?

Meissner: We tested the antibody with ELISA and it was always of the same quality. For the other products, I cannot say at the moment.

Cho: While you were transfecting, your cells were changed to DMEM medium. DMEM has a high calcium concentration and aggregated the cells. Was this to help transfection efficiency?

Meissner: We used this medium because it is optimal for calcium phosphate transfection. We transfected the cells in suspension before the large aggregates were formed.

Bernard: Could you give us a personal impression on how far we can go with scale in this? The reason I am asking is that my personal prejudice is that we are going to have problems of pH control at larger scale. You are talking about controlling pH very accurately from 7.4 to 7%. Within a large bioreactor you are going to have these variations anyway.

Meissner: If you are working on medium optimisation, maybe we do not have this problem because we can remove the precipitate, without reducing the pH, by altering the feeding strategy. This is one solution. If we can reduce the amount of DNA, then there is no limitation to scale-up to, let us say, 100L.

Bergemann: Can you please tell us about the quality of the plasmid DNA which you need to get such high transfection efficiency?

Meissner: The quality is not the main problem. We are not really testing if it is LPS-free, or for freedom from toxicity.

Bergemann: So you can use standard kit-form preparation of DNA and not use caesium chloride centrifugation?

Meissner: Yes.

A METHOD TO DETERMINE THE OPTIMAL TIME TO INFECT INSECT CELLS WITH THE BACULOVIRUS EXPRESSION SYSTEM

J. LJUNGGREN, M. ALARCON, A-K. RAMQVIST, A. WESTLUND, AND L. ÖHMAN
Karo Bio AB, NOVUM, S-141 57 Huddinge, SWEDEN
E-mail: jan.ljunggren@karobio.se

1. Abstract

A set of batch cultures infected with the Baculovirus expression system were carried out to find out the optimal cell density, when to infect the insect cells (Sf-9), and how this correlates to the cell size and specific growth rate (μ, h^{-1}). By measuring the cell size before infection it is possible to predict the optimal time for infection as cell size changes, increases, during the non-infected phase of the culture. From the experiments it was found that the protein expression (production) increases as long as the cell density, at the time of infection (MOI=5), was increased up to a "critical density" where the protein expression dramatically terminated (implying a loss of "infectability" of the cells). Above this cell density level it was no longer possible to infect the insect cells and get a subsequent protein production without replenish the old medium to new fresh cell culture medium.

2. Materials and methods

Sf9 (*Spodoptera frugiperda*) insect cells were used in this work. Suspension cultures of these cells were maintained in SF900II medium supplemented with Gentamycin (15µg/L). The cells were maintained in Erlenemeyer shaker flasks at 27°C, 130 rpm before the start of the bioreactor. A three litre bioreactor was used to grow inoculated cells for the "different cell density infection experiment". The bioreactor temperature was controlled at 27°C, the DOT was set to 40% and maintained by surface airation and sparging of oxygen. The pH was not controlled during the culture. The SF900II medium used for the bioreactor was supplemented with 11ml/L of 10% pluronic F68, 4ml/L of 0.3% antifoam C and 15µg/L Gentamycin. Initial inoculum density for the bioreactor was 0.5 x 10^6 cells/ml. Aliquots (drawn twice daily) from the bioreactor were used to seed Erlenemeyer shaker flasks at different cell densities. At each time point duplicate aliquots were removed and one of the duplicates was kept in the old medium whereas the other one was pelleted for 5 minutes at 1000 g and resuspended in

A. Bernard et al. (eds.), Animal Cell Technology: Products from Cells, Cells as Products, 359–361.

new SF900II medium. The cells were added to shaker flasks and the resulting suspensions were infected at MOI =5 with recombinant baculovirus (ß-galactosidase). Samples from the bioreactor (drawn 3 times per day) were taken to determine the cell density and cell size. In addition, samples from the shaker flask cultures were taken at 0 h, 24 h and 48 pi in order to measure cell density, cell size and ß-gal concentration. Cell density and cell size were quantified by the CASY®1 TTC instrument (cell distribution range 7.5-30 μm) and ß-gal concentration was determined enzymatically.

3. Results and discussion

Examination of the uninfected bioreactor culture showed that the specific growth rate, μ (h^{-1}), increased very rapidly after inoculation and reached its maximum value before 50 hours of cultivation, followed by a rapid decline (Figure 1). This decrease in specific cell growth indicates that the cells are not in an optimal medium already after 50 hours of cultivation and is possibly due to the loss of nutrients (unknown factors) from the medium (or cells) and/or the release of some growth inhibiting metabolite into the medium.

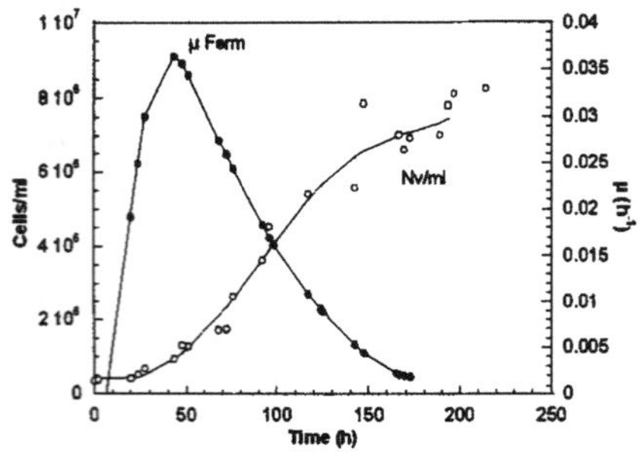

Figure 1. Viable cell density CASY (o), and specific growth rate (●) in the bioreactor.

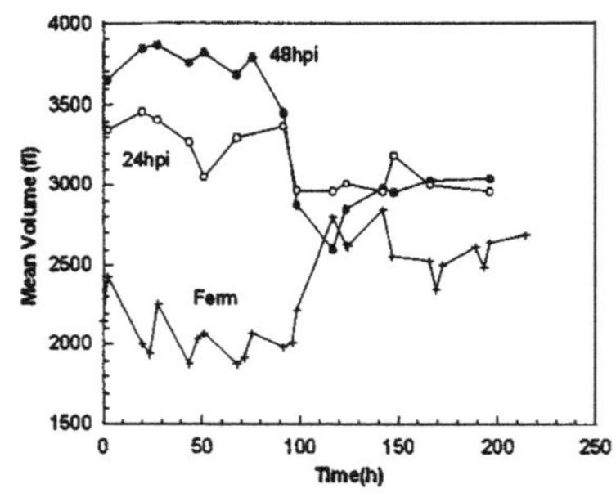

Figure 2. Mean cell volume in the bioreactor (+) and mean cell volume in the Erlenmeyer shaker flasks with old medium at 24pi (o) and 48pi (●).

The mean volume of the uninfected cells in the bioreactor increased from a stable value around 2000 fl (femto litre) to a value slightly above 2500 fl after 100 hours of cultivation (Figure 2). This indicates a change in cell growth, where the cell population probably enters a different phase in the cell cycle. The effect of infection can be seen by the large increase in cell volume at both 24hpi and 48hpi in the Erlenemeyer shaker flasks (Figure 2). However, after 100 hours of cultivation in the bioreactor the cells did not become swollen when they were infected. This is clearly seen by the loss of increase in cell volume at 24hpi and 48hpi in the Erlenmeyer shaker flasks (Figure 2).

When analysing the decrease of ß-gal production (Figure 3), it is evident that it coincides with the loss of cell swelling at 48 hours post infection. In Figure 3 it can also be seen that the loss of ß-gal production also coincides with the time point when uninfected cells increase in volume.

The total ß-galactosidase production in the Erlenmeyer shaker flasks grown in both the unchanged and replenished medium at 48hpi was plotted against the total cell concentration in the bioreactor at the time of infection (TOI). The data demonstrates that cells grown in the replenished medium produce ß-galactosidase at higher cell densities compared to cells grown in the unchanged medium (Figure 4).

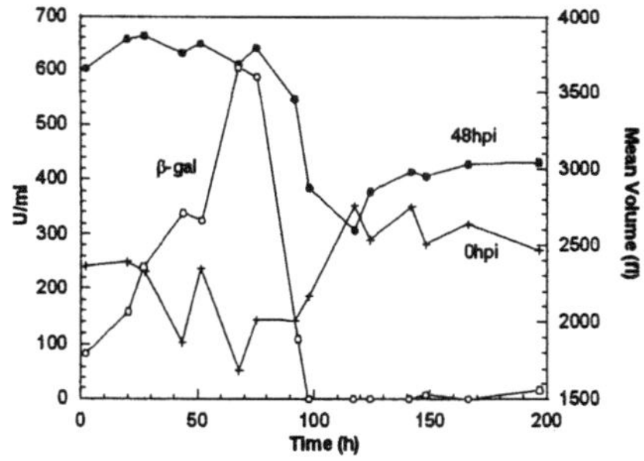

Figure 3. Total ß-galactosidase production in the Erlenemeyer shake flasks with old medium at 48h (o). Mean cell volume in the Erlenmeyer shaker flasks with old medium at 0pi (☛) and 48pi (●).

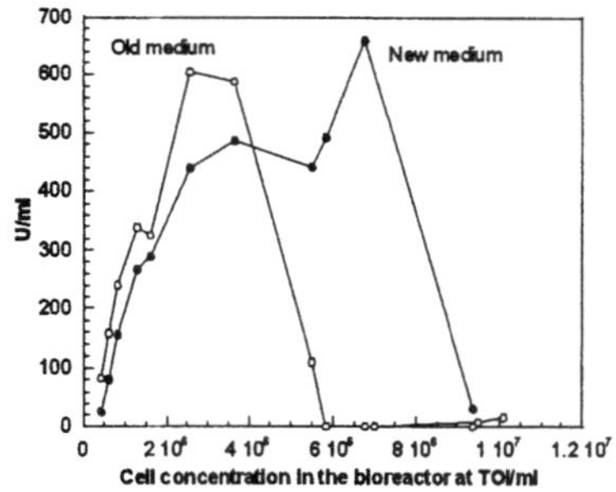

Figure 4. Total ß-galactosidase production in the Erlenmeyer shake flasks with old (o) and new (●) medium at 48hpi plotted against total cell concentration in the bioreactor at TOI.

4. Conclusions

Monitoring cell size before infection can be used as a powerful way to determine the best time for infecting the Sf-9 insect cells in order to harvest large amounts of the expressed recombinant protein. The simple parameter of cell size indicates, somewhat surprisingly, how susceptible the cells are to infection and consequently to the uptake and expression of the recombinant baculovirus. If the cells have passed the time point where they start to increase in volume it is no longer advisable to infect the cells since the infection will result in very low protein yields. This phenomenon may be due to the fact that the cell population has entered a different phase in the cell cycle and is therefore no longer readily susceptible to infection.

Cell size determination is also valuable for determining how often to subculture the insect cells in order to insure that the optimal growth conditions are maintained.

ANALYSIS OF BACULOVIRUS INFECTED SF9 CELLS USING FLOW CYTOMETRY

G. R. PETTMAN
Microbial and Cell Culture Sciences, SmithKline Beecham, Harlow, UK

Abstract
The insect cell-baculovirus expression system is a widely used system for the expression of recombinant proteins and is the first choice system for many laboratories. There are many published reports on the optimisation of the process discussing cell lines, growth medium, etc, however, there are relatively few reports covering the analysis of insect cells for the identification, and quantitation, of infected cells within culture.
This study set out to develop a rapid and sensitive method for the identification and quantitation of insect cells infected with baculovirus in order to titrate virus and/or monitor the viral infection process within production cultures. By using a monoclonal antibody specific to a baculovirus coat protein (gp64) and analysing samples using flow cytometry, a rapid (2 hour) and reproducible method was developed. Initial studies gave relatively poor separation between the infected and non-infected cells. Optimisation of the labelling procedure led to a 50 fold separation which enabled identification and quantitation of the two populations of cells.

Materials and methods
Shake flask cultures of Sf9 cells were routinely grown in IPL-41 serum free medium (Maoirella et al), maintained at a cell concentration of between 2-20x10e5 cells/ml, shaken at 120 rpm and incubated at 27°C. Cultures were infected with recombinant baculovirus by the direct additon of virus to a culture at 1x10e6 cells/ml using a multiplicity of infection (moi) of 10 virus particles/cell.
The hybridoma cell line expressing the gp64 monoclonal antibody was a kind gift from Professor L. Volkman (University of California). Conditioned medium from the cultures was purified on Protein A sepharose and concentrated by ultrafiltration. The antibody was labelled with the fluorophore Alexa 488 (Molecular Probes) according to the manufacturers instructions.
Samples of baculovirus infected and non-infected Sf9 cells (2x10e6 cells/sample) were prepared in the following manner, all incubations were on ice. Samples were centrifuged (895g x 5 min), the cell pellets resuspended in PBS containing 1% BSA and incubated for 20 minutes. The samples were centrifuged, the cell pellets resuspended in 200ul of Alexa 488 labelled gp64 mAb (5ug/ml in PBS/BSA buffer) and incubated for 20 minutes, with gentle mixing approximately every 10 mins. The cells were then fixed using the fixation/permeabilisation kit (Coulter), by the addition of 200ul of reagent 1 and incubated for 15 minutes. The samples were made up to 1ml with ice cold PBS/BSA buffer and centrifuged. The cells were resuspended in 200ul of reagent 2 (Coulter kit) for permeabilisation, incubated for 5 minutes after which time 200ul of labelled gp64 mAb was added. The samples were incubated for a further 20 minutes after which time 1ml of PBS/BSA buffer was added and the cells centrifuged. Finally the cells were resuspended in 1ml of PBS/BSA buffer and stored on ice prior to analysis by flow cytometry using excitation at 488 nm. Routinely samples were analysed within 30-60 minutes of preparation.

A. Bernard et al. (eds.), Animal Cell Technology: Products from Cells, Cells as Products, 363–365.

Prior to the inclusion of the fixation/permeabilisation procedure, these stages were omitted along with the additional antibody incubation.

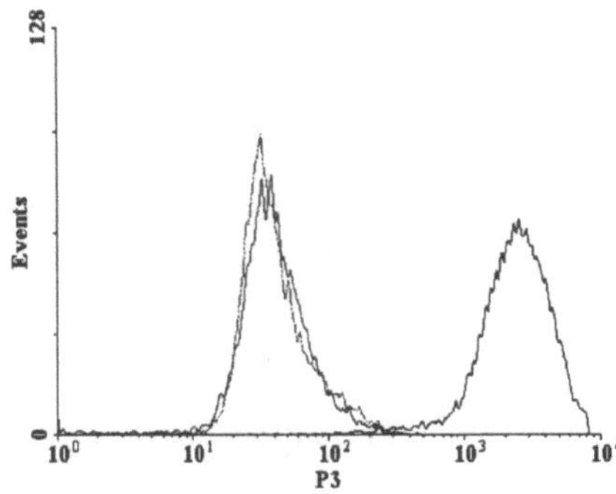

Figure 1
Flow cytometry analysis of Sf9 cells, non-infected, unlabelled (red line), non-infected (blue line) and infected with baculovirus (black line), both labelled with antibody. Samples were taken at 24 hour post infection and prepared as described in material and methods.

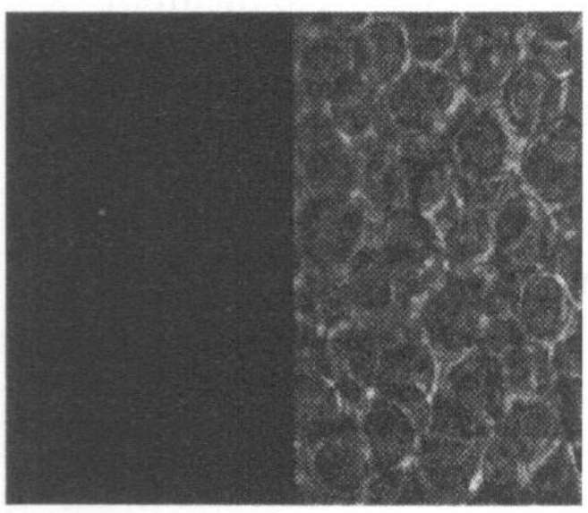

Figure 2.
Transmission and fluorescence image of labelled, uninfected Sf9 cells.

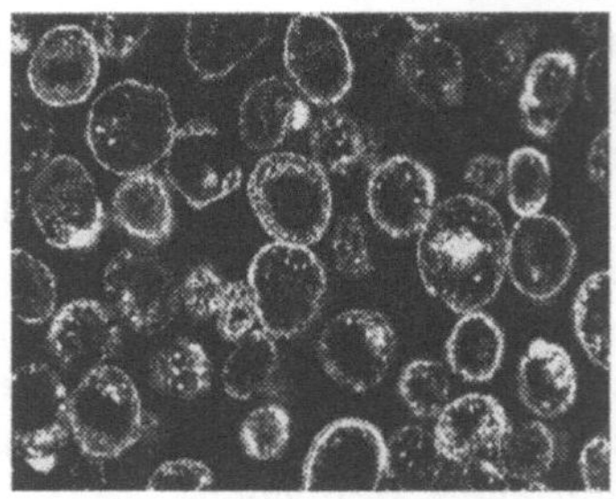

Figure 3.
Fluorescence image of fixed and permeabilised, labelled, infected Sf9 cells

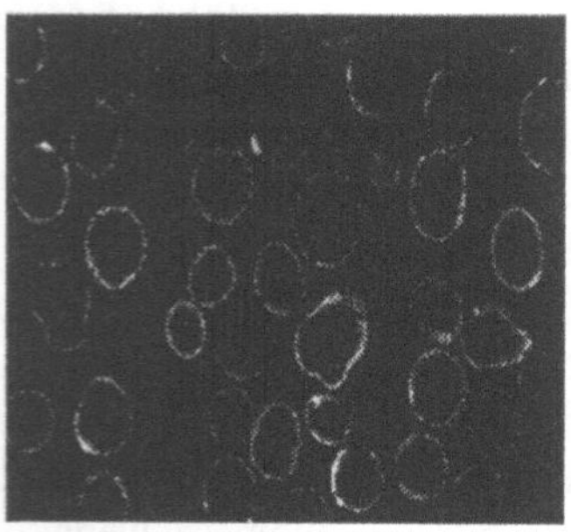

Figure 4.
Fluorescence image of unfixed, non-permeabilised, labelled, infected Sf9 cells.

Labelling/culture conditions	Fold increase in mean fluorescence
0 hours pi	1.1
4.5 hours pi	7.4
16 hours pi	14.3
24 hours pi	18
48 hours pi	9.3
Fixed/permeabilised, 24 hours pi	50

Table 1.
Mean fluorescence increase of infected cells relative to non-infected control. Samples of Sf9 cells were prepared at various time points, as described in the methods section. For the time course samples no fixation/permeabilisation was performed.

Results and Discussion

Initial labelling studies were performed using a crude, partially purified preparation of the gp64 mAb followed by labelling with a secondary FITC detection antibody (data not shown). The results from these analyses gave relatively poor separation (3-5 fold) between infected and non-infected cells and also gave very high non-specific binding of the detection antibody. Therefore, antibody was purified and labelled directly with an improved fluorophore, Alexa 488. In addition, other parameters were optimised, such as antibody concentration and additional washing steps (data not shown). The inclusion of BSA in the buffers reduced non-specific binding to zero (Figure 1). The fixation/permeabilisation of the cells improved the fluorescent signal significantly, up to 50 fold over control cells (Figure 1.). This separation enabled clear identification of infected and non-infected cells and, by specific gating, also allowed enumeration of the two cell populations. Samples were analysed over a time course and showed peak fluorescence (for the time points analysed) at approximately 24 hours post infection (Table 1).

In addition to flow cytometry, samples were analysed using confocal microscopy to examine the distribution of labelling (Figures 2-4). In unfixed, non-permeabilised cells staining of the cell membrane was evident (Fig.4), whereas, with fixed, permeabilised cells, intracellular staining was observed (Fig. 3) with a concomittant increase in fluorescence signal.

This study has provided a novel, rapid method for the analysis and quantitation of infected cells in culture. It is envisaged that this method will prove a useful additional tool for analysing and monitoring the baculovirus infection process

ADAPTATION OF RECOMBINANT HEK-293 CELLS TO GROWTH IN SERUM FREE SUSPENSION

R. MCALLISTER, C. SCHOFIELD, G. PETTMAN and C. MANNIX.
Microbial and Cell Culture Sciences, SmithKline Beecham, Harlow, UK.

1. Abstract

HEK-293 cells are a popular host for the generation of stable cell lines for recombinant protein production, and particularly cloned membrane receptors which are required continuously for screening of drug targets. The purpose of this work was to adapt cell lines to growth in suspension culture in order to increase efficiency. This objective was achieved with three recombinant HEK-293 cell lines expressing 5HT receptors, which were successfully adapted for robust growth in suspension culture using a new serum-free medium, Ex-Cell 520 (JRH Biosciences). The adaptation process was cell line dependent and the time scale of this process varied between 4 and 8 weeks. Cells were observed to grow as a mixture of single cells and loose aggregates and could be repeatedly passaged in the range of 6 x 10^4 cells/ml to 1 x 10^6 cells/ml with a doubling time of between 24 and 35 hours (cell viability of 95%). In summary, the adaptation process resulted in an increased specific cell productivity, and cell growth rates equivalent to standard cultures and all three cell lines have now been successfully scaled to growth in 50L and 100L bioreactors. In addition, the process achieved an estimated 5-fold reduction in consumable costs.

2. Introduction

Recombinant HEK-293 cells expressing membrane receptors are continuously required for high throughput screening. The standard method for production of HEK-293 cells was as adherent cultures in medium containing serum and therefore scale-up was both time consuming and expensive. Previous adaptation experiments in standard medium containing serum showed that cells preferentially formed large aggregates and showed a decrease in receptor expression. In addition, these aggregates are problematic for cell number determination and subculture. Three recombinant HEK cell lines expressing 5HT receptors were adapted to growth in suspension in a new serum-free medium, Ex-Cell 520 (JRH Biosciences). This was initiated by a direct switch into serum-free suspension culture but the subsequent rate and method for full adaptation was different for each cell line. In general this required complete medium replenishment initially every 2 days. The three fully adapted cell lines had growth rates consistent with the adherent control cultures and receptor expression per cell was increased. The data presented shows representative 293_5HT2C growth curves, receptor expression and glucose/lactate metabolism.

A. Bernard et al. (eds.), Animal Cell Technology: Products from Cells, Cells as Products, 367–369.

3. Materials and Methods

Prior to adaptation cells were routinely grown in T-flasks in EMEM supplemented with 10% foetal bovine serum, 1% non essential amino acids and the selective agent G-418 as appropriate. The serum-free medium used was Ex-Cell 520 (JRH Biosciences). Cells used for the adaptation process were taken from cryostorage (293_5HT2C and 293_5HT2A), or from exponentially growing cultures (293_5HT2B). In both cases, cultures were initially inoculated into Ex-Cell 520 medium (supplemented with 4mM L-glutamine) at approximately 2 x 10^5 cells/ml and stirred at 70 rpm. Subsequently, cultures were maintained in the range 2-6 x 10^5 cells/ml in either spinner vessels (Techne) or shake flasks (Corning) until adaptation was complete. Cultures were counted using a haemocytometer (Neubauer) and Erythrosin B dye exclusion stain to determine viable cells. Samples, as appropriate, were taken for glucose/lactate analysis (YSI) and cell pellets prepared for receptor expression analysis. Large scale growth of the 293_5HT2C cell line was performed in a 100L (working volume) stirred tank bioreactor (LH).

4. Results and Discussion

The adaptation of the 293_5HT2C cell line initially showed high viability and only a slight decrease in doubling time (Td) in spinner flask culture but by day 14 the cell viability had decreased to 58% and the growth rate had slowed (Td=69h). By day 33 cell viability was 91% with a satisfactory growth rate (Td=35h). Fig.1 shows a growth curve in which cells grew to a maximum cell concentration of 6.7 x 10^5 cells/ml but recent data has demonstrated cell growth to 1 x 10^6 cells/ml. This cell line has been scaled to growth in a 100L (working volume) bioreactor (Fig.2).

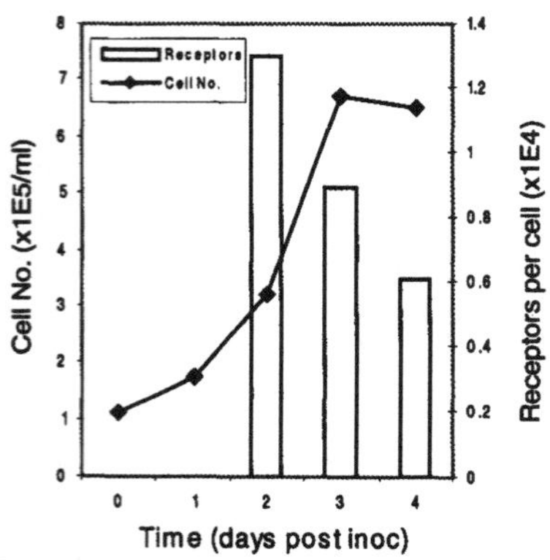

Fig 1. Growth and receptor expression of 293_5HT2C cells. (0.6L culture in 1.5L Techne).

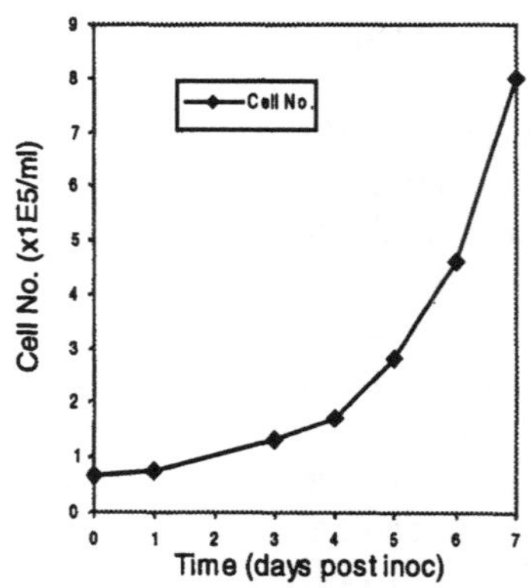

Fig 2. Growth of 293_5HT2C cells in a 100l bioreactor. (28L culture).

The 293_5HT2A cell line showed little growth in the first 14 days in spinner flask culture with a Td between 100 to 120h and cell viability dropped to 75%. By day 49 the rate of growth had improved significantly (Td=34h) and cell viability had increased to 98%.

The 293_5HT2B cell line also showed little growth during the first 14 days in spinner flask culture and during this time cell viability decreased to 70%. This viability remained constant and growth continued slowly (Td=60 to 90h). A cell suspension was transferred to a roller bottle and the viable population adhered. The attached cells were returned to shake flask culture and had a viability of 90%, a greatly improved growth rate (Td=30h) and were grown routinely between 1.0 x 10^5 cells/ml and 1.5 x 10^6 cells/ml.

The adapted cell populations were screened to ensure that the selection events did not result in loss of or decrease in recombinant protein expression. Receptor expression was increased 4-8 fold and 2-3 fold in the 293_5HT2A and 293_5HT2C cell lines respectively compared to that of adherent cultures (see table 1). The different cell lines utilised and produced different amounts of glucose and lactate respectively during growth in Ex-Cell 520 medium (see table 2).

Table 1: Comparison of receptor expression from adherent (850cm^2 roller bottle) and suspension cultures

Growth method	293_5HT2A		293_5HT2C	
	Cell concentration (xE5/ml)	Receptors per cell (xE6)	Cell concentration (xE5/ml)	Receptors per cell (xE5)
Adherent	4.2[a]	0.3	3.6[a]	0.5
Suspension	3.7[a]	2.1	3.2[a]	1.3
Adherent	5.0[b]	0.3	6.6[b]	0.3
Suspension	5.3[b]	1.2	6.7[b]	0.9

(a,b) denote two representative cell concentrations

note: cell concentration equivalents for adherent cultures were calculated from roller bottles.

Table 2: Glucose utilisation and lactate production by 293_5HT2A, 2C and 2B cells during suspension culture growth in Ex-Cell 520.

Days post inoc.	293_5HT2A			293_5HT2C			293_5HT2B		
	Cell conc^n xE5/ml	Glucose (mM)	Lactate (mM)	Cell conc^n xE5/ml	Glucose (mM)	Lactate (mM)	Cell conc^n xE5/ml	Glucose (mM)	Lactate (mM)
0	1.1	15.9	4.8	1.1	14.8	6.5	1.0	18.1	1.0
2	2.8 (2.2)	10.3 (12)	14.9 (12)	1.7	13.5	8.9	3.6	16.2	3.5
3	3.7	8.0	18.8	3.4	10.8	13.0	6.2	14.6	5.8
4	5.3	4.9	23.2	6.7	9.2	15.9	11.0	9.6	11.6
5	6.7	2.3	25.8	-	-	-	-	-	-

( ) Denotes change in parameter due to dilution of culture.

GROWTH ON MICROCARRIERS AND NUTRITIONAL NEEDS OF HIGH DENSITY INSECT CELL CULTURES

L. IKONOMOU[1], G. BASTIN[2], Y.-J. SCHNEIDER[3], and S.N. AGATHOS[1]
[1]*Unit of Bioengineering, Catholic University of Louvain, Place Croix du Sud 2/19, B-1348, Louvain-la-Neuve, Belgium*
[2]*Centre for Systems Engineering and Applied Mechanics, Av. George Lemaître 4, Catholic University of Louvain, B-1348, Louvain-la-Neuve, Belgium*
[3]*Laboratory of Cellular Biochemistry, Catholic University of Louvain, Place L. Pasteur 1, B-1348, Louvain-la-Neuve, Belgium*

1. Introduction

The insect cell/baculovirus (ICB) system has become popular both for the production of recombinant proteins and biopesticides. The development of serum-free media, advances in baculovirus vector construction and the relative simplicity of insect cell culture make the ICB quite powerful and versatile for recombinant protein expression. We report here experimental results on the growth of insect cells in bioreactor. The evaluation of insect cell growth on Fibra-Cel® microcarriers in suspension is also presented.

2. Materials and Methods

Sf9 and High Five™ cells were a gift from SmithKline Beecham, Belgium and from the Laboratory of Virology, Agricultural University of Wageningen, Netherlands, respectively. The SF-900 II and Insect XPRESS media were purchased from Life Technologies and BioWhittaker Europe. Metabolite levels were determined using enzymatic kits from Boehringer Mannheim. Amino acid quantification was performed by HPLC. The bioreactor used in this study was Celligen Plus™ (New Brunswick Scientific), with working volume of 1 l and Insect XPRESS medium. Oxygen level was set at 50% of air saturation and aeration was performed via the headspace. Fibra-Cel® disks from Bibby Sterilin were employed. They were evaluated in siliconised 250-ml disposable Erlenmeyer flasks containing 50 ml of medium and kept at 100 rpm and 27°C. Half of the medium was changed on day 3, 6 and 8 for Sf9 and on day 3 and 6 for High Five™ cells. All of the medium was changed on day 8 in the flasks containing High Five™ cells. Disk samples were incubated for 2 h with crystal violet (0.1% w/v) dissolved in citric acid (0.1 M) and Triton X-100 (0.1% v/v).

A. Bernard et al. (eds.), Animal Cell Technology: Products from Cells, Cells as Products, 371–373.

3. Results and discussion

3.1. BATCH CULTURE OF SF9 CELLS IN THE CELLIGEN PLUS™

Figure 1A shows the viable cell concentration and the viability of Sf9 cells during a batch culture in the Celligen Plus™ bioreactor. By the time glucose was exhausted

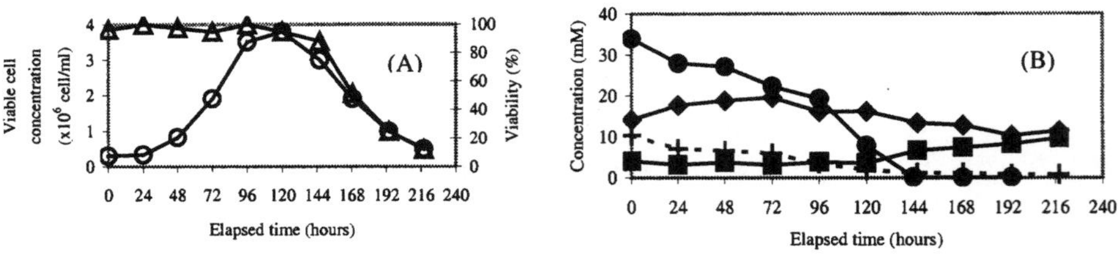

Figure 1. (A) Time course profiles of viable cell concentration (○) and viability (△) of Sf9 cells in bioreactor (B) Consumption of glucose (•), glutamic acid (▲), glutamine (+) and formation of ammonia (■) during batch culture of Sf9 cells in bioreactor (medium : Insect XPRESS).

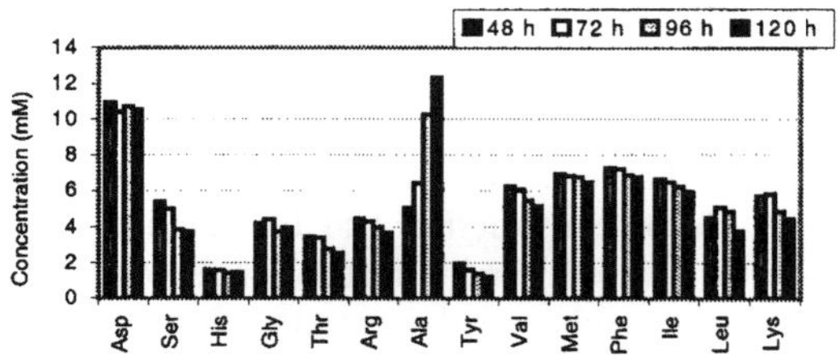

Figure 2. Profile of amino acids during growth of Sf9 cells in bioreactor (medium : Insect XPRESS).

(Figure 1B), cells entered the death phase. The glutamate accumulate at the beginning of growth and was consumed thereafter. Alanine accumulate during the exponential growth phase (Figure 2) and ammonia at the late stage of the culture. The above are in agreement with the metabolism scheme proposed by Bédard *et al.* (1993). The profiles of amino acids in the time period from 48h to 120h, show that they were all present in significant proportions in the medium even at the beginning of the stationary phase. Therefore potential nutrient limitation could not attributed to their exhaustion. It seems that the initial amino acid content can be reduced without any effects on cell growth, as has been suggested by Ferrance *et al.* (1993). The specific consumption rates for glucose and glutamine for the exponential phase of growth (24-96 h) are shown in Table 1. Both values are similar to values reported in the literature (Rhiel *et al.*, 1997). The determination of consumption rates could facilitate the design of feeding strategies.

TABLE 1. Specific consumption rates of key metabolites for Sf9 cells

Metabolite	Specific consumption rate (mol/cell.s)
Glucose	$2.59*10^{-17}$
Glutamine	$1.01*10^{-17}$

3.2. GROWTH ON FIBRA-CEL® MICROCARRIERS

Although Fibra-Cel® disks are used mainly in packed bed configurations, there is at least one reference where they had been utilised in suspension culture (Racher *et al.*, 1995). As seeding at a density of $2.5*10^5$ cells/cm^2 of projected area resulted in absence of growth for both Sf9 and High Five™ cells we tested two higher densities namely

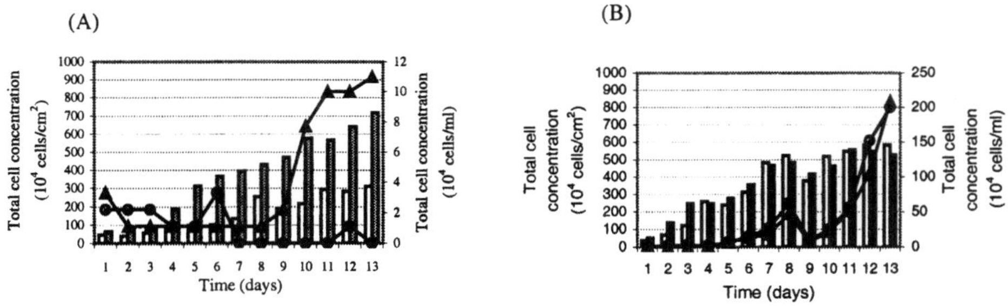

Figure 3. Time course profiles of total cell concentration in disks and in bulk liquid for Sf9 cells (A) and High Five™ cells (B): i) seeding density of $5*10^5$ cells/cm^2, disks (open bars), liquid (○) ii) seeding density of $7.5*10^5$ cells/cm^2, disks (full bars), liquid (△)

$5*10^5$ cells/cm^2 and $7.5*10^5$ cells/cm^2. As can be seen in Figure 3A, cell growth was dependent of the seeding density as far as the Sf9 cells are concerned. That was not the case with High Five™ cells where both seeding densities led to similar final densities (Figure 3B). Cell growth in the microcarriers was accompanied by cell leaking in the medium after the 8th day for Sf9 cells and the 7th for High Five™. Nevertheless, the further increase in cell concentration in liquid is probably due to the proliferation of free cells rather than to continuous leaking of cells from disks. This is supported by the relatively stable or even increasing total concentration in the disks (Figure 3A and B) and the increase of cell viability in the bulk liquid (data not shown). Leaking cells from the microcarriers to the bulk medium act as an inoculum for the liquid leading to a mixture of growing free cells with cells in microcarriers. Thus, use of Fibra-Cel® disks in suspension cultures is an easy way to reach higher cell densities with lower inoculation densities than usual.

4. References

Bédard, C., Tom, R. & Kamen, A. (1993). Growth, nutrient consumption, and end-product accumulation in Sf-9 and BTI-EAA insect cell cultures: Insights into growth limitation and metabolism. *Biotechnol. Prog.* **9**, 615-624.

Ferrance, J. P., Goel, A. & Ataai, M. M. (1993). Utilization of glucose and amino acids in insect cells cultures: Quantifying the metabolic flows within the primary pathways and medium development. *Biotechnol. Bioeng.* **42**, 697-707.

Racher, A. J., Fooks, A. R. & Griffiths, J. B. (1995). Culture of 293 cells in different culture systems: cell growth and recombinant adenovirus production. *Biotechnol. Tech.* **9**, 169-174.

Rhiel, M., Mitchell-Logean, C. M. & Murhammer, D. W. (1997). Comparison of *Trichoplusia ni* BTI-Tn-5B1-4 (High Five™) and *Spodoptera frugiperda* Sf-9 insect cell line metabolism in suspension cultures. *Biotechnol. Bioeng.* **55**, 909-920.

USE OF FLOW CYTOMETRY IN THE EVALUATION OF HYDROXYAPATITE FOR THE PURIFICATION OF PLASMID DNA

Jason Wright, Pascal Batard, Johanna van Adrichem, Ruth Freitag and Florian M. Wurm.
LBTC, EPFL, 1015 Lausanne, Switzerland.

Keywords: Flow Cytometry, hydroxyapatite, plasmid DNA

Introduction

Flow cytometry is traditionally used to analyse and or sort cells in terms of diffusion and fluorescent characteristics, whereby the analyser detects specific properties of cells such as the expression of a fluorescent protein, (e. g. GFP). Solid particles have been used in a number of applications, including the monitoring of phagocytosis (1) and immunoassay (2). We have analysed ceramic hydroxyapatite particles and investigated the binding of plasmid DNA (in the presence of ethidium bromide) to this material under different conditions. We hereby demonstrate the effectivity of cytometry in the analysis of the binding characteristics of molecules on a chromatographic stationary phase.

Materials and Methods

DNA containing solutions were mixed with 10mg of hydroxyapatite (ceramic type II, 20µm, BioRad laboratories, California) in 1.5 ml eppendorf reagent tubes at room temperature by gentle inversion. Binding was carried out either in a 10mM calcium chloride solution (strong DNA binding) or a 10mM sodium phosphate buffer, pH 7.2 (weak DNA binding), either in the presence or absence of ethidium bromide. Control binding assays to monitor the background signal were performed without DNA in the presence and absence of ethidium bromide, under exactly the same conditions and in parallel to binding assays containing DNA. 12.5µg of purified plasmid DNA were used in each binding assay. Binding for 1 min was followed by centrifugation at 13,000 rpm for 2 minutes at room temperature, supernatants were removed and the hydroxyapatite pellet resuspended in the wash buffer, 1mM PO_4^{3-} pH 7.2. Hydroxyapatite particles were further separated by centrifugation (as before) and then washed a further two times in the wash buffer. Resuspension in the same buffer was followed by a 1 in 10 dilution before analysis by cytometry, i. e. the sample used in particle analysis contained 1mg / ml hydroxyapatite.

A. Bernard et al. (eds.), Animal Cell Technology: Products from Cells, Cells as Products, 375–377.

Particles were analysed with the Pas III, Partec, Munster, Germany. Sample volumes were typically 1 to 2 ml, flowrates were relatively high 5 – 20 µl / s to prevent particle sedimentation. Particles were excited by an argon ion laser at 488nm, an emission filter >620nm was used to monitor the fluorescence emitted by ethidium bromide on particles. Data acquisition was performed using Pas software, Partec, subsequent analysis was performed using WinMDI software. The intensity of fluorescence was quantified as the mean of fluorescent events in the defined region of morphology, corresponding to 20 micron Hydroxyapatite particles.

Results and Discussion

The histograms and accompanying fluorescence mean values presented in figures 1(a) to 1(d) represent all fluorescence events in the defined morphological region (figure 1(e)), the fluorescence events outside of this region are excluded from the analysis. The mean fluorescence of DNA bound to particles in the presence of ethidium bromide is 95 fold greater than that of DNA bound to particles in the absence of ethidium bromide (see Figures 1(a) and (b)). In the presence of ethidium bromide, a 38 fold difference in particle fluorescence is observed under strong binding conditions, where DNA is bound in a calcium containing solution as opposed to a phosphate buffer (Figures 1(b) and (d)). Ethidium bromide exposed to particles in the absence of DNA revealed that no non specific binding of the dye to the particles had occurred and that the fluorescence exhibited by particles was due only to ethidium bromide – DNA interactions. Suprisingly, the binding of plasmid DNA in the absence of ethidium bromide resulted in an enhanced particle fluorescence. The mean fluorescence intensities of figures 1(a) and 1(c), reveal an approximate six fold increase in the fluorescence on particles exposed to DNA only in the presence of calcium in comparison to binding in the phosphate buffer. The observed shift in relative fluorescence of particle bound DNA - ethidium bromide complexes appears to be a useful property, which can be used to optimise DNA binding on HA particles. Allowing a large number of binding conditions, i. e. buffer, pH the presence of additional compounds to be screened in a short time. This may help in the development of purification procedures. We are currently constructing standard curves to enable us to quantify accurately the amount of bound DNA, based on the mean fluorescence values obtained for different amounts of bound DNA.

Whether the methods and data presented herein are directly applicable to the binding behaviour of DNA on chromatography columns is debatable, as during the binding assays the particles undergo resuspension and centrifugation several times. However, analysis of the laser light scattering pattern from these particles shows that they are relatively homogeneous, (Figure 1(e)). Hence, the binding assays and the subsequent cytometric analysis had little effect on the particle size distribution. One of the major limiting factors thus far has been the relatively high flow rates used in the analysis; these are required to prevent particle settling in the sample tube during injection. To optimise our protocol further we are looking at ways to keep the hydroxyapatite particles in suspension longer prior to injection on the partec analyser. The particles involved in the binding assays were exposed to light at certain points in the binding assay and during analysis, approximately 30 - 60 minutes. Prolonged exposure to natural light has an

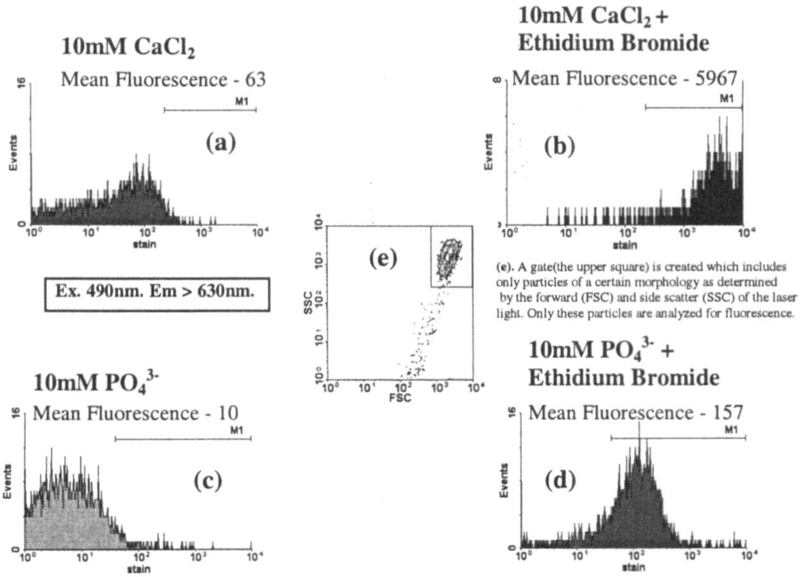

Figure 1. The enhanced fluorescence displayed by Hydroxyapatite (HA) particles upon DNA binding by flow cytometry. A comparison of binding in the presence of calcium and phosphate.

effect on the fluorescent signal emitted ethidium bromide; this is therefore a parameter which needs to be carefully monitored in order to achieve reproducible data.

From here, we envisage applying cytometry further to model the binding of molecules on particles. We are particularly interested in quantifying the fluorescent signal emitted by DNA specific dyes on plasmid DNA bound to particles. The working range for the quantification of DNA bound to particles in the presence of ethidium bromide is currently under evaluation. We envisage taking this work further to look at competitive binding assays, e. g. using a combination of protein and DNA specific dyes as a preliminary step in evaluating the binding properties of DNA in more complex mixtures. In addition to monitoring fluorescent events on particles (a representation of the amount of molecules bound to a population of particles), flow cytometry also allows the morphology of particles to be monitored. Thus the homogeneity in the morphology of a large population of particles from a specific chromatographic stationary phase can be readily detected in a short time frame.

References

1. Steinkamp J.A., Wilson J.S., Saunders G.C. and Stewart C.C. - Phagocytosis: flow cytometric quantitation with fluorescent microspheres. *Science.* **215** (1982), 64 - 66.

2. Bishop JE and Davis KA – A flow cytometric immunoassay for beta2 – microglobulin in whole blood. *Journal of Immunological methods.* **210** (1997), 79 – 87.

REGULATORY AND STABILITY ISSUES IN ANIMAL CELL CULTURE FOR PHARMACEUTICAL PRODUCTION

Chapter VI

REGULATION AND STANDARDISATION OF GENE TRANSFER PRODUCTS IN THE EUROPEAN UNION

Klaus Cichutek
Dept. of Medical Biotechnology
Paul-Ehrlich-Institut
Paul-Ehrlich-Str. 51-59
D-63225 Langen
E-mail to: cickl@pei.de

1 Introduction

In Germany gene therapy is used as a general term for human somatic gene and (genetically modified) cell therapy. It encompasses the use of recombinant drugs in humans. These drugs are on the one hand naked DNA, viral or non-viral vectors in vivo, or, on the other hand, genetically modified cells in vivo. The active ingredient is the gene product made in vivo, and thus, indirectly, also the gene to be delivered. The vector, if used, may be considered as a component of the drug. The formulation may contain additional ingredients.

Definition: Gene therapy products as defined above and used in vivo are medicinal products according to the German Drug Law (AMG; „Arzneimittelgesetz"). They include DNA, viral or non-viral vectors and genetically modified autologous, allogeneic or xenogeneic cells (used in vivo) and are often summarised under the term „gene therapy drugs" in Germany. No official definition of the term „gene therapy drug" has been included in the AMG (up to the 9th revised version).

Generally, proprietary medicinal products are used in clinical trials of phase I to III to learn about their safety, efficacy and potential environmental risks associated with their use. The data collected form the basis for an application for marketing authorisation which, if granted, allows the standard use of the drug. In contrast, special preparations for individual patients do not require marketing authorisation in order to be used in humans. For example, autologous modified cells are generally only used in the patient himself. Nevertheless, current German practice, as proposed in the GT guidelines of the German Medical Association and in the final report of the working group "Bund/Länder-Arbeitsgemeinschaft Somatische Gentherapie", is to also use individually prepared drugs under the provisions of § 40 AMG, i.e. in clinical trials.

Regulation of gene therapy drug use in humans prior to marketing authorisation is mainly provided by the AMG and the professional law of physicians. Application of GMOs and therefore of gene therapy drugs in humans is not regulated by the German Gene Technology Law (GenTG). Approval of deliberate release according to the

A. Bernard et al. (eds.), Animal Cell Technology: Products from Cells, Cells as Products, 381–387.

GenTG is not required. Regulations are identical for gene therapy drugs and other drugs.

2 Preclinical Research

Experimental work in gene therapy including the construction, use, storage and inactivation of vectors, genetically modified bacterial or mammalian cells or animals has to be conducted according to the German Law on Gene Technology (GenTG; „Gentechnikgesetz"; transformation of Council Directives 90/219/EEC and 90/220/EEC). Basically, all experiments therefore are to be performed in gene-technology laboratories under contained use. If only risk group 1 organisms are used, this only involves notification of the competent authority, which is different for each Land in Germany. This is true for the use of naked DNA or non-viral vectors and genetically modified cells assuming that genetic sequences void of any potential pathogenicity for humans or animals would be used. Work with risk group 2 organisms such as adenoviral and retroviral (including lentiviral) vectors has to b performed in safety level 2 laboratories which have to be registered and approved. Each line of experiment involving the use of GMOs also has to be approved. Similar pre-conditions are given for work with risk group 3 GMOs, which could for example include hybrid vectors derived from adeno- and lentiviruses.

Generally in Germany, experiments involving the use of genetically modified organisms (GMOs) have to be performed in laboratories or animal facilities of one of four safety levels (S1 to S4), which are accordingly equipped. Laboratory approval is given by the competent authority of the Land („Bundesland") for the GenTG. As experiments in safety level 1 laboratories only have to be documented and the competent authority has to be notified, the experiments can be started as soon as the authority has received the notification. As experiments falling under higher safety levels need additional approval by the same authority, experiments normally can be started about 3 months or less after application.

The Central Commission for Biological Safety (ZKBS; „Zentrale Kommission für die Biologische Sicherheit", secretariat located at the Robert-Koch-Institut, Berlin) provides a list containing the safety level classifications of „standard" vectors or plasmids and GMOs and is in some cases (e.g. approval of safety level 3 operations) to be consulted by the competent authority of the Land for the GenTG. Thus, advice about the approval of gene laboratories and genetic experiments is given by the competent authority of the Land where the contained use facility is located.

Other laws and regulations that may apply (which are executed by different competent authorities of the Land where the laboratory is located) include the law on epidemics („Bundesseuchengesetz", to be replaced by the „Infektionsschutzgesetz"), the law on animal protection („Tierschutzgesetz"), the law on human embryo protection („Embryonenschutzgesetz"), the radiation protection ordinance („Strahlenschutzverordnung") and the ordinance on the use of hazardous substances („Gefahrstoffverordnung").

3 Manufacture

Drugs for clinical use have to be produced according to Good Manufacturing Practice (GMP). GMP has to be implemented as defined in the „Operation Ordinance for Pharmaceutical Entrepreneurs (PharmBetrV; „Betriebsverordnung für pharmazeutische Unternehmer"). Also generally relevant are the GMP guidelines issued by the WHO, the European Community and PIC. Good Laboratory Practice (GLP) implemented by the German law on the use of chemical substances (ChemG; „Chemikaliengesetz") is required in certain cases, e.g. for analyses relevant to the safety of the drugs, according to the German law on the use of chemical substances (ChemG; „Chemikaliengesetz"). Manufacturing authorisation is necessary for those manufacturers who intend to commercially or professionally distribute the drug (and/or the active ingredient) to others. It is granted by the authority of the Land, where the facility is located, competent for the AMG (§ 13 AMG). This authority is also responsible for inspections and supervision (§ 64 AMG). Notification of the competent authority of the Land for the AMG is necessary beforehand for companies and establishments (also clinical departments) which develop, manufacture, test, package drugs or subject them to clinical trials (§ 67 AMG).
Manufacturing authorisation for drugs which are blood products (like genetically modified blood stem cells) or vaccines (like genetically modified tumour cells), is given under consultation with the Paul-Ehrlich-Institut. This involves a site visit from a member of the competent authority together with a member of the Paul-Ehrlich-Institut before authorisation is granted. According to the drug law in its newer version (§ 14 AMG), the production and testing of the drug intended to be used in the clinic has to be performed according to the established standards of science and technology. Therefore, respective prescriptions for production and testing have to be included in the application and will be reviewed.

4 Clinical Trial („Klinische Prüfung")

In the guidelines „Richtlinien zum Gentransfer in menschliche Koerperzellen", published by the German Medical Association the requirement to use gene therapy drugs during clinical trials only is stressed. Clinical trials can only be conducted, if certain requirements are met (see §§ 40, 41 AMG). Positive appraisal of a local, independent ethics committee formed according to the law of the Land, where the trial is performed, is required before initiation of a clinical trial (see § 40 (1) AMG for exception). If a multicenter trial is going to be performed, all relevant ethics committees have to give their appraisal. No IND approval by a higher federal authority is necessary except for the submission of a complete set of certain documents (see below).

Notification of the competent authority of the Land (AMG) and deposition of the clinical study plan is required according to § 67 AMG. This authority is also responsible for inspections and supervision of the trial according to § 64 AMG. If necessary, samples of the material produced or tested can be taken.
Submission of documents (presentation according to § 40 AMG; forms available by Internet: http://www.dimdi.de/germ/amg/klifo.htm") including the positive appraisal of

the local ethics committee(s) and the pharmacological-toxicological data (see German Directive „Arzneimittelprüfrichtlinien" for content) to the competent federal higher authority (AMG) are required. The competent authority is either the Paul-Ehrlich-Institut, Langen, for gene therapy drugs which are vaccines or blood preparations or the Federal Institute for Drugs and Medical Devices (BfArM), Berlin, for other gene therapy drugs (see § 77 AMG). The trial can only be initiated after written confirmation of the competent higher authority that all documents required have been received.
A positive appraisal has to be obtained from the Commission for Somatic Gene Therapy of the German Medical Association (KSG-BÄK, „Kommission für somatische Gentherapie", Koeln. This appraisal is required according to the professional law of physicians (see „Richtlinien zum Gentransfer in menschliche Koerperzellen", Deutsches Aerzteblatt 92, Heft 11, B-583-B588 (1995)). The Commission is giving advice to the local ethics committees about issues related specifically to gene therapy. Members of the Commission are currently scientific experts in the fields of clinical gene therapy and so-called vectorology as well as an expert in ethics.
Registration of the clinical trial and patients involved with the German Gene Therapy Register (DGTR, „Deutsches Gentherapie-Register") is requested at the German Working Group for Gene Therapy (DAG-GT) and the publication of the protocol is recommended. The registration is important in order to keep records of the gene therapy trials going on.

5 Contacts and further information:

The competent authorities of the Land are in charge of authorising and supervising the drug manufacture and the clinical trials. They can give advice on related questions.
The competent higher federal authorities involved in gene therapy can be contacted for further information. The author is a member of the Commission for Somatic Gene Therapy and was thus recruited from the Department of Medical Biotechnology of the Paul-Ehrlich-Institut to give advice on questions of drug safety and pharmacology (Prof. Dr. K. Cichutek, Paul-Ehrlich-Str. 51-59, D-63225 Langen, Tel. +49-6103-77-5307, Fax +49-6103-77-1255, e-mail cickl@pei.de). The Paul-Ehrlich-Institut also arranges informal discussion on the design of clinical gene therapy trials as well as questions of preclinical testing of such drugs and the biological monitoring with a view to drug licensing. The expert of the Federal Institute for Drugs and Medical Devices (BfArM), Dr. U. Kleeberg, can also be contacted for further information Seestr. 10-11, D-13353 Berlin, Tel. +49-30-4548-3356, Fax +49-30-4548-3332, e-mail u.kleeberg@bfarm.de.
The secretariat of the Commission for Somatic Gene Therapy (KSG-BÄK, Wissenschaftlicher Beirat der Bundesaerztekammer, Herbert-Lewin-Str. 1, D-50931 Koeln, Tel. +49-221-4004-0, Fax +49-221-4404-386, e-mail dezernat6@baek.dgn.de) can be asked for advice on questions related to the application. The secretariat of the Commission can also arrange hearings during which general quesstions related to gene therapy can be discussed. For general infromation, the German Working Group for Gene Therapy, (contact: Dr. M. Hallek, Muenchen, e-mail hallek@lmb.uni-muenchen.de) is available.

6 Overviews on German gene therapy regulations

Overviews on and introduction to German gene therapy regulations can be found in:

- Guidelines for the design and implementation of clinical studies in somatic cell therapy and gene therapy, A. Lindemann et al., J. Mol. Med. 73, 207-211 (1995);
- Abschlußbericht der Bund/Länder-AG „Somatische Gentherapie", in: Eberbach/Lange/Ronellenfitsch (Hrsg.), Recht der Gentechnik und Biomedizin, GenTR/BioMedR, Teil II, F., Loseblattwerk, Heidelberg, Stand: 19. Erg.Lfrg., Dezember 1997).
- Gene therapy in Germany and in Europe: Regulatory Issues, K. Cichutek and I. Krämer, Qual. Assur. J. 2, 141-152 (1997).

A brief summary of German gene therapy regulations is also available on the Euregenethy website (http://193.48.40.240/www/euregenethy/reg/Germanfront).

Discussion (Cichutek)

Bavand: I have a question on autologous tumour vaccines - how would they be regulated? As a product with a centralised procedure, and if it is a centralised procedure, how would it work compared to the known procedures now established?

Cichutek: If I understand the thrust of the question, it is that a lot of the tumours are autologous products at the moment - in the laboratory samples from tumour tissue have been grown and genetically modified, or have been mixed with other cells, and given back to patients. There are products which are not regulated at this point as the process is for individually prepared products which cannot be licensed as they are just prepared for individual patients. They are not finished products and are not going to be licensed until after safety, efficacy and quality has been shown in clinical trials. The second point is that a lot of companies take cell lines which they have genetically modified, and provide them to the patient as a tumour vaccine. Those are products that are going to be licensed through the central licensing procedure. That is going to be similar to the one we have, ie a single file is sent to the EMA and experts are selected and a reporter is appointed. The experts will examine the file, questions will go back and forth with the applicant and the product eventually licensed, or not. So there is a difference between finished products and those that have been prepared locally for individuals.

STUDY OF STABILITY OF EXPRESSION OF A HUMANIZED MONOCLONAL ANTIBODY FROM A TRANSFECTED NSO MYELOMA CELL LINE USING DIFFERENT CULTURE MEDIA AND SERUM CONCENTRATIONS.

CASTILLO A. J.[(1)], VICTORES S.[(1)], MARISON I. W[(2)].
[(1)] Center of Molecular Immunology (CIM), P.O. Box 16040, Havana 11600, Cuba,
[(2)]Institute of Chemical Engineering and Bioengineering, Chemical Department, Swiss Federal Institute of Technology, CH-1015, Lausanne, Switzerland.

Introduction

The use of mammalian cell lines for expression of recombinant proteins with potential therapeutic use has become increasingly prevalent.
Optimising productivity by recombinant cells relies partly on understanding relationships between regulation of protein expression and culture conditions, e.g., composition of cell culture medium.
Stability of expression of recombinant cell lines is a major problem for many researchers and manufacturers.
A humanised monoclonal antibody against EGF-receptor has been expressed in NSO myeloma cells (1). After successive selection and cloning procedures the clone R3T/16 showed the higher expression level and was selected for further development. However this clone decreases antibody productivity after 20 days in culture using DMEM/F-12 medium supplemented with 5 % of foetal calf serum.
In attempt with that clone R3T/16 was adapted to growth in different cell culture media in order to study the stability of these cells in the respective media.
The aim of this work was to answer the question: How to obtain a good producer stable NSO recombinant cell line expressing an humanised monoclonal antibody?.

Results

1.- Stability studies of R3T/16 in different cell culture medium:
Five media composition were tested:
- DMEM/F-12 with 5 % of foetal bovine serum (FBS): Control medium.
- RPMI-1640 medium with 5 % (v/v) of FBS.
- RPMI-1640 medium with 5 % of Myoclone FBS: In order to eliminate the effect of serum batch.
- RPMI-1640 medium with 1 % (v/v) of FBS.
- Turbodoma HP-1 supplemented with 1 % (v/v) of FBS (THP1 medium).

Dr. F. Messi Cell Culture Technologies kindly provided Turbodoma medium. All other media and serum were purchased from Gibco.
Cells were sub-cultured every 3 or 4 days in 25 cm^2 T-flasks by duplicates. Periodically cells were taken from each flask and seeded by triplicate in 24-well cell culture plates at 1 x 10^5 cells/ml and one milliliter per well. After 9-11 days supernatant was tested by an anti-human IgG sandwich ELISA in order to control the productivity of the cells during the culture. From this results (Fig. 1) we can conclude that R3T/16 cell line in THP-1 medium with 1 % of FBS had a stable productivity during the first 50 days. On the contrary cultures in DMEM/F-12 and RPMI-1640 supplemented both with 5 % of FBS decreased their immunoglobulin production levels from the same beginning.

A. Bernard et al. (eds.), Animal Cell Technology: Products from Cells, Cells as Products, 389–391.

<u>2.- Kinetic parameters of R3T/16 before and after long term culture:</u>

The decrease in the antibody production could be a consequence of reduction of specific antibody production rate or due to a decrease in the maximal viable cell concentration.

Some kinetic parameters were determined at the end of the long term cell cultures in order to elucidate the source of reduction of IgG production levels in DMEM/F-12 and RPMI-1640 media supplemented with 5 % of FBS. Results from these experiments are showed in Table 1.

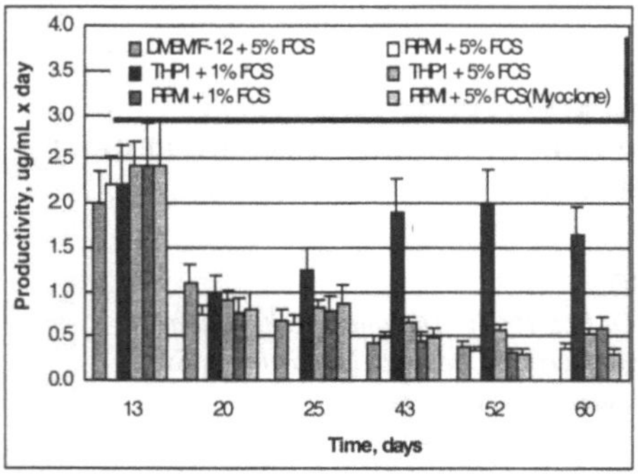

Figure 1: Stability of immunoglobulin production rates for R3T/16 cell line in different cell culture media

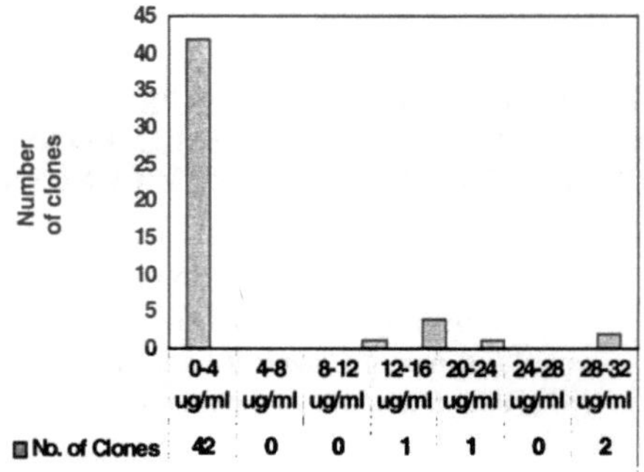

Figure 2: Clone distribution by production levels for a cloning of R3T/16 cells in RPMI-1640 medium with 5 % FBS after 50 days in culture. Only one clone produces between 1 and 4 μg/ml

Table 1: Kinetic parameters for R3h3/t16 cell line in DMEM/F-12 media with 5 % of FBS, RPMI-1640 supplemented with 5 % of FBS and THP1 with 1 % of FBS cell culture media. Cells were seeded by triplicate in 24 wells cell culture plates at 10^5 cells/ml per well. μ–specific growth rate, q_{ab}-specific production rate, Xv_{max}-maximal viable cell concentration, IgGmax-maximal Ig concentration.

Culture medium	Xv_{max} x10^6 cells/ml	μ_{max} h^{-1}	IgG_{max} μg/ml	q_{ab} pg./cell x h
DMEM/F-12 (5% FBS)	1.47	0.103	1.8	0.007
RPMI-1640 (5% FBS)	1.21	0.082	2.8	0.017
THP/1 (1% FBS)	1.05	0.086	10.7	0.056

Growth parameters (μ, Xv_{max}) were very similar for all tested conditions, however the production parameters (q_{ab}, IgGmax) were higher for THP1 medium with 1 % of FBS. From these results we can conclude that the decreases in production levels after long term cell culture in DMEM/F-12 and RPMI-1640 media supplemented with 5 % of FBS are induced by a reduction in the specific production rates.

<u>3.- Cloning of R3T/16 after long term cultures.</u>

Emerging non-producing populations could induce the decrease in specific antibody production rate. In order to elucidate this point, R3T/16 cell line was cloned in THP-1 medium with 1 % of FBS and RPMI-1640 medium with 5 % of FBS at the end of the long-term cell culture. Non growth was observed for THP-1 medium with 1 % of FBS, indicating that the sub-cloning of these cells at low serum levels (below 5% v/v) is not possible. For RPMI-1640 medium with 5 % of FBS it was observed cell growth in 50 wells, however IgG production was detected only for 9 clones (Fig. 2), indicating that most of this cells became non-producing after long term culture in RPMI-1640 with 5 % of FBS.

4.- Stability studies of clones selected after long term culture of R3T/16 cell line in RPMI-1640 medium with 5 % of FBS.

From the sub-cloning of R3T/16 cell line in RPMI-1640 medium with 5 % of FBS we selected two clones in order to characterise them of point of view of stability of productivity: a high producer one (R3H/7; 31.6 µg/ml) and medium producer one (R3D/8; 14 µg/ml). The stability studies were carried out following the protocol above mentioned for R3T/16 clone. The results obtained for these clones (Fig. 3) showed that both cell lines were stable for 55 days in culture.
Cell line R3H/7 was sub-cloned in RPMI-1640 medium with 5 % of FBS after this stability study (Fig 4). There was observed cell growth in 68 wells and most of them were producers. These results contrast with those obtained for the sub-cloning of R3T/16 cells, after the first stability study.

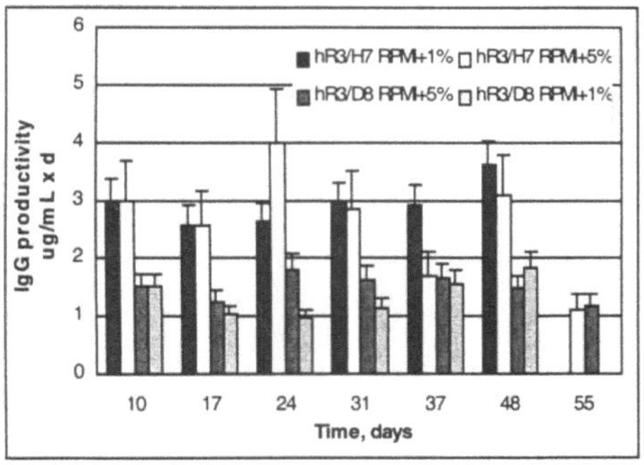

Figure 3: Stability study for R3H/7 y R3D/8 clones in RPMI-1640 medium supplemented with 5 % and 1 % of FBS.

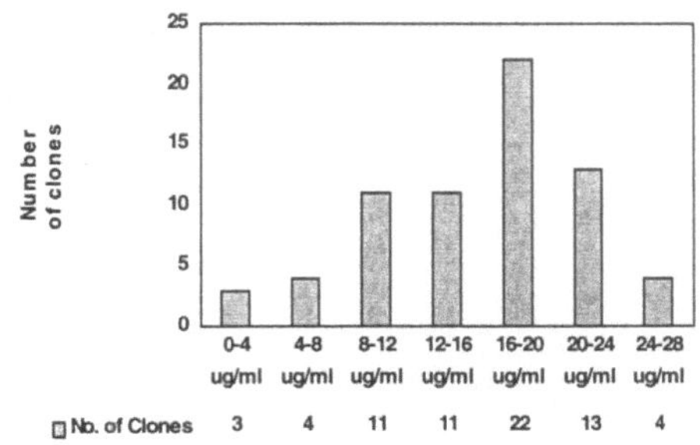

Figure 4: Clone distribution by production levels after cloning R3H/7 cells in RPMI-1640 medium supplemented with 5 % FBS after 55 days in culture.

Conclusions

From obtained results (Fig. 1) we can conclude that basal medium composition or level of serum supplement could have a great influence upon the stability of expression level for transfected mammalian cell lines. In certain cases the increase in serum supplement induces a reduction of production levels for R3T/16 cell line in long term cell cultures. This decrease of IgG productivity is due to a reduction in specific production rates (Table 1), but seems not be correlated with measured growth parameters (maximum of viable cells and specific growth rate).
Changes in the culture medium could induce heterogeneity in cell population and appearance of non-producing or low producing cell sub-populations (Fig. 2).
On the other hand clones obtained after 50 days in culture are stable in the pre-determined medium (Fig. 3). This result indicates that one possible way to obtain stable producing clones is to adapt transfected cells to desired medium, to passage them for a defined number of cell generations and to select by sub-cloning stable producing clones that remains in this heterogeneous cell population.

References

1.- Mateo C., Moreno E., Amour K., Lombardero J., Harris W. and Pérez R. (1997). Humanization of a mouse monoclonal antibody that blocks the epidermal growth factor receptor: recovery of antagonistic activity. Immunotechnology 3; 71-81.

Conclusions

References

ANALYSIS OF ALTERATIONS IN GENE EXPRESSION AFTER AMPLIFICATION OF RECOMBINANT GENES IN CHO CELLS

OTMAR HOHENWARTER, JOHANNES GRILLARI, KLAUS FORTSCHEGGER, REINGARD M. GRABHERR AND HERMANN KATINGER
Institute of Applied Microbiology, University of Agricultural Sciences, Vienna, Austria

Abstract

Dihydrofolate reductase based amplification of recombinant genes using increasing concentrations of methotrexate is a common method to establish CHO cell lines which secrete high amounts of the desired protein.
We have investigated alterations in gene expression which occur beside amplification of the recombinant genes, because we wanted to define genes which are up- or downregulated in a cell line that shows high secretion rates. For this purpose we have used the suppression hybridization method (SSH) to create a cDNA library which is enriched for differentially expressed sequences in the original and a recombinant antibody producing CHO cell line. Differential expression was confirmed by Northern Blotting and "Northern ELISA".
In addition to the expected recombinant genes we have identified 5 genes which are upregulated in the recombinant cell line. One sequence has not been found in existing data bases, the others revealed to be genes involved in protein synthesis and regulation of transcription.
Beside an improved characterization of altered gene expression in a recombinant CHO cell line the results prove the high efficiency of the SSH technique for abundant as well as for low copy number mRNAs.

Introduction

Differential gene expression occurs in the process of development (differentiation, quiescence, senescence), injury and apoptosis on cellular level as well as on the level of tissues and organisms. The two most common methods to detect and characterize these changes on the level of the mRNA transcription are differential display and subtractive hybridization. In order to evaluate the efficiency of these methods, subtractive suppression hybridization (Diatchenko, 1996) was performed on two different Chinese Hamster Ovary (CHO) cell lines. Whereas the first cell line was producing a recombinant antibody against HIV-1(CHO-2F5) the second cell line was the non transfected control (DHFR- CHO). By subtracting mRNA of CHO DHFR- from CHO

A. Bernard et al. (eds.), Animal Cell Technology: Products from Cells, Cells as Products, 393–395.

2F5 a cDNA library was created and differential expression of isolated cDNAs was confirmed by non radioactive Northern Blotting.

Materials and Methods

CHO DHFR- and CHO 2F5 cell lines were cultivated in DMEM/Ham's-F12 Medium with 2% FCS and 9,6 μM Methotrexat.
RNA from CHO-2F5 (tester) and CHO-DHFR- (driver) was prepared and ds cDNA was synthetized. Adaptors 1 and 2 were ligated to the tester cDNA. Subtractive Hybridization and Suppression PCR was performed according to the scheme of Fig. 1. PCR-products were cloned. Northern Blots were hybridized with DIG labelled probes to confirm differential expression by chemiluminescent detection (Fig. 2), and positive cDNAs were sequenced by the Dideoxy method.

Tab 1: Genes characterized by Northern Analysis and by sequencing by the didesoxy method and their frequency

Characterized gene	Number of clones
2F5 light chain	8
2F5 heavy chain	3
DHFR	8
others	4
not differential	10
no signal on Northern Blot	5
differential in Northern ELISA	1

Results and Discussion

35 individual cDNA-clones were characterized by restriction analysis, identifying 7 unique sequences(Tab 1). Northern Blot analysis revealed that 23 clones were differentially transcribed, while 5 clones gave no detectable signal (Fig. 2). These clones were analized by the more sensitive "Northern ELISA".
In addition to the expected genes (2F5 heavy chain, 2F5 light chain, dihydrofoate reductase) we isolated TAXREB107 (probably identical with ribosomal protein L6), Aspartic acid aminotransferase, lysyl tRNA synthetase and an unknown gene. Analysis with the "Northern ELISA" reveald one gene with 80% higher expression in 2F5 cells: TAF II 30 (TATA-Box binding associated factor). Thus, five different mRNAs coding for transcriptional or translational activity have been identified, suggesting an upregulation of the protein synthesis machinery of the antibody producing cell line.
These results prove the high efficiency of the Subtractive Suppression Hybridization Technique for abundant as well as for low copy number mRNAs and ist value for further investigations on differential gene expression. More hybridization experiments for

elucidating the quantitative differences in gene expression in these as well as in other cell lines are in course.

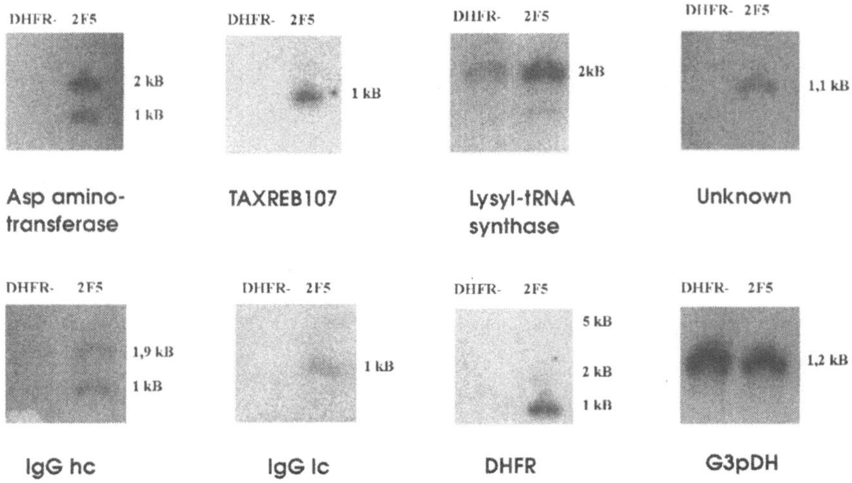

Fig. 2:Northern Blots of CHO 2F5 and CHO DHFR- 10 µg total RNA/lane, hybridized with DIG labelled probes obtained by Subtractive Suppression Hybridization Technique.

References

Diatchenko, L., Y. C. Lau, et al. (1996). "Suppression Subtractive Hybridization: A Method for Generating Differentially Regulated or Tissue Specific cDNA Probes and Libraries." Proc. Natl. Acad. Sci. USA 93: 6025-6030

Acknowledgements

We thank Renate Kunert, Willi Steinfellner and Sonja Preis for providing us generously with CHO cells.

STABLE RECOMBINANT EXPRESSION AND FUNCTIONAL IDENTITY OF THE ANTI HIV-1 MONOCLONAL ANTIBODY 2F5 AFTER IGG3/IGG1 SUBCLASS SWITCH IN CHO-CELLS

R. KUNERT, W. STEINFELLNER, A. ASSADIAN AND H. KATINGER
Institute of Applied Microbiology, University of Agricultural Sciences, Nussdorfer Lände 11, A-1190 Vienna, Austria
E-mail: R.Kunert@iam.boku.ac.at, http://www.boku.ac.at

Abstract

The human monoclonal antibody 2F5 (Buchacher et al., 1994) is a potent candidate for immunotherapy of HIV-1 (Katinger et al., 1995). The hybridoma derived humAb is of IgG3 kappa isotype. Since the IgG1 isotype has a longer half-life in human than IgG3, we switched the subclass-type to IgG1 by ligation of the 2F5 heavy chain variable region to an IgG1 constant region and expressed the IgG1 kappa molecule in CHO-cells. Stable recombinant 2F5-IgG1 expressing CHO-cells were isolated and the recombinant protein compared to the hybridoma derived immunoglobulin. Here we confirm, that specificity and affinity of different isotypes has not changed. Stability assays of the recombinant 2F5 IgG1 clone were performed with and without selection pressure.

1. Expression vectors for recombinant antibody 2F5-IgG1

2F5 light chain was amplified by RT-PCR and cloned into the plasmid pRC/RSV, carrying the rous sarcoma virus long terminal repeat (RSV-LTR) -promoter, the bovine growth hormone-terminator and the neomycin expression cassette. The variable region of the 2F5 heavy chain was ligated to the constant region of a different human antibody of IgG1 specificity and also cloned in pRC/RSV. The plasmid pdhfr is able to express the dihydrofolate-reductase under control of the SV40 early promoter and SV40 small-t intron and polyA sequence.

2. Clonal selection of 2F5 IgG1 recombinant CHO-cells

Dihydrofolate-reductase deficient CHO-cells (Urlaub and Chasin, 1980) were co-transfected with p2F5LC, p2F5IgG1 and pdhfr by Ca-phosphate coprecipitation method and selected 24 hours post transfection with dialysed FCS and 500 µg/ml G418. High level expressing clones were established by stepwise increasing the methotrexate (MTX) concentration (starting with $1x10^{-7}$ M) two fold. Subcloning with decreasing numbers of cells per 96-well plate stabilizes a monoclonal recombinant cell-line. The final recombinant 2F5 IgG1 expressing CHO-line (2F5/1F8/4F11/9E6/4G11) was subcloned four times, with a final MTX concentration of $9.6x10^{-6}$ M.

A. Bernard et al. (eds.), Animal Cell Technology: Products from Cells, Cells as Products, 397–399.

3. Specificity test of different 2F5 preparations

We tested the reactivity of different human antibodies with the synthetic epitope ELDKWA coated onto an ELISA plate. Cell culture supernatants were incubated at serial dilutions (1:2) starting with an IgG concentration of 200 ng/ml; the specific binding of IgG to the epitope was detected with anti-human-IgG horseradish-peroxidase conjugate (1:1000) (fig. 1).

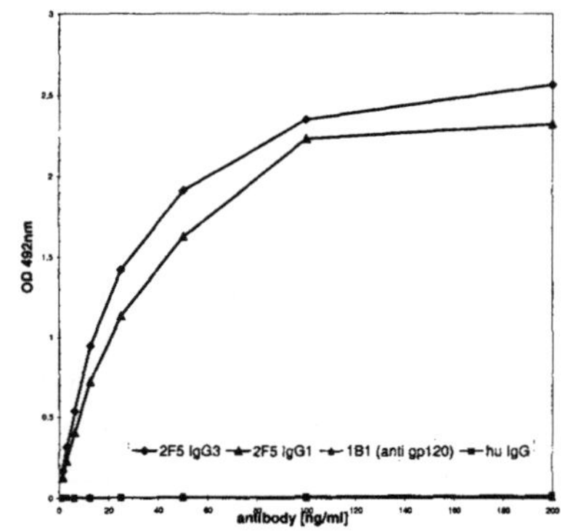

Fig.1.: 2F5 IgG3 and IgG1 specificity compared to antibodies not recognizing ELDKWA

4. Determination of antibody affinity

We have determined the affinity of three different antibody preparations (two culture supernatants 2F5 IgG3 and 2F5 IgG1 and the purified 2F5 IgG1, C2F5 IgG1) by indirect ELISA and Scatchard analysis.

Competitive binding assays were performed with constant amounts of antibody, incubated with various amounts GST-ELDKWA fusion protein. The percentage of free antibody was estimated after reaching equilibrium. The apparent affinity constant (aK) is defined as the reziprokal value of molar concentration of the epitope in the liquid phase resulting in 50% inhibition of the antibody binding to the antigen in the solid phase. We measured affinities of 1-2x10^7 l/mol, typically for monoclonals after affinity maturation *in-vivo* (fig. 2).

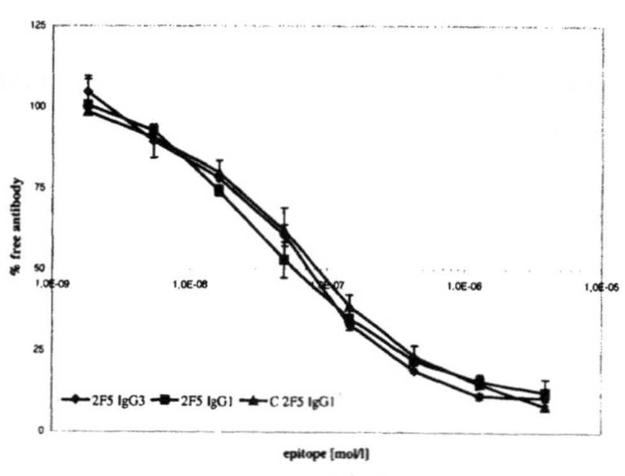

Fig. 2: Affinity determination of different 2F5 preparations

5. Long time cultivation of recombinant 2F5 IgG1

The final CHO-recombinant 2F5 IgG1 subclone (2F5/1F8/4F11/9E6/4G11) was tested for its productivity in long time cultivation in 25 cm^2 T-flask bottles. Three different cultivation series were analysed: one experiment with selection medium comprised of DMEM with 4mM glutamine, 10% dialysed FCS, $9.6x10^{-6}$ M MTX and 500 µg/ml G418. The second medium was without G418 and the third medium was without both selection criteria, G418 and MTX. Each culture was splitted twice a week 1:3 and the productivity was determined every two weeks by ELISA (fig. 3).

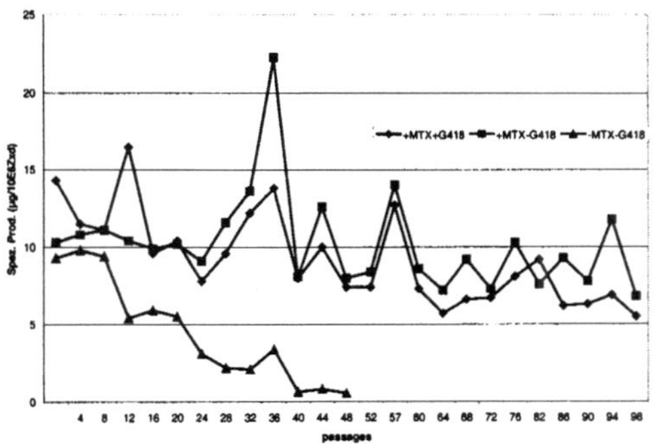

Fig. 3: IgG productivity of the recombinant 2F5 IgG1 in three different cultivation media over several months.

6. Conclusion

The human monoclonal antibody 2F5 has been subclass switched from IgG3 to IgG1. The quality of therapeutic mAbs is determined by characters such as binding specificity, affinity and half-live. Both antibody isotypes, expressed in different host cells, show identical patterns with respect to their specificity, affinity and also *in-vitro* function (data not shown). Stable high level expression in CHO-cells enables the mass production of the IgG1 isotype for in vivo studies in humans. The expectation that the isotype switch from IgG3 to IgG1 shall prolong the circulation of the mAb in the patients serum remains to be investigated in clinical trials.

7. References

Buchacher A., Predl R., Strutzenberger K., Steinfellner W., Trkola A., Purtscher M., Gruber G., Tauer C., Steindl F., Jungbauer A., and Katinger H. 1994, AIDS Research and Human Retroviruses 10:359-369.

Katinger H., Purtscher M., Muster T., Steindl F., Döpper S., Vetter N., Armbruster Ch., Gelbmann H. 1995, Dixime colloque des cent Gardes; 291-295

Urlaub G. and Chasin L. 1980, Proc Natl Acad Sci USA 77(7): 4216-4220

ANALYSIS OF GENOMIC DNA AND RNA-TRANSCRIPTS OF A HETEROHYBRIDOMA SIMULTANOUSLY EXPRESSING IGG3 AND IGG1 WITH IDENTICAL SPECIFICITY

R. KUNERT, S. WOLBANK, A. HÜSER AND H. KATINGER
Institute of Applied Microbiology, University of Agricultural Sciences,
A-1190 Vienna, Nussdorfer Lände 11, Austria
E-mail: R.Kunert@iam.boku.ac.at, http://www.boku.ac.at

1. Introduction

During the development of B-lymphocytes, the immunoglobulin heavy chain (IgH) gene undergoes two types of DNA recombinations. Joining of VDJ pieces of variable-region exons upstream of the Cμ determines - beside somatic mutations - the specificity of the antibody (Ab). This first recombination event induces IgM specific immunogobulin expression. Further differentiation of B-cells leads to the production of other classes of IgH molecules through isotype switch on the IgH locus. The class switch recombination takes place between a pair of sites, one site on the intron between the JH and Cμ, the other site in a region upstream of the later expressed CH gene. These recombination sites are variable and fall in or near the so-called switch (S)-regions. S-regions do not contain any conserved recombination signals, but are rich in repetitive sequences. Different models have been proposed to explain the subclass switch: recombination between homologs, unequal sister chromatide exchange, and looping out and deletion (Harrison et al., 1993). The deletion model is the simplest and most convincing explanation for class switch recombination. Circular DNA is excised from the chromosome, bringing subclass exons of gamma, alpha and epsilon specificity in proximity to the VDJ exons (fig.1).

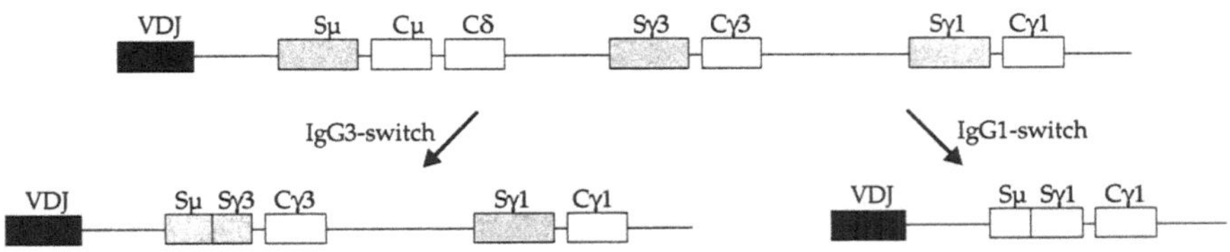

Fig.1: Rearrangement of human IgH constant region locus on chromosom 14 for IgG3 and IgG1. S designates the switch regions, VDJ and C the heavy chain exons.

2. Expression of the humAb 4E10 and determination of subclass

The human/mouse heterohybridoma 4E10 (Buchacher et al., 1994) was generated by fusion of human peripheral blood lymphocytes of HIV-1 infected asymptomatic patients

A. Bernard et al. (eds.), Animal Cell Technology: Products from Cells, Cells as Products, 401–403.

with the human mouse heteromyeloma cell-line CB-F7 (Grunow et al., 1988). It was stabilized by four subcloning steps and production of antibody over more then one year confirmed monoclonality. Immunoglobulin subclass of anti-HIV-1 antibody was determined by four step ELISA, using subclass specific monoclonals as catcher antibody and also after applying the 4E10 sample we used a different subclass specific monoclonal antibody. The detection was performed with horseradish-peroxidase conjugated antiserum directed against the last specific antibody. The ELISA showed only minor reactivity in the IgG1 specificity test, and so we assumed that the monoclonal 4E10 is IgG3 isotype.

3. cDNA preparation and amplification humAb 4E10 hinge region

mRNA from $1x10^6$ hybridoma cells was prepared by hybridizing poly-A tailed mRNA to oligo-dT cellulose. First strand cDNA was synthesized using heavy chain CH3 3'-antisense-primer. PCR amplification was performed with a sense primer annealing to the beginning of CH1 and an antisense-primer recognizing the last 20 nucleotides of CH2. The hinge region of IgG1 displays a length of 44 basepaires while in the hinge region of IgG3 this segment is quadruplicated. After ampilfication of 4E10 CH1-hinge-CH2 we compared the two fragments to the database (Altschul et al., 1997) and confirmed the presence of IgG3 and IgG1 specific mRNA.

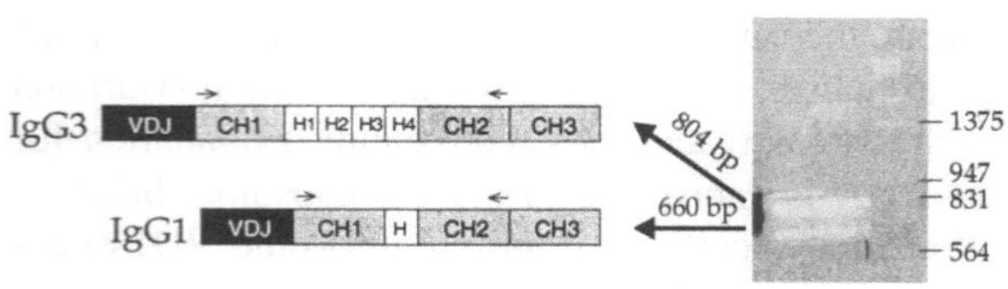

Fig.2: Amplification of 4E10 mRNA with constant region primers, gave two fragments indicating the 660 bp long IgG1 and the 804 bp long IgG3 specific CH1-hinge-CH2 PCR product.

4. Amplification of genomic DNA and characterization of switch regions

Genomic DNA was isolated by DNAzol-BD reagent (MRC,Inc.) according to the manufacturer instructions. For elucidation of recombination events taken place in the IgH locus of 4E10, we amplified chromosomal DNA fragments displaying the intron between rearranged VDJ heavy chain exon and Ig gamma constant region CH1 exon. The primers were designed in sense in the JH4 region found to be used in 4E10. Antisense-primer bound at the beginning of CH1 being specific for both, IgG1 and IgG3. Amplification gave two fragments; a larger one with approximately 7000 bp and a smaller one with 5858 bp. The 5858 bp fragment was subcloned via TA-cloning kit and sequenced on the ABI DNA sequencer 373 with the ready reaction dideoxy terminator cycle sequencing kit. In figure 3 the sequencing data were related to comparable units on the IgH locus.

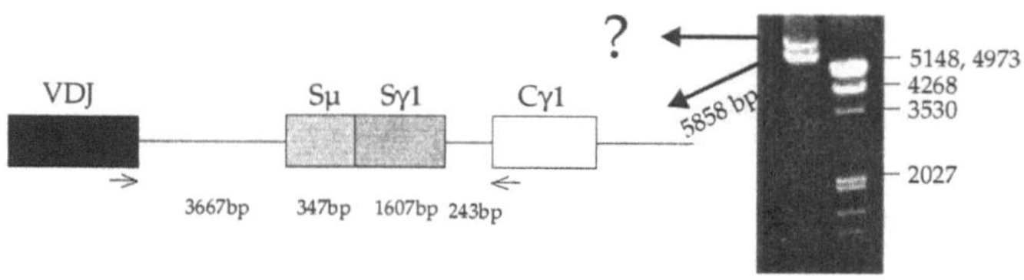

Fig. 3: PCR amplification with JH4 sense primer and CH1 antisense primer gave two different fragments. The shorter fragment (5858 bp) was sequenced and identified as rearrangement between Sμ and Sg1. The second larger fragment has not been sequenced yet.

5. Discussion

We have stabilized a human monoclonal antibody 4E10 recognizing the ectodomain of gp41 and neutralizing laboratory strains as well as primary HIV-1 isolates. 4E10 hybridoma cells are monoclonal and all characterizations, discussed in this poster, were carried out on 4E10 cells from early and late subcloning steps. All clones had the same sequences so we can exclude, that the recombination has occured do to in vitro cultivation events. Subclass determination by ELISA showed IgG3 specificity and also in immunofluoreszence we couldn't detect IgG1 in comparing amounts.

The characterization of mRNA sequences gave clear evidence for IgG3 and IgG1 transcripts (fig. 2). We tried to explain these results by analysing the genomic rearranged IgH locus. As shown in fig. 3 two different fragments were amplified from the chromosomal locus. The shorter fragment (fig. 3) displays typical rearrangement of an IgG1 switch via looping out. These findings are contrary to expectations, since 4E10 expresses predominantly IgG3 subtype. Rearrangement of both constant regions can occur by recombination of homologous chromosomes, but literature provides some strong arguments against this as major mechanism of subclass switch. Following this theory, in hybridoma 4E10 the sequenced chromosome expresses the IgG1 molecule, while the longer fragment is the homologous active allele expressing IgG3. The second possibility is a mechanism in which coexpression occurs by discontinuous transcription (Nolan-Willard et al., 1992). In this case IgG1 and IgG3 are expressed from the same allele. The long genomic PCR-fragment corresponds to the rearranged IgH locus including both, C□1 and C□3 exons. Recent experiments support this theory.

6. References

Harrison W, Völkl H, Defrnoux N, Wabl M, 1993, Annu. Rev. Immunol. 11: 361-84

Altschul S, Stephen F, Madden T, Schäffer A, Zhang J, Zhang Z, Miller W, Lipman D, 1997, Nucleic Acids Research: 25: 3389-3402

Grunow R, Jahn S, Porstmann T, Kiessig SS, Steinkellner H, Steindl F, Mattanovich D, Gürtler L, Deinhardt F, Katinger H, Von Baehr R, 1988, J Immunol Methods, 106: 257-265.

Buchacher A, Predl R, Strutzenberger K, Steinfellner W, Trkola A, Purtscher M, Gruber G, Tauer Ch, Steindl F, Jungbauer A, Katinger H, 1994, AIDS Res hum Retrov. 10(4), 359-369

Nolan-Willard M, Berton MT, Tucker P, 1992, PNAS, 89,1234-1238.

NOVEL THERAPEUTIC AND PROPHYLACTIC APPROACHES BASED ON CELLS AND NUCLEIC ACIDS

Chapter VII

TISSUE ENGINEERING OF A BLADDER WALL PATCH

I.BISSON, J. HILBORN*, F. WURM°, P. FREY

Experimental Pediatric Surgery , CHUV, Departments of Chemistry° and Material Science, EPFL, Lausanne Switzerland*

Introduction

Congenital bladder disorders, such as bladder and cloacal exstrophy as well as acquired pathologies such as traumatic neurogenic bladder, chronic interstitial cystistis, and possibly bladder cancer, require the increase in size or the replacement of the bladder. This to improve or to create adequate storage capacity, and/or to definitively treat the bladder disease.

Several techniques using intestinal or gastric tissues have been proposed for the reconstruction of the lower urinary tract. Applying these techniques complications such as severe metabolic disturbances, mucus secretion, bacterial infection, stone formation and even malignant transformation may be induced.

A new strategy, for bladder augmentation or replacement, may be found in tissue engineering techniques enabling us to create an autologous functional bladder wall graft derived from its components, the urothelial and the smooth muscle cells expanded in culture.

The cells are harvested by endoscopic biopsy of the patient's own bladder tissue. The tissue fragments are dissociated into individual cells to then be cultured *in-vitro*. Once an adequate amount of cells is obtained they are seeded on a compliant biocompatible polymer matrix acting as scaffold. Once cell adherence has been guaranteed the artificial bladder wall is further cultured *in-vitro* for a limited time, to then be reimplanted into the patient. Thereafter the reimplanted polymer scaffold degrades slowly, while spontanous differentiation and regeneration into functional tissue takes place.

A. Bernard et al. (eds.), Animal Cell Technology: Products from Cells, Cells as Products, 407–411.

I. *In-vitro* urothelial cells culture

Although we are able to culture bladder smooth muscle cells in a reproducible way, we first concentrated our research on urothelial cells.
We established urothelial cell cultures, using a serum free keratinocyte medium, in the rodent, the porcine and the human system (Zhang *et al.*, 1999, Ludwikowski and Frey 1999). The urothelial cells are qualified by multiple monoclonal anti-cytokeratin antibodies and in particular antibodies against cytokeratin (CK) 17, confirming the urothelial phenotype of the cells.

II. *In-vitro* induced, stratified urothelial construct

By altering the composition of the culture medium, in particular of the bi-cations, the urothelial cells can be induced to form a stratified construct, resembling normal urothelium. (Southgate *et al.*, 1994) We have characterized these urothelial constructs and compared them with native urothelium.

II.1. MORPHOLOGY OF THE UROTHELIAL CONSTRUCT

In a first step morphology and epithelial polarity of the construct were evaluated. By transmission electron microscopy the apical cell membrane was found to form micro-vilosities and glycocalyces as in the native tissue.
Polarity of the stratified urothelial construct could be evaluated by the analysis of the expression of cytokeratins 7, 8, 13, 17, E-cadherin and symplekin, characterising each cell layer. Similar expression of these proteins could be found in the *in-vitro* engineered construct as well as in the native urothelium. No expression of CK 20, CK 18 and uroplakin was observed in the construct. As in native urothelium, these proteins are preferentially expressed in the apical membrane, being in contact with urine, we can speculated that the lack of their expression in the construct might be due to the abscence of the acidity of urine in culture. Further experiments to evaluate this speculation have been initiated.
The basement membrane, of normal urothelium, is connected with the *lamina propria* composed of extra cellular matrix (ECM) proteins. Collagen type IV and laminin are the major ECM proteins that are localized in the proximity of the urothelial basement membrane. Collagen type I and III and fibronectin are the ECM proteins mostly found in the submucosa, area of the *lamina propria* facing the smooth muscle cells. (Wilson *et al.*, 1996) The interaction between cells and ECM protein ligands are realized by hemidesmosomes that are composed of different cell-receptor proteins. The interation of the cell-receptor integrin $\alpha6\beta4$ and the ligand laminin type 5 seems to be important for cell adhesion (Jones *et al.*, 1998). By transmission electron microscopy we could observe hemidesmosome formation in the urothelial constructs.

II.2. FUNCTIONALITY OF THE UROTHELIAL CONSTRUCT

We further assessed the functionality of the *in-vitro* engineered urothelial construct with permeability studies, and compared the results with the ones of native human urothelium. Mammalian urinary bladders are characterized by having a relatively impermeable urothelial barrier. Permeability of the construct for urea, water and ammoniac was analyzed and compared to the one of native human urothelium. A similar very reduced permeability could be demonstrated (Sugasi *et al.*, 1999).

III. Cell seeding on surface modified, protein grafted, polymer scaffold

Cell isolated from the patient's tissue will be, once expanded *in-vitro*, placed on a three-dimensional scaffold surface that holds them close enough together to organize themselves and to form functional tissue. The scaffold surface of the scaffold must have the right surface chemistry to guide and let reorganize the cell mass. Unfortunately polymer scaffold surfaces commonly used and FDA approved are not appropriate for urothelial cells seeding. In this case, the surface of these polymers must be first modified by the immobilization of cell-adhesive proteins such as ECM proteins or peptide cell-binding sequences.
We first grafted the ECM proteins collagen type I and III onto the polymer surfaces.

III.1. COLLAGEN IMMOBILIZATION ON PET/PAA

For surface tailoring of polymer a new technology (patent Hilborn/Frey pending) was developed. We performed surface modification initially using a non-degradable polymer poly(ethyleneterephtalate) (PET) that does not possess any sites for protein immobilization. Polyacrylic acid (PAA) is therefore grafted on the polymer scaffold surface to provide functional groups for protein attachment. Thereafter a mixture of type I and type III collagen, optimal for cell adherence is immobilized on the PAA grafted surface.
To verify the presence of collagen on polyacrylic acid grafted PET surfaces, X-ray photoelectron spectroscopy was used to determine the quantity of nitrogen, expressed in percentage of atoms, before and after protein immobilization. An increase of nitrogen, originating from the protein molecules, from 1% to 14% is observed after protein immobilization. The results of this screening method can be a preliminary indicator of successful collagen grafting on PET/PAA surfaces. Further detection methods of the proteins grafted, including optical thickness measurement, Ninhidrin reaction, and radioactive marking, will be performed.

III.2. CELL CULTURE ON PETT/PAA GRAFTED WITH COLLAGEN

Onto non-degradable polymers grafted with collagen, human urothelial cells were successfully seeded. Excellent cell adherence was observed and confluence was obtained within 7 days of culture. In contrast, cells seeded on PET or PET/PAA, acting as controls, showed morphological deterioration with vacuole formation and detachment of the cells from the polymer. This senescence may be due to the lack of ECM-cell interaction.

Although the preliminary results with collagen grafting is very encouraging, future research should investigate other ECM proteins grafted onto non-degradable and degradable polymers.

Conclusions

The challenge to tissue engineer an artificial bladder wall must be approached step by step.

We have first established reproducible urothelial cell cultures. Thereafter we constructed an *in-vitro* induced multilayered human urothelial construct of which the morphological and functional behavior are very similar to the ones of the native human urothelium.

In a further step a novel technology to immobilize ECM protein ligands was developed to allow selective cell seeding on non-degradable polymer surfaces.

In a third step the protein grafting is applied to degradable polymers, and in a final step a knitted, three-dimensional, compliant and degradable polymer scaffold will be seeded with urothelial and smooth muscle cells *in-vitro*, to then be implanted *in-vivo* to allow final tissue differentiation and regeneration.

References

Hilborn, J. and Frey, P. UK. Pat. Appl. 9824562.4

Jones, J.C.R., Hopkinson, S.B., and Goldfinger, L.E. (1998) Structure and assembly of hemidesmosomes, *Bioessays* **20,** 488-494.

Ludwikowski, B., and Frey, P. (1999) Long-term culture of porcine urothelial cells and induction of urothelial stratification, *Brit. J. Urol.* (in press).

Southgate, J., Hutton, K.A., Thomas, D.F., and Trejdosiewicz, L.K. (1994) Normal human urothelial cells in vitro: proliferation and induction of stratification, *Lab. Invest.* **71,** 583-594.

Sugasi, S., Lesbros, Y., Larbi, N., Bisson, I., Kucera, P., and Frey, P. (1999) In vitro engineering of human stratified urothelium: analysis of its morphology and functionality. (in review: J.Urol)

Wilson, C.B., Leopard, J., Cheresh, D.A., and Nakamura, R.M. (1996) Extracellular matrix and integrin composition of the normal bladder wall, *World J. Urol.* **14,** S30-37.

Zhang, Y., Leuba, A.N., and Frey, P. (1999) Long-term culture and characterization of normal rat urothelial cells in vitro. (in review: J.Urol

Discussion (Bisson)

Baetge: Have you taken any of your bladder patches and transplanted them into nude mice, for instance, to see if they spot?

Bisson: No, not yet.

INTEGRATED OPTIMIZATION OF RETROVIRUS PRODUCTION FOR GENE THERAPY

P.E. CRUZ[1], D. GONÇALVES[1], J. ALMEIDA[2], J.L. MOREIRA[1], M.J.T. CARRONDO[1, 3]
1 - IBET/ITQB, Apartado 12, P-2780 Oeiras, Portugal
2 - ITQB, Apartado 127, P-2780 Oeiras, Portugal
3 - Laboratório de Engenharia Bioquímica, FCT/UNL
P-2825 Monte da Caparica, Portugal

1. Introduction

Currently, a large number of gene therapy applications make use of retroviruses as vehicles for gene delivery [1], since retroviruses provide permanent integration of foreign genes in cellular DNA [2] and, in contrast to other viral delivery systems, stable helper cell lines for their production are available [3]. However, several factors are limiting the commercial production procedures and wider clinical applications of retrovirus mediated gene delivery. The most relevant are the low titers obtained in culture, which seldom rise above 10^6 infective particles per milliliter and the low retrovirus stability not only during culture but also during downstream processing and storage [4]. This stresses the need for integrated optimization since degradation will occur in all steps of the production process. Nonetheless, few protocols and systematic optimization methods have been published. For this purpose in the present work the production of retroviruses was mathematically modeled in order to optimize the controllable intermediary steps between helper cell growth and final retrovirus titer.

2. Model Description

The production of retroviruses using perfusion cultures can be described by a mathematical model composed of a set of differential equations relative to cell growth, retrovirus production and degradation in a bioreactor (vessel A). This bioreactor is operated in perfusion mode with the harvested broth being recovered in another vessel (B) maintained at low temperature to minimize product degradation before downstream processing. In this vessel B, total and infective retrovirus concentration and vessel volume are also described by a set of differential equations. These most relevant equations are presented in Table 1. In these equations, δ is the cell death rate, $X_{V,max}$ is the maximum achievable cell concentration, β is the specific productivity, V_A and V_B are the vessel volumes, F is the perfusion flow rate and γ_A and γ_B are the retrovirus degradation rates.

A. Bernard et al. (eds.), Animal Cell Technology: Products from Cells, Cells as Products, 413–420.

TABLE 1. Relevant equations for system modeling.

Variable	Equation
Viable cells	$dX_V/dt = \mu X_V - \delta X_V$
Specific growth rate	$\mu = \mu_{max} (1-X_V/X_{V,max})$
Retrovirus concentration (vessel A)	$dRv_{V,A}/dt = \beta X_V - F / V_A Rv_{V,A} - \gamma_A Rv_{V,A}$
Retrovirus concentration (vessel B)	$dRv_{V,B}/dt = Rv_{V,A} F/V_B - Rv_{V,B} F/V_B - \gamma_B Rv_{V,B}$
Vessel B volume	$dV_B/dt = F$

For optimization purposes, the objective function to be maximized was viable retrovirus productivity:

$$f_{objective} = (Rv_{V,A} V_A + Rv_{V,B} V_B) / (\Delta t V_A)$$

This model can also be extended to include the downstream processing with the final goal of overall process optimization.

3. Materials and Methods

3.1. CELL GROWTH AND RETROVIRUS PRODUCTION

A PA317 packaging cell line producing recombinant retrovirus MoMLV derived was kindly provided by Dr. Hansjoerg Hauser (GBF, Braunschweig, Germany). The recombinant retrovirus encoded two genes: the puromycin resistance gene for selection and the Lac Z gene as a marker. Cells were grown in 125 cm^3 stirred vessels (Wheaton, Millville, NJ) at 37°C in a 5 % CO_2 atmosphere using DMEM medium + 10% FBS (Life Technologies, Paisley, UK). Cultispher S porous supports (Percell, Astorp, Sweden) at a concentration of 3 g/l were used. These stirred vessels were inoculated with approximately $3.8x10^7$ cells. Perfusion was usually initiated 3 days after innoculation. Similar experiments were performed adding dexamethasone (Sigma, St. Louis, MO) and sodium butyrate (Merck, Darmstadt, Germany) to the culture medium. Infective viruses were determined by using standard β-gal assay.

3.2. PARAMETER DETERMINATION AND PROCESS OPTIMIZATION

The growth and production parameters were determined by fitting the non-linear differential equation system to the experimental data. For this purpose, the software Scientist® v2.0 (MicroMath, Inc, USA) was used. Optimization was performed using the software MatLab® v5.0 (MathWorks, Inc., USA).

4. Results and Discussion

The results of the stability tests carried out at 37, 32 and 4°C indicated that the degradation constant at 37°C (0.23 h^{-1}) is nearly two fold higher than that at 32°C (0.12 h^{-1}). As expected, at 4°C the degradation rate is significantly smaller (0.07 h^{-1}), in agreement with previously published results for retroviruses [5].

Table 2 presents the results of the determination of other model parameters. These results were obtained by fitting the viable cell concentration and infective retrovirus titer under different conditions of temperature and additive usage (see Materials and Methods).

TABLE 2. Cell growth and retrovirus production parameters for batch cultures at 37°C and for batch and perfusion cultures at 32°C with and without additives.

	Experiment		
Parameters	$\mu_{max} - \delta$ (h^{-1})	Xv_{max} (10^6 cells/ml)	β (virus/cell-h)
37°C, no additives	0.041	2.7	0.017
32°C no additives	0.031	3.1	0.011
32°C with additives	0.020	2.0	0.033

In all experiments the glucose concentration was above 1 g cm^{-3} to avoid nutrient limitation that would affect cell growth and retrovirus production. The effect of the temperature reduction from 37 to 32°C upon cell growth is evident. The significant decrease of (μ_{max} - δ) clearly indicates that the reduction of temperature leads to a reduction in cell growth. The change in the value of $X_{V,max}$ was not statistically significant. These results are consistent with the findings of Lee et al. [4] and Kotani et al. [6] for both PA317 and ψ–CRIP cells.

The use of additives caused an even greater reduction in the growth parameters. In fact, dexamethasone, while not toxic to the cells, may exert a growth-retarding influence [7]. Sodium butyrate has also shown some growth inhibitory properties in SV40 transformed human embryonic lung fibroblasts [8]. The use of additives had a positive effect upon retrovirus productivity, with β increasing 3 fold in comparison with that of the experiment performed at the same temperature without additives (Table 2). The observed increase in productivity was expected since the addition of sodium butyrate and synthetic glucocorticoids such as dexamethasone stimulates the production of retroviruses, namely MoMLV [9]. Although the results indicated that the model could describe the experimental data with accuracy, this does not prove that the model can predict changes in the operational conditions.

For this purpose, an experiment involving a batch growth phase without additives and a perfusion phase with additives and a perfusion rate of 0.066 h^{-1} (well outside the range used for parameter determination) was performed. This experiment was performed at 32°C since the retrovirus concentration results were higher in this case. For simulation purposes, the parameters presented in Table 2 were used. The experimental results and the model predictions are shown in Figure 1.

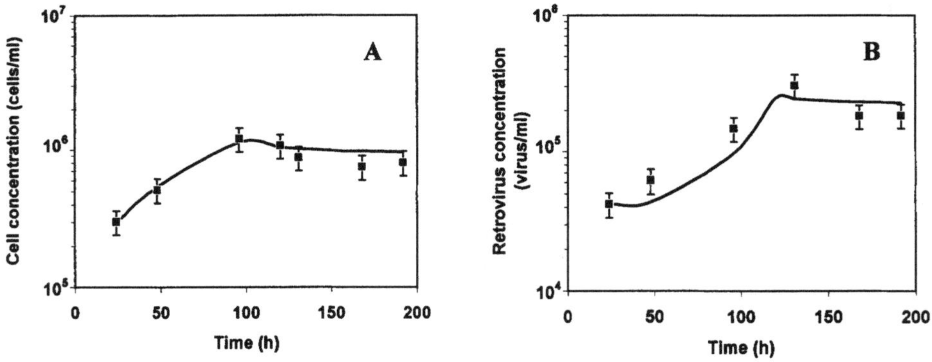

Figure 1. Model validation: variation of cell concentration (A) and retrovirus concentration (B) with time in a culture with a batch phase without additives and a perfusion phase (0.066 h^{-1}) with sodium butyrate (5 mM) and dexamethasone (0.1 μM) as additives at 32°C. The solid lines represent the model predictions based on the parameters presented in Table 2.

Statistical analysis performed by the software used indicated, for the simulation presented in Figure 1, a correlation between the model predictions and the experimental data of 0.91 for the viable cell concentration and of 0.85 for the viable retrovirus concentration. This clearly shows that the model predictions simulated these trends and proves the robustness of the model under changing conditions.

The final step of the modeling of this system was the determination of the batch and perfusion duration and conditions (temperature and additives) as well as the perfusion rate that maximizes productivity and simultaneously satisfies the titer constraint. For optimization described in this work a constant perfusion rate will be used for simplification of the computational methods; however the batch and perfusion times may vary.

To compare batch and perfusion operation modes, the constraint used was the titer obtained for a batch culture at 32°C without additives, i.e., 9.3 x 10^4 viable virus/ml. This corresponds to a productivity of 750 viable virus/h-ml of bioreactor. The optimization results are shown in Table 3.

The first conclusion to be drawn from these results is that a perfusion step should be performed, since in all cases the perfusion time output was different from zero. As a consequence, the productivities obtained for the same titer were higher than those obtained in batch cultures. As for the different scenarios, it is clear that additives have to be used during perfusion, since the overall productivity was at least doubled in comparison with the scenarios without additives. In the batch phase the temperature had little effect on the overall productivity, in contrast with the use of additives that, due to their negative effect upon cell growth, actually reduced the productivity.

Whatever system is used, there will obviously exist limitations in the achievable titer and quality pairs. Whether or not a particular system meets the final constraints will depend on the final product requirements and storage and downstream processing characteristics. Nevertheless this model will be able to determine the limit values and, for these, the optimal operational conditions.

TABLE 3. Process optimization: determination of the model operational variables that maximize productivity using as constraints a final titer of 9.3 x 10^4 viable virus cm^{-3}.

Scenario	Model Outputs Productivity (virus $cm^{-3}h^{-1}$)	t_{batch} (h)	$t_{perfusion}$ (h)	F ($cm^3 h^{-1}$)
batch 37°C – na[1]				
perfusion 32°C - a[2]	2682	53	23	13.4
batch 32°C – na				
perfusion 32°C – na	951	67	18	2.5
batch 32°C – na				
perfusion 32°C - a	2618	64	22	15.7

[1] na: no additives, [2] a: with additives

5. Conclusion

The model described in this work could explain and predict the production process based on a limited number of experimental runs. For a given titer constraint, the model could define the optimal operational conditions that maximize productivity and satisfy this requirement. The optimization output indicated advantages in promoting cell growth in a batch phase and productivity during perfusion.

However, bioprocess optimization will not be global unless this model is extended to include downstream processing equations and parameters to account for the factors that affect each step of the process. This will consist on a major advantage for the development of a tailor-made approach to each gene therapy programme.

6. References

1. Havenga, M., Hoogerbrugge, P., Valerio, D. and van Es, H. (1997) Retroviral Stem Cell Gene Therapy, *Stem Cells* **15**, 162-179.
2. Hodgson, C.P. and Solaiman, F. (1996) Virossomes: Cationic Lipossomes Enhance Retroviral Transduction *Nat. Biotechnol.* **14**, 339-342.
3. Ory, D.S., Neugeboren, B.A. and Mulligan, R.C. (1996) A Stable Human-Derived Packaging Cell Line for Production of High Titer Retrovirus/Vesicular Somatitis Virus G Pseudotypes *Proc. Natl. Acad. Sci. USA* **93**, 11400-11406.
4. Lee, S.-G., Kim, S., Robbins, P. D. and Kim, B.-G. (1996) Optimization of Environmental factors for the Production and Handling of Recombinant Retroviruses. *Appl. Microbiol. Biotechnol.* **45**, 477-483.
5. Shen, B. Q., Clarke, M. F. and Palsson, B. O. (1996) Kinetics of Retroviral Production for the Amphotropic ΨCRIP Murine Producer Cell Line *Cytotechnol.* **22**, 185-196.
6. Kotani, H., Newton, P., Chiang, Y. L., Otto, E., Weaver, L., Blease, R. M., Anderson, W. F. and McGarrity, G. J. (1994) Improved Methods of Retroviral Vector Transduction and Production for Gene Therapy. *Hum. Gene Ther.* **5**, 19-28.
7. Tchekneva, E. and Serafin, W. E. (1994) Kirsten Sarcoma Virus-Immortalized Mast Cell Lines. Reversible Inhibition of Growth by Dexamethasone and Evidence for the presence of an Autocrine Growth Factor. *J. Immunol.* **152**, 5912-5921.
8. Goldberg, Y. P., Leaner, V. D. and Parker, M. I. (1992) Elevation of Large-T Antigen Production by Sodium Butyrate Treatment of SV40-Transformed WI-38 Fibroblasts. *J. Cell Biochem* **49**, 74-81.
9. Pagès, J.-C., Loux, N., Farge, D., Briand, P. and Weber, A. (1995) Activation of Moloney Murine Leukemia Virus LTR Enhances the Titer of Recombinant Retrovirus in ΨCRIP Packaging Cells *Gene therapy* **2**, 547-551.

Discussion (Cruz)

Forestell: One of the critical parameters is the half-life of the virus. Can you speculate on what mechanism the retroviruses are inactivating, and what your additive is doing to extend that half-life?

Cruz: The mechanism of inactivation has been deeply discussed at our project meetings because it is unclear whether shedding of the envelope occurs, or some kind of proteolytic degradation. We have found proteolytic activity in our supernatants, but it is difficult to see if this is the reason behind it as there are a number of degradation mechanisms for proteins. The additives which we used are dexamethasone and sodium butyrate which, although not toxic to the cells, exert some growth-retarding influence. I think dexamethasone is related to the promotor enhancement.

Noé: It is not clear to me whether this system is of general application or a basis for optimising a virus process. Is it a general model?

Cruz: As you build a model, you proceed from lower complexity to higher complexity. At the beginning you do not have, for instance, glucose limitation, then you can simplify the model. These were not the final equations - we have far more complete equations. The computer selects what parameters are relevant or not. In that case we have a general model but, as I pointed out, there may always be something extra that we have forgotten and then we have to introduce that into the model to see if it is relevant or not.

Noé: In general you have to use this iterative process for each different process?

Cruz: Yes, but even so you can reduce your workload 4-fold.

Kempken: For such systems you usually need a very good, nice and complex knowledge base. Can you give us a hint as to how many bioreactor runs, or months, of optimisation that were needed to set up this capability?

Cruz: Parameter determination can be done in one month. The computer processing is almost irrelevant - it is overnight. Then you have to test the model output to see if it is working or not. You do not need too long for adapting this general model to your system, and we did it quite quickly for the cell tube.

Kempken: I was more interested in how much experience you need to get to this predictive capability of the model. I think it is easy to establish a model and to prove that it is correct for a certain application but, until you get a predictive status, it is a very time consuming effort.

Cruz: It may well be but if the parameters are determined and they have good degrees of confidence, and if the model is correct, it will work. That is what we have seen.

TOXICOLOGICAL EVALUATION OF SURFACE WATER SAMPLES IN SENSITIZED CULTURED FISH CELLS, AND COMPARISON WITH THE MICROTOX® METHOD

P.J. DIERICKX[a], C. VAN DER WIELEN[b] and N. FRANCOIS[a]
Institute of Public Health, Afdeling Toxikologie, Wytsmanstraat 14, B-1050 Brussel, Belgium[a] and ISSeP, Section Environnement, rue du Chéra 200, B-4000 Liège, Belgium[b]

The previously described neutral red uptake inhibition assay on cultured fathead minnow (FHM) fish cells revealed a good correlation as well with fish toxicity data for a series of 50 test chemicals [1] as with *Daphnia magna* toxicity data for inorganic metal compounds [2]. The total protein content could also be used as an endpoint [3]. The major drawback was the lower sensitivity of the cytotoxicity assay. Aiming at a higher sensitivity the assay was adapted by reducing the cell number, by a longer treatment period, and by simultaneous treatment with sodium dodecyl sulfate and buthionine sulfoximine. The fluorimetrically measured protein content was chosen as the endpoint [4]. This assay revealed to be sensitive enough to measure the toxicity of environmental water samples [4]. The results for 93 surface water samples are reported here. They are compared with toxicity data obtained by the Microtox method using the bacteria *Vibrio fischeri*.

The surface water samples were taken in Southern Belgium between March and May 1998. The samples were sterilized by filtration on a Millex-GS filtre (0.22 μm). The toxicity was measured on sensitized FHM cells [4] by diluting the water samples with an equal volume of twice concentrated culture medium, which is called the 100% water sample in this report. The toxicity of each sample was measured at 50 (results not shown) and 100%. The control cells received medium prepared in the same way, but diluted with ultra pure water. The Microtox results were obtained within the ecotoxicological monitoring program of surface waters in Wallonia (Belgium).

FHM cells were considered to be intoxicated when the total protein content was reduced by at least 20% as compared to control cultures. The results are summarized in Table 1. In borderline cases the cytotoxicity was measured on a serial dilution of the water sample, as illustrated for the toxic water sample 31 in Fig. 1. On this basis the water samples 21, 63, 64 and 75 were evaluated as nontoxic, and the samples 35, 41, 76, 79 and 89 as toxic. An increased protein content was observed for samples 23, 25 and 50, what can be explained by the presence of nutrients. All but one (sample 6) toxic samples in the Microtox assay were also toxic in FHM cells. Moreover, 37 other samples were toxic in the FHM assay. However, no linear relationship was found for the toxicity results obtained with both methods. This is not surprising since *V. fischeri* bacteria and FHM cells showed also different sensitivities for 3 chemicals. An EC_{50} of 0.19 mg/l for $CuSO_4$, 34.65 mg/l for $K_2Cr_2O_7$ and 43.4 g/l for NaBr was observed in the Microtox assay, against respectively 0.51 mg/l, 1.05 mg/l and 7.95 g/l in the FHM assay.

A. Bernard et al. (eds.), Animal Cell Technology: Products from Cells, Cells as Products, 421–423.

Table 1. Toxicity results for the whole series of water samples. The cytotoxicity results in FHM cells are expressed as the protein quantities in % of the control cells.

Sample number	Toxicity		
	% Proteins	Evaluation[a]	Tu20[b]
1	90 ± 6	-	-
2	98 ± 6	-	-
3	121 ± 21	-	-
4	104 ± 4	-	-
5	108 ± 7	-	-
6	104 ± 9	-	34
7	99 ± 12	-	-
8	100 ± 15	-	-
9	108 ± 13	-	-
10	98 ± 7	-	-
11	100 ± 8	-	-
12	97 ± 8	-	-
13	96 ± 7	-	-
14	98 ±10	-	-
15	101 ± 12	-	-
16	92 ± 14	-	-
17	92 ± 7	-	-
18	58 ± 19	+	-
19	54 ± 8	+	-
20	55 ± 5	+	pos.
21	79 ± 11	-	-
22	98 ± 15	-	-
23	127 ± 20	-	-
24	104 ± 32	-	-
25	167 ± 37	-	-
26	55 ± 72	+	-
27	95 ± 86	-	-
28	59 ± 15	+	-
29	60 ± 8	+	pos.
30	52 ± 12	+	-
31	0 ± 9	+	83
32	38 ± 16	+	89
33	59 ± 17	+	-
34	96 ± 16	-	-
35	78 ± 15	+	-
36	52 ± 3	+	-
37	67 ± 5	+	-
38	61 ± 23	+	-
39	42 ± 21	+	-
40	44 ± 22	+	-
41	72 ± 11	+	-
42	67 ± 12	+	-
43	76 ± 11	+	-
44	57 ± 22	+	-
45	31 ± 18	+	-
46	40 ± 15	+	-
47	90 ± 17	-	-
48	103 ± 22	-	-
49	98 ± 11	-	-
50	128 ± 25	-	-
51	58 ± 13	+	pos.
52	10 ± 8	+	120
53	55 ± 16	+	9.7
54	103 ± 12	-	-
55	75 ± 13	+	pos.
56	65 ± 7	+	-
57	93 ± 0	-	-
58	61 ± 10	+	-
59	86 ± 20	-	-
60	114 ± 25	-	-
61	56 ± 47	+	-
62	111 ± 41	-	-
63	74 ± 23	-	-
64	75 ± 25	-	-
65	34 ± 16	+	-
66	9 ± 17	+	5.0
67	32 ± 19	+	-
68	88 ± 17	-	-
69	59 ± 28	+	-
70	85 ± 24	-	-
71	0 ± 15	+	-
72	0 ± 18	+	-
73	49 ± 25	+	-
74	18 ± 25	+	-
75	80 ± 20	-	-
76	75 ± 12	+	-
77	68 ± 15	+	-
78	63 ± 17	+	23
79	74 ± 9	+	-
80	107 ± 9	-	-

Table 1. (continued)

Sample number	Toxicity		
	% Proteins	Evaluation[a]	Tu20[b]
81	69 ± 6	+	-
82	69 ± 10	+	-
83	68 ± 10	+	-
84	57 ± 15	+	-
85	59 ± 11	+	-
86	58 ± 12	+	-
87	88 ± 18	-	-
88	88 ± 26	-	-
89	73 ± 25	+	-
90	98 ± 11	-	-
91	85 ± 5	-	-
92	81 ± 14	-	-
93	100 ± 17	-	-

[a]Cytotoxicity in FHM cells, after further evaluation of the borderline cases as described in the text,- = nontoxic,+ = toxic.
[b]Microtox® results, Tu20=the number of times the sample has to be diluted to obtain the EC20, pos.= light inhibition, lower than the quantification limit of the Tu20, -=nontoxic.

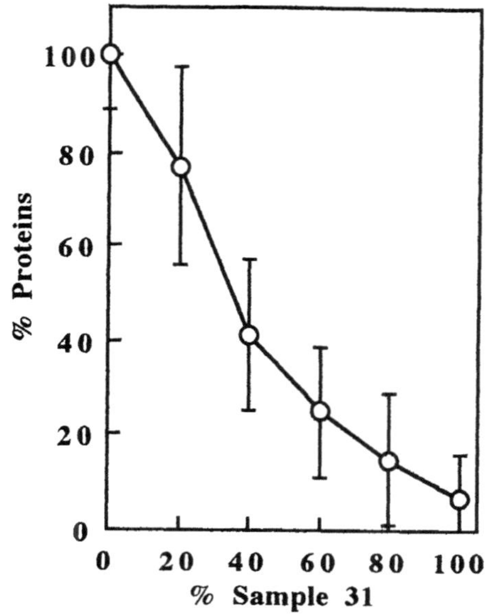

Figure 1. Influence of the dilution of a toxic water sample on the total protein content in FHM cells.

Conclusion

The results show that the sensitized FHM assay allows to measure the toxicity of environmental water samples. Further research to reduce the variability of the results and to explain the significance of the toxic results in the samples which are not toxic in the Microtox assay will increase the reliability of the FHM assay.

1. Brandao, J.C., Bohets, H.H.L., Van De Vyver, I.E. and Dierickx, P.J. (1992). Correlation between the in vitro cytotoxicity to cultured fathead minnow fish cells and fish lethality data for 50 chemicals. *Chemosphere* **2 5**, 553-562.

2. Dierickx, P.J. and Bredael-Rozen, E. (1996). Correlation between the in vitro cytotoxicity of inorganic metal compounds to cultured fathead minnow fish cells and the toxicity to Daphnia magna. *Bull. Environ. Contam. Toxicol.* **5 7**, 107-110.

3. Dierickx, P. J. (1993). Correlation between the reduction of protein content in cultured FHM fish cells and fish lethality data. *Toxicol. in Vitro* **4**, 527-530.

4. Dierickx, P.J. (1998). Increased cytotoxic sensitvity of cultured FHM fish cells by simultaneous treatment with sodium dodecyl sulfate and buthionine sulfoximine. *Chemosphere* **3 6**, 1263-1274.

ELECTROLYZED AND NATURAL REDUCED WATER EXHIBIT INSULIN-LIKE ACTIVITY ON GLUCOSE UPTAKE INTO MUSCLE CELLS AND ADIPOCYTES

M. Oda, K. Kusumoto, K. Teruya, T. Hara, T. Maki, S. Kabayama, Y. Katakura, K. Otsubo[1], S. Morisawa[1], H. Hayashi[2], Y. Ishii[3] and S. Shirahata
Graduate School of Genetic Resources Technology, Kyushu University, Fukuoka, Japan; [1]Nihon Trim Co. Ltd., Osaka, Japan; [2]Water Institute, Tokyo, Japan; [3]Hita Aqua Green Co. Ltd., Hita, Japan)

Abstract

In the type 2 diabetes, it has become clear that reactive oxygen species (ROS) cause reduction of glucose uptake by inhibiting the insulin-signaling pathway in muscle cells and adipocytes. We demonstrated that electrolyzed-reduced water (ERW) scavenges ROS and protects DNA from oxidative damage[1)]. Here we found that ERW scavenges ROS in insulin-responsive L6 myotubes and mouse 3T3/L1 adipocytes. Uptake of 1-deoxy-D-glucose (2-DOG) into both L6 cells and 3T3/L1 cells was stimulated by ERW in the presence or absence of insulin. This insulin-like activity of ERW was mediated by the activation of PI-3 kinase, resulting in stimulation of translocation of glucose transporter GLUT4 from microsome to plasma membrane. These results suggest that ERW may be useful to improve insulin-independent type 2 diabetes.

1. Introduction

Reactive oxygen species (ROS: $^1O_2$, $O_2^{\cdot-}$, H_2O_2, $\cdot OH$ *etc.*) are known to cause irreversible damage to macromolecules such as nucleic acids, proteins and lipids. Since ROS are produced in many processes in many tissues, there are many diseases caused by ROS. Cancer, diabetes and arteriosclerosis are representative those.

Diabetes is mainly grouped into two types; IDDM (insulin-dependent diabetes mellitus) and NIDDM (insulin-independent diabetes mellitus). Insulin stimulates blood glucose uptake into muscle and adipocytes which are main tissues in the body. IDDM is caused by deficiency of insulin secretion from pancreas. NIDDM is caused by lowered responses of cells against insulin (insulin-resistance). Recently participation of ROS has been noted in both IDDM and NIDDM. Since oxidative damage in insulin signaling pathway (hyperoxia[2)] , high glucose[3,4)]) has been reported, we tried to improve glucose uptake by shifting redox state of muscle cells and adipocytes to more reduced one by reduced water.

A. Bernard et al. (eds.), Animal Cell Technology: Products from Cells, Cells as Products, 425–427.

Devices to reform tap water by way of electrolysis were produced in Japan before half a century ago and now very popular in Japan. Daily intake of electrolyzed alkaline reduced water produced near cathode by electrolysis is believed to be beneficial for health. The ministry of Health and Welfare In Japan authorized in 1965 that the intake of electrolyzed reduced water is effective for restoration of unusual fermentation of intestinal flora. However, the action mechanism of electrolyzed reduced water was unknown for a long time. Recently we found electrolyzed reduced water (ERW) contains a lot of hydrogen, scavenges ROS and protects DNA from oxidative damage[1]. We proposed active hydrogen water hypothesis that active hydrogen in reduced water may be ideal radical scavenger to scavenge ROS[1]. Cancer cells cultured in electrolyzed reduced water exhibited suppressed growth, drastic morphological changes, telomere-shortening, lowered activities of telomere binding proteins, suggesting that cancer cells lost tumor phenotypes in reduced water[5]. Since sooner decline in blood sugar level in diabetic patients by daily intake of electrolyzed reduced water or of natural reduced water drawn from deep underground in some districts has been reported, we examined effects of those reduced water on glucose uptake into muscle and adipocytes.

2. Materials and Methods

Preparation of reduced water Electrolyzed reduced water (ERW) was obtained from ultrapure water containing 0.01% NaCl by using an electrolyzing device (type TI-7000S, Nihon Trim Co., Osaka). Natural reduced water drawn from 1000 m underground was provided from Hita Aqua Green Co. Ltd.

Cell Culture Rat skeletal muscle L6 cells were differentiated in DMEM containing 2% FBS and experiments on myotubes were usually performed between days 9 and 11 after the initiation of differentiation[6]. Differentiation from 3T3/L1 fibroblasts into adipocytes were accomplished as previously described[7,8]. Mature 3T3/L1 adipocytes were used between days 10 and 12 after the initiation of differentiation.

Assay of intracellular redox state Intracellular redox state levels were measured using a fluorescent dye, 2',7'-dichlorofluorescein-diacetate(DCFH-DA).

2-Deoxyglucose uptake After incubation in serum-deprived DMEM for 5 h at 37 ℃ prior to incubation with or without insulin, cells were rinsed twice with HEPES-buffered saline, followed by determinations of transport of 2-deoxy-D-[^{3}H]glucose(1μCi/ml) for 10 min in the same solution, and the associated radioactivity was determined by a liquid scintillation counter.

3. Results and Discussion

First we observed intracellular redox state of L6 myotubes treated with ERW. After incorporation into cells, DCFH-DA changes to fluorescent substance by oxidation with H_2O_2 accumulated in cells. We found ERW-treated cells shifted to more reduced redox state compared to control cells. Even when the cells were more oxidized in the presence of a high concentration of glucose, ERW could reduce the oxidized state (data not shown).

Both ERW and Hita natural reduced water could stimulate the 2-DOG uptake into L6 myotubes in the absence of insulin, as shown in Figure 1. In the presence of reduced water, the action of insulin seemed to be more effective. Since the insulin-like effect of ERW was inhibited by wortmannnin, a PI-3 kinase specific inhibitor, the action mechanism of reduced water on the stimulation of glucose uptake was suggested to be similar to that of insulin. After binding to insulin receptor, insulin activate PI-3 kinase which then accelerate the translocation of a glucose transporter, GLUT-4. Actually, ERW could accelerate the translocation of GLUT4 into plasma membrane from microsome (data not shown). Similar results were obtained in the case of adipocytes (data not shown).

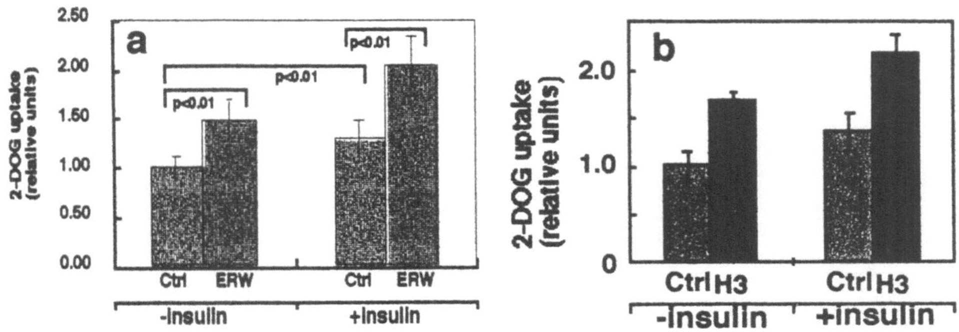

Figure 1. Stimulation of 2-deoxy glucose uptake into L6 muscle cells by ERW (a) and Hita natural reduced water (b). L6 myotubes were incubated with or without 100 nM insulin after cultivation in serum deprived DMEM or ERW-DMEM for 5 hours. Ctrl, control milli Q water; ERW, electrolyzed reduced water; H3, Hita reduced water.

This study revealed that reduced water stimulated glucose intake into muscle and adipocytes. Natural reduced water also protected DNA from oxidative damage. Further intensive researches on the relationship between oxidative damage of cell functions and anti-oxidative effects of reduced water will be needed to clarify the molecular action mechanism of anti-diabetes effect of ERW and natural reduced water.

5. References

1) Shirahata,S. *et al. Biochem. Biophys. Res. Commun.*, **234**, 269-274(1997)
2) Bashan N. *et al. Am J Physiol*, **262**, C682-90(1992)
3) Antonio Ceriello *et al. Diabetes*, **45**, 471-477(1996)
4) Muller H *et al. Diabetes*, **40**, 1440-1448(1991)
5) Shirahata, S. *et al.* Animal Cell Technology, 11, in press (1999).
6) D. Elizabeth Estrada *et al. Diabetes*, **45**, 1798-1804(1996)
7) Student AK *et al. J. Biol. Chem.*, **255**, 4745-50(1980)
8) Sargeant R *et al. Am. J. Physiol.*, **269***(Cell Physiol.* 38*)*, C217-C225(1995)

STUDY OF DDS FOR A CANCER THERAPY APPLYING LIPID VESICLE IMMOBILIZING *EUCHEUMA SERRA* LECTIN.

K. KATO, T. SUGAHARA, Y. MARUYAMA, N. YOSHIMURA
A. YOSHIHIRO, N. TATEISHI, Y. SUZUKI, A. KAWAKUBO
AND T. SASAKI
Applied Chemistry, Faculty of Engineering, Ehime University
3 Bunnkyou cyo Matuyama Ehime 790-8577, Japan

Abstract

The investigations on a specific adsorption between a lipid vesicle and a cancer cell were carried out. The vesicles were prepared by means of a two-step emulsification technique. The main component of the lipid membrane of the vesicle was Span80 (sorbitan monooleate). The attention was given to *Eucheuma serra* (algal lectin: abbreviated as ESA), which showed the specific affinity to a high-mannose. The ESA-immobilized vesicle was found to be specifically adsorbed to the cancer cells, such as human breast adenocarcinoma (MCF7), colon adenocarcinoma (COLO201), and human-human hybridoma (HB4C5). The vesicle flew smoothly in the superior artery of rabbit. Then, the immobilized vesicle is expected to be useful for a cancer therapy in a DDS.

1. Introduction

A drug delivery system (DDS) by use of a lipid vesicle as a drug carrier has been expected to be applied for a cancer therapy. However, there are many problems for the practical therapy use. The purpose of this work is the preparation of the vesicle, which can be specifically adsorbed by the targeting cancer cell. It is well known that a growth rate of either glycolipid or a glycolprotein of cancer-cells increases compared with a normal cell. From our investigations, the *Eucheuma serra* (ESA) has been found to inhibit the growth of a cancer cell. The *Eucheuma serra* (ESA) was tried to be immobilized to a lipid vesicle as a drug carrier. Thus, the fundamental experiments on the ESA-immobilized vesicle toward an application of DDS for a cancer therapy.

2. Materials and method

Large lipid vesicles were prepared by the two-step emulsification technique by use of a homomixer [1,2,3]. The main components of the vesicle were Span80 (sorbitan monooleate). Small lipid vesicles were also prepared by means of supersonification. Span80 was mixed in Tween80 (the weight ratio = 2:1) by use of a homomixer. Then the PBS buffer solution was added and the mixed solution was supersonicated for about 3 minutes. From the TEM-microscopic observations, the vesicle diameters are ranged from about 50nm to 300nm. Hunan-human hybridoma (HB4C5) cells were fusion product of a

A. Bernard et al. (eds.), Animal Cell Technology: Products from Cells, Cells as Products, 429–431.

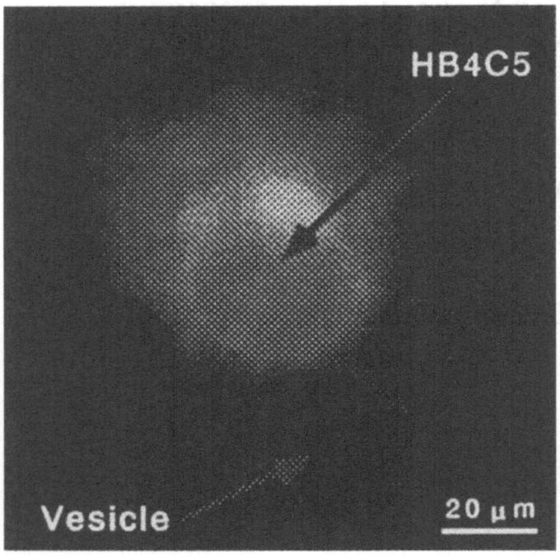

Fig. 1. Phase-contrust microscopic analysis of the binding between lipid vesicle and hybridoma HB4C5 cells

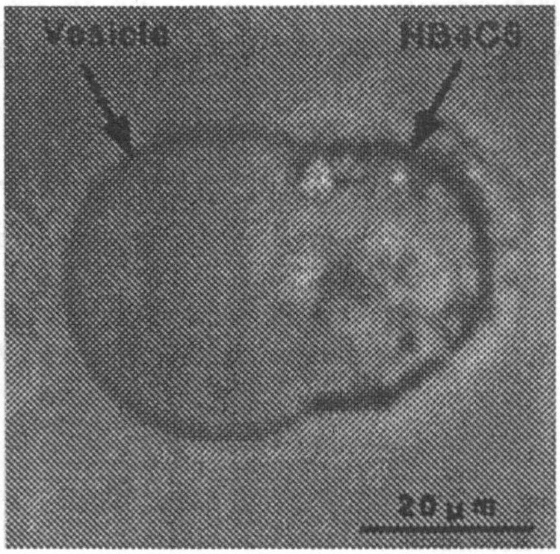

Fig. 2. Phase-contrust microscopic analysis of the fusion between lipid vesicle and hybridoma HB4C5 cells

human lymphocyte from lung cancer patient and a human lymphoma line,NAT-30 cell [4]. Cells were counted by use of hemocytometer. Additional agents to a cell were used after 0.22μm-membrane filtration. Protein A (adhering type to cell membrane) was preliminary purified with a gel clomatography. The 1.5mg/ml-ProteinA was prepared being mixed in Tween80 solution and sonificated for 5min. Then, the protein A was immobilized to a vesicle by the second-stage emulsification. The cells were cultured in ERDF medium supplemented with FCS using plastic dishes. The rate of ESA-induced erythrocyte aggregation was measured with low share rheoscope which consisted of a transparent 0.8" cone and a glass plate. Measurements of flow dynamics of the vesicle were carried out, using a mesentery isolated from anesthetic rabbit.

3. Results and discussion

3.1 Characteristics of ESA

The various lectins were successfully separated from *red alga*. The lectins were *Eucheuma serra* (ESA), *Eucheuma cottonii* (ECA) and *Eucheuma amakusaensis* (EAA). The lectins were examined from the viewpoint of the specific affinity to a sugar chain. The ESA was found to have an excellent specifically affinity to a high-mannose [5]. Therefore, the ESA was chosen as a ligand immobilized to a lipid vesicle.

We observed the flow dynamics of the vesicles in microciculatioin and measured the flow resistance. As a result, the vesicles were found to have high fluidity in a superior mesentric artery of rabbit.

3.2 Effect of ESA on cancer cells

The ESA had an excellent heat-stability and the inhibition of the growth of a cancer cell was observed. The free ESA of various concentrations was contacted with the cells, such as COLO201, MCF7 and MCF10 cells. The contact of 23μg/ml-ESA with the cancer cells of both COLO201 and MCF7 caused the death of the cells during a day after the contact, while MCF10 cells as a normal cell didn't die. This is probably due to the difference of the amount of the high-mannose between the cancer cells and MCF10 cell: the ESA was preferentially adsorbed to the cancer cells, which has lager amount of high mannose compared with that of the MCF10 cells. Thus, ESA is expected to either the targeting ligand or the inhibitor of cancer cells.

3.3 Immobilization of ESA to a vesicle

ESA mixed in the Tween80 solution was used to immobilize ESA to a vesicle at the second-stage emulsification. The driving force of the immobilization is the hydrophobicity of the ESA surface. ESA could not be easily immobilized to a vesicle due to the strong hydrophilicity. The various immobilization methods of the ESA were tried. The methods tried were as follows: give an osmotic pressure at the immobilization to incorporate ESA into the inner lipid membrane; mixing of Sphingomyelin with Span80 to support of the ESA in the lipid membrane and etc. From the many examinations on the immobilization, the labeling of XRITC of fluorescent probe was found to exchange the ESA surface hydrophobic and very effective for the immobilization of ESA about 30wt% amount of ESA used were immobilized to a vesicle.

3.4 Both the microscopic observations and the fluorometric measurements of the adsorption binding between the ESA-immobilized vesicle and various cells

The phase-contrast fluorescent microscopic observations of the cell adsorption between the ESA-immobilized vesicles and HB4C5 were carried out taking into account the convenient observation. Many vesicles were observed to be adsorbed by the HB4C5 cells as shown in the micrographs in **Fig.1**. **Figure 2** also shows the fusion between the vesicle and the cell. On the other hand, such binding shown in Fig.1 could be hardly observed in the case of normal cell. From these results, the immobilized vesicle was found to be specifically adsorbed by HB4C5 cell.

Nextly, the florometric observations of the adsorption binding between the ESA-immobilized vesicles and cancer cell. In the case of the florometric experiments small vesicles (about 50~300nm) were applied. Fluorescent probes of either FITC or XRITC were labeled to the immobilized ESA. The observations by use of a phase-contrast florescent microscopic and flowcytometric measurements were carried out to the adsorption binding. From the results, the ESA-immobilized vesicles were specifically adsorbed by the cancer cells, such as COLO201 and MCF7cell. Moreover, the vesicle adsorption gave a serious lesion to the cancer cells and the rate was inhibited.

Thus the ESA-immobilized vesicles were very hopeful for a cancer therapy in DDS.

References

1. Kato, K, Ikeda,T. and Shinozaki,M.:(1993),Lipid-Membrane Characteristics of large lipid-vesicles prepared by two-step Emulsification Technique and enzymatic NAD^+ -recycling in the vesicles, *J.Chem.Eng.Japan*, **26**, No.2,212-216
2. Kato, K, Tsutanaga,A. and Shinozaki,M.:(1997), Entrapment of enzymes in a giant lipid-vesicle as a biomimetic membrane, *Solvent Extraction Research and Development, Japan* , **4**, 51-61
3. Kato, K, Hirata,K.: (1996),Water Permeability through the Lipid Membrane of a Giant Vesicle Prepared by a Two-Step Emulsification Technique, *Solvent Extraction Research and Development, Japan* , **3**, 62-78
4. Murakami, F, Sasaki, T. and Sugahara, T. (1997) Lysozyme stimulates immunoglobulin Production by human-human hybridoma and human peripheral blood lymphocytes., *Cytochnology* **24**,177-182
5. H.Kawakubo, *Journal of Applied Phycology*, **9**, 331-338 (1997)

STUDY OF THE SPECIFIC BINDING BETWEEN LIPID VESICLES AND HUMAN-HUMAN HYBRIDOMA TOWARD EITHER DDS OR GENE TRANSFECTION

K. KATO, T. SUGAHARA, S. KAWASHIMA, Y. HAYASHI, A. YOSHIHIRO AND T. SASAKI
Applied Chemistry, Faculty of Engineering, Ehime University
3 Bunnkyou cyo Matuyama Ehime 790-8577, Japan

Abstract

The lipid vesicles prepared were composed of such as artificial lipids of Span80 and cationic lipids (DDAB and $N^{+}C_5Ala2C_{16}$). Firstly, the cationic vesicles were prepared by the addition of the cationic lipids to Span80. The addition enhanced the affinity between the vesicles and the cells, such as MCF7 (human breast adenocarcinoma), COLO201 (colon adenocarcinoma) and the human-human hybridoma (HB4C5 cell), which was a fusion product of a human lymphocyte from lung cancer patient and a human lymphoma line, NAT-30. Nextly, the anti-IgM antibody was immobilized to vesicle (immunovesicle). The targeting antigen was the monoclonal IgM produced at the surface of the HB4C5 cells. The binding and the fusion between the vesicles and HB4C5 cells were confirmed and the addition of the cationic lipids to Span80 enhanced the fusion between the immunovesicles and HB4C5 cells. Moreover, the membrane fluidity and the HLB value of the lipids was also found to strongly effect on the vesicle fusion.

1. Introduction

An immunovesicle, which is an antibody-immobilized vesicle, is expected to be applied in either DDS or gene transfection. However, many problems to be solved have remained for the practical use. The driving force of the binding between an immunovesicles and cells is due to an antigen-antibody affinity. The immobilized antibody to a vesicle should be orientated to the outer direction of a vesicle. The surface of many cells is charged to be anionic. Then, the cationic vesicle becomes to have an stronger affinity between vesicles and cells. The prepared vesicles with cationic lipids have, however,been known to be unstable. Thus, the vesicle stability and the immobilization method of an antibody are the important factor of the preparation of an immunovesicle.

In this work, the human-human hybridoma (HB4C5 cell) [1] was mainly used as an targeting cell. Then, the binding affinity between the HB4C5 cells and the immunovesicles, at which surface the IgG was immobilized, was studied.

2. Materials and method

Materials: The Cationic lipids used here were Di-lauryl-dimethylammonium bromide (DDAB) and *N,N*-dihexadecyl-N^{α}-[6-(tri-methylammonio)hexanoyl]-L-alaninamide-bromide

A. Bernard et al. (eds.), Animal Cell Technology: Products from Cells, Cells as Products, 433–435.

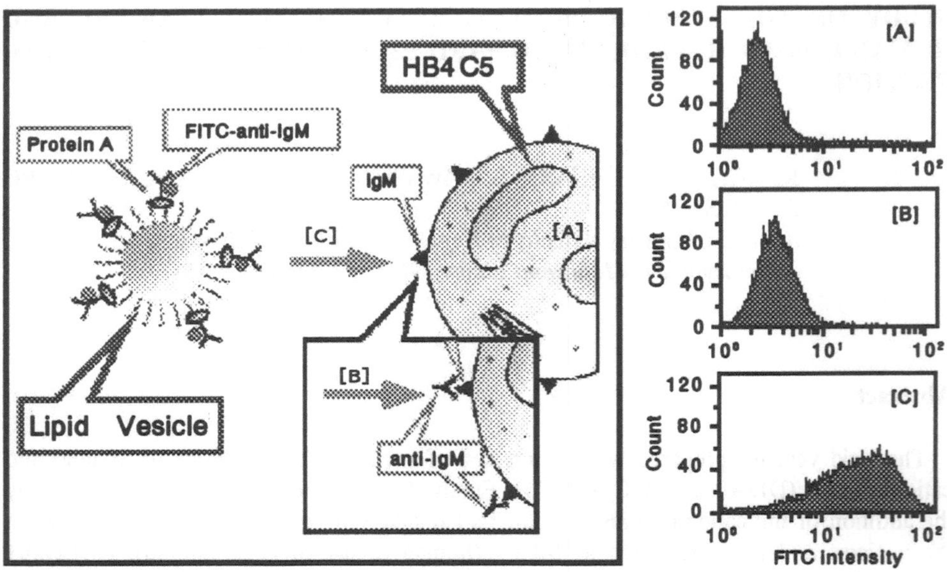

Fig.1. Schematic view and the Flowcytometric measurements of the binding between HB4C5 and the immobilized-anti-IgM vesicle: [A] control (HB4C5 without contacting vesicle), [B] the binding in the case of masking IgM on HB4C5 surface with free anti-IgM antibody, [C]the binding without masking. Non-immobilized vesicle was not adsorbed to HB4C5

($N^{+}C_5Ala2C_{16}$). As an antibody immobilized to a vesicle, human anti-IgM antibody (IgG) was chosen. The cells used here were HB4C5, MCF7 and COLO201 cells. As a fluorescent probe, FITC and Propidium iorido (PI, Ex=493nm, Em=638nm) were used .

Method: Vesicles were prepared by means of a two-step emulsification technique [2,3,4]. The W/O emulsions were prepared by adding PBS buffer solution to Span80-dissolved hexane emulsifying with a homomixer at the first-step emulsification. After removal the hexane by evaporation, the residual lipid-PBS mixture was mixed with Tween80 solution. The mixed solution was also emulsified at the second-step emulsification and vesicle suspension was prepared . Then, the suspension was filtrated with Millipore membrane. The cells were cultured in ERDF medium supplemented with FCS using plasticdishes .and the binding experiments between the vesicles and cells were carried out. The observations of the cells were by use of a fluorescent phase-contrast microscope (Olympus, BH-2), confocal-laser fluorescent microscope (Meridian) and flowcytometer (Becton Dickinson, Facscaliber, K21).

3. Results and discussion

3.1 The binding affinity between cationic vesicles and cancer cells

The vesicle surface composed of only Span80 has already been found to be anionic from our electrophoresis experiments. It is well known that the surfaces of many cancer cells are strongly anionic. Then, the electrostatic repulsion between the Span80 vesicle and a cancer cell weakens the binding affinity. We tried the preparation of the cationic vesicle by adding cationic lipids to Span80 of a vesicle to enhance the affinity.

From our many experiments, the vesicles were most stable at about 30% weight-ratio of cationic lipid related to the total lipid. The fluorescent probe of PI as a tracer was entrapped in the cationic vesicles, of which membrane was composed of either mixture of DDAB and Span80 or that of CPL and Span80. Flowcytometric measurements revealed that the vesicles composed of only Span80 (Span80 vesicle) could hardly be adsorbed by the cancer cells, such as MCF7, COLO201 and HB4C5 cells. On the other hand, the cationic vesicles could be adsorbed by the cells and the entrapped PI in the vesicles could be successfully introduced in the cells. Thus, the mixing of the cationic lipid with Span80 enhanced the binding affinity between the vesicles and the cancer cells.

It was also found that the cationic vesicles composed of the mixture of Span80 and CPL (30wt%-CPL vesicle) were more stable than those composed of the mixture of Span80 and DDAB (30wt%-DDAB vesicle).

3.2 The preparation of an immunovesicle

The HB4C5 cells produce the monoclonal antibody of IgM. Therefore, the IgM at the HB4C5 surface was chosen as a targeting antigen. A proteinA, which has a binding site to Fc fragment of IgG, was preriminaly immobilized to the vesicle by means of the two-step emulsification. The human anti-IgM antibody (IgG) was immobilized to a vesicle by incubation method. It was fluorometrically confirmed that the IgG was successfully immobilized orientated to the outer direction of the vesicle surface and the antigen-antibody binding of the immobilized IgG well functioned.

3.3 The binding affinity between the immunovesicle and a cancer cell

Firstly, the function of the immnovesicle composed of only Span80 (Span80 immunovesicle) was studied. The immobilized human anti-IgM antibody (IgG) was preriminally labeled with FITC. The fluorescent probe of PI as a tracer was also entrapped in the vesicle. As shown in **Fig.1**, the flowcytometric measurements of the FITC revealed that the Span80 immunovesicles could be specifically adsorbed by the HB4C5 cells due to the antigen-antibody affinity (see the legend in Fig.1). However, the entrapped PI was found not to be introduced in the cells from the PI observations by use of either confocal-laser fluorescent microscope or flowcytometer (data not shown). This indicates that the fusion between the Span80 immunovesicles and the HB4C5 cells does not happen.

Nextly, the immunovesicles were prepared by mixing cationic lipids, such as either DDAB or CPL, with Span80 (cationic immunovesicle). The fluorometric measurements also revealed that the cationic immunovesicles could be specifically adsorbed by the HB4C5 cells and the entrapped PI could be successfully introduced in the cells. The above results indicate that the mixing of the cationic lipid with Span80 enhances the fusion between the immunovesicles and the HB4C5 cells.

References

1. Murakami,F. Sasaki,T. and Sugahara,T.(1997) Lysozyme stimulates immunoglobulin Production by human-human hybridoma and human peripheral blood lymphocytes., Cytochnology **24**, 177-182.
2 Kato,K. Hirao,H. Sagawa,M. and Tamura M. (1997)Preparation and function of a new nano-size W/O/W emulsion modified by a monoclonal antibody of CD45 for a DDS to a human neutrophil, Proc. of Asia-Pacific Biochemical Engineering Conference '97, 592-596
3. Kato,K. and Hirashita,J.(1997),Preparation and function of a hybrid-type vesicle modified by Concanavalin A for a DDS directed toward a cancer therapy, Proc. of Asia-Pacific Biochemical Engineering Conference '97, 592-596
4. Kato, K. Tsutanaga, A. and Shinozaki, M. (1997), Entrapment of enzymes in a giant lipid-vesicle as a biomimetic membrane, *Solvent Extraction Research and Development, Japan*, **4**, 51-61.

SUPPRESSION OF APOPTOSIS IN UV-DAMAGED HUMAN MELANOMA CELLS BY A FERMENTED MILK, KEFIR

T. NAGIRA, J. NARISAWA, K. TERUYA, K. KUSUMOTO, Y. KATAKURA, D. W.BARNES[1], S. TOKUMARU[2], and S. SHIRAHATA
Graduate School of Genetic Resourses Technology, Kyushu University, Fukuoka, Japan; [1]American Type Culture Collection, Manassas, USA, [2]Nihon Kefir Co.Ltd., Fujisawa, Japan.

Abstract : Accumulation of DNA damage caused by UV irradiation which could not be excluded by DNA repair or apoptosis mechanisms may cause skin cancer. We studied DNA repair enhancing factors in Kefir, a fermented milk by detecting thymine dimer formed by UV irradiation with FITC labeled anti-thymine dimer antibody. The aqueous extract of Kefir remarkably enhanced the repair of thymine dimer and the cells treated with Kefir enhanced unscheduled DNA synthesis in the process of DNA repair of UV-damaged cells. The Kefir extract exhibited potent anti-oxidative activity and shifted the redox state of HMV-1 cells irradiated with UV from oxidative to more reduced state. The cells treated with Kefir exhibited high colony forming ability than control. Since apoptosis of HMV-1 cells caused by UV irradiation was also suppressed by the Kefir extract, suggesting that DNA repair enhancing factors in the Kefir extract could decrease DNA damage by UV irradiation and suppress apoptosis.

1. Introduction

It has been known that ultraviolet ray (UV) produces free radicals or reactive oxygen species in cells, which damage chromosomal DNA. This DNA damage results in genetic mutation, cell death or cellular carcinogenesis. Kefir is a fermented milk drink originating in Caucasus mountains and is known to have numerous benefits including anti-cancer effect and stimulation of the immune system. We have demonstrated that the interferon-β production stimulating effect of Kefram–Kefir derived sphingomyeline.[1] Here we report the possibility that Kefir contained DNA repair enhancing factors which could decrease thymine dimer, a kind of DNA damage formed by UV irradiation.

2. Material and Method

The human melanoma cell line HMV-1 cells were provided by RIKEN Cell Bank. This

A. Bernard et al. (eds.), Animal Cell Technology: Products from Cells, Cells as Products, 437–439.

cell relatively resistant to irradiation.

Kefram–Kefir powder was supplied by Nihon Kefir Co. Ltd. The aqueous extract of the Kefram–Kefir powder was used as a Kefram–Kefir extract sample.

After UV irradiation, HMV-1 cells were incubated with Kefir, we aspirated medium, washed cells, and changed medium to fresh one. Then we analyzed the effects of Kefir to UV-irradiated cells. UV irradiation was performed using 254nm-UVC.

We detected thymine dimer formed in chromosomal DNA by mouse monoclonal anti-thymine dimer antibody (KYOWA). Fixed cells were treated with the antibody, and labeled with FITC-conjugated anti-mouse IgG antibody.[2]

HMV–1 cells were incubated with Kefir and bromodeoxyuridine (BrdU) for 24 h. Unscheduled DNA synthesis (UDS) was measured by incorporation of BrdU in DNA and detecting it by specific antibody. The cells were arreated with 50mM of hydroxyurea to minimize replication of DNA as a result of cell division.[3]

The redox state in HMV-1 cell was measured by using a fluorescence pigment,2',7'-dichlorofluorescein-diacetate (DCFH-DA). DCFH-DA incorporated in cells werehydrolyzed by peroxidase and then oxidized to fluorescent DCFH by hydrogen peroxide acculmulated in cells.[4]

We used Apoptosis Detection System, Fluoresce Kit (Promega) to detect apoptosis. Apoptotic cell was detected by TUNEL method, which detects 3'OH end of chromosomal DNA breakage by apoptosis.[4]

After UV irradiation HMV-1 cells were incubated with the Kefir extract for 5 hours. Then, after washing, the cells were cultured for 1 week. The number of colonies were counted and the colony formation ration was calculated as % of control.[5]

3. Result and Discussion

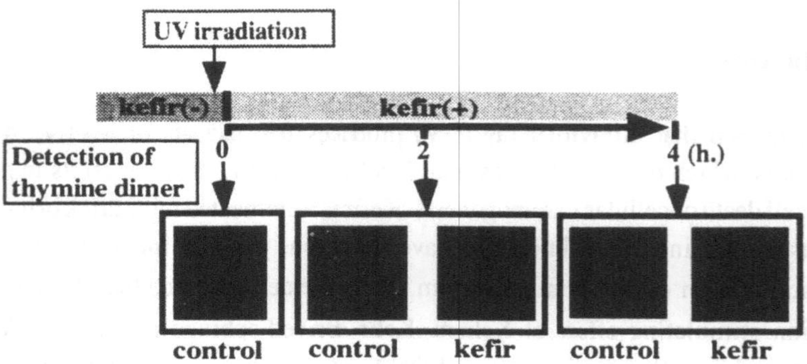

Figure 1. Suppressive effect of Kefir against thymine dimer formation

we examined the effect of Kefir on UV-induced thymine dimer formation, which is the main cause of cell death by UVC irradiation. The cells treated with Kefir had less thymine dimer than control. The number of thymine dimer in cells treated with Kefir decreased with time, suggesting that Kefir stimulated the DNA repair activity in cells. (Fig.1)

The cells treated with Kefir had high quantity of UDS than control after UV-irradiation. It means that the cells treated with Kefir enhanced DNA synthesis in the process of DNA repair of UV- damaged cells.

The Kefir extract shifted the redox state in UV-irradiated cells from oxidative to more reductive state in short time. These results suggest that Kefir may protects DNA, protein and lipid from the oxidative-damage by reactive oxygen species.

The Kefir extract suppressed UV-induced apoptosis of the cells compared to control. This suggests that DNA damage was repaired by DNA repair enhancing factor contained in Kefir and apoptosis was suppressed as a result.

UV-damaged cells treated with Kefir exhibited high colony forming ability than control. This means that Kefir could repair UV damage of DNA to the extent that cells can survive and proliferate without inducing apoptosis.

Intracellular oxidative stress causes wide ranges of damage to DNA, proteins and cell memberane. Defect of DNA repair system is known to extremely elevate the probability of suffering from cancer. Kefram-Kefir can not only decrease intracellular oxidative stress but also enhance DNA repair activity of UV-damaged cell, suggesting that Kefir may be benefical to prevent skin cancer caused by UV irradiation.

4. References

1. Osada, K., Nagira, K.,Teruya, K., Tachibana, H., Shirahata, S., and Murakami, H. (1994) Enhancement of interferon-β production with spingomyelin from fermented milk, *Biotherapy*, **7**,
2. Koji, T. and Nakane, P.K. (1990) Localization in situ of specific mRNA using thymine-thymine dimerized DNA probes. Sensitive and reliable non-radioactive in situ hybridization., *Acta Pathol JPN*, **40**, 783-807(Review)
3. Anna, T., Jeno, M., and Matyas, G, J. (1994) Monitoring of benzene-exposed workers for genotoxic effects of benzene:improved-working-condition-related decrease in the frequencies of chromosomal aberrations in peripheral blood *Mutat. Res.*, **304**: 159-165
4. Gavrieli Y, Sherman Y, Ben-Sasson SA (1992) Identification of programmed cell death in situ via specific labeling of nuclear DNA fragmentation, *J.Cell.Biol*, **119**,493-501
5. Ursula, K., Leon, H.F.M., and Andrea, H. (1997) Cobat(II) inhibits the incision and the polymerization step of nucleotide excision repair in human fibroblasts, *Mutat. Res.*, **383**,81-89

LARGE SCALE GMP PRODUCTION OF RETROVIRAL GENE THERAPY VECTORS IN VERY LOW SERUM MEDIUM

N. MOY, M. DILLINGHAM, S.MCGOWAN, M. ROTHERHAM, T.M. CLAYTON
GlaxoWellcome R&D, Biotechnology Development Laboratories
South Eden Park Road, Beckenham, Kent. BR3 3BS UK.

1. Abstract

Approximately 800 l of retrovirus containing supernatant were produced using the Costar Cell Cube™ system. Vector titres of approximately 2 to 8 x 10^6 infective units per ml were achieved and a total of over 10^{12} virus particles were recovered over the campaign. Initial growth was performed in serum containing medium and production was performed using a version of our proprietary protein free medium supplemented with 0.5 % FCS.
The production of the vector was enhanced by reducing the culture temperature during the production phase and by performing repeat harvesting of supernatant from the cell cube.

2. Introduction

The production of retroviral gene therapy vectors is difficult because of the low viral titres and instability of the virus (). The viruses are temperature sensitive and pH sensitive which makes purification difficult and the downstream processing of the virus includes a concentration stage to produce high enough concentrations of vector to allow the virus to be used therapeutically. Consequently the purification process used for retroviruses are likely to co purify high molecular weight medium components. It is essential to reduce the chances for contamination as much as possible by eliminating protein altogether or by using a very low protein medium. We had to produce a large quantity of virus in a short time and there was no time to develop a serum free medium formulation for the production of retrovirus. We developed a two stage system where the cells were grown in culture medium containing serum and then the serum content was reduced for the production phase of the work.

3. Materials and Methods

3.1. Cell Culture

The producer cell line was a PA317 derived cell line obtained from B. Huber's group.

A. Bernard et al. (eds.), Animal Cell Technology: Products from Cells, Cells as Products, 441–443.

Cells were revived and scaled up in tissue culture flats and roller bottles. The medium was DMEM supplemented with 1m/l of essential amino acids and 5% FCS. Cell were seeded at 2x 10^4 cells/cm^2 and harvested by trypsinisation when 70 to 95% confluent (approximately 2x 10^5 cells/cm^2). Cells were seeded into the cell cube to give a density of 1 to 2 x10^4 cells/cm2 and grown in normal culture medium for 3 days. The medium was then exchanged for a version of our propriatory serum free medium supplemented with 0.5% FCS and the temperature dropped to from 37 to 32 °C. Glucose and lactate analyses were performed using a YSI 2700 analyser. Viable cell counts were obtained using Erythrosin B staining and the virus was measured by a colony formation in indicator cells caused by the transfer of G418 resistance.

3.2. Cell Cube™ Operation

Four 100 stack cell cube units were manifolded together using sterile connectors and tubing and seeded with cells. The Cell Cube™ was seeded over a period of approximately 4 hours with alternate sides of the cubes being seeded in turn. All manipulations were performed in a Class 100 environment within the main culture area that was Class 10,000. All additions and removals of material were made using single use sterile connectors that were manifolded onto the outlet ports. The liquid circulation rate was set to maximum required for the process to prevent sloughing of cells if the flow rate was increased in the later stages of the process. Dissolved oxygen tension was measured on the inlet and outlet streams for the cell cube and the aeration rate of the culture medium was modified to keep the outlet DOT above 20%. Virus containing supernatant was harvested by completely emptying the cell cube system into a sterile plastic bag and refilling the system with fresh medium. The harvest was filtered through Sartorious Sartobran filters (0.45 micron prefilter and 0.2 micron final filter) prior to downstream processing.

4. Results

Revival and scale up of the cells gave consistent growth with the cell doubling rate being constant and the production seed was ready 21 days after the revival of a vial of cells from the cell bank. The culture of the cells in the cell cube could be followed by monitoring the dissolved oxygen tension and glucose utilisation (figure 1). As the cultures progressed the cell sheet started to shed cell debris into the culture medium and this made the filtration of the harvest progressively harder. In one case we saw a partial loss of biomass by sloughing of the cell sheet followed by recovery of the biomass. In most cases the culture was ended by total loss of cells from the cell cube.
The titres obtained during the production phase were mostly between 1x10^6 and 1x 10^7 and the virus production appeared to be relatively stable in most cultures as is demonstrated by the example in figure. The one exception was a culture where the cells appeared to grow quickly and the titre followed the oxygen utilisation curve.

Figure 1

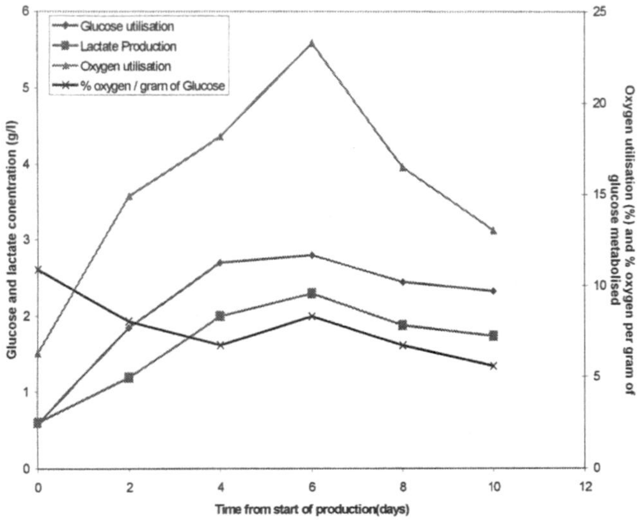

Following the growth of a retroviral vector producing culture indirectly by glucose and oxygen utilisation

5. Discussion

The Cell Cube™ proved itself a useful tool for production of quantities of retroviral vectors from attachment dependent cells. The system was installed and operated with a minimum of development and the cell cultures grown in the cell cube produced virus. The ability to completely drain the system meant that it was easy for every harvest to be defined as a batch. The system can be scaled up by adding further units but does not have the flexibility of suspension and microcarrier culture.

VACCINES AND IMMUNOLOGICALS

Chapter VIII

CHIMERIC LYSSAVIRUS GLYCOPROTEIN: NEW VECTOR FOR MULTIVALENT VACCINES

E. DESMEZIERES, Y. JACOB, M-F. SARON, F. DELPEYROUX
N. TORDO AND P. PERRIN.
Lyssavirus Laboratory
Pasteur Institute
25-28 rue du Dr. Roux
75724 Cedex 15 Paris, France

1. Abstract

We have developed a multivalent vaccine prototype using the DNA technology and chimeric lyssavirus glycoproteins to carry foreign virus epitopes. Lyssaviruses (rabies and rabies-related viruses) induce a fatal encephalomyelitis. They are divided in 7 genotypes (GT) and two principal groups according the cross-reactivity of virus neutralizing antibody (VNAb); group 1: GT 1, 4, 5, 6 and 7; group 2: GT2 and 3. Currently available vaccines belong to GT1. They induce protection against rabies (GT1) and are more or less efficacious against the other members of the group 1. They do not induce protection against group 2 viruses. Lyssavirus glycoprotein (G) is involved in the induction of both VNAb and protection. Rabies G molecule can be divided in two parts separated by a flexible hinge: the NH2 half and the COOH half containing the VNAb-inducing antigenic site II and III respectively. Injection of chimeric plasmid containing the COOH half of Pasteur Virus (PV: GT1) and the NH2 half of GT5 or GT3 G induced VNAb and protection against parental viruses but also enlarged to the other genotypes.
We have taken into account the flexibility of the site II-site III junction to insert foreign epitopes with the view to construct a multivalent vaccine prototype. The inserted sequences corresponded to two well characterized epitopes: the C3 B cell epitope of the poliovirus VP1 protein and the $CD8^{+}$ T cell epitope of the lymphocytic choriomeningitis virus (LCMV) nucleoprotein. Under these conditions, injection of mice with chimeric G genes carrying the foreign epitopes induced antibodies against poliovirus and protection against LCMV whereas VNAb production against parental lyssaviruses was maintained. Therefore, chimeric lyssavirus glycoproteins can be proposed as new vector for multivalent vaccines not only against lyssaviruses but also against other pathogens.

A. Bernard et al. (eds.), Animal Cell Technology: Products from Cells, Cells as Products, 447–453.

2. Introduction

Vaccines conferring protection against both a number of serotypes and various pathogens would have obvious advantages (4). Protection against viral diseases is mediated by both antibodies and CD8$^+$ cytotoxic lymphocytes (CTL) induced frequently by only a few immunodominant epitopes. Such a multivalent vaccine generating a complete immune response without the potential risk of pathogenicity as for live-virus vaccines could be developped using DNA-based immunization. Direct injection of DNA into animals is a promising and versatile method (3) and the use of chimeric lyssavirus glycoproteins enlarges the protection against rabies and rabies related viruses from the genotype 1 (GT1) to the other genotypes (2 to 7) (1, 5, 7). The protection against the lyssaviruses is mainly induced by the glycoprotein (G) which is associated as trimers and constitutes the spicules of the virus. G induces both virus neutralizing antibodies (VNAb) and CTL if injected as live attenuated virus or naked DNA (1). We have recently shown than G molecule can be divided in two parts separated by a flexible hinge composed of the NH2 half and the COOH half containing the VNAb-inducing antigenic site II and III respectively (1, 7). Thus with the view to construct a prototype of a multivalent vaccine various combinations of two well characterized epitopes were inserted in a chimeric lyssavirus G gene (NH2 half of the EBL1 G -GT5- and COOH half of the PV -GT1- G molecules). The inserted foreign sequences corresponded to: the C3 B cell epitope of the poliovirus type 1 capsid VP1 protein (2) and the H2^d CMH I-restricted CD8$^+$ T cell epitope of the lymphocytic choriomeningitis virus (LCMV) nucleoprotein (11). The C3 poliovirus epitope induces the synthesis of virus neutralizing antibodies (2) and the LCMV epitope is involved in both the induction of CTL and protection against LCMV challenge (11). The DNA-based immunization of mice with these genes induced antibodies against poliovirus and protection against LCMV as well as VNAb against parental Lyssaviruses. Therefore, chimeric lyssavirus glycoproteins can be proposed as new vector for multivalent vaccines not only against lyssaviruses but also against other pathogens.

3. Materials and Methods

H2^d BALB/c mice were used for immunological and protection studies. Chimeric EBL1-PV G gene was inserted in the eukaryotic expression vector pCIneo which possesses the CMV promotor as described elsewhere (1, 7). The different sequences inserted in the hinge region (a.a. 253) of chimeric EBL1-PV G gene correspond to (CTL and B represent LCMV and poliovirus epitope respectively):

GEBL1-(CTL)-PV: EBL1--a.a. 253- NLERPQASGVYMGNLTAQ-a.a.254--PV
GEBL1-(CTL-B)-PV: EBL1--a.a. 253- NLERPQASGVYMGNLTAQ NLDNPASTTNKDK-a.a.254--PV
GEBL1-(B-CTL)2-PV: EBL1--a.a. 253- (NLDNPASTTNKDKLFAVPQASGVYMG)2-a.a.254--PV

Mice were injected intramuscularly with 50 μg of different plasmids. T helper cells induction was evaluated according interleukin-2 (IL-2) production by splenocytes after *in vitro* stimulation with lyssavirus antigens as previously reported (8). Lyssavirus neutralizing antibodies were assayed by the RFFIT method (12) and poliovirus by ELISA (2). Immunized mice were also challenged intracerebrally with a lethal dose of LCMV (Arm/53b strain) (11).

3. Results

DNA-based immunization of mice with the chimeric EBL1-PV G protein carrying various B and CTL foreign epitopes induced a similar production of T helper cells (Table 1). This clearly indicates that foreign sequences can be inserted in EBL1-PV G and expressed under an immunologically potent form able to induce T helper cells.

TABLE 1. T helper cells induced by DNA-based immunization with various plasmids: production of IL-2 after in vitro stimulation of splenocytes from immunized mice.

Plasmid injected	IL-2 production (cpm) after stimulation with:		
	Medium	IPLV	ConA
pCIneo	5000	4 500	5 050
pEBL1-PV	3500	70 000	200 000
pEBL1-(CTL)-PV	800	65 000	210 000
pEBL1-(CTL-B)-PV	600	68 000	190 000
pEBL1-(CTL-B)2-PV	500	40 000	195 000

The insertion of B and CTL epitopes into the chimeric EBL1-PV G protein was also studied for the induction of antibodies against poliovirus and VNAb against PV and EBL1 lyssaviruses (Table 2). All plasmids containning the poliovirus B cell epitope induced antibodies against the poliovirus. The insertion of foreign epitopes decreased anti-lyssavirus VNAb production. However antibody levels remained at an acceptable level. Thus the chimeric G protein can carry and express *in vivo* B cell epitopes.

TABLE 2. Antibody production induced by the chimeric EBL1-PV G protein carrying foreign B epitope.

Plasmid injected	VNAb (IU/ml) against:		Antibody against Poliovirus
	PV	EBL1	
pCIneo	0	0	0
pEBL1-PV	6	22	0
pEBL1-(CTL)-PV	2.6	5.2	0
pEBL1-(CTL-B)-PV	1.8	8	1 515
pEBL1-(CTL-B)2-PV	1	1	1 100

As the LCMV nucleoprotein $H2^d$ $CD8^+$ T cell epitope is involved in the protection against an intracerebral challenge (11), chimeric EBL1-PV G proteins carrying this epitopes were tested for their protective activity in $H2^d$ BALB/c mice (Table 3). The chimeric GEBL1-(CTL-B)-PV plasmid induced a significant 70% protection against a lethal intra cerebral challenge.

Table 3. Protection induced by the EBL1-PV G protein carrying foreign CTL epitope.

Plasmid injected	Survival (%)
pCIneo	1/25 (4)
pEBL1-(CTL-B)2-PV	2/15 (13)
pEBL1-(CTL-B)-PV	7/10 (70)

These results show that both B and CTL epitopes can be inserted in the chimeric plasmid without destroying their immunological properties. Therefore, the chimeric G permits a correct presentation of these epitopes.

4. Discussion

Vaccine generating various immune responses against various serotypes and pathogens would be of great value (4). Among current vaccines, only those displaying viral replication (attenuated or recombinant viruses) generate CTL. In contrast, DNA-based immunization including that against lyssaviruses is able to generate a complete immune response without the potential risks of pathogenicity as for live viruses. Thus, as we have previously shown that chimeric lyssavirus G proteins enlarged the protection against lyssaviruses, immunization with plasmids has great potential for the development of safe multivalent vaccines.

Combined vaccines including a vaccine against rabies have been used for the immunization of dogs ((leptospirosis, distemper, hepatitis, and parvo- canine viruses), cats (panleukopenia, calici- and parvo-feline viruses) and cattle (foot and mouth disease virus) (10). Thus, we investigate the development of a multivalent DNA vaccine « prototype » including lyssaviruses using the versatile plasmid DNA technology. We have recently shown that (1, 7): i) plasmids encoding lyssavirus glycoproteins induced both humoral and cell-mediated immune responses including CTL; ii) chimeras between different lyssavirus genotypes can be used to broaden the spectrum of protection against rabies and rabies-related viruses; iii) the COOH half containing the antigenic site III of the rabies PV glycoprotein induces Th cells. Consequently, we tried to investigate the potential of a multivalent DNA vaccine prototype based on the use of a chimeric lyssavirus G protein (EBL1-PV) to carry non-lyssaviral B and CTL cell epitopes as fragments of foreign antigens. The B and CTL cell epitopes corresponded respectively to the C3 epitope of poliovirus type 1 VP1 protein (involved in VNAb induction) and to the $CD8^+$ T cell epitope of LCMV nucleoprotein (involved in the protection of $H2^d$ mouse against LCMV).

The chimeric EBL1-PV plasmid encoding the foreign B and CTL cell epitopes was efficient in inducing antibodies, Th cells and protection against LCMV challenge, demonstrating that foreign epitopes are expressed under an immunologically potent form and correctly presented by both MHC II and MHC I molecules. Virus neutralizing antibody titers against lyssaviruses were maintained at a protective level. Indeed we have previously shown that when VNAb titers in mice sera were above 1.5 IU/ml after DNA-based immunization, all animals survived an i.c. challenge (7). In addition, levels of antibody against the poliovirus peptide induced by pGEBL1-(CTL-B)-PV was similar to that obtained by a single injection of the same epitope on hybrid hepatitis B surface antigen particles (2). Moreover, high levels of protection against LCMV were obtained showing that the efficacy of the chimera was similar to that of the gene encoding the full LCMV nucleoprotein, or to that induced by recombinant bacteria or bacterial toxin carrying the same epitope (6, 14).

We demonstrated that it is possible to insert various epitopes presented by different MHC alleles in order to obtain CTL production in non-congenic animals or in human because: i) at least 49 aa residues can be inserted with satisfying immunological responses; ii) the effective delivery of multiple CTL epitopes involved in protection against many viruses, tumors, intracellular bacteria and parasites has been reported; iii) degenerated CTL epitopes can be recognized in the context of multiple MHC I molecules reducing the number of epitopes to be used (13, 14). Therefore, multiple B and CTL epitopes could presumably be inserted and expressed with the view of human and/or veterinary use.

The chimeric lyssavirus glycoproteins carrying antigen fragments may be a new prototype for the development of multivalent vaccines against various zoonoses including rabies. DNA-based immunization was used for this demonstration but, other systems able to induce CTL might be accepted: in particular, recombinant vaccine virus which has been widely used in Europe for oral vaccination of foxes

5. References

1. Bahloul, C., Y. Jacob, N. Tordo, and P. Perrin. (1998). DNA-based immunization for exploring the enlargement of immunological cross-reactivity against the lyssaviruses. Vaccine. 16:417-425.
2. Delpeyroux, F., Van Wezel, E., Blondel, B. and Crainic, R. 1990. Structural factors modulate the activity of antigenic poliovirus sequences expressed on hybrid hepatitis B surface antigen particles. J. Virol. 64:6090-6100.
3. Donnelly, J., Ulmer., J., Shiver, J. and Liu, M. 1997. DNA Vaccines. Annu. Rev. Immunol. 15:617-648.
4. European Commission COST/STD-3. 1996. Advantages of combined vaccines. Vaccine. 14:693-700.
5. Fekadu, M., J. Shaddock, D. Sanderlin, and J. Smith. 1988. Efficacy of rabies vaccines against virus isolated from European house bats (*Eptesicus serotinus*), classic rabies and rabies-related virus. Vaccine. 6:533-539.
6. Goossens, P. Milon, G., Cossart, P. and Saron, M-F. 1995. Attenuated *Listeria monocytogenes* as a live vector for induction of CD8$^+$ Tcells *in vivo*: a study with the nucleoprotein of the lymphocytic choriomeningitis virus. Internation. Immunol. 7: 797-805.
7. Jallet, C., Jacob, Y., Bahloul, C., Drings, A., Desmezières, E. Tordo, N. and Perrin, P. 1999. Chimeric lyssavirus glycoproteins with increased immunological potential. J. Virol; 73:225-233.
8. Joffret, M-L., Zanetti, C., Morgeaux, S., Leclerc, C. Sureau, P. and Perrin P. 1991. Appraisal of rabies vaccine potency by determination of *in vitro*, specific interleukin-2 production. Biologicals. 19:113-123.
9. Lodmell, D., Ray, N., Parnell, M., Ewalt, L., Hanlon, C., Shaddock, J., Sanderlin, D. and Rupprecht, C. 1998. DNA immunization protects non-human primates against rabies virus. Science Med. 4:949-952.
10. Pastoret, P-P., Brochier, B., Aguilar-Setién, A., Blancou, J. 1997. Vaccination against rabies. In «Veterinary Vaccinology» Eds. Pastoret, P-P. *et al.* (Elsevier): 616-628.
11. Saron, M-F., Fayolle, C., Sebo, P., Ladant, D., Ullmann, A. and Leclerc, C. 1997, Anti-viral protection conferred by recombinant adenylate cyclase toxins from *Bortella pertussis* carrying a CD8+ T cell epitope from lymphocytic choriomeningitis virus. Proc. Natl. Acad. Sci. USA. 94: 3314-3319.
12. Smith, J., P. Yager, and G. Baer. 1996. A rapid fluorescent focus inhibition test (RFFIT) for determining virus-neutralizing antibody. In "Laboratory techniques in rabies", Fourth edition (Eds Meslin, F-X; Kaplan, M and Koprowski, H) WHO, Geneva.:181-189.
13. Thomson, S., Sherritt, M., Medveczky, J., Elliott, S., Moss, D., Fernando, G., Brown, L. and Suhrbier, A. 1998. Delivery of multiple CD8 cytotoxic cell epitopes by DNA vaccination. J. Immunol. 160: 1717-1723.
14. Yokoyama, M., Zhang, J. and Whitton, L. 1995. DNA immunization confers protection against lethal lymphocytic choriomeningitis virus infection. J. Virol. 69:2684-2688.

Discussion (Desmezieres)

Burger: Can you tell me something about your immunisation schedule? How often did you give the DNA and where (intramuscular), and what was the time between the last immunisation and the challenge?

Desmezieres: 50μg of DNA is injected. There is an inhibition of the response if too much DNA is given. The immunisation is by intramuscular injection and the challenge is 21 days after immunisation.

Burger: Did you check a later time point?

Desmezieres: No.

Renner: Did you compare the levels of neutralising antibody that you achieved with the DNA vaccine, with those from a conventional vaccine?

Desmezieres: Yes, the titre levels are similar but the DNA vaccine is better because the production of antibodies is longer in time. After 160 days after immunisation we still have a high level of antibodies (as shown in the slide).

THE SPP BIOTECH MODULE 1: PROTEINS FOR MEDICAL APPLICATIONS
A MULTIDISCIPLINARY EFFORT FOR THE DEVELOPMENT OF A RECOMBINANT ANTI-RHD ANTIBODY PREPARATION FOR THE PREVENTION OF HEMOLYTIC DISEASE OF THE NEWBORN

H. AMSTUTZ, S. MIESCHER, R. MOUDRY, S. DEJARDIN, M. IMBODEN, J.-J. MORGENTHALER
ZLB Zentrallaboratorium Blutspendedienst SRK, Bern, Switzerland
M. ZAHN, M. KOBR, I. FISCH, B. CORTHÉSY, N. MERMOD
Laboratoire de Biotechnologie moleculaire, Université de Lausanne, Switzerland
L. BERRUEX, M. ABDELLALI, O. BRÜGGEMANN, D. FRABOULET, R. FREITAG *Laboratoire de Biotechnologie cellulaire, Ecole Polytechnique Fédérale de Lausanne, Switzerland*
E. KRAGTEN, S. NAHRGANG, P. DUCOMMUN, P. PUGEAUD, I. MARISON, U.VON STOCKAR
Laboratoire de Genie Chimique et Biologique, Ecole Polytechnique Fédérale de Lausanne, Switzerland
M. DE JESUS, M. JORDAN, M. BOURGEOIS, L. HUNT, S. RADICE, L. BALDI, A. KULANGARA, P. GIRARD, R. RIBICKI, F. M. WURM
Laboratoire de Biotechnologie cellulaire, Ecole Polytechnique Fédérale de Lausanne, Switzerland

1 Introduction

Hemolytic Disease of the Newborn (HDN) can be a life threatening condition in Rh D positive babies born from Rh D negative mothers. For several decades, prophylactic treatment with immunoglobulins derived from human blood plasma from immunized donors has been available. Because of concerns for donor safety and limited supply of raw material, alternatives for production of specific immunoglobulins were investigated. Since 1994 ZLB has, in collaboration with a group at the University Hospital in Bern, started to develop recombinant anti-D antibodies. In 1997, a multidisciplinary team with expertise ranging from molecular biology to biochemical engineering for large scale operations was constituted, co-funded by the Swiss National Science Foundation and ZLB, as the industrial partner. The aims of this team are to establish the prophylactic principle and the manufacturing process for anti-D therapy based on recombinant antibody preparations produced from Chinese hamster ovary (CHO) cells.

A. Bernard et al. (eds.), Animal Cell Technology: Products from Cells, Cells as Products, 455–458.

2 Strategy

A phage-display library was constructed with immunoglobulin gene fragments amplified from PBMC-RNA of an immunized anti-D donor. From this library Rh D specific phage/Fab fragments were isolated (Vox Sang 75, 278, 1998) and used to construct complete antibodies.

3 Results

3.1. FINE SPECIFICITY

The fine specificities of four Fab fragments and corresponding complete antibodies were compared in a hemagglutination test (Figure 1). IgG1-14 shows weak reactivity with the C/E Protein on Rh negative cells. IgG1-3 strongly agglutinates all Rh D cells tested including variant DVI and therefore was chosen for further development.

Antibody	**Erythrocytes**												
ID-Geltest	CCDDee	ccddee	CCddEE	Ccddee	ccddEe	D III	D IVa	D IVb	D Va	D VI(J.)	D VII	Rh 33	DFR
Fab LD2-14	++++	-	-	nd	nd	++++	+++	+++	+++	-	+++	+++	-
IgG1-14 (Sup)	++++	-	+	+	±	++++	++++	++++	++++	++	++++	++++	++++
Fab LD1-52	+++	-	-	-	-	+++	-	-	+++	-	++	-	-
IgG1-52 (Sup)	++++	-	-	-	-	++++	+++	++++	++++	-	++++	+	+++
Fab LD1/2-6-3	+++	-	-	-	-	+++	+++	+++	+++	+	+++	+++	±
IgG1-3 (Sup)	++++	-	-	-	-	++++	++++	++++	++++	++++	++++	++++	++++
Fab LD1/2-6-33	+++	-	-	-	-	+++	+++	+	+++	+	+++	±	±
IgG1-33 (Sup)	++++	-	-	-	-	++++	++++	++++	++++	++++	++++	+	++++

Figure 1. Comparison of fine specificities of Fab fragments and corresponding complete antibodies.

3.2. FACS ANALYSIS

The antibody IgG1-3 was produced in the cell lines MDJ1 and MDJ8 (stable, transfected CHO DG44 cells). They were cultivated serum free in roller bottles as well as in batch fermentation up to the 100 liter scale. A large-scale recovery and purification process based on three chromatographic steps was developed. Binding of purified IgG1-3 to Rh D positive red blood cells (RBC) was analyzed by FACS analysis (Figure 2). It compared very favorably with polyclonal anti-D immunoglobulin.

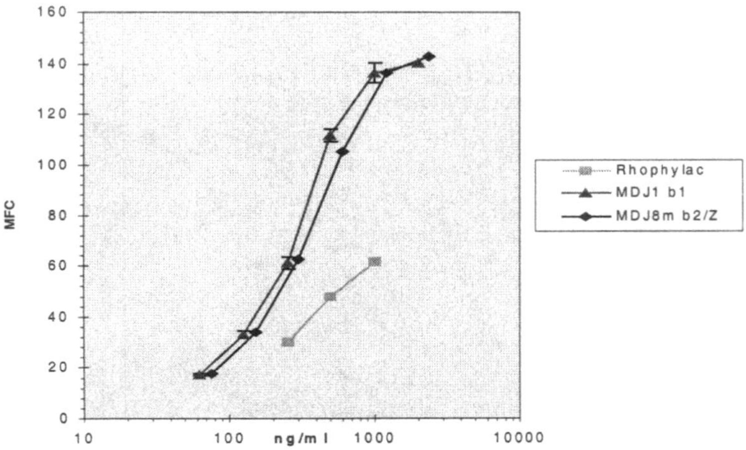

Figure 2: FACS analysis: Binding of purified mAb IgG1-3 and of polyclonal anti-D immunoglobulin (Rhophylac®) to Rh D pos. red blood cells (RBC); MFC, mean fluorescence channel.

3.3. BIOLOGICAL ACTIVITIES

In order to test the Fc-dependent activities of purified IgG1-3, activation of human macrophage by sensititized RBC (Figure 3) and antibody dependent cellular cytotoxicity (ADCC) of human lymphocytes (Figure 4) was measured. Macrophage activation was monitored by measuring oxygen radical production in a luminol based assay. Lytic activity in the ADCC assay was obtained by comparing the total release of a fluorescent dye from RBC mediated by detergent with antibody dependent release mediated by human lymphocytes.

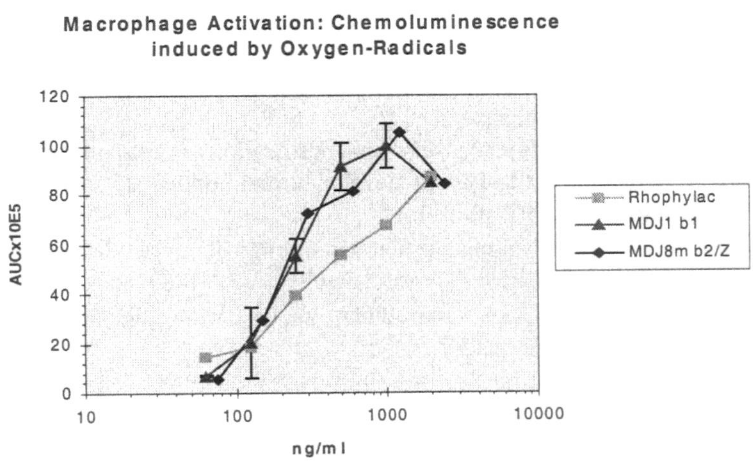

Figure 3. Activation of human macrophages with sensitized RBC shown in Fig. 2.

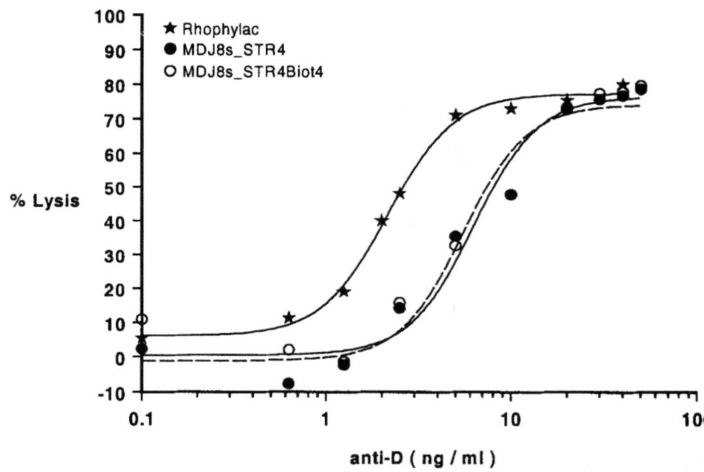

Figure 4. ADCC: Antibody dependent lysis of Rh D pos. RBC by human lymphocytes mediated by purified IgG1-3 (MDJ8s, batch STR4), biotinylated IgG1-3 (MDJ8s_STR4Biot4) and polyclonal anti-D immunoglobulin (Rhophylac•).

4 Conclusions and achievements

- High affinity recombinant human antibodies have been constructed and expressed in CHO cells. Antigen specificity and Fc-mediated activities were comparable to approved polyclonal anti-D immunoglobulin (Rhophylac®).
- High expression vectors for mammalian cells based on MAR sequences and transactivation of promoter sequences were developed and tested in transient and stable expression systems.
- High yielding, stable (1 year without selection) and clinically acceptable, suspension cell lines were obtained.
- A 10 to 100 liter scale, serum free and animal protein free production system (stirred tank) was established.
- A large scale three-step recovery and purification process towards therapeutics grade recombinant antibody was developed and applied up to the 20 liter scale (overall yield 75%, >99% purity).
- Analytical method development and monitoring of glycosylation using high pH anion-exchange chromatography validated by LC-ES/MS was performed.
- High throughput ELISA and other assays were developed for product characterization and for quality control.
- New fast on-line quantification and characterization of key cell culture components were developed.

LOW SERUM AND SERUM-FREE CULTIVATION OF MAMMALIAN CELLS USED FOR VIRUS PRODUCTION APPLICATIONS.

D.W. JAYME, P.J. PRICE, M.Z. PLAVSIC and D.A. EPSTEIN
Life Technologies, Inc., Grand Island, New York 14072 USA

1. Introduction.

Various cell lines are routinely used to produce veterinary and human vaccines, including kidney-derived lines from multiple species. This study focused on VERO (primate), PK-15 (porcine), MDBK (bovine), and HEK 293 (human). Classically, virus production is performed in roller bottle culture using relatively simple basal nutrient formulations supplemented with animal serum and various hydrolysates. Technical and regulatory issues encourage more labor-sparing bioreactors for process scale-up and reduction of concerns associated with adventitious agent contamination.

The challenges [Shah, 1999] associated with serum-free nutrient optimization of cells may be broadly-simplified into four categories: (a) Ability to promote attachment of anchorage-dependent cells in microcarrier culture in the absence of serum or other costly, unstable, animal-derived attachment factors (e.g., vitronectin, fibronectin); (b) Optimization of nutrients to maximize high density generation of biomass per unit time of culture incubation; (c) Optimization of nutrients to generate infective virus, recognizing that proliferation does not necessarily correlate with active virus production; and (d) Achievement of the previous three objectives without introducing constituents that might adversely impact virus titer or contamination risk.

2. Results and Discussion.

2.1. SERUM TREATMENT

Gamma irradiation [Daley *et al.*, 1998] and heat treatment [Danner *et al.*, in press] were examined independently and in tandem to evaluate eradication of a challenge panel of viruses. Treated sera were evaluated for persistent ability to support multiple subcultures of relevant cell types.

2.1.1. *Gamma Irradiation.*

A high titer panel of five representative model viruses was spiked into multiple lots of fetal bovine serum (FBS) (and other biological materials with similar results – data not shown). Spiked serum samples were exposed to gamma irradiation and the kinetics of viral inactivation are summarized in Table 1. All virus titers were reduced to below detectable limits by 35 kGy, resulting in 6-7 logs of virus reduction [Daley *et al.*, 1998].

Table 1. Effect of Gamma Irradiation on Residual Titer of Model Viruses Spiked into Fetal Bovine Serum.

Radiation Dose (kGy)	Bovine Reovirus [a]	Porcine Parvovirus [a]	Canine Adenovirus [a]	Infectious Bovine Rhinotracheitis [b]	Bovine Viral Diarrhea Virus [b]
Control	9.8×10^7	9.2×10^7	9.3×10^8	1.3×10^8	2.8×10^7
15	3.9×10^6	1.4×10^7	5.66×10^4	5.0×10^7	3.3×10^6
20	6.1×10^4	2.8×10^5	2.86×10^4	2.9×10^5	2.2×10^5
25	7×10^0	≤ 0.5	7.42×10^2	2.3×10^0	1×10^0
35	≤ 0.5	≤ 0.5	≤ 0.5	≤ 0.5	≤ 0.5
45	≤ 0.5	≤ 0.5	≤ 0.5	≤ 0.5	≤ 0.5
Log Reduction	≥ 6.78	≥ 7.09	≥ 7.13	≥ 6.78	≥ 6.78

[a] Quantitated by $TCID_{50}$ Assay; [b] Quantitated by Plaque Assay; n = 3

A. Bernard et al. (eds.), Animal Cell Technology: Products from Cells, Cells as Products, 459–461.

2.1.2. *Heat Inactivation.*

A high titer panel of six representative model viruses was spiked into FBS at room temperature and samples were placed in a 56°C water bath. Residual virus was quantitated [Danner *et al.*, in press] once the sample reached bath temperature and at fifteen minute intervals. BVD and PI-3 viruses were highly sensitive to heat inactivation, approaching baseline detection during the equilibration phase. Reo-3, IBR and CAV viruses exhibited intermediate sensitivity, but were reduced to undetectable levels after 30 minutes at 56°C. PPV was relatively insensitive to heat inactivation.

2.1.3. *Biological Performance.*

Samples of horse serum or fetal bovine serum were exposed to <u>both</u> heat treatment (56°C for 30 minutes) and maximal dose gamma irradiation (45 kGy). Biological performance was monitored relative to sham-treated samples from the same serum lot. Growth of 4-5 representative cell lines was monitored over multiple sequential subcultures. All treated samples performed equivalently to sham-treated controls, except in the adherent CHO assay. These results, combined with the virus titer reduction data above, indicated that a cumulative 10-12 log reduction of model virus may be achieved by these two processes without material impact on biological performance for many cell culture applications [Daley *et al.*, 1998; Danner *et al.*, in press].

2.2. CONSTITUENT REPLACEMENT

Common ingredients of cell culture medium (e.g., insulin, transferrin, serum albumin and various non-protein constituents) have traditionally been obtained from animal sources [Shah, 1999; Jayme, 1999]. The risk of introducing adventitious agents (e.g., virus, prion, mycoplasma) into the production environment led to qualification of non-animal-derived substitutes with equivalent biological performance. Re-engineering of classical media and design of new nutrient formulations permit production of vaccines and biologicals in culture environments completely free of animal-sourced raw materials [Jayme, 1999]. Serum-free media containing human transferrin used to grow HEK 293, MDBK, PK-15 or VERO cells were re-engineered to eliminate transferrin and other animal origin raw materials and to achieve a culture environment with reduced risk of adventitious contaminants. Figure 1 illustrates results obtained with an improved OptiMEM® I prototype, designed for serum-free, animal origin-free cultivation of adherent cell lines for virus production. These data indicate the differential ability of various iron chelates to sustain growth of MDBK and PK-15 cells over multiple passages.

Figure 1. Growth of MDBK and PK-15cells in OptiMEM II using iron chelates as transferrin substitutes.

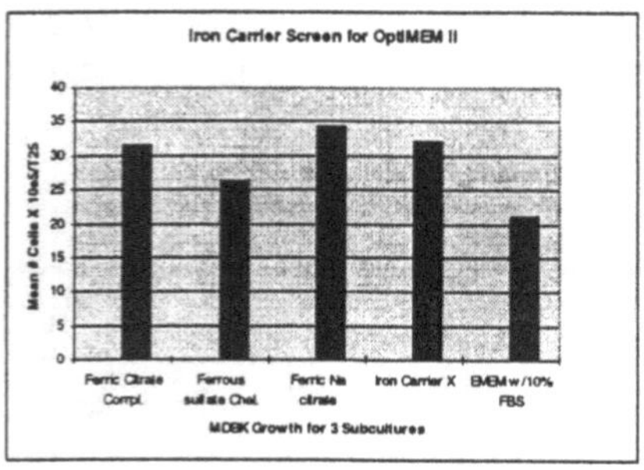

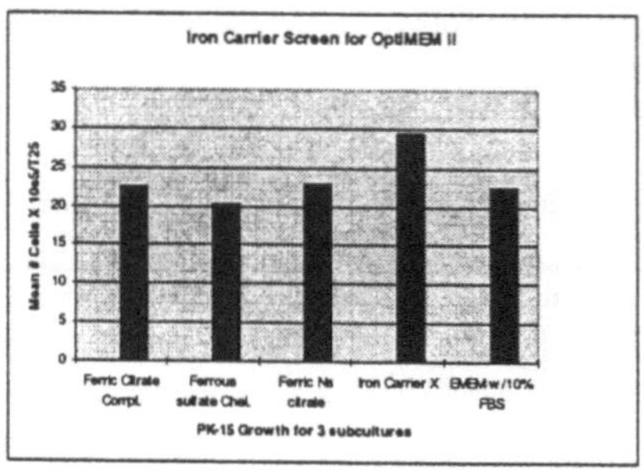

2.3. SERUM-FREE MEDIUM DEVELOPMENT

OptiMEM® I, developed a decade ago as a reduced serum medium, has been widely used to produce veterinary and human vaccines. However, depending upon the cell type and bioreactor system, it generally required supplementation by animal sera (0.1 - 4%). Evolving applications have taken our <u>serum-free</u> medium development efforts in three unique directions:

- VP SFM – A medium free of all components of animal origin and containing only a single recombinant protein (10 ug/ml) which supports VERO cell applications.
- OptiMEM® II – A prototype serum-free formulation designed for virus production applications using a broad range of kidney-derived lines.
- 293 SFM – A serum-free medium for virus production applications of HEK 293 and similar cell lines.

Each nutrient formulation required kinetic analysis of nutrient consumption (e.g., amino acids, vitamins, carbohydrates, lipids, etc.) under high density cultivation conditions and optimization of constituents to promote performance in suspension or adherent culture virus production. Data in Tables II and III were derived from collaborative studies with BioReliance. Table II compared virus titers obtained from VERO cells adapted to serum-free VP-SFM or maintained in FBS-supplemented control medium. Titers of three model viruses (sindbis, polio I and pseudorabies) were equivalent in the two test systems. Table III examined adenovirus production by HEK 293 cells cultivated in serum-free 293 SFM vs. FBS-supplemented control medium. Studies were performed both in 100 ml shaker culture and 2 L bioreactors. Specific adenovirus production was unchanged in serum-free medium.

Table II. Virus Production by VERO Cells in VP-SFM.

Test Medium	Sindbis[a]	Polio I	Pseudorabies
EMEM + 2% FBS	8.1	7.8	8.1
VP -SFM	8.0	7.6	8.1

[a]Virus titers expressed in Log_{10} PFU/ml.

Table III. 293 Cell Adenovirus Production.

Medium	FBS (%)	Scale	$TCID_{50}$/cell (x 10^3)[a]
293 SFM	0	100 ml	2.0[b]
DMEM	2	100 ml	2.2
DMEM	5	100 ml	2.2
293 SFM	0	2 L	4.1

[a]Spec. virus production; [b]48 hrs post-infection

3. Conclusion

Serum processing by gamma irradiation or by heat treatment may be controlled to reduce virus risk while retaining biological performance. Serum-free media support proliferation of selected suspension and anchorage-dependent cell lines. Virus titers obtained in serum-free media compare favorably with titers obtained in serum-supplemented control media. All required nutrient constituents may be obtained from non-animal sources, yielding culture environments of reduced risk of adventitious agent contamination.

4. Acknowledgments

We gratefully acknowledge the collaborative contributions of Paddy Iyer, Dominick Vacante and Jeffrey Ostrove (BioReliance). Technical support and scientific consultation from colleagues, Brian Long, John Daley, Doug Danner, David Judd, Ethel Evege, Shawn Smith and Steve Gorfien are acknowledged with appreciation.

5. References

- Daley, J.P., Danner, D.J., Weppner, D., and Plavsic, M.Z.: Virus inactivation by gamma irradiation of FBS, *Focus* **20** (1998) 86-88.
- Danner, D.J., Smith, J., and Plavsic, M.Z.: Inactivation of viruses in fetal bovine serum using heat inactivation at 56°C, *BioPharm* (in press).
- Jayme, D.W.: An animal origin perspective of common constituents of serum-free medium formulations, *Developments in Biological Standardization* **99** (1999) 181-187.
- Price, P.J. and Evege, E.K.: Serum-free medium without animal components for virus production, *Focus* **19** (1997) 67-69.
- Shah, G.: Why do we still use serum in the production of biopharmaceuticals?, *Developments in Biological Standardization* **99** (1999) 17-22.

DEVELOPMENT OF A SERUM-FREE MEDIUM FOR MRC-5 CULTURE

C. MAGGETTO[1], S. PIROTTON[2], I. KNOTT[1], B. AERTS[1], E. EVRARD[1], M.M. GONZE[1], M. RAES[2] and L. FABRY[1]

1- SmithKline Beecham Biologicals, B-1330 Rixensart, Belgium

2- Lab. Biochimie et Biologie Cellulaire, FUNDP, B-5000 Namur, Belgium

1. INTRODUCTION

Serum-free media are more and more used for different cell lines. We report here the preliminary results obtained with serum-free prototype media (SF1 and SF2, property of SB Bio) in order to cultivate MRC-5 cells without serum: i) various serum replacement components were mixed to medium and tested on MRC-5 growth which was evaluated by the MTT method; ii) the impact on MRC-5 growth was confirmed by DNA measurements; iii) the prototype media were also tested on the cell growth of MRC-5 in T-flasks (175 cm^2).

2. MATERIAL AND METHODS

In 96-well plates, MRC-5 cells were seeded in medium supplemented with 2% or 10% FBS (BM + 2%FBS; 10% M) or in a basal medium (BM) containing additives or combinations of additives. Cells were kept at 37°C for maximum 7 days. Viability and cell growth at specific days, were measured by MTT and Propidium Iodide assays.

In T-175 cm^2 flasks, MRC-5 cells cultivated in a medium supplemented with 2%FBS (BM + 2%FBS) were seeded and cultivated for 1 passage in the prototype serum-free media.

3. RESULTS

The growth and viability of MRC-5 cells with different prototype media were first studied by a colorimetric method (MTT assay) in 96-well plates (Figure 1). The cells did not proliferate when incubated in a basal medium containing no additives (BM), but displayed improved proliferation with the SF1 and SF2 prototype media, whatever the incubation time (from day 1 to 7). The positive controls were MRC-5 cells either cultivated in medium containing 10% FBS (10%M) or in a basal medium supplemented with 2% FBS (BM + 2%FBS), both controls giving comparable growth profiles.

MRC-5 growth was also evaluated using the fluorescent DNA dye propidium iodide. Cells cultured in the presence of SF1 and SF2 prototype media displayed improved cell growth compared with the BM +2%FBS positive control (Figure 2). Moreover, the pro-proliferative effect of SF1 and SF2 was dose-dependent (results shown for SF2 in Figure 3).

One passage in T-flask (175 cm^2) was performed with the prototype media SF1 and SF2. The density at confluence with the SF2 medium was higher than with the control media containing 2% or 10% FBS (254.000 cell/cm2 vs 180.000 cell/cm2). With the SF1 medium, this density was lower than for the controls. The population doubling levels were 2.5/4days for SF2 versus 2PDL/4days for controls.

A. Bernard et al. (eds.), Animal Cell Technology: Products from Cells, Cells as Products, 463–465.

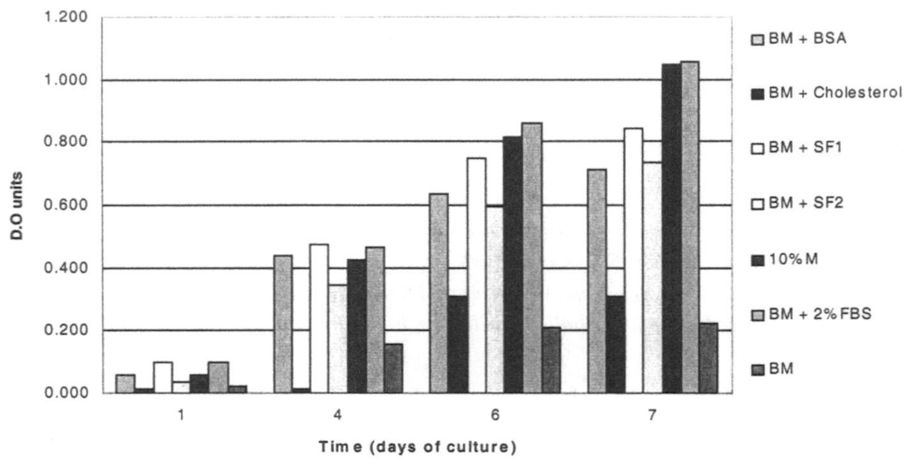

Figure 1 : MRC-5 growth in response to serum free prototype media.
MRC-5 cells were seeded in a 96-well plate (4000 cells/well) and incubated in a basal medium (BM) implemented with additives or in serum-free prototypes media (SF1 and SF2). The viability/growth was measured at day 1, 4, 6 and 7 by MTT assay. Results are expressed as mean of triplicate determinations.

4. CONCLUSION

Two serum-free prototype media were demonstrated to promote MRC-5 cell growth. The prototype media will be now evaluated for several consecutive passages in T-flasks.

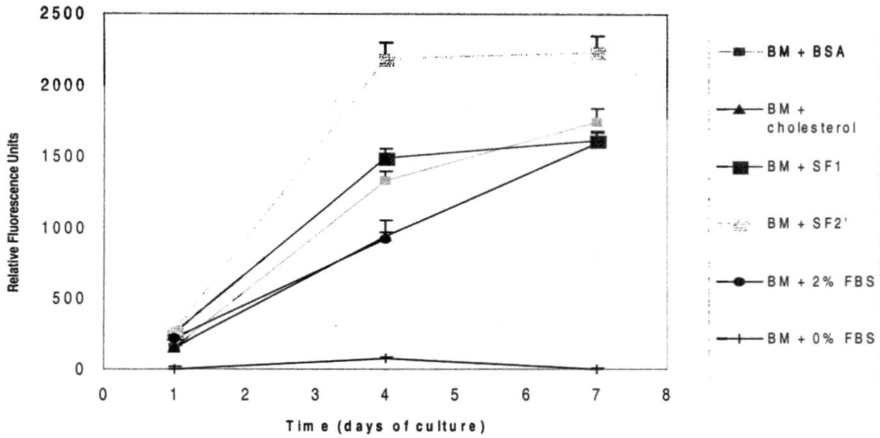

Figure 2 : MRC-5 growth in response to serum free prototype media.
MRC-5 cells were seeded in a 96-well plate in the same conditions as described in figure 1. The cell growth was evaluated by DNA quantification (Propidium Iodide assay) at day 1, 4 and 7. Results are expressed as mean +/- SD of triplicate determinations.

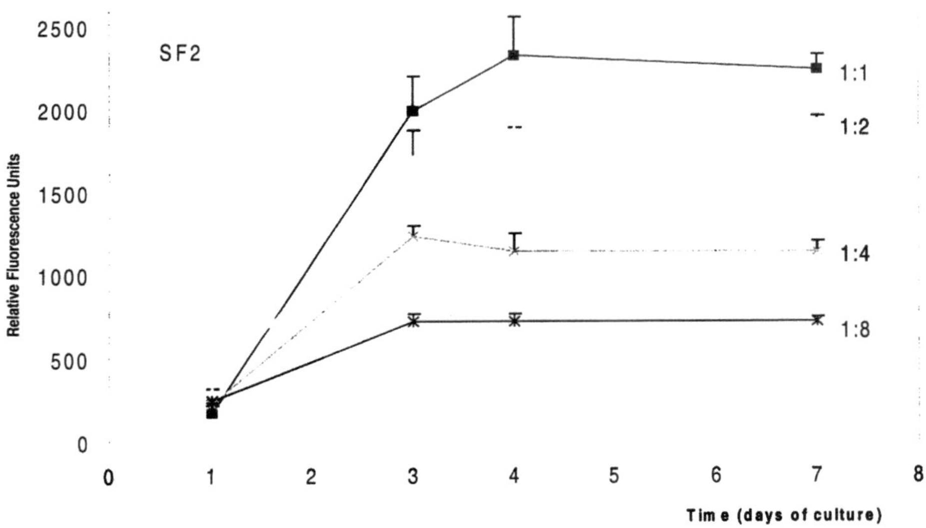

Figure 3 : Effect of different dilutions of the SF2 medium on the proliferation of the MRC-5 cells. The cells were seeded in a 96-well plate (4000 cells/well) and incubated with different dilutions of serum free prototype medium SF2. The cell growth was evaluated by DNA quantification (Propidium Iodide assay) at day 1, 3, 4 and 7. Results are expressed as mean +/- SD of triplicate determinations.

ADAPTATION OF VERO CELLS TO A SERUM FREE MEDIUM FOR THE PRODUCTION OF RABIES VIRUS

S. MAJOUL[1], H. KHARMACHI[2] , M. SAADI[2], A. CHOUAIB[1] and H. KALLEL[1].
[1] *Service de production des vaccins virologiques.*
[2] *Laboratoire de la rage.*
Institut Pasteur de Tunis. 13, Place Pasteur. BP. 74. 1002 Tunis. Tunisia.

1. Introduction

The large scale production of inactivated vaccines such as polio and rabies vaccines, using continuous cell lines (CCLs) such as Vero cell line, is still based on a serum-containing media (SCM), at least for the production of biomass (Duchene et al., (1990)). However, the use of serum during the cell proliferation step of the production process, has several disadvantages such as quality variation among lots and high cost. Furthermore, sera are also a potential source of contamination by virus, mycoplasma, BSE agent, etc. For those reasons, the adaptation of animal cells to grow in serum free media (SFM), allows a more controlled and a less risky bioprocess.
In this work, we described the adaptation of Vero cells to three serum free media. The results of the growth of Vero cells in these media, are also presented. In addition, we investigated the production of rabies virus by Vero cells in T-flasks and roller bottles in a serum free medium (MDSS2). The results obtained are compared to the standard medium which is DMEM supplemented with foetal calf serum.

2. Materials and Methods

Cell lines : Vero cells, used for the production of a human rabies vaccine, were used for this study.

Virus strain : Pasteur 2061/Vero -strain was used to infect the cells.

Culture Medium : DMEM (Gibco BRL, ref. 52100-039)+5%FCS (Hyclone, ref. A-1115-L), MDSS2 (Axcell Biotechnologies, ref. 34601), DMEM/F12 (Bio media, ref. DM12NAB2052) and CCM5 (Hyclone, ref. SH30100) were used. Prolifix (Bio media, ref. PROLI22012) was used as a serum substitute.

Growth assay : Cell culture assays were performed at 37°C, in 25 cm^2 flasks, at an initial concentration of $1x10^5$ cells/ml. Cell concentration was determined daily. The assays were performed in duplicate.

Virus production : Vero cells were infected by the strain Pasteur 2061/Vero of rabies virus at a cell concentration ranging from $0.8x10^5$cells/cm^2 to $1.4x10^5$cells/cm^2 with a MOI of 0.1/cell. Virus production was performed at 34°C, in 25 cm^2 flask or in roller bottle (850cm^2). 3 days after the infection, the medium was removed and replaced by

A. Bernard et al. (eds.), Animal Cell Technology: Products from Cells, Cells as Products, 467–469.

DMEM+0.3% human albumin (LFB, ref. R5144-26CSP). Harvests were performed at 8, 11 and 14 days post infection. Samples were taken to determine virus titre and total glycoprotein content. The assays were performed in duplicate.

Cell counting : Cells cultivated on CCM5 and DMEM+5%FCS, were washed with PBS then trypsinized at 37°C for 5 to 10 minutes and counted in a Mallassez chamber. For cells grown on DMEM+10%Prolifix, they were first washed with a trypsin substitute called SPLITIX (Bio media, ref. SPLIT142052), treated at 37°C with the same solution for 15 to 20 minutes then counted.

Rabies virus titration : Virus titres were determined according to a modified RFFIT method (Perrin et al. (1995)) and expressed in Fluorescent Focus Units per ml (FFU/ml).

Glycoprotein titration : rabies glycoprotein content was determined by ELISA using polyclonal antibodies (Perrin et al. (1995)).

3. Results and discussion

3.1 Kinetics of Vero cells growth in serum free media :
To study the growth of Vero cells, we tested the following media : CCM5, DMEM/F12 supplemented with 10% of Prolifix and MDSS2. DMEM+5% FCS was used as a control. Before conducting kinetic studies, cells were first adapted to the serum free media tested. To adapt the cells to MDSS2 and DMEM/F12+10%Prolifix, the amount of serum added was reduced progressively. Concerning the adaptation to CCM5, cells previously grown in DMEM+5%FCS, were transferred directly to CCM5.

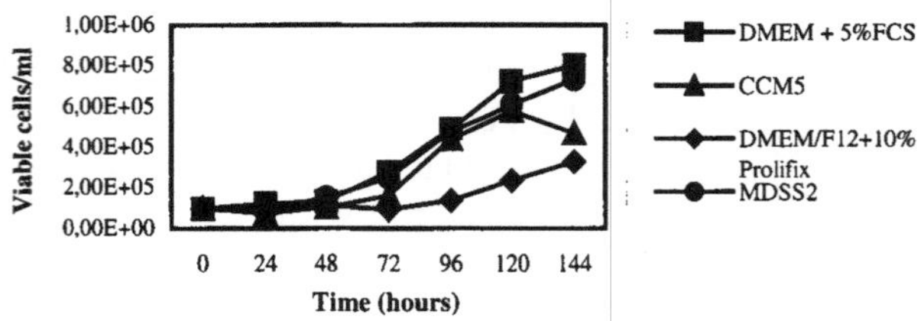

Figure 1 : Growth of Vero cells in DMEM+5% FCS, CCM5, MDSS2 and DMEM/F12 + 10% Prolifix

The results of the kinetic studies are shown in figure 1. They indicate that the maximal cell density of Vero cells obtained were $8x10^6$, $7.31x10^6$, $5.82\ 10^6$ and $3.25x10^6$, respectively in DMEM + 5% FCS, MDSS2, CCM5 and DMEM/F12 + 10% Prolifix.
With regard to the average specific growth rate, we obtained 0.0144, 0.0138, 0.0107 and 0.0098 h^{-1} in DMEM + 5% FCS, MDSS2, CCM5 and DMEM/F12 + 10% Prolifix, respectively.
The comparison of Vero cells growth in the different serum free media, shows that the growth of Vero cells in MDSS2 was better than the other media tested. However, the maximal cell density and the average specific growth rate of Vero cells obtained in MDSS2 were slightly lower than those obtained in DMEM+5% FCS. Nevertheless, MDSS2 was chosen to study the production of rabies virus by Vero cells.

3.2 Rabies virus production :
To study the production of rabies virus by Vero cells in MDSS2, cells were infected by the strain Pasteur 2061/Vero. The studies were conducted in T-flasks and in roller bottles. The results obtained in T-flasks, shown in table 2, indicate that for all the

harvests, virus titres obtained in MDSS2 were always lower than those obtained in DMEM+5%SVF. The maximal virus titre obtained in T-flasks and in MDSS2, was 1.82 10^6 (FFU/ml). In DMEM+5%FCS, virus titre was 24 (harvest N°1) to 2.8 (harvest N°3) time higher. On the other hand, the levels of total glycoprotein obtained in MDSS2, were higher than those obtained in DMEM+5%FCS (table 2).

TABLE 2 : Rabies virus production by Vero cells, in T-flasks and in roller bottles

		T-flasks			Roller bottles		
		Harvest number			Harvest number		
		1	2	3	1	2	3
MDSS2	Total glycoprotein level (ng/ml)	6543.6	6258.7	9321.3	6183.9	8074.4	12921.4
	Virus titre (FFU/ml)	2.53 10^5	4.74 10^5	1.82 10^6	5.67 10^5	9.04 10^5	4.17 10^5
DMEM+5%FCS	Total glycoprotein level (ng/ml)	5064.6	4115.2	9554.4	5317.9	7392.4	13326.6
	Virus titre (FFU/ml)	6.04 10^6	5.25 10^6	5.25 10^6	2.76 10^6	5.25 10^6	5.77 10^6

The results related to the study of the production of rabies virus by Vero cells, in MDSS2 and in roller bottles, presented in table 2, indicate that the level of virus titres obtained in MDSS2 and in roller bottles, were always lower than those obtained in DMEM+5%FCS. However, the production of glycoprotein, in MDSS2 and in DMEM+5%FCS were similar.

These results indicate that the kinetic of rabies virus production by Vero cells in MDSS2 and DMEM+5%FCS, were similar to those obtained in T-flasks. These results also indicate that MDSS2 seems to be less appropriate than DMEM+5%FCS, for rabies virus production by Vero cells. This is probably due to the non adaptation of the virus strain used to cells grown in serum free medium and to the new environment (SFM). In fact, rabies virus can be adapted to a new environment by successive passages (Perrin et al. (1995)). Nevertheless, the evaluation of the optimal medium relies upon accurate assessment (percentage) of the soluble and the non soluble glycoprotein levels, and not only the total glycoprotein level, for each harvest of Vero cells in MDSS2 as compared to the control (DMEM+5%FCS).

4. Conclusion

We compared the growth of Vero cells in three serum free media (MDSS2, CCM5, DMEM/F12+10%Prolifix) to the growth in the standard medium (DMEM+5%FCS). Growth of Vero cells in MDSS2 was similar to that observed in DMEM+5%FCS.

Rabies virus production by Vero cells in MDSS2 in both T-flasks and roller bottles was also compared to that performed in DMEM+5%FCS. MDSS2 seems to be less appropriate than the classical DMEM supplemented with 5%FCS. This is probably due to the non adaptation of the virus strain used to cells grown in MDSS2 and to the new environment.

5. References

Duchene, M., Peetermans, J., D'Hondt, E., Harford, N., Fabry L. and Stephenne J. (1990) Production of polio virus vaccines : past, present and future, *Viral Immunology* **3**, 243-272.

Perrin, P., Shampur, M., Gontier-Jallet, C., Petres, S., Tordo, N. and Merten, O.-W. (1995) An experimental rabies vaccine produced with a new BHK-21 suspension cell culture process : use of serum-free medium and perfusion-reactor system, *Vaccine* **13**, 1244-1250.

INCREASED PRODUCTIVITY OF A RICKETTSIA VACCINE: ENHANCEMENT OF ADHESION PROPERTIES OF INFECTED ENDHOTHELIAL CELLS BY TGFβ

MARIQUE T., NDONGO D.E., TANGA M., RASCHELLA A., HENRICK V., DECLERQ B. and WERENNE J.
Laboratory of Animal Cell Biotechnology, Université Libre de Bruxelles. Belgium

Abstract

We showed here using human or mouse cellular model system that by addition of TGFβ, at a preconfluent stage of cell growth, cell adhesion is greatly improved and rickettsia productivity therefore much higher. This observation allowed us to propose a two step process for the vaccine production based first on cell biomass accumulation up to preconfluence stage, followed by infection with *Cowdria ruminantium* in presence of " conditioned medium" in which inactive pre TGFβ produced by the endothelial cells themselves has been activated.

1 Introduction

The tick-born rickettsia *Cowdria ruminantium* causes a common deadly disease in cattle south of Sahara. *In vitro*, this obligate intracellular bacteria grows exclusively in endothelial cells. However, these slow growing cells readily detach from their substratum. The development of an appropriate vaccine thus necessitates better control to allow long-term (1-2 weeks) culture maintenance.

2 Materials and methods

BUEC (bovine umbilical endothelial cells), PSV1 (human established endothelial cell line) or Mouse immortalised endothelial cells (MIEC) are grown in TC-flasks (1). Cytodex 3 microcarriers are then allowed to stand over the cell monolayer and colonised. When confluent, microcarrier beads are introduced in spinner flasks containing normal medium or ACM (acid-treated supernatant from endothelial cell culture). Cell population is counted every day with hematocytometer chamber.

A. Bernard et al. (eds.), Animal Cell Technology: Products from Cells, Cells as Products, 471–473.

3 Results:

TGFβ and *Cowdria* vaccine in endothelial cells

Agitated BUEC and PSV1 show bad adhesion in normal medium. Good adhesion can be restored in presence of activated conditioned medium (ACM) containing TGF-β (as shown by inhibition with antibody).

Figure 1 : Properties of BUEC and PSV1

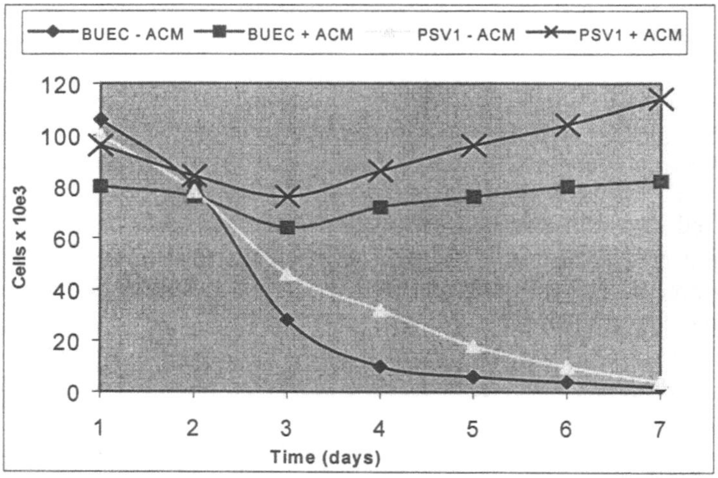

Moreover with MIEC Cells we showed that adhesion is better maintained using 25% of activated conditioned medium.

Figure 2: Properties of MIEC

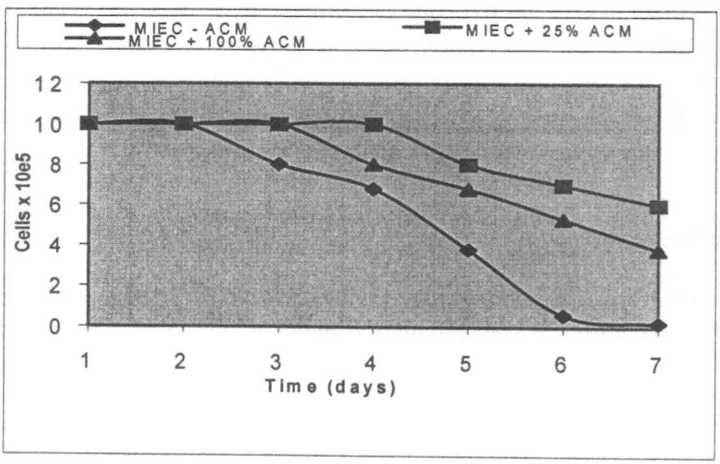

4 Discussion

The rickettsia *cowdria ruminantium* causing a deadly disease in cattle which is at present one of the main causes of economical losses for breeders south of Sahara, could spread also on the American continent. Possible tick vectors are indeed present there and the disease has appeared already in the Caribbean region. The development of an appropriate vaccine at an affordable cost is therefore essential.

A major difficulty to produce such a vaccine efficiently nowadays, arises from the fact that obligate intracellular maturation process of the rickettsia is long (it requires 1-2 weeks) and infected cells become fragile and detach readily from the substratum on which they have been grown, before completion of the bacteria production process. Scale-up of the production in bioreactors using microcarriers is therefore very inefficient.

We treated our cultures with TGF-β or with activated conditioned medium (ACM) to improve cell adhesion. A two phase procedure was developed:

1. Biomass production
2. *Cowdria* infection and production in presence of TGF-β (ACM).

Small percentage (25%) of ACM can give better results then high ones (100%) for MIEC cells. This could be due to the negative effect of TGF-β on the specific growth rate.

Vaccine production will necessitate control of cell adhesion through cytokine action.

We showed here that adhesion and growth of endothelial cells could be restore by activated conditioned medium ; pure TGFβ (data not show) is equally active. This leads us to propose a procedure for rickettsia vaccine production. Such single procedure should permits efficient cattle protection until recombinant or DNA vaccine can be produced (2).

5 References

(1) Marique T.,Blankaert,D., Hendrick,V., Raschella,A., Declerck,B.,Alloin,C.,Teitera-Guerra,I.,Sandron,D.,Cherlet, M., Parent, D., Kirkpatrick, C., Van Vooren, J.P. and Wérenne J.,1997, *Biological response of endothelial cells and its modulation by cytokines prospect for therapy and bioprocesses,.,* cythotechnologie, 1997: **25**:183-189

(2) Nyika, A., Mahan, S.M., Burridge, M.J., McGuire, T.C., Rurangirwa, F. and Barbet, A.F.. 1998, *A DNA vaccine protects mice against the rickettsial agent Cowdria ruminantium.,* Parasite Immunology, 1998: **20**: 111-119

COMPARATIVE STUDY OF CELL CULTURES PERSPECTIVE FOR RUBELLA VIRUS PRODUCTION

T.N. GETMANOVA, A..A. CHEPURNOV, N.M. MAKHOVA,

E.A. NECHAEVA, T.D. KOLOKOLTSOVA

Institute of Cell Cultures, State Research Center of Virology and Biotechnology «Vector», Koltsovo, Novosibirsk region, Russia

1. Introduction

Rubella is a widespread infectious disease, often causing serious health problems in both adults and children. Teratogenic effect of rubella virus and the resulting inborn rubella syndrome bring forth the demand for development of both vaccinal and diagnostic preparations. No rubella vaccine is produced in Russia now, although the rubella virus vaccine strain Orlov was deposited in 1977. Development of technology for producing diagnostic preparations requires additional studies on selection of both the viral producing strain and the cell system providing high yield of rubella virus, relatively simple cultivation, and availability of the substrate. The goal of the work was to study the reproduction of rubella virus strains in sensitive cell cultures and assess the potential of human diploid cell culture L-68 for rubella virus production.

2. Materials and methods

Continuous cell cultures Vero, RK-13, and BHK-21 were obtained from Flow Laboratories. Diploid human cell culture L-68 was produced from lung of 10-week

A. Bernard et al. (eds.), Animal Cell Technology: Products from Cells, Cells as Products, 475–477.

embryo at the Institute for Viral Preparations, Russian Academy of Medical Sciences, certified according to the WHO's requirements, and used at passages 21 and 30. Rubella virus strain P-1965 was obtained from the collection of the Institute of Virology, Russian Academy of Sciences, and has undergone 18 passages in cell culture MT-4, 18 passages in the primary culture of baby rabbits kidney, and 4 passages in newborn rabbits. Intermediate variant of the strain Orlov was provided by the Pasteur Institute of Epidemiology and Microbiology. It has undergone 36 passages in primary culture of newborn rabbits and 2 passages in the cell culture L-68. Cells were grown in a monolayer at a temperature of 37°C in modified Eagle medium supplemented with 10% bovine serum (fetal bovine serum was used in case of L-68 and RK-13 cultures). Cells were infected with rubella virus at a dose of 0.01 $TCID_{50}$/cell on day 2. Virus reproduction was studied by light and electron microscopy in monolayer cultures and cell suspensions, produced using trypsin. The reproduction of rubella virus strain P-1965 in Vero, RK-13, and BHK-21 cell cultures was studied on day 7 post infection in cell suspension sedimented by centrifugation. The monolayers of diploid cells and Vero cells were grown on cover glasses and fixed daily starting from day 1 to day 8 post infection in parallel with the control. Monolayers of L-68 at passages 21 and 30 were infected with the rubella virus strain Orlov. The dynamics of the infection was studied starting from day 1 to day 8 post infection. The diameter of microfoci was measured in toluidine blue-stained semithin sections of monolayer at 21th passage number.

3. Results and discussion

Study of the virogenesis of rubella virus strain P-1965 in Vero, RK-13 and BHK-21 has demonstrated that the viral replication is accompanied by characteristic changes in the cell ultrastructure, causing vacuolation of the cytoplasm, hypertrophy of the Goldgi complex, and an increased number of phagosomes. Spherical-shaped rubella virions with a diameter of about 60 nm were found within small vacuoles and cisterns of the lamella complex. In the intercellular space, the virions often contact with cell detritus and membranes. The amounts of both inter- and intracellular virions are highest in Vero and RK-13 cells. The cytopathic effect of the virus on monolayer of all the cultures studied is weakly expressed, which is often observed in case of rubella virus infection (3) and hinders determination of its specific activity through titration. The reproduction of rubella virus strain P-1965 in L-68 diploid cell culture is accompanied by changes in the cytoplasm ultrastructure similar to those described above. However, L-68 is inferior to Vero cells in the number of virions and their accumulation.

No distinctions in morphogenesis of rubella virus strain Orlov in the L-68 diploid cells as compared to strain P-1965 were recorded. The virions were detected both outside and inside the cells and displayed a typical rubella morphology. Thus, the L-68 diploid cell culture at passages 21 to 30 is capable of maintaining the replication of the virus in question. However, the cytopathic effects of the virus on cell monolayer of different passages are dissimilar. Microfocus-type cell aggregates with a typical morphology described earlier (1) appear in the monolayer of passage 21 cells starting from day 4 post infection. The diameter of the microfoci varies in the range of 180 to 250 μm. Unlike the cells of passage 21, the cytopathic effect of the virus on the monolayer of passage 30 cells manifests itself in the diffuse affection with cell rounding and detaching from the glass. The distinctions found in the cytopathic effect of the virus on the cell monolayer of different passages require a more detailed investigation with regard to cells properties and conditions for virus production.

4. Conclusion

The performed study of rubella virus reproduction has demonstrated that the virogenesis in the cell cultures studied displays a typical character accompanied by release of virions into intercellular space. The results obtained confirm that rubella virus strain P-1965 can be used for producing the antigen for diagnostic test kits. Vero cell culture proved most optimal substrate, as it is capable of maintaining a high level of virus reproduction without addition of expensive fetal serum. The diploid cell culture L-68 is sensitive to rubella virus infection, but the yield of virions is inferior to Vero cells. However, employment of diploid cells as a substrate for production of immunobiological preparations is promising (2); therefore, diploid cell culture L-68 may be recommended for developing production technology of rubella vaccine.

5. References.

1. Kouri; G., Aguilera, A., Rodriguez, P., Korolev, M. (1974) A study of microfoci and inclusion bodies produced by rubella virus in RK-13 cell line. J. General Virol. 22, 73—80.

2. Just, M., Berger, R., Gluck, R., Wegman, A. (1986) Evaluation of a combined vaccine against Measles–Mump–Rubella produced on human diploid cells. 19th IABC Congr. Use and Stand. Comb. Vaccines, Proc. Symp., Amsterdam 1985 - Basel e.a., 25—37.

3. Frey, T.K. (1994) Molecular biology of rubella virus. Advances in Virus Res. 44, 69—160.

BIOGENESIS OF EPITHELIAL PHENOTYPE DURING LONG-TERM CULTIVATION OF MDCK CELLS IN SERUM-FREE ULTRA-MDCK MEDIUM

N.KESSLER, G.THOMAS, M.AYMARD
Laboratoire de Virologie, Faculté Medecine Grange-Blanche, Lyon, France

1. Introduction

The Madin Darby Canine Kidney (MDCK) cell line is an ideal model for studying questions pertaining the influence of the environment on the epithelial phenotype biogenesis. Wild-type MDCK cells, are heterogeneous in terms of morphology and functions and consist of two genetically different subpopulations : Principal (P) cells and Intercalated (IC) cells, whose respective amounts can vary as a function of culture conditions (culture medium, substrate, passages number...). Our previous experiments showing that MDCK cells can be easily cultivated over more than 150 passages under serum-free conditions , (Kessler et al, 1999) prompted us to follow up the genotypic and phenotypic characteristics of cells as well as the biogenesis of the epithelial phenotype during long term cultivation in such an environment.

2. Results and discussion

2.1. MDCK CELL CULTIVATION

Wild type MDCK cells at passage 73 in standard medium were either transferred to and then subcultivated in Ultra-MDCK medium from BioWHITTAKER (**U-MDCK cell series**) or subcultivated for twenty additional passages in EMEM (**EMEM cell series**).

U-MDCK series : p73 EMEM pX U-MDCK (X= up to 160 passages)
EMEM series : p (73+X) EMEM (X= up to 20 passages)

2.2. KARYOTYPING ANALYSIS OF EMEM AND U-MDCK CELLS

The monitoring of MDCK karyotype during subcultivation of cells in either EMEM [p(73+X) EMEM cells with X=0, X=20] or Ultra-MDCK [p73 EMEM pX U-MDCK cells with X=20, X=40, X=75, X=160] showed that 1) Starting cells (p73 EMEM) contained two distinct populations with chromosome modal number 78 and 90, which were assumed respectively as Intercalated (IC) cells and Principal (P) cells according to Wünsch et al (1995) 2) After an additional ≥20 passages period in either EMEM or Ultra-MDCK, the "P" cells population disappeared and a unique population of IC cells was still found , exhibiting a modal number at 78 chromosomes in EMEM cells and 76 to 80 chromosomes in U-MDCK cells. These results show that a similar

A. Bernard et al. (eds.), Animal Cell Technology: Products from Cells, Cells as Products, 479–481.

karyological evolution mechanism proceeds with subcultivation of MDCK cells, independently of the culture medium formulation.

2.3. EVOLUTION OF PHENOTYPIC MARKERS DURING LONG-TERM CULTIVATION OF EMEM CELLS AND U-MDCK CELLS

2.3.1. Markers of the junctional complex

Tight junctions

Trans-epithelial resistance (TER) was chosen as a marker of tight junctions. Transfer of MDCK cells to and further subcultivation in, Ultra-MDCK were responsible for a huge increase in TER, especially when using Transwell clear inserts and a low cell seeding density (EMEM cells : 100 ohms/cm^2, U-MDCK cells 1000-2000 ohms/cm^2).

Desmosomes and uvomorulin

Interesting differences were identified when comparing EMEM and U-MDCK cells. EMEM cells (p73 EMEM, 73+20) EMEM) were shown to exhibit a segregation between clusters of positive cells and clusters of negative cells, when labelled in terms of desmosomes as well as uvomorulin and the positive to negative clusters ratio varied with the support and cell density. These results are highly surprising ; indeed, Collares-Buzato et al. (1998) reported a similar aggregation phenomenon when mixing P cells and IC cells and an uvomorulin labelling restricted to P cells . On the contrary, U-MDCK cells were shown to exhibit a homogeneous pattern of positive cells, irrespective of support, cell seeding density and passages number. These results reinforce those we obtained previously when studying TER, and are in favour of a better development of the junctional complex in cells grown in Ultra-MDCK.

2.3.2.Markers of intercalated cells

Carbonic anhydrase (CA) : α and β intercalated cell marker

Levels and cell pattern of CA activity were shown completely different when comparing both cell series. EMEM cells, independently of the passage number, exhibited a "cluster" pattern, with colonies of negative cells surrounded with highly stained cells. On the contrary, U-MDCK cells which were shown to have the same "intercalated type" genotype as p(73+20) EMEM cells consistently gave a "mosaic-like" pattern of CA activity, with a regular dispersion of negative cells and cells exhibiting different degrees of CA activity.

PNA binding : β intercalated cell marker

As previously mentioned with CA activity, completely different patterns of PNA binding were found when studying EMEM cells and U-MDCK cells. The percentage of positive cells and amount of PNA binding were both consistently lower in U-MDCK cells than in EMEM cells and the "cluster" pattern was observed in the latter only. In addition, PNA binding to U-MDCK cells was shown to be dependent on support (T.col > T.clear), cell seeding density (high density > low density) and cell aging (4 days > 3 days) in contrast to EMEM cells. The presence of a high amount of PNA negative cells in U-MDCK cell series could suggests two hypothesis : the presence of α intercalated cells or a lower determinism of the apical pole of U-MDCK β intercalated cells when compared with EMEM cells.

2.3.3. Products of the extra-cellular matrix (ECM)

Both EMEM and U-MDCK cells were shown to synthesise fibronectin, CSPG, and HSPG and significant differences were observed as a function of the culture support in terms of number of secreting cells (Cytodex 3 > Biosilon > culture plates) and pattern of localisation of synthesised products ("orange-like" pattern with Cytodex 3 only). In addition differences appeared as a function of culture medium, with a regular lower synthesis of fibronectin and HSPG in U-MDCK cells than in EMEM cells. No significant differences were observed in ECM product synthesis in U-MDCK cells as a function of passages (up to 160 passages).

3. Conclusion : influence of culture medium on biogenesis of MDCK epithelium

Uvomorulin–mediated cell-cell contact is critical to formation of the baso-lateral domain since it is required for the formation of tight junctions, zonula adherens, and desmosomes, while integrin-mediated substratum interactions are responsible for the orientation of the apico-basal axis and the development of apical membrane characteristics. (Eaton and Simons, 1995). Our data regarding the phenotypic markers of MDCK cells show without any ambiguity that culture medium plays a major role in the biogenesis of the polarised epithelium.

EMEM cell series was characterised by both an excellent development of the apico-basal axis (heavy PNA binding to the apical membrane) due likely to a good development of the basement membrane in presence of FBS (high synthesis of ECM products) and a poor development of the baso-lateral axis (poor formation of cell-cell junctions) due likely to an increased paracellular permeability in response to the presence of serum in the culture medium (Chang et al, 1997).

U-MDCK cell polarisation was found as the mirror image of EMEM cells, with a poor development of the apico-basal axis as demonstrated by a low PNA binding in only a few cells, and an excellent development of the baso-lateral axis with an excellent formation of the junctional complex (tight junctions –TER-, desmosomes and adhesive junctions –uvomorulin-). It is interesting to notice that an inverse polarisation schedule was observed in U-MDCK cells when changing the culture medium daily; so, using serum-free Ultra-MDCK medium give us the possibility to turn on demand the polarisation of MDCK cells to either the baso-lateral axis or the apico-basal axis

1. Chang, C.W., Wang, X., Caldwell, R.B.(1997) Serum opens tight junctions and reduces ZO-1 proteins in retinal epithelial cells, *J. Neurochem.* **69**, 859-867.

2. Collares-Buzato, C.B., Jepson, M.A., Mc Ewan, G.T.A., Hirst, B.H., Simmons, N.L. (1998) Co-culture of two MDCK strains with distinct junctional protein expression : a model for intercellular junction rearrangement and cell sorting, *Cell Tissue Res.* **291**, 267-276.

3. Eaton, S. and Simons, K. (1995) Apical, basal and lateral cues for epithelial polarization, *Cell*, **82**, 5-8.

4. Kessler, N., Thomas-Roche, G., Aymard, M. (1999) Suitability of MDCK cells grown in serum-free medium for influenza virus production, *Dev. Biol. Stand.* **98**, 13-21, Basel Karger Publishers.

5. Wünsch, S., Gekle, M., Kersting, U., Schuricht, B., Oberleithner, H (1995) phenotypically and karyotypically distinct Madin-Darby canine kidney cell clones respond differently to alkaline stress, *J. Cell. Physiol.* **164**, 164-171.

SANDWICH ELISA TODAY:
LIMITS FOR SENSITIVITY, SPEED, PRECISION AND THROUGHPUT

M. JORDAN, D. FRABOULET, G. FOURMESTRAUX, F. M. WURM AND R. FREITAG

Centre of Biotechnology UNIL – EPFL, Swiss Federal Institute of Technology Lausanne
CH-1015 Lausanne, Switzerland

1. Introduction

Optimization of cell culture and process parameters requires accurate assays for the quantification of the product. One of the most widely used techniques for measuring recombinant proteins is the ELISA assay in multiwell plates. Using grandfathers approved protocol, it yields valid results within one working day. Since the establishment of the widely accepted standard ELISA protocols, many factors have been improved: Better detection antibodies and more potent detection substrates are available, ELISA plates are made to strongly bind proteins, and most importantly, many steps can be automated. Computers are used for sophisticated data acquisition and handling. Below we outline how a combination of these parameters can be used to improve ELISA throughputs to a point where one person can handle over 800 samples per day with the high precision requested by the biopharmaceutical industry.

2. Procedures

Improvements	**Grandfather's approach**	**Modern protocol**
	Coating of plates: over night at 4 ° C	30 minutes at 37°C
Assay time	Blocking: up to 4 hours	None (casein in dilution buffers)
	Incubation with samples: 2 hours	30 minutes
	Incubation detection antibody: 2 hours	30 minutes
	Substrate reaction: 30 minutes	Substrate reaction: 15 minutes
Throughput	Manual forms for samples and results	Intranet for sample description/report of results
	Primitive data analysis software	Software controls data acquisition and analysis
Workload	Max. samples per plate: 6 (as serial dilutions)	Up to 88
	Total assay time: 24 hours	Less than 2 hours

A. Bernard et al. (eds.), Animal Cell Technology: Products from Cells, Cells as Products, 483–485.

3. Results and Discussion

After optimizing each step of the original 24-hour protocol, the total assay time was reduced to 2 hours. In particular, coating plates at 37°C for 30 minutes instead of coating at 4°C and eliminating the blocking step saved time without biasing the results. Shortening sample/detection antibody incubation times reduced the absolute OD-values by a constant factor of 0.5, thereby actually extending the dynamic range of the assay. Reproducibility and accuracy for the long and the short protocol was found to be comparable.

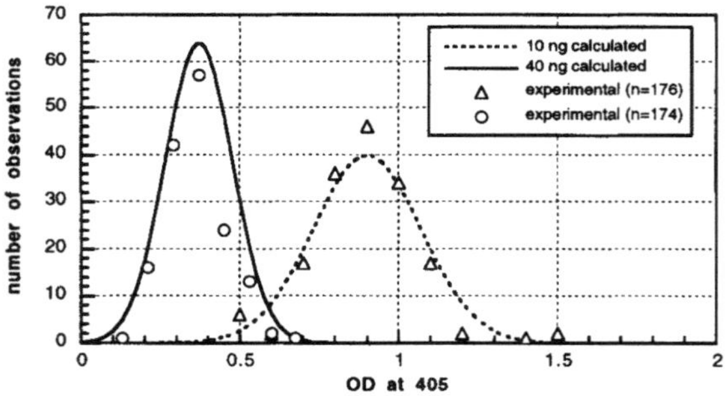

Figure 1: Distribution of absolute OD values obtained during a period of 3 months in comparison with a calculated normal distribution. The y-axis shows the frequency of the experimental values.

For the 2-hour protocol, day to day variations were estimated. We found considerable variations of the absolute OD values when repeatedly testing prediluted aliquots stored at – 80°C (Fig. 1). A probability distribution of 176 values fitted perfectly to a calculated normal distribution (mean value and standard deviation from experimental data). The relatively wide distribution and even overlapping OD readings between 40 and 10 ng values shows that an absolute precision can be not obtained by manual operations, even if protocols are strictly followed. In order to correct for such variations, an internal standard is needed on every plate.

Using multiples increases the probability of precisely hitting the real value and reduces the risk of errors (Fig.2).

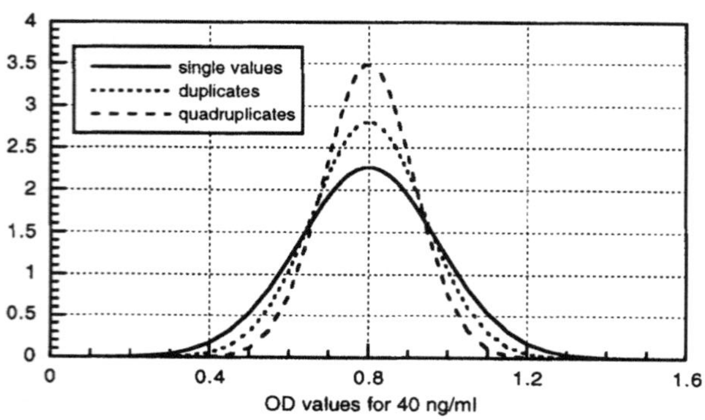

Figure 2: Multiples and precision as described by a normal distribution. Depending on the precision of the assay, individual points (single values) will distribute with a certain probability around the average value.

While all of the above measures helped to reduce the total assay time, no major improvement in throughput was possible that way. A major handicap was the limited detection range imposed by the calibration curve (linear or quadratic fit, see Fig 3). This made serial dilutions for each sample mandatory. By simply using a 4 parameter curve fit, a function with a linear increase for small values simulating the saturation of higher concentrations, we could increase our range to cover amounts from 0.4 – 140 ng/ml, i.e. 3 orders of magnitude. Serial dilutions are no longer necessary and up to 88 samples can be treated per plate. The relative standard deviation is minimal between 20 and 40 ng/ml.

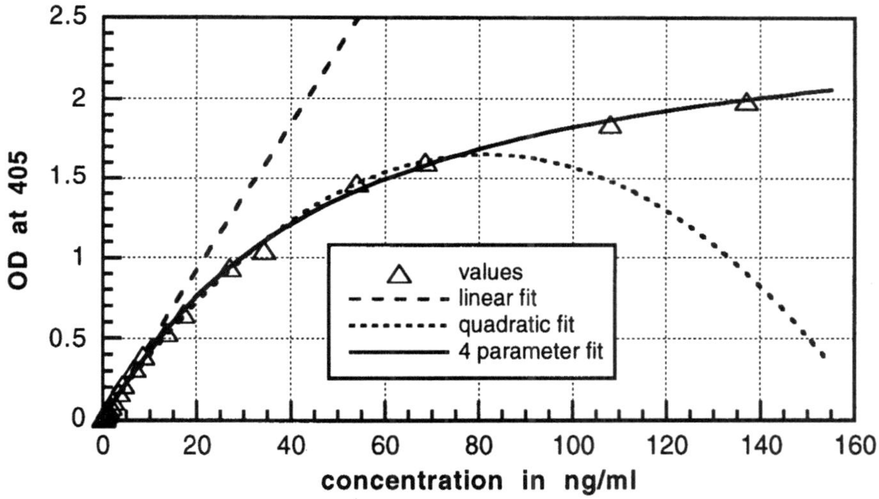

Figure 3: Various curve fits can be used to describe the standard curve. A linear fit is useful for small values up to an OD of 0.4 or 10 ng/ml, a quadratic fit up to 70 ng/ml and the 4 parameter curve fits perfectly even beyond 150ng/ml.

4. Conclusion

Modern software can treat and analyse results quickly and offers a good choice for curve fits, statistics and documentation. With protocols reduced in length to two hours and an optimised template for loading the samples, the ELISA assay can be fast, accurate and still provide a high throughput.

This work is supported by research funds from the Swiss National Science Foundation, Biotechnology Priority Program.

Chapter IX. CELL CHARACTERIZATION, VALIDATION, SAFETY, ADVENTIOUS AGENTS

IMPACT OF ICH GUIDELINES ON PHASES OF DEVELOPMENT OF BIOTECHNOLOGY PRODUCTS

Anthony S. Lubiniecki, Sc.D.
SmithKline Beecham Pharmaceuticals
King of Prussia, Pennsylvania 19406, USA

International Conferences on Harmonization (ICH) began in 1991 and have attempted to harmonize the requirements and guidance for new product approvals in the European Union, Japan, and the United States. Families of documents on quality, safety, and efficacy were created for all drugs, and some special documents were prepared for biotechnology products when special considerations existed. These documents address fundamental issues of harmonizing biotechnological/biological product regulations. In addition to creating foundational documents, the goal of this effort was to prevent divergence of biotechnological regulations before it occurred, and to set the stage for future discussions on important topics in biotechnology. At present, six biotechnology documents are completed (Q5A, Q5B, Q5C, Q5D, Q6B, and S6). In addition, there are numerous ICH documents covering efficacy determination in clinical trial which cover all drugs and, of course, cover biotechnology products as well. These documents influence the conduct of R&D at various points, some earlier than others (Table 1). To maximize flexibility during development, it is generally better to apply the ICH documents earlier rather than later. By the time the dossier is about to be filed, it is generally too late to apply them to R&D activities.

The impact of some of the specific documents is easier to judge than others, the impact will be greater on some manufacturers than others. The virus safety guideline (Q5A) provides a clear common understanding of methods and standards for virus detection and clearance. Manufacturers with highly developed programs before the guideline became official will not notice much difference afterwards, while manufacturers with less well developed programs will need to catch up quickly. The guideline on evaluation of the expression construct (Q5B) establishes the need to characterize important properties of the predominant nucleic acid sequences encoding rDNA-derived products, but also establishes the utility of protein analysis for some facets of the evaluation. Q5B will avoid wasted resources devoted to sequencing artifacts of cDNA cloning, both in the industry that would perform the studies and in the regulatory community that would

A. Bernard et al. (eds.), Animal Cell Technology: Products from Cells, Cells as Products, 489–493.

attempt to interpret them. The product stability (Q5C) guideline introduced the concept of the adequately characterized product as the foundation of a product stability program initiated on representative pilot plant material and carried through to marketed product. This will allow manufacturers to gather data in support of shelf life without investment in a commercial plant, and without having to wait until the plant is built, allowing more rapid development. The cell substrate (Q5D) guideline sets forth criteria for the preparation and characterization of cell banks. It eliminates the need to rebank cells after medium formulation changes if product consistency is unaffected, provides for mutual acceptance of various appropriate tests for detection of bioburden and mycoplasma, and creates harmony with WHO standards. The safety assessment guideline (S6) reaffirms the utility of a science-based case-by-case approach. This will clearly lead to reduction in the use of animals in essentially duplicative studies which provided little new information.

Finally, the new specifications (Q6B) guideline, which became official in March, 1999, provides a common understanding of the scope and criteria for assessment of biopharmaceutical quality, and allows manufacturers to know how regulatory bodies expect specifications to be established. It also establishes several important principles, including the concept that specifications are both product- and process-specific, that specifications should account for the stability of both drug substance and drug product, that specifications are linked to data obtained for materials used in preclinical and clinical studies which should be representative of commercial supply, and that specifications are linked to analytical procedures. Among its most significant impacts are the establishment of a harmonized scientific basis for establishment of specifications based on characterization of product and process, and on data for commercial and representative pilot batches.

The impact of these documents is felt in several ways. In a tangible sense, all of these can eliminate duplicative studies and permit resources to be utilized more efficiently to develop additional new products that might not be funded without the resources saved by ICH activity. In an intangible sense, the documents can lessen resistance to science-based change in regulations and standards, create an atmosphere of mutual trust and respect based on scientific discussions and data, and increase the availability of novel medicines to patients who are waiting for their approval. In addition to forming a basis for a common technical dossier, the documents may set the stage for discussions of additional important topics such as comparability/manufacturing variations and other issues in the future.

Industry believes that ICH represents the first time that the industry which created virtually all the data regarding regulated products was invited to use these data to devise science-based regulatory guidance in collaboration with the regulatory community. Thus, industry has a positive opinion about the process, the results, and their impact.

New ICH activities that are underway include a Good Manufacturing Practices guideline and a guideline for the Common Technical Document. When completed, these will provide guidance for all drugs including biotechnology products. These and other matters will be discussed at a public meeting scheduled for the week of 6 November 00 in San Diego, USA. A maintenance procedure to update official guidance documents was established which should provide for thorough review by expert working groups and approval by the ICH Steering Committee by inexpensive communications media without the necessity of travel.

Current copies of all ICH guidelines can be obtained from the following Internet sources:

Europe	http://www.eudra.org/emea.htm
Japan	http://www.nihd.go.jp/dig/ich/ichindex.htm
USA-CDER	http://www.fda.gov/cder/guidance/index.htm
USA-CBER	http://www.fda.gov/cber/guidelines.htm

TABLE 1. Phases of development

Document	Discovery	Preclinical Devel	Phase I/II	Phase III	MAA Filing	Post-Approval
Q5D Cell Substrates	X	X	X	X	X	X
Q5B Expression Construct		X	X	X	X	
Q5A Virus Safety		X	X	X	X	X
S6 Preclin Safety		X	X	X	X	
Q5C Prod Stability				X	X	X
Q6B Specificat				X	X	X

Discussion (Lubiniecki)

Anon: I have a question about comparability manufacturing changes for the future which are regulated by ICH documents - when do you anticipate these guidelines will come into force? It is a lengthy process and would hit some imminent (say 2-3 years) things like Biogenics. How would that be regulated by the ICH?

Lubiniecki: This is not yet an official topic before the ICH. There have been some preliminary discussions that it might be. If, and when, it became an official topic using the other existing 6 documents as historical standards, then expect it to take 3-4 years from when they start to when it became a step 4 document. I repeat, this is only an example of what might be in the future and is by no means read that it will be so.

THE MICROBIAL RISKS OF XENOTRANSPLANTATION

Daniel N Galbraith and David E Onions
Q-One Biotech Ltd, Todd Campus, West of Scotland Science Park
Glasgow, SCOTLAND

Since its initial problems, human to human transplantation has resulted in the successful clinical management of a number of life threatening conditions. This success has recently resulted in an acute shortage of organs and tissues. The lack of available organs has caused patients to remain on transplant lists for a number of years with severely debilitating conditions.[1] It has been estimated in Europe and the USA that 30 individuals die each day while waiting for a transplant.

Xenotransplantation is seen by some scientists and clinicians as a future solution to this problem. The use of animal organs and tissues for transplantation is rare and has been generally discontinued owing to the lack of success. The best result was recorded over 30 years ago where a patient survived nine months after having received a chimpanzee kidney. The difficulty in the use of animal organs is the prevention of acute organ rejection. The transplanted material is seen as foreign and the immune response to the animal tissue (xenograft) is severe, resulting in devastating necrosis of the tissue transplanted. Near-human primates have been investigated as possible sources of tissue for transplant. Baboons have been the primate of choice, however the use of these animals presents difficulties because of public disquiet at the use of primates and also because of the problems of maintaining animal stocks. With respect to the microbial risk, primates are known to harbor a number of viruses which are potentially pathogenic for humans including *Retroviruses*, *Simian herpesviruses* and *Simian papovaviruses*. This is especially relevant as laboratory animals are sometimes only one generation removed from wild caught animals. The recently presented data suggesting *Human immunodeficiency virus (HIV)* was transmitted from a primate to human[2] highlighted the risk of transmission of agents from primates to humans and would indicate that if primates were proposed as a source of tissue for transplantation this would raise major public health concerns. Recently, pigs have been investigated as an alternative source of organs. Pigs are seen by the public as more acceptable to use in the production of tissues for transplant as these are an animal also used in the production of food. The pathogens of pigs are well characterized and may not pose as great a risk to humans as primate viruses, pigs are also more capable of being produced pathogen free. The use of pigs, however, has been hampered by the acute rejection which occurs when the tissues from these animals are used. This problem can be ameliorated by making the pigs used for transplant transgenic for one or more human genes which prevent the activation of complement. The use of modified pigs and the development of improved immunosuppressive drugs regimens may now advance the use of these tissues for treatment of humans.

A. Bernard et al. (eds.), Animal Cell Technology: Products from Cells, Cells as Products, 495–499.

Given that the rejection problems can be overcome another important consideration is the transmission of infectious disease to humans following transplant of the xenograft (zoonosis). A number of agents are known to infect both humans and animals such as, *Influenza virus* which causes lung infection in both humans and pigs as well as a number of other species. It would be logical that tissue for use in transplants should have been tested and shown to be free of zoonotic agents in the same way to human donors are assessed for the presence of *HIV*, *Cytomegalovirus* and hepatitis causing viruses[3] (B and C). In addition to infectious agents which are common to humans and animals, many animal species harbor specific pathogens which do not cause infection in non-related species, such as *Porcine circovirus*, a pig virus which does not infect humans.[4] There is a risk that if such agents are transmitted during transplantation they may then become pathogenic in the recipients of the xenograft. Alternatively, it is possible that such agents while not pathogenic for the human xenograft recipient may infect the transplanted organ. This infection may be exacerbated given the immunosuppresive regime under which the patients will be maintained.

Although infection of the xenograft recipient is of great importance the principal public health concern is the potential for the transmission of an agent from the xenograft recipient to the wider community. Xenotransplantation provides a novel mechanism for the potential transmission of pathogenic agents from one species to another. Viruses, in particular have been known to spread using clinical interventions as a vector. For example the spread of HIV, Hepatitis B and Hepatitis C was facilitated by the widespread use of blood transfusions. Such infections are difficult to treat using traditional drugs, and are quick to mutate and avoid both immune system clearance and drug treatment.

Owing to the risks of transmitting disease to human xenograft recipients it is important that tissue for xenografts is from animals free of known agents. This would necessitate production animals being monitored regularly for levels of infection. There are no compulsory guidelines for the monitoring of animals. However, best practice would be to screen for the maximum number of known agents. Given that these animals are transgenic then consideration should also be given to human viruses which may infect the production animals owing to the presence of human cell receptors on the cells of the transgenic animals. The complement mediating receptors for which the pigs are transgenic are known receptors for *Measles virus* and *Human Picornaviruses*. These human viruses may gain entry into porcine cells expressing these human cell receptors. It is therefore possible that transgenic pigs may be susceptible to viruses which were previously unable to replicate in the porcine host. The possibility of new recombination between porcine viruses and human viruses exists as these viruses can now co-infect the same host for the first time. This possibility should be exhaustively investigated *in vitro* before xenotransplantations are common place. The pigs and recipients of these xenografts should be monitored for these agents.

One virus which has recently caused concern is *Porcine endogenous retrovirus (PoERV)*. The virus is present in the genome of all pigs, is genetically transmitted and importantly has been shown to infect human cells.[5] The greatest risk for the xenograft recipient would be the release of the virus from the xenograft cells and the subsequent infection of host cells. Infection of cells with retroviruses may result in viral integration and subsequent production of lymphomas. However, the wider community is also at risk as there is the possibility PoERV may undergo recombination with the *Human endogenous retrovirus* resulting in a novel retrovirus. This novel recombined virus is giving greatest concern to Virologists as this may result in a virus which is capable of infecting the wider community. As PoERV is present at around 50 copies in the porcine genome removal of the virus from the pig germ line using traditional breeding techniques would be difficult. Gene knockout technology has been proposed as a means of removing the virus but this would require considerable time to accomplish. If clinical trials, as have been proposed, are to start in the near future then the challenge to scientists and clinicians is to assess the risk of retroviral infection and to monitor the patients as closely as possible during their treatment. Monitoring such patients requires testing regimes for detection of virus and antibody, in a similar manner to those used for monitoring of HIV infected individuals. At Q-One Biotech we have developed and validated an assay to detect the presence of antibody to PoERV proteins. This assay is designed to monitor the antibody response of patients to PoERV and enable early detection of infection. However, serological analysis have the disadvantages that antibody detection lags behind viral expression by a number of weeks and antibody response can be heterogenious. A second method of monitoring for PoERV in human patients is the nucleic acid amplification techniques, polymerase chain reaction (PCR) and reverse transcriptase - PCR. These methods allow the monitoring of viral genome present as RNA or integrated proviral DNA. All detection techniques are validated to ensure the sensitivity and specificity of the assay. By monitoring the patients during clinical trials of xenotransplants this will demonstrate the PoERV risks associated with the procedure giving clinicians and patients the necessary information required for patient management judgements.

In conclusion, xenotranplantation appears to be nearing a position where it may be a major breakthrough in clinical management. The rejection problems appear for the main part to have been offset by the use of genetically modified animals and the development of immuno-suppresive drug therapy. All that remains is an assessment of the microbiological safety of the procedure. When this has been clearly established and the risks noted, it then remains for the public to decide if this technology should be used.

1. Annual Report of the U.S. Scientific Registry for Transplant Recipients and the Organ Procurement and Transplantation Network - Transplant Data 1988-1994. (1995). *UNOs and the U.S.A Department of Health Human Services.*

2. Cohen J. (1999). AIDS virus traced to chimp sub-species. *Science*, **283**; 772-3.

3. Moor AC, Dubbelman TM, Van Stevenick J, Brand A. (1999). Transfusion transmitted diseases: risk, prevention and perspectives. *European Journal of Haematalogy*, **62**; 1-18.

4. Tischer I, Gelderblam H, Vettermann W and Koch MA. (1982). A very small porcine virus with a circular single stranded DNA. *Nature*, **295**: 64-66.

5. Wilson CA, Wong S, Muller J, Davidson CE, Rose TM, Burd P. Type C retrovirus released from primary porcine peripheral blood mononuclear cells infects human cells.

Discussion (Galbraith)

Rhiel: Dr Hayflick reminded us in his talk on the potential of latent virus. Suppose such a virus exists, what is the probability now that it would slip through the screening methods? Will a complete screening only be available when gene therapy will be borne?

Galbraith: Are you asking about the porcine endogenous retrovirus?

Rhiel: In general, ie any latent virus not known at present.

Galbraith: For latent viruses, such as porcine CMV, we can use PCR techniques to exclude those animals that have the virus from the breeding population. By this means we can exclude it from the cohort intended for xeno-transplantation.

Rhiel: But if the sequence of the virus is unknown?

Galbraith: How can you screen for the unknown? If you have an idea, such as a Herpes virus, we can use degenerate primers to see if there is a retrovirus there. We can look for differential expression of normal nucleic acids.

Rhiel: Are prions not an issue with the pig?

Galbraith: As far as I am aware there are no known TSE in pigs. Again, it is unknown

Anon: I would like to challenge your views on the risk of porcine endogenous viruses. In nature we have seen transmission from monkeys to humans, when blood has been mixed. This could have happened with pork all the time. When you infect human cells they only get transiently infected at a very low level, and one has never seen a recombinant between a human endogenous retrovirus. What is your data?

Galbraith: It is because these are transgenic animals producing retrovirus. As you are aware when the retrovirus buds from the surface it will carry the Hdark receptors. Therefore, it will not be inactivated by complement so one of the most important ways to retro-viral inactivation is unavailable.

Anon: But it only transiently infects human cells at a very low level. Do you have data for recombination and cross-packaging between human endogenous viruses and porcine viruses?

Galbraith: No.

CRYOPRESERVATION UNDER PROTEIN-FREE MEDIUM CONDITIONS: A RELIABLE WAY TO SAFE CELL BANKING

F. HESSE, K. SCHARFENBERG, R. WAGNER
Cell Culture Technology Department
Gesellschaft für Biotechnologische Forschung mbH
Mascheroder Weg 1, Braunschweig, Germany

1. Introduction

Cell banking systems serve as source for the producer cell line throughout the lifetime of the manufacture of a biopharmaceutical product. Therefore, a high quality of the cell banks has to be guaranteed leading to an extensive validation effort. In particular, processes depending on cryopreservation media which contain serum or other components of animal origin suffer from a high contamination risk and must be excessively validated to ensure a high safety standard of the product.

We have developed a protocol which enables the establishment of cell banking systems under protein-free medium conditions. This protocol combines well-defined procedures for freezing and thawing of the cells with the application of a special protein-free cryopreservation medium.

Various cell lines (e.g. BHK, CHO, MDCK) were successfully cryopreserved by applying this method after being adapted to protein-free medium conditions according to Scharfenberg and Wagner (1995). After revitalization, the cells showed the same growth characteristics as those before cryopreservation. The data were also comparable to those obtained from control experiments performed with serum-containing cryopreservation medium.

2. Materials and Methods

CELL CULTIVATION

CHO cells were cultivated using the protein-free SMIF 6 medium. For BHK-21 and MDCK cells the protein-free SMIF 7 medium was used. All cultivations were performed in a CO_2-incubator at 37 °C, 12% CO_2, and 80% humidity. Samples were collected from the culture vessels once a day for determination of the culture parameters.

CRYOPRESERVATION MEDIA

All cryopreservation media used were based on the protein-free SMIF medium line. For the preparation of protein-free cryopreservation media, SMIF media were supplemented

A. Bernard et al. (eds.), Animal Cell Technology: Products from Cells, Cells as Products, 501–503.

with 10.0% DMSO and 0.1% methyl cellulose as cryopreserving substances according to Merten *et al.* (1995). The serum-containing cryopreservation media consisted of 80.0% fetal calf serum (FCS), 12.5% SMIF medium, and 7.5% DMSO. Cryopreservation and revitalization were performed according to very detailed protocols under GLP conditions. These protocols are available under http://www.gbf.de/bereiche/uk/div_bvt/zkt/Research_Activities/Culture_Media_Development/culture_media_development.html.

3. Results and Discussion

We cryopreserved and revitalized a wide range of cell lines adapted to protein-free SMIF media, including BHK, CHO, and MDCK, successfully according to the described procedure and compared it with the standard serum-dependent procedure. For this purpose, cells were taken from stem cultures and cryopreserved according to the freezing protocol given above using protein-free and serum-containing cryopreservation media, respectively. Cells cryopreserved under protein-free and serum-containing medium conditions were revitalized in parallel and cultivated using protein-free SMIF media. Revitalization parameters were comparable for cells cryopreserved in protein-free and serum-containing media. Total cell recovery was about 70% in both cases with greater variations when using serum-containing media. The viability of the cells after revitalization was even slightly higher when using protein-free media.

TABLE 1. Comparison of revitalization parameters of cells cryopreserved in FCS-containing media with revitalization parameters of cells cryopreserved in protein-free media for CHO, BHK and MDCK cells. The generation doubling times for the stem cultures from which the cells were taken for cryopreservation were 22.2 h for CHO cells, 34.0 h for BHK cells, and 41.1 h for MDCK cells, respectively.

	CHO		BHK		MDCK	
	cryopreserved with FCS	protein-free cryopreserved	cryopreserved with FCS	protein-free cryopreserved	cryopreserved with FCS	protein-free cryopreserved
Viability before cryopreservation [%]	97.6	97.6	96.7	96.7	99.5	99.5
Viability after revitalization [%]	80.7	90.3	89.2	90.6	93.3	97.0
Total cell recovery [%]	58.1	68.8	70.4	73.3	76.3	66.3
Generation doubling time [h]	22.9	22.9	29.1	30.4	44.3	45.4

Growth rates and viability of cells cryopreserved using protein-free media were the same as for cells cryopreserved using serum-containing media (see figure 1 and 2). The generation doubling times were nearly identical for cultures started with cells cryopreserved in protein-free or serum-containing media (see table 1). These data clearly demonstrate that the presented procedure is just as suitable for cryopreservation of cells adapted to protein-free culture media as the standard method depending on FCS-containing media. No significant differences could be observed when cryopreserving cells in protein-free or serum-containing media using the described method.

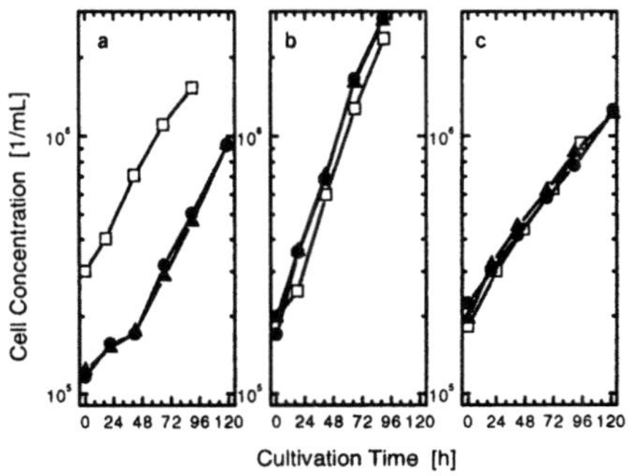

FIGURE 1. Cultivation diagrams of BHK (a), CHO (b), and MDCK (c) cultures. The diagrams show the comparison between the stem cultures from which the cells were taken for cryopreservation (□), the cultures started with cells cryopreserved in FCS-containing media (●), and the cultures started with cells cryopreserved in protein-free media (▲).

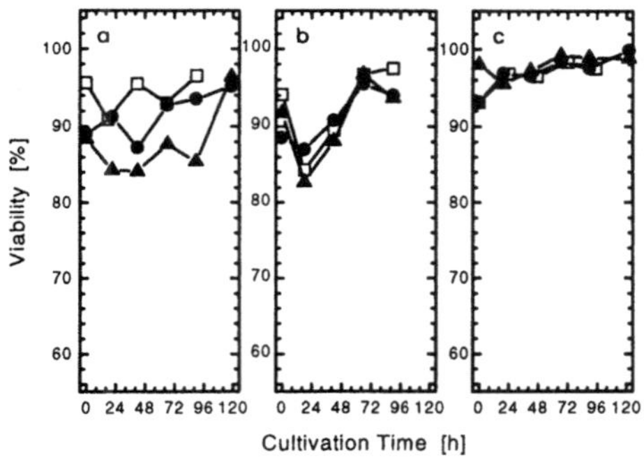

FIGURE 2. Viability diagrams of BHK (a), CHO (b), and MDCK (c) cultures. The diagrams show the comparison of viabilities between the stem cultures from which the cells were taken for cryopreservation (□), cultures started with cells cryopreserved with FCS-containing media (●), and cultures started with cells cryopreserved using protein-free media (▲).

The described method was successfully applied for the establishment of several cell banks during the EU-Demonstration Project BIO4-CT97-2140.

The protocols we developed also contain several documentation steps which lead to a high documentation standard. This enables an easy integration of the procedure into processes which are performed according to Good Manufacturing Practice.

4. Conclusions

The reported procedure enables cryopreservation of cells adapted to protein-free cultivation conditions and therefore the establishment of master and manufacturer´s working cell banks under defined medium conditions, free of proteins and any components of animal origin. This will reduce the validation effort for cell banking systems used as cell sources for production processes of biopharmaceuticals using mammalian cell cultures.

5. References

Scharfenberg, K. and Wagner, R. (1995) A Reliable Strategy for the Achievement of Cell Lines Growing in Protein-free Medium, in: E.C. Beuvery, J.B. Griffiths, W.P. Zeijlemaker (eds.), *Animal Cell Technology: Developments towards the 21st Century*, Kluwer Academic Publishers, Dordrecht, pp. 619-623.

Merten, O.W., Petres, S., and Couve, E. (1995) A simple serum-free freezing medium for serum-free cultured cells. *Biologicals* **23**, 185-189.

6. Acknowledgements

Part of his work was performed during the EU-Demonstration Project BIO4-CT97-2140 and were supported by the Biotechnology Programme of the European Commission.

CREATION OF CERTIFIED CELL BANKS SUITABLE FOR PRODUCTION OF HEALING AND IMMUNOBIOLOGICAL PREPARATIONS

T.D.KOLOKOLTSOVA, N.D.YURTCHENKO, A.A.ISAENKO, E.A.NECHAEVA, O.V.SHUMAKOVA, T.N.GETMANOVA.
Institute of Cell Cultures, State Research Center of Virology and Biotechnology "Vector", Koltsovo, Novosibirsk Region, 633159, RUSSIA

1. Introduction

The basis for successful development of virological, biotechnological and gene-engineering studies and production is the creation of cell banks certified in accordance with general requirements and maintained in quantities that allow a long-term provision of developing research directions with standard cells. The aim of this work was the creation of certified seed, master and working banks from the most promising cell cultures suitable for production of healing and immunobiological preparations and perspective for transplantation.

2.Material and methods

CELL CULTURES. **MT-4** the human T-lymphoblastoid cell line contain a genome of Human T-Cell leukemia virus (HTLV-I). **L-68** diploid cell culture was obtained from the lung of an eleven-week-old human embryo[1]. **BHK TK-** and **CHO TK-** are the thymidine kinase deficient mutants, derived from the syrian hamster kidney cell line (BHK) and from the chinese hamster ovary cell line (CHO), respectively. **CHOpE** cell culture, a stable producer of recombinant human erythropoietin was produced by transfection of CHO TK- cells with the recombinant plasmid containing human erythropoietin gene[2]. **FL-16, FK-16** diploid cell cultures were obtained from the lung and skin of a seven-week-old human embryo.
CERTIFICATION. Certification of the cell cultures was carried out by assays of morphological and cultural properties, karyological characteristics, species identification, control of contamination, virus susceptibilities and tumorogenicity according to the WHO's requirements.

3. Results and discussion

A. Bernard et al. (eds.), Animal Cell Technology: Products from Cells, Cells as Products, 505–507.

The use of cell cultures in production of preparations for human presents a certain risk caused both by biochemical and genetical abnormalities. The cell cultures may be contaminated with human's pathogenic viruses, mycoplasma, bacteria, yeast and fungi or might contain potential oncogenity. All the problems may be resolved due creation of sertified seed, master and working cell banks. Taking into account the Recomendations of the National Research Institute of Standardisation and Control of Medical and Biological Preparations (Russia) and latest requirements of WHO we created and attested banks of most promising cell cultures. At present 17 master and working banks of the promising diploid and heteroploid cell cultures have been formed and attested in accordance with the WHO's standards.

Human diploid cells are capable to long time maintaining in culture and are characterised with high viability and diploid karyotype. According to the WHO's classification, diploid cells belong to the group of substrates of the least risk. Diploid cells L-68, FK-16 and FL-16 cell cultures does not contain mycoplasmas, viruses and bacteria, are not tumorogenic and have stable cytogenetic characteristics The conducted studies showed a high stability and absence reconstruction chromosomes in the caryotype of the L-68 cell culture and morphological properties were also unchangeable during 45 passages. L-68 is highly receptive to the measles virus and now L-68 is used in production of live measles vaccine in our Centre [3]. On the other hand, this cells are susceptible to a wide range of human viruses. Therefore diploid cell lines are suitable for the production of viral vaccines and have been useful for screening of antiviral chemicals. The diploid cells L-68, FK-16 and FL-16 also can be used for transplantation and for healing of burn wounds [4].

It is known that T-limphoblastoid cells have been used for large scale production of biological products. In addition to, since essential component of the receptor for the AIDS retrovirus is antigen CD4-antigen of population of T-helpers, T-lymphoblastoid cell lines may by used for secretion human immunodeficiency virus (HIV) and for screening assays of potential anti HIV chemicals [5]. T-lymphoblastoid cell line MT-4 is promising for investigation of antiviral substances screening on cell cultures [6]. Our researches revealed the stability of karyotype and the other culture's properties in cell line MT-4 during 30 passages. The seed and working banks of MT-4 cell culture have been formed and officially allowed for the works with the HIV-I, HIV-II viruses.

Mammalian cell lines carrying nonsense suppressor are comfortable model for selection of cell strains, which may be potential sources for production recombinant human proteins. As BHK TK- and CHO TK- cell lines are the starting material for genetic engineering approaches the main and working banks of the each culture were formed. The strain of cell culture CHOpE producing recombinant human erythropoietin (rhEPO) have been developed using methods of genetic and cellular engineering. Cells CHOpE were derived from CHO TK- cell line. Conducted studies showed that CHOpE cells characterised by stable karyotype, morphology and productivity of rhEPO during 25 passages [7]. On the basis of the control and certification of the master and working banks of CHOpE cell cultures and, first of all, due to stability of characteristics the National Research Institute of Standardisation and Control of Medical and Biological Preparations NISK approved the use of the CHOpe cells in erythropoietin production technology.

4. Conclusion

Thus, seed, master and working banks of the promising diploid and heteroploid cell cultures have been formed and attested in accordance with the WHO standards. 10 of them are certified by the National Research Institute of Standardisation and Control of Medical and Biological Preparations. These certified cell cultures can be used for production of immunobiological and pharmaceutical preparations, for genetic engineering experiments and for healing of burn wounds.

5.References

1. Stepanova, L.G., Alekseev, S.B., Zgursky, A.A. (1986) Isolation and characterization of a new strain of diploid cells from human embryonic lung tissue, *Cytologiya* **28**(12), 1373-1376, The Russia.
2. Kipriyanov, S.M., Tsareva, A.A.,et al. (1995) *Strain of ovarian cells-producer of human erythropoietin*, Patent N93029235 SU, 28.03.93., The Russia.
3. Sandakhchiev, L.S., Tsareva, A.A., Nechaeva, E.A. et al.(1996) *Methods of production of live measles vaccine*, Patent of RF, N95117008 from 07.12.95. The Russia.
4. Kolokoltsova, T.D., Yurchenko, N.D., Kolosov, N.G. et al (1998) Prospects of sertified human fetal fibroblasts in the treatment of wounds of various etiology, *Vestnik RAMN* **3**, 32-35.
5. Dalgleish, A.C., Bevorley, P.C.L., Clapham, P.K., et al.(1984) The CD4(T4) antigen is an essential component of the receptor for the AIDS retrovirus, *Nature*, **312**, 763-767.
6. Harada, S., Koyanagi, Y., Yamamoto, N.(1985) Infection of HILV-III/LAV in HTLV-I. Carrying cells MT-2 and MT-4 and application in a plaque assay, *Science*, **229**, 563-566.
7. Adamson, J.W., Spivak, J.I. (1994) Physiologic basis for the pharmacologic use of recombinant human erythropoietin in surgery and cancer treatment, *Surgery*, **115**(1), 7-15.

ENDOGENEOUS RETROVIRUS PARTICLES AND THEIR EFFECTS ON THE GROWTH BEHAVIOR OF HYBRIDOMA CULTIVATION PROCESSES

ANDREA HAWERKAMP, D. LÜTKEMEYER, A. FALKENHAIN, F. GUDERMANN, H. BÜNTEMEYER AND J. LEHMANN
Institute of Cell Culture Technology, P.O. Box 100131, 33501 Bielefeld, Germany

1. Abstract

Murine hybridoma cell lines used for the production of monoclonal antibodies are known to produce retrovirus-like particles such as infectious c-type murine leukemia viruses (MuLV) and/or intracisternal a-type particles. Froud et al. (1997) showed that in batch fermentation the level of murine c-type retrovirus particles followed the viable cell density. It appeared that the maximum titres of virus particles were found in the exponential growth phase of the cells at an optimal nutrient supply. The cell's consumption of the nutrients caused a decrease in the virus titres. In contrast to these results we examined the virus expression and a potential growth inhibition effect of these virus particles in a perfusion fermentation process. Therefore, a special reactor setup was developed which guaranteed a sufficient nutrient supply and the simultaneous removal of waste products. Even in the perfusion process a decrease of viability could be observed under optimal nutrient supply. To analyse whether there is a potential relationship between the observed decrease in viability caused by an increased retrovirus expression during this cultivation process, we examined the harvest after a decrease of cell growth for existance of retrovirus particles. Monitoring the reverse transcriptase-activity, a rapid increase of the sig-nal in the supernatant with a coinciding decrease in viability could be observe. The examination of concentrated and fractionated supernatant from this period shows growth inhibitory effects in the biological assay (MTT-assay). Investigations of respective fractions demonstrate retrovirus particles with reverse transcriptase-activity. Based on RT-PCR data we can show that only inhibitory fractions contain retrovirus particles which are E-MuLV origin. Therefore, we assume a correlation between the retrovirus-expression in continuous fermentation process and the observed growth modulating effect.

2. Materials and Methods

2.1 Cultivation:

- Cell line: rat-mouse hybridom [187.1 (HB 58) ATCC]
- Bioreactor: 2 litre reactor, Biostat BF2 (B. Braun International) with bubble free membrane aeration and membrane perfusion

A. Bernard et al. (eds.), Animal Cell Technology: Products from Cells, Cells as Products, 509–511.

- Medium: DMEM/F12 (1:1) supplemented with human transferrin (5 mg/L), bovine insulin (5 mg/L) and amino acids. To avoid high ammonia levels, glutamine (100 mmol/L) was added according to its consumption by the cells.

2.2 Concentration and purification of the cell free harvest:

- Concentration: Ultrafiltration: with Sartocon Micro (50 kDa, Sartorius)
- Separation: Sephacryl S-1000 (fractionation range: up to 10^8 Da, Amersham Pharmacia Biotech); Elutionbuffer: 20 mM phosphate, 150 mM NaCl, pH 7.2

2.3 Analytical methods:

- Monitoring of substrates: The extracellular amino acid concentrations were measured daily by using an automated reversed phase HPLC system (Kontron). Glucose and lactate were measured with an automatic analyzer (Yellow Springs Instruments). Ammonium was measured by an ammonia-sensitive sensor (WTW).
- MTT-Bioassay: This assay system determines the activity of the dehydrogenases in the mitochondria by colorimetric reaction.
- Reverse transcriptase-activity: The reverse transcriptase-activity was tested by using the Retro-Kit/RTA-Assay (Retro-Tech).
- RT-PCR: For the detection of c-type retrovirus (ecotrope murine leukemia virus = E-MuLV) specific primer were used for the RT-PCR (Perkin Elmer).

3. Results:

The perfusion process was started after inoculation with a dilution rate of $D=1.75d^{-1}$. At a viable cell density of $1x10^6$ cells/ml the dilution rate was reduced ($D=0.26d^{-1}$) and the pulsed glutamine feed (100 mmol/L) was started (Fig. 1) and always kept on a concentration above 250 µmol/L (K_S=30-70 µmol/L). The grafic shows a continuous decrease of the viability from day 4 to day 8 under optimal nutrient supply. It is known from other investigations (data not shown) that only harvest after a decrease in viability contains growth inhibitory substances. To analyse whether there is a potential relationship between the observed decrease in viability caused by the increased retrovirus expression, we examined the harvest after the decrease in viability for the existance of retrovirus like particles.

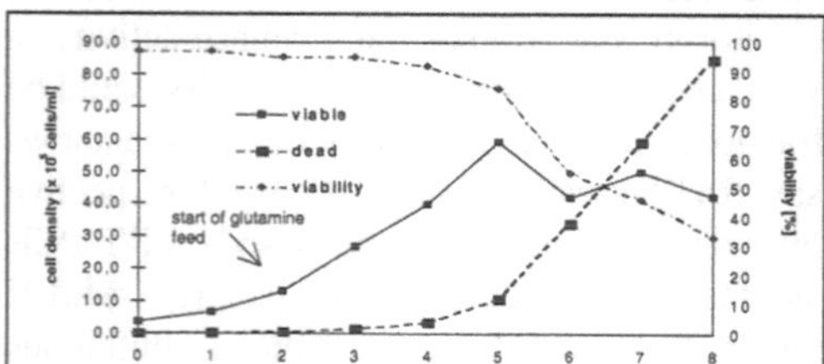

Fig. 1: Time course of perfusion process with glutamine feed

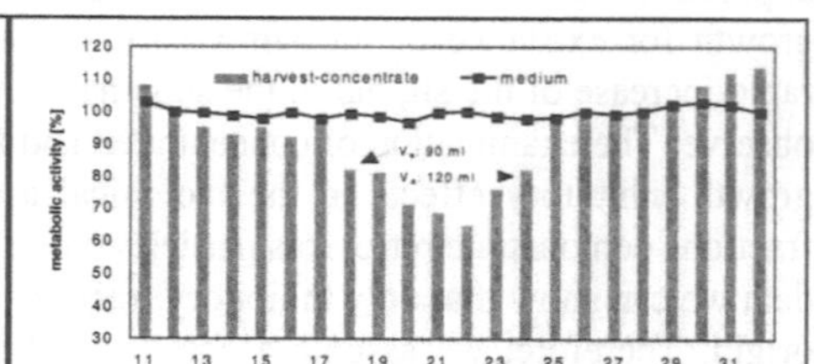

Fig. 2: Activities of Sephacryl S-1000 fractions in the MTT-assay (V_e:signs the elution range of the inhibitory substance)

Therefore cell free supernatant was collected over a period of 24 h and harvested (500 ml per day) for further purification (Fig. 2) and examinations of retrovirus particles with reverse transcriptase-activity (Fig. 3 and 4) and RT-PCR (Fig. 5).

To determine the molecular weight range of the inhibitory substance, a Sephacryl S-1000 column was used. After the decrease in viability cell free harvest was first concentrated, than fractionated on a Sephacryl S-1000 column and the resulting fractions were examined in the MTT-assay (Fig. 2). The fractions 18 to 24 showed an inhibitory effect only for the harvest-concentrate, not for medium. The elution range of this substance, between 90 to 120 ml, corresponds to the elution volume of blue dextran with a molecular weight of 2000 kDa. The figure 3 shows a comparison of the viability and the reverse transcriptase-activity during the perfusion process, demonstrated in figure 1. The graphical representation shows an increase in the reverse transcriptase-activity when the viability decreased.

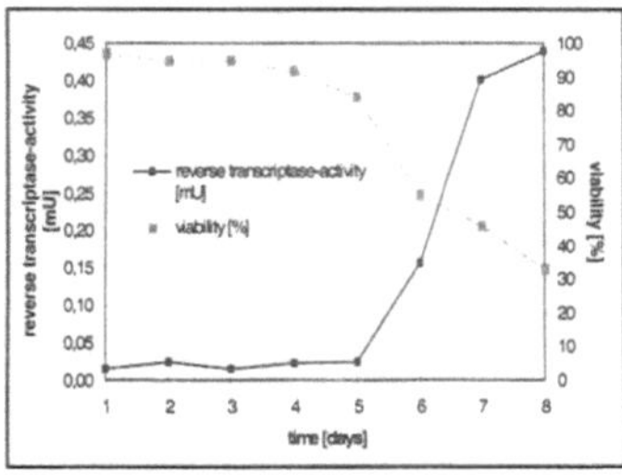

Fig. 3: Time course of reverse transcriptase-activity during cultivation

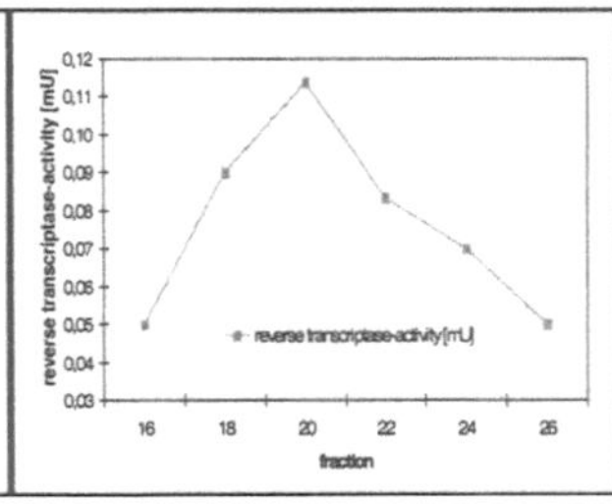

Fig. 4: Reverse transcriptase-activity in the fractions 16 to 26 of the Sephacryl S-1000

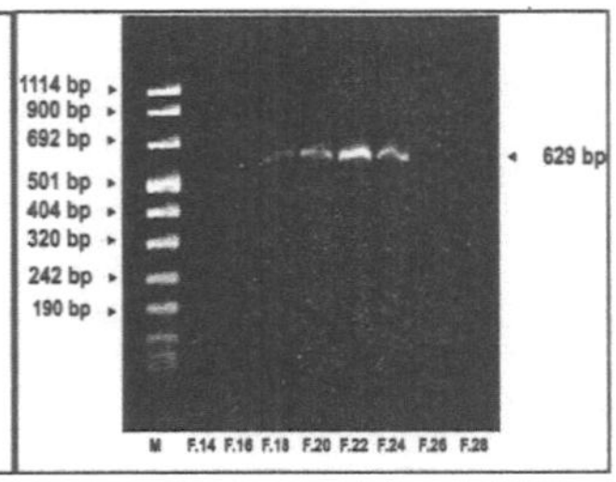

Fig. 5: RT-PCR analysis with the fractions 14 to 28 of the Sephacryl-column for detection of E-MuLV

Furthermore, we examined the growth inhibitory fractions for existance of retrovirus particles. Figure 4 demonstrates the results of the reverse transcriptase-activity assay with the fractions 16 to 26 of the Sephacryl-column. All inhibitory fractions show reverse transcriptase-activity and the highest signal could be measured for fraction 20. The RT-PCR analysis (Fig. 5) with specific primers for detection of E-MuLV suggest the presence of retrovirus particles which are E-MuLV origin. Moreover, we can show that only growth inhibitory fractions contain E-MuLV particles.

3. Conclusion

By using a special reactor setup we can show, that the viability of the hybridoma cells decrease in spite of optimal nutrient supply. The fractionation of cell free harvest after a decrease in viability demonstrates an inhibitory substance with a molecular weight of approximately 2000 kDa. The monitored reverse transcriptase-activity showed a rapid increase of the signal in the supernatant with a coinciding decrease in viability during the fermentation process. Further investigations suggest the presence of retrovirus particles which are E-MuLV origin. Based on RT-PCR and reverse transcriptase-activity analysis we can show that only growth inhibitory fractions contain E-MuLV particles. Therefore, we assume a correlation between the retrovirus-expression in perfusion cultivation processes and the observed growth inhibiting effect.

4. References

Froud, S.J., Birch, J., McLean, C., Shepherd, A.J. and Smith, K.T. (1997) Viral contaminants found in mouse cell lines used in the production of biological products, in R.E. Spier, J.B. Griffiths and W. Berthold (eds.), *Animal Cell Technology: products of Today Prospects for Tomorrow*, Kluwer Academic Publishers, Dordrecht, 681.

TESTING FOR ADVENTITIOUS RETROVIRUSES USING PCR-BASED REVERSE TRANSCRIPTASE ASSAYS

LOVATT, A., BLACK, J., GALBRAITH, D., DOHERTY, I., MORAN, M. W., WILSON, N., SHEPHERD, A. J. AND K. T. SMITH.
Q-One Biotech Ltd., West of Scotland Science Park, Glasgow, G20 OXA, UK

Introduction

Product enhanced reverse transcriptase (PERT) assays, also termed Amp-RT or polymerase chain reaction-based reverse transcriptase (PBRT) assays (1), are extremely sensitive tests for the detection of reverse transcriptase (RT). They have been reported to be up to 10^6-fold more sensitive than conventional RT assays for detecting the presence of retroviruses (2,3,4,5). The tests are also the first choice for detection of RT in live viral-vaccines, gene therapy preparations and the screening of animals and patients for *Porcine endogenous retrovirus* (PoERV) thought to be of concern in xenotransplantation. The assays involve converting an RNA template to cDNA and then amplifying the cDNA using product-specific primers. Since no exogenous RT activity is added to the reaction, cDNA will only be generated if the sample itself contains RT activity.

Materials and Methods

CELL CULTURES AND RETROVIRUSES: Cell lines and retroviruses used were maintained by Q-One Biotech Ltd. in fetal calf serum DMEM or RPMI 1640 medium.

REVERSE TRANSCRIPTASE ASSAYS: Virus stocks and cell line supernatants were clarified and virus particles pelleted by ultracentrifugation. The fluorescent PERT assay (F-PERT) used a Perkin Elmer Corp./Applied Biosystems ABI 7700 sequence detection system. During amplification with PCR primers, a specific fluorogenic oligonucleotide probe with both a reporter and quencher dye attached anneals specifically to the amplified product between the amplimers. The detection reaction uses the 5' nuclease activity of AmpliTaq Gold™ DNA polymerase to cleave the reporter dye from the probe, which results in an increased fluorescence when reverse transcriptase activity is present (6). The C_T, or threshold cycle value is calculated in real time and represents the PCR cycle at which an increase in reporter fluorescence above the baseline signal can first be detected. During validation of the F-PERT assay, analysis carried out at the limit of detection and on negative control reactions indicates that C_T values equivalent to or below 32 should be considered positive or suspect for reverse transcriptase activity. C_T values greater that 32 are in the lower boundary of the limit of detection, and therefore considered RT negative. The conventional PERT (C-PERT) assay was carried out in a single, two-compartment tube. The finished reactions were electrophoresed and analysed by southern blotting. The conventional RT assay (C-RT) measured activities of RT with poly(rA)-oligo(dT) and the

A. Bernard et al. (eds.), Animal Cell Technology: Products from Cells, Cells as Products, 513–515.

activities of contaminating DNA polymerase with poly(dA)-oligo(dT) with incorporation of ^{3}H-TTP.

Results

SUPRESSION OF CELLULAR DNA POLYMERASE SIGNALS

One major problem with PERT assays is the potential for misinterpretation of test results, leading to the conclusion that a sample contains retroviral RT. This arises from the sensitivity of the test, coupled with the ability of DNA polymerases, telomerases and other cellular enzymes to utilize RNA templates, thereby showing RT-like activity (7,8). The use of activated calf thymus DNA (aCT) to "mop up" signals generated by DNA polymerase δ and *Taq* polymerase demonstrates that the F-PERT assay that has the ability to discriminate between RT-like DNA polymerase and retroviral RT activity (Table 1).

TABLE 1

Sample	α	β	δ	*Taq*	TdT	SMRV
Minus aCT	37-40	40-40	26-28	22-31	40-40	22-29
Plus aCT	40-40	40-40	39-40	40-40	40-40	26-28

Table 1. Suppression of RT-like DNA polymerase activity. Abbreviations: DNA polymerase α (calf thymus), β (human), and δ (calf thymus). *Taq*; native *Taq* DNA polymerase. TdT; terminal deoxynucleotide transferase. SMRV; 10^3 squirrel monkey retrovirus particles Numerical values show 95% confidence limits of C_T values.

ASSAY SENSIVITY AND DETECTION OF RETROVIRAL ACTIVITY IN CELL CULTURE SUPERNATANTS

Both F-PERT and C-PERT assays were able to detect at least 10^2 *Murine leukaemia virus* (MLV) retroviral-like particles (rVLPs) and 10^1 *Squirrel monkey retrovirus* (SMRV) rVLPs. Analysis of PoERV, indicates that F-PERT was able to detect 10^4 rVLPs and lower. Purified RT enzyme was detected at 10 molecules for *Avian myeloblastosis virus* (AMV) and MLV using F-PERT. The detection limit of C-PERT was 10^3 and 10^2 molecules for MLV and AMV respectively (data not shown). We assessed the sensitivity of the conventional RT and PERT and compared this with the F-PERT assay. Depending on the retroviral type, both C-PERT and F-PERT are between 3 and 6 logs more sensitive than conventional RT assays and can detect a wide range of retroviruses, including representatives of the Deltaretroviruses (SMRV, SRV-1) and Gammaretroviruses (MLV, PoERV), Lentiviruses (SIV), and Spumaviruses (SFV) (data not shown).

No RT activity was detected in non-induced Vero, Raji, MRC-5, BT, *Mus dunni* and C127I cell supernatants. Induction of *Mus dunni* and C127I with 5-iodo-2-deoxyuridine (IDU) resulted in a cell supernatant that was positive for RT activity. RT activity was detected in non-induced Sf-9, K-Balb, NS0, SP2/0-Ag14 and BHK-21 cell supernatants. Induction of the endogenous retrovirus from the K-Balb cell line resulted in a significant increase in RT activity in the cell supernatant. Furthermore, the F-PERT assay inhibited the RT-like activity produced by cellular DNA polymerases of non-induced MRC-5 *Mus dunni* Vero and cells, further supporting the suppression of false positive signals by activated calf thymus DNA (data not shown).

Discussion

Several laboratories have employed methods that suppress the RT-activity associated with cellular DNA polymerases (7,9). Chang *et al.*, 1997 have reported that authentic RT activity has a broader pH range than cellular polymerase RT-like activity. They report that lowering the pH of the RT reaction to pH 5.5 suppresses non-retroviral RT activity. However, they demonstrate no RT activity in CHO and BHK-21 cell lysates at the lower assay pH 5.5. Interestingly, the F-PERT assay was able to detect endogenous retroviral activity from BHK-21, C127I, *Mus dunni* and Sf-9 cell supernatants. Endogenous retroviral particles with reverse transcriptase activity have been detected in numerous species including invertebrates (10), Chinese hamsters (11), murine (12) and *Homo sapiens* (13). In order to characterise the nature of these viruses it is important to interpret results from PERT assays in combination with induced and non-induced test cells with detector cell co-cultivation assays. Thereby, allowing an assessment of potential infectivity of the viruses.

References

1. Maudru and Peden (1997). Elimination of background signals in a modified polymerase chain reaction - based reverse transcriptase assay. *Journal Virological Methods*, **66**, 247-261.
2. Silver J., Maudru T., Fujita K. and Repaske R. (1993). An RT PCR assay for the enzyme activity of reverse transcriptase capable of detecting single virions. *Nucleic Acid Research*, **21**, 3593-3594.
3. Pyra H., Böni J. and Schüpbach J. 1994. Ultrasensitive retrovirus detection by a reverse transcriptase assay based on product enhancement. *Proceedings National Academy Sciences, U.S.A.*, **91**, 1544-1548.
4. Heneine W., Yamamoto S., Switzer W. M., Spira T. J. and Folks T. M. 1995. Detection of reverse transcriptase by a highly sensitive assay in sera from persons infected with human immunodeficiency virus type 1. *Journal of Infectious Diseases*, **171**, 1210-1216.
5. Roberston J. S., Nicolson C., Riley A. M., Bentley M., Dunn G., Corcoran T., Schild G. C. and P. Minor. 1997. Assessing the significance of reverse transcriptase activity in chick cell-derived vaccines. *Biologicals*, **4**, 403-414.
6. Arnold, B. A., Hepler, R. W. and P. M. Keller. 1998. One step fluorescent probe product enhanced reverse transcriptase assay. *Biotechniques*, **25**, 98-106.
7. Lugert R., König H., Kurth R. and Tönjes R. R. 1996. Specific suppression of false positive signals in the product enhanced reverse transcriptase assay. *Biotechniques*, **20**, 210-217.
8. Boni, J., Pyra H. and J. Schupbach. 1996. Sensitive detection and quantification of particle-associated reverse transcriptase in plasma of HIV-1 infected individuals by the product enhanced reverse transcriptase (PERT) assay. *Journal of Medical Virology*, **49**, 23-28.
9. Chang, A., Ostrove J. M. and R. E. Bird. 1997. Development of an improved product enhanced reverse transcriptase assay. *Journal of Virological Methods*, **63**, 45-54.
10. Gabriel, A. and J. D. Boeke. 1993. Retrotransposons and reverse transcription. *In* Reverse transcriptase. Edited by Skalka, A. M. and S. P. Goff. Cold Spring Harbour Laboratory Press 1993. USA.
11. Anderson, K. P., Low, M. A., Lie Y. S., Keller, G. A. and M. Dinowitz. 1991. Endogenous origin of defective retrovirus-like particles from a recombinant chinese hamster ovary cell line. *Virology*, **181**, 305-311.
12. Wolgamot, G., Bonham L., and A. D. Miller. 1998. Sequence analysis of *Mus dunni* endogenous virus reveals a hybrid VL30/gibbon ape leukaemia virus-like structure and a distinct envelope. *Journal of Virology*, **72**, 7459-7466.
13. Berkhout, B., Jebbink, M. and J. Zsiros. 1999. Identification of an active reverse transcriptase enzyme encoded by a human endogenous HERV-K retrovirus. *Journal of Virology*, **73**, 2365-2375.

RAPID RNA QUANTITATION FROM XENOTROPIC SP2/0 RETROVIRUSES BY COMPETITIVE RT-PCR ELISA

W. HANSEN, L. SCHUMACHER, M.C. JAEGGI, H.P. KNOPF, M. WIRTH
Dept. of Regulation and Differentiation, GBF, Braunschweig
Novartis AG[1], Basle

1. Abstract

Xenotropic retroviruses are common contaminants of mouse hybridomas and myeloma cell lines. Virus load differs widely within individual cell lines commonly used for antibody production, making a fast and reliable method for virus quantitation desirable. We have developed a quantitative competitive RT-PCR ELISA which enables exact and specific determination of infectious as well as noninfectious Sp2/0 xenotropic retrovirus in cellular supernatants or in probes spiked with the respective retrovirus. The method is based on cDNA synthesis and subsequent DNA amplification with digoxigenin-labelled specific xenotropic retroviral primers in the presence of an internal, competitive RNA standard (IQS). cDNAs of probe and reference are identical in length and nucleic acid composition to achieve corresponding amplification efficiencies during PCR. However, they differ in a 27 b region which is permutated in the IQS standard for later distinction. This step is followed by hybridization and immobilization with capture probes for specific RNA and IQS and colorimetric detection of the digoxigenin labelled DNA.

2. Results

2.1. OPTIMIZATION

The method outlined in Fig.1 has been optimized with respect to different parameters. Different cycle numbers (26, 28, 30) were applied. 28 cycles gave the best results with respect to sensitivity and proportionality and were chosen for further experiments. Maximal amount of capture probe concentration was calculated and optimal concentration of capture probes were determined empirically therafter. DNA denaturation conditions were investigated (heat, NaOH). Alkali naturation was found to be superior with respect to handling (data not shown).

A. Bernard et al. (eds.), Animal Cell Technology: Products from Cells, Cells as Products, 517–519.

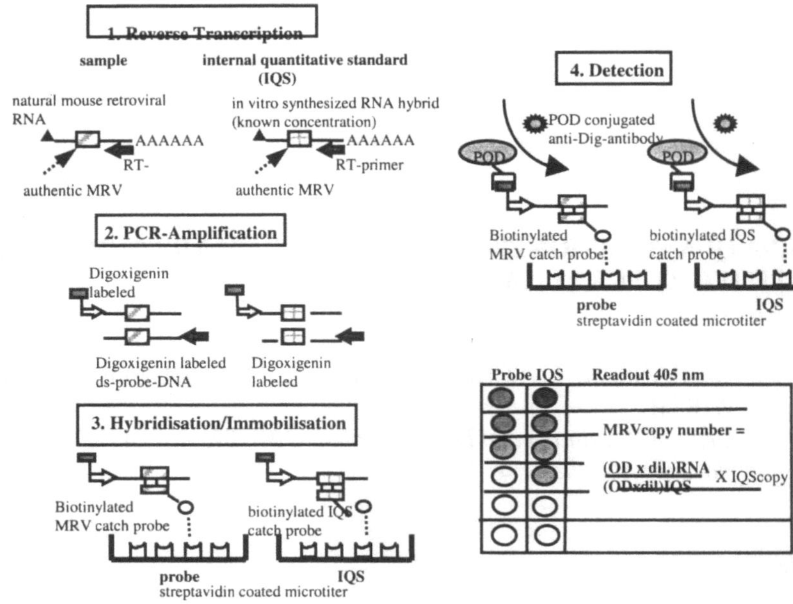

Fig. 1: RNA quantitation of xenotropic Sp2/0 retroviruses consists of a reverse transcription step (1), a PCR amplification (2) and a hybridisation-ELISA (3/4). RNA probe and IQS RNA were reverse transcribed in the same reaction tube using the same primer (1), which is specific for xenotropic Sp2/0 retrovirus. A 107bp region is amplified with specific primers binding in the capsid coding region, whereby one of the primers is digoxigenin-labelled for later detection. (2). Amplified, labelled DNA is hybridized to a specific, biotinylated primer and the whole complex is immobilized on a streptavidin-labelled microtiter plate (3). The final step is the immunological detection of digoxigenin-labelled DNA in a colorimetric assay (4).

2.2. INDEPENDENCE OF RNA AND IQS AMPLIFICATION

To assess the influence of probe and IQS RNA on each other, experiments were performed using dilution experiments with constant amount of IQS and altering amounts of RNA and vice versa (Fig.2). We could clearly demonstrate that probe RNA and IQS standard RNA amplify independently of each other within the concentrations investigated. This allows us to determine the RNA amounts with a confidence over a wide range of concentrations.

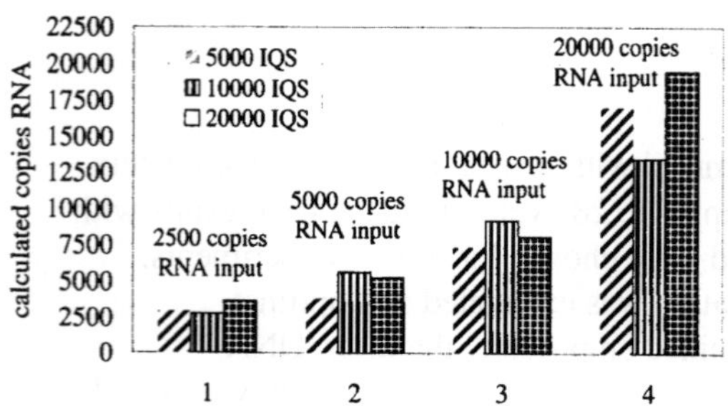

Fig. 2:Independenc of RNA and IQS amplification

To determine the effects of coamplification of xenotropic MLV RNA on the internal quantitation standard 2500 - 20000 copies of RNA probe were mixed with 5000, 10000 and 20000 copies of the IQS.

2.4. SPECIFITY

Signals may arise from unspecific binding to surfaces or specific binding to related viral RNAs and DNAs present in the supernatant or as endogenous provirus in the cell. The quality of the twofold purification procedure of the in vitro transcribed RNA (DNAse I treatment and polyA affinity purification) was ascertained by omitting a reverse transcription step in the assay (data not shown). As the method is based on a validated qualitative RT-PCR procedure for detection of C-type mouse retroviruses [3] cross-reactions with other retroviruses can be excluded. Primer mismatches in two out of the three primers used are prominent enough to allow discrimination between the xenotropic Sp2/0 retrovirus and related retroviruses. No signals were obtained examining unspecific binding of primer/probe or amplified DNA to each other or to surfaces (Fig.3).

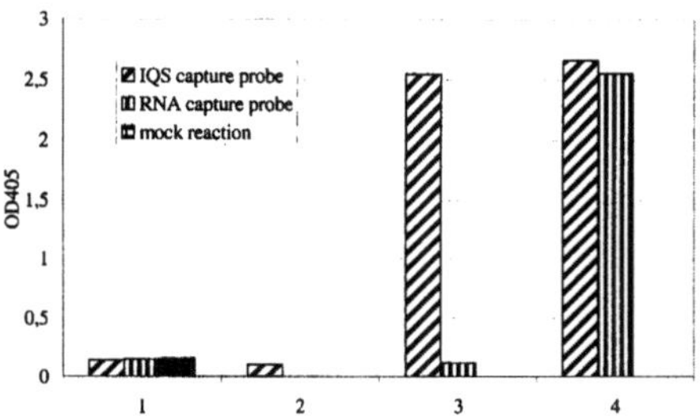

Fig.3: Specificity
To exclude unspecific bindings of amplicons to surfaces or to the capture probes, DIG - labelled unspecific cDNA derived from bacteriophage MS2 RNA was used as template for hybridisation with IQS capture probe, RNA capture probe and mock reaction (1). No template was added in (2). IQS RNA was included in the hybridisation reactions with IQS- or xenotropic MLV capture probe (3). As a positive control a mixture of IQS RNA and xenotropic MLV RNA was used as template (4).

3. Discussion

A competitive RT-PCR-ELISA for quantitation of RNA from xenotropic mouse retroviruses of the Sp2/0 cell line is described. Due to the wide distribution of xenotropic retroviruses in hybridomas and myelomas the competitive RT-PCR-ELISA is valuable for following the viral load during antibody production processes and establishing conditions to reduce or even suppress virus occurance. Furthermore, the RNA quantitation procedure is usefull for virus clearance studies, as infectious and noninfectious can be quantitated. This makes the competititive RT-PCR-ELISA the method of choice in such validation studies. The assay is specific for xenotropic mouse retroviruses, reliable, avoid radioactivity and can be completed within one day. Future aspects aim to the reduction of total assay time and may make use of ultra-fast real time quantitation methods (Taqman, Lighcycler) reducing RNA quantitation to 30 minutes and increasing the assays feasability.

4. References

1. Deo, Y., Ghebremariam, H., Cloyd, M. (1994) Detection and characterization of murine ecotropic recombinant virus in myeloma and hybridoma cells, *Hybridoma* **13**(1), 69-76
2. Froud, S.J., Birch, J., McLean, C., Shepherd, A.J., Smith, K.T. (1997) Viral contaminations found in mouse cell lines used in the production of biological products, in M.J.T. Carrond, B. Griffiths and J.L.P. Moreira (eds.), *Animal Cell Technology*, Kluwer Academic Publishers, Netherlands, pp. 681-686.
3. Heinemeyer, T., Klingenhoff, A., Hansen, W., Schumacher, L., Hauser, H., Wirth, M. (1997) A sensitive method for the detection of murine C-type retroviruses, *J. Virol. Meth.* **63**, 155-165.

HUMAN AMNIOTIC EPITHELIAL (HAE) CELLS EXPRESS ERYTHROPOIETIN (EPO) AND EPO-RECEPTOR

KEIKO MATSUURA and AKINORI HOSHIKA
Department of Pediatrics, Tokyo Medical University
6-1-1, Shinjuku, Shinjuku-ku, Tokyo, 160-8402, Japan
SATOSHI TERADA
Department of Applied Chemistry and Biotechnology, Fukui University
3-9-1, Bunkyo, Fukui, 910-8507, Japan
NORIO SAKURAGAWA
Department of Inherited Metabolic Disorders, NCNP
4-1-1, Ogawahigashi, Kodaira, Tokyo 187-8502, Japan

1. Introduction

Amnion is the membrane around fetus and the amniotic fluid is thought to play an important role in the process of parturition. Amniotic fluid contains erythropoietin (Epo) (1). Epo enhances erythropoiesis and it is synthesized by the kidney and by liver. It is reported that there is correlation between the level of erythroblastosis and the amount of Epo in amniotic fluid (2). As the infant and adult, erythropoiesis in the fetus is regulated by Epo (3). Recently, other functions of Epo have been reported. Several groups observed that the brain has a paracrine Epo/Epo-R system (4). It is also reported that Epo stimulates proliferation of human endothelial cells (5) and angiogenesis of rat thoracic aorta (6). The origin of Epo in amniotic fluid has not been identified yet. Finne *et al.* reported that Epo did not cross the placenta into the fetus (7), and the results suggests that erythropoiesis in the fetus would be regulated by Epo produced from sites within the fetus. Recently Epo mRNA was detected in the human placenta by RT-PCR (8) and their immunostaining supports expression of Epo in human placenta. But it has not been investigated whether placenta is origin of Epo in amniotic fluid. Additionally the role of EPO in amniotic fluid is still unknown. Therefore we used human amniotic epithelial cells in this study to investigate these questions.

2. Materials and Methods

A human placenta was obtained from an uncomplicated elective Caesarean section. The human amniotic epithelial (HAE) tissue was mechanically peeled free from the chorion of the placenta. HAE-SV40 cells, HAE cells transfected with SV40 large T antigen, were also examined. HepG2 cells and KMT-2 cells were served as a positive control for EPO expression and for Epo-R expression, respectively.

A. Bernard et al. (eds.), Animal Cell Technology: Products from Cells, Cells as Products, 521–523.

3. Results and Discussion

3.1. DETECTION OF EPO AND EPO-R mRNA

In order to understand better the role of amnion, we investigated the expression of Epo and Epo-R genes by using RT-PCR. The PCR products were electrophoresed and shown in *Fig.1*. Because the band derived from contaminating genomic DNA was not be observed in RT (-), genomic DNA did not contaminated. In all RT (+) samples, β-actin band was observed and the density of each band is apparently same (*Fig. 1-a*). Thus, the amount of mRNA in each preparation should be almost same. In *Fig.1-b*, the band in HepG2 lane is denser than in HAE lane and this result suggests that HepG2 cells produce much more Epo mRNA than HAE cells. Epo-R also detected in HAE cells (*Fig. 1-c*). Though both bands in HAE cells and in KMT-2 cells as positive control are not so clear in this figure, the Epo-R bands are observed. The electroeluted Epo and Epo-R RT-PCR fragments were purified and cloned and the sequencing analysis revealed that the amplified product is derived from Epo mRNA.

Fig. 1 Detection of Epo and Epo-R mRNA in human amnion epithelial (HAE) cells
β-actin (a), Epo (b), and Epo-R (c) mRNA were analyzed by RT-PCR.
1: HAE cells RT(+), 2: HAE cells RT(-), 3: HepG2 RT(+), 4: KMT-2 RT(+).

3.2. WESTERN BLOTTING ANALYSIS

It is well known that Epo production is stimulated by hypoxia. Western blot studies were performed testing for presence of Epo. HAE cells transfected with SV40 large T antigen (HAE-SV40) and HepG2 cells as positive control were cultured for 5 days. As shown in *Fig. 2-a*, hypoxia increased Epo production of HepG2. In HAE-SV40 lanes, Epo bands apparently observed and the band in hypoxia lane is denser than in ambient oxygen lane. But considering the background, hypoxia did not enhance Epo production of HAE-SV40 cells. Finne *et al.* reported that hypoxia increased Epo concentration in amniotic fluid (9). Amnion would only contribute base level of Epo and stimulated Epo by hypoxia is not derived from amnion. The similar result about uterine endometrium was reported (10).

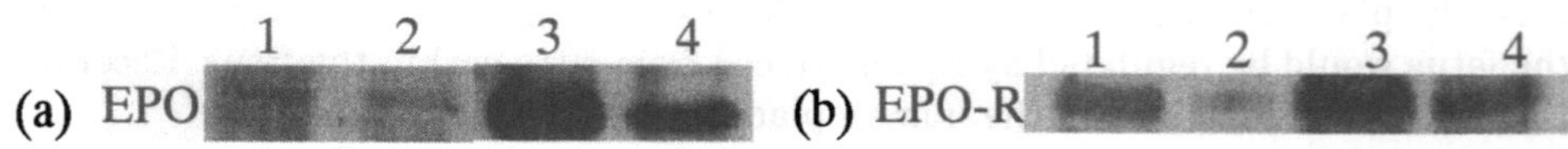

Fig. 2 Detection of Epo and Epo-R by western blot analysis.
(a) Detection of Epo. Culture supernatant of HAE cells in hypoxia (lane 1), in ambient oxygen (lane 2), and HepG2 cells in hypoxia (lane 3), ambient oxygen (lane 4) were loaded.
(b) Detection of Epo-R. Lysate of HAE (lane 1, 2) and HepG2 (lane 3, 4) were loaded. 12 μl of lysate was loaded in lane 1 and 3, and 4 μl of lysates were loaded in lane 2 and 4.

Epo-R was also detected in the lysate of HAE cells (*Fig. 2-b*). The density of the band in HAE cell lane is fainter than in KMT-2 lane. Considering that both Epo and Epo-R expression were observed, Epo paracrine system might exist in HAE cells.

3.3 DETECTION OF EPO ACTIVITY IN HAE CELL CULTURE SUPERNATANT

Culture supernatant HAE-SV40 cells was condensed by centrifugal filter device. F36E cells, dependent on Epo, were incubated in medium supplemented with the condensed supernatant. As shown in *Fig.3*, treatment of purified Epo increased the proliferation of F36E and addition of anti-Epo antibody decreased this effect. As positive control, culture supernatant of HepG2 cells increased the proliferation of F36E cells. Treatment of the condensed culture supernatant of HAE increased the proliferation and this increase was reduced by addition of the neutralizing antibody. This result suggests that HAE culture supernatant contains Epo activity.

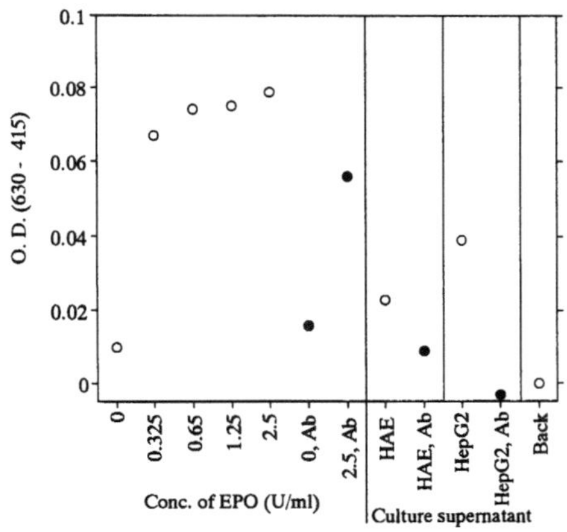

Fig.3 Detection of Epo activity in supernatant
HAE and HepG2 culture supernatant was condensed. F36E cells, dependent on Epo, were incubated with the condensed culture supernatant, rh-Epo and anti-Epo neutralizing antibody (Ab). Proliferation of F36E was measured with TetraColor ONE (Seikagaku, Japan).

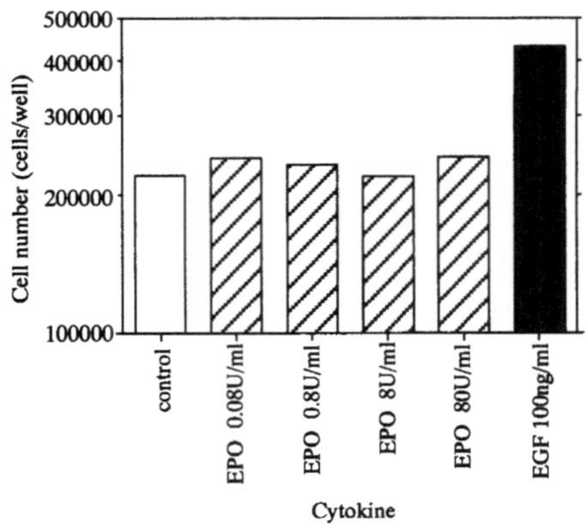

Fig. 4 Effect of Epo on proliferation of HAE cell
1.1 × 10^5 of HAE cells were harvested in 24 well plate with 600 µl of RPMI medium supplemented with 10 % FBS and EPO or EGF as positive control. After 5 days culture, cell number was determined by counting in a hemacytometer using trypan blue exclusion.

3.4. EPO EFFECT ON HAE CELL PROLIFERATION

We studied Epo effect on HAE cell. HAE cells treated with Epo were cultured for 5 days. As shown in *Fig.4*, treatment of 80 U/ml concentration of Epo did not effect on proliferation of HAE cells. Therefore Epo is neither autocrine nor paracrine growth factor. If Epo in amniotic fluid has functions during pregnancy, the functions would be problem. Because fetus drinks amniotic fluid and urinates, amniotic fluid is not distinguished from urine. Thus, Epo in amniotic fluid circulated in fetus. It could be imagined that human amnion is involved in some important fetus function, for example erythropoiesis or anything, by supplying Epo.

4. References

1. Buscher U.et al. (1996) *Geburtshilfe Frauenheilkd* **56** 243
2. Halvorsen S. et al. (1963) *Br. Med. J.* **27**, 1132
3. Finne P, H et al.. (1972) R *Arch Dis Childh.* **47**, 683
4. Masuda, S. et al. (1993) *J. Bio.Chem.* **268** 11208
5. Anagnostou A. et al. (1990) *Proc Natl Acad Sci.* **87**, 5978
6. Carlini, R. G. et al. (1995) *Kidney Int* . **47** 740
7. Finne, P.H. (1966). *Acta Paediat Scand.* **55**, 609
8. Conrad, K.P. et al. (1996) *FASEB J.* **10** 760
9. Finne, P. H. (1966). *Acta Paediat Scand.* **55** 478
10.Yasuda, Y. et al. (1998) *J. Biol. Chem* .**273** 2538

THE NORDENAU-PHENOMENON – FACTS AND HYPOTHESES

Results from a prospective follow-up study of 515 patients

GADEK, Z., M. D.,
Seifert, D. J., Dipl.-Psych.,
Zentrum für Ganzheitsmedizin und Naturheilverfahren
57392 Schmallenberg-Nordenau, Talweg 14, Germany
Phone +49 2975 / 9622-190
Fax +49 2975 / 9622-200

The reaction shown by the 515 test-subjects (318 female, 197 male) to the so-called "Nordenau-Phenomenon" (meaning an old slate tunnel with an underground source of spring-water, which are said to have a healing effect) was verified. Chronically ill and incurable people have been pilgrimaging there for 6 years (a few hundred daily). The average age of the test-subjects was 66, the length of stay was 6 days (taking 2 tunnel-walks at 30 min. each and consuming 2 litres of water daily). The spring water was seen as a possible active substance, as it possesses the quality of "reduced water" (triple analysis in the Laboratory of Cellular Regulation Technology Kyushu University, Japan). In order to record a picture of the symptoms, a list of complaints based on a complaint questionnaire (Höck. K. and Hess, H., Berlin 1975) was modified and presented in form of a rating-scale in 10 steps. The test-subjects had a choice of 47 different statements to describe the current experienced extent of their respective symptom at the beginning of their stay (Measure Point MP1), at the end of their stay MP2 and after three months MP3, divided into steps of intensity. The most important point concerning the evaluation of this survey is the following: at no time were the examinees asked if they had felt any improvement or deterioration in their condition. Furthermore, it was technically impossible for the testees to draw comparisons between the three test results as they had no access to previous pathological findings. Consequently the examinees had no point of orientation to set the placebo mechanism in motion.

In addition to the description of the participant-data or basic information, the details of the complaints questionnaire were subject to a statistical analysis. Firstly, the data was checked against a norm-distribution in order for us to decide on a statistical choice of procedure. (KOLMOGOROFF-SMIRNOFF-test, see SACHS 1997). Since this could not be confirmed for all variables, the examination data was also analysed using the parameter-free method of the WILCOXON-test next to the t-test in order to get correlating samples.For this, the computer-backed programme-pack BiAS (Ackermann 1997) was used.

Results with $p < 0.05$ were judged to be an indication for the presence of a secured effect. Correspondingly, $p < 0.01$ were valid as very significant and $p < 0.001$ as highly significant. The statistic evaluation resulted in highly significant improvement with regard to all 15 diagnostic collective terms (p-value ≤ 0.01 to 0.001). From the outset of our investigations and our proposed line of study, we were always open to

A. Bernard et al. (eds.), Animal Cell Technology: Products from Cells, Cells as Products, 525–527.

and willing to accept the theory that purely psychological reactions or mechanisms could be the root cause of the Nordenau Phenomenon. In order to include the psychopathological dimension which could influence the credibility, manoeuvrability or self-assessment of the testee and thus create a certain personal prejudice, the MMPI-procedure was applied in each individual case. The results taken from all the patients can be defined as follows: It is impossible to proof of any of the personality structures in the testees likely to result in above average suggestibility, manoeuvrability or other shortcomings within the field of self-assessment and judgement of situations in conjunction with aberrations of thought-patterns or bizarre ideas. Observing the individual figures, we found that among the 515 testees, there were 108 people with above average characteristics as potential causes of psychosomatic complaints. These distinctive characteristics only serve to illustrate to which extent the possibility or grade of mental strain might have an effect upon the physical phenomena. And in addition, of course, the somato-psychical influences in the form of a reaction to the suffering of an illness must likewise be taken into account. There were definitely no significant signs which could have triggered off either hallucinations or delusions. To summarise, it may be said that the testees, on average, represented a random population sample and thus it follows that the influences and changes which took effect in the tunnel cannot, according to the data, be described as personality phenomena. Nevertheless, we shall be comparing those testees with unobtrusive personality structures with testees with increased neurotic personality traits (the so-called "neurotic trias") as soon as we have a sufficient number of patients in both groups at our disposal. In this survey we have already made comparisons between groups of different numbers in order to be at least able to pass a rough judgement on the tendencies. At present it appears that the patients with neurotic personality traits reveal either the same or a smaller change in their condition. An analysis of the survey figures shows a highly significant improvement with regard to symptoms as with regard to diagnostic generic terms. There was slight regression after the patient's return home, but seen within the time-span of three months, the final result must be judged to be a stabilisation as against the original complaints. The changes which have been assessed according to their subjective appearance reflect certain changes in conditions whose somatic correlations we will not be able to discuss before the following survey stage. A significant improvement in the patient's condition with reference to all 15 diagnostic generic terms points to a universal factor being responsible for an activation of self-repair mechanisms of the body.

These observations correlate with the results of the research on “reduced water” by Shirahata, S. (Biochemical research communications, Vol.234, No. I, May 8, 1997, New developments – New applications by O. Merten et al. Kluwer academic publishers, 1998, in press). Shirahata, S., Fernandes, G. et al. Harrison M., Yoshikawa, T. (JAACT/ESACT '98 Meeting, July 26-30, 1998, Kyoto, Japan). Further research on a different level is necessary to find out if antioxidative operating mechanisms of the spring water are at the only thing at the root of the Nordenau phenomenon or if, in addition, a form of energy yet unknown exists, which converts this water into microwater.

References

Ackermann, H. (1997). BiAS. Biometrische Analyse von Stichproben, Epsilon Verlag, Darmstadt.

Hennig, B., Mehl, J. (1974). Untersuchungen zur Tauglichkeit eines Kurzverfahrens zur Psychodiagnostik von Neurosen, Deutscher Verlag der Wissenschaften, Berlin.

Höck, K., Hess, H. (1975). Beschwerdenfragebogen (BFB), Deutscher Verlag der Wissenschaften, Berlin.

Kincannon, J.C. (1968). The prediction of the scale scores of the standard MMPI with 71 items, J. clin. Psychol. 32, 3.

Huber, H. (1982). Deutsche Kurzform für Handauswertung, Bern, Göttingen, Toronto, Seattle.

Sachs, L. (1997). Angewandte Statistik, Springer-Verlag,. Berlin, Heidelberg, New York.

Shirahata, S. et al. Electrolyzed reduced water scavenges active oxygen species and protects DNA from oxidative damage. Biochemical research communications, Volume 234, No. 1, 1997.

Shirahata, S. et al. (1998). Electrolyzed reduced water which can scavenge active oxygen species supresses cell growth and regulates gene expression of animal cells. Animal cell technology: New developments – New applications“ ed. by O. Merten et al. Kluwer academic publishers, 1998, in press.

Shirahata, S., Fernandes, G., et al., (1988) JAACT/ESACT ´98 Meeting, July 26-30, Kyoto, Japan. Harrison, M., Yoshikawa, T. (1998) JAACT/ESACT ´98 Meeting, July 26-30, Kyoto, Japan.

CONVERSION OF A PROCESS DEVELOPMENT PILOT PLANT INTO A CLINICAL PHASE I/II MANUFACTURING FACILITY FOR RECOMBINANT PROTEINS EXPRESSED IN MAMMALIAN CELLS

Tom Klamer, Jørn Meidahl Petersen and Randi Skovgaard
Novo Nordisk Engineering and Novo Nordisk Biologics Development, Cell Culture
Novo Allé, DK-2800 Bagsvaerd, Denmark

1. Introduction

A process development pilot plant designed and established more than ten years ago according to fermentation standards rather than biopharmaceutical standards has been rebuilt. The purpose of the rebuilding project was to establish a clinical phase I/II manufacturing facility for the new biological entities discovered through functional genomics. The engineering project has been carried out under a number of constraints:

- Budget
- Available area for expanding the building
- Shape of the building
- The major equipment already existed.

Therefor solutions that exactly fit the purpose have been sought throughout the project.

2. User requirements

A user requirement statement was developed based on an analysis of the current standard in the industry and the original design of the plant. The analysis focused on flows of people and materials, classification of rooms, clean-ability of process equipment and quality requirements for process utilities.

Active Pharmaceutical Ingredients derived form mammalian cells for clinical use must be produced under conditions where:

- Risk of contamination from the surroundings is minimised
- Risk of contamination form people is minimised
- Means are taken to separate the process from research and development activities
- Medium preparation is separated from cell cultures
- Cell propagation is separated from cell samples from ongoing batches.

The flow patterns of people and materials should if possible be unidirectional and room classifications as stated in table 1.

A. Bernard et al. (eds.), Animal Cell Technology: Products from Cells, Cells as Products, 529–531.

Room	Classification	Air changes per hour	Temperature	Relative humidity	Pressure
Wash	100.000/D	20	N.A.	N.A.	+++
Autoclave	100.000/D	20	N.A.	N.A.	+++
Medium prep. At rest	10.000/C	20	15-21°C	30-70%	++++
Medium prep. At work	100.000/D				
Seed culture lab.	10.000	20	15-21°C	30-70%	++++
Fermentation	Controlled	20	N.A.	N.A.	+
Inventories	Controlled	10	N.A.	N.A.	++
Air-lock	Controlled	10	N.A.	N.A.	+

Table 1: Room classifications.

Water for final rinse of equipment and medium preparation should comply with WFI specifications and be non-toxic to cells. Process gasses should comply to the class 100 specification with respect to viable and non-viable particle counts and be dry ($H_2O(g) \leq 800$ ppm) and oil free (hydrocarbon ≤ 0.1 mg/Nm^3). Product contact process equipment should be designed for Clean In Place and Steam In Place procedures.

3. Engineering solutions

Based on the user requirement statement engineering solutions were developed. A building layout that suits the requirements to flows of people and materials has been developed and HVAC systems that supply the rooms with the specified air quality have been designed. Fig. 1 shows the lay-out of the core of the building. Because of the shape and area constraints it was impossible to fulfil the requirements regarding unidirectional flow. The design, however, secure sufficient separation of the activities. Table 2 shows the choices of room finish for the various rooms.

All product contact stainless steel surfaces will be mechanically polished to Ra $\leq$ 0.6 μm, and pipe work must be sloped for drainage and free of dead legs. A water system that is fed with deionised water of boiler feed quality and includes the unit operations Continuos Deonisation and double Ultrafiltration on ceramic filters. The circulation loop will be kept cold during normal operation and sanitised by heating to > 80°C daily for two hours. The CIP system that has been chosen is a portable re-circulating system with a process vessel. The tanks are equipped with spray balls and connected to the CIP unit with flexible tubes.

3. Validation plan

The main contractor was responsible for the engineering quality assurance activities. The IQ/OQ will be conducted as a documented review of the quality documentation handed over by the main contractor. The PQ plan focuses on providing documentation for the quality of room environment and water, steam and gases.

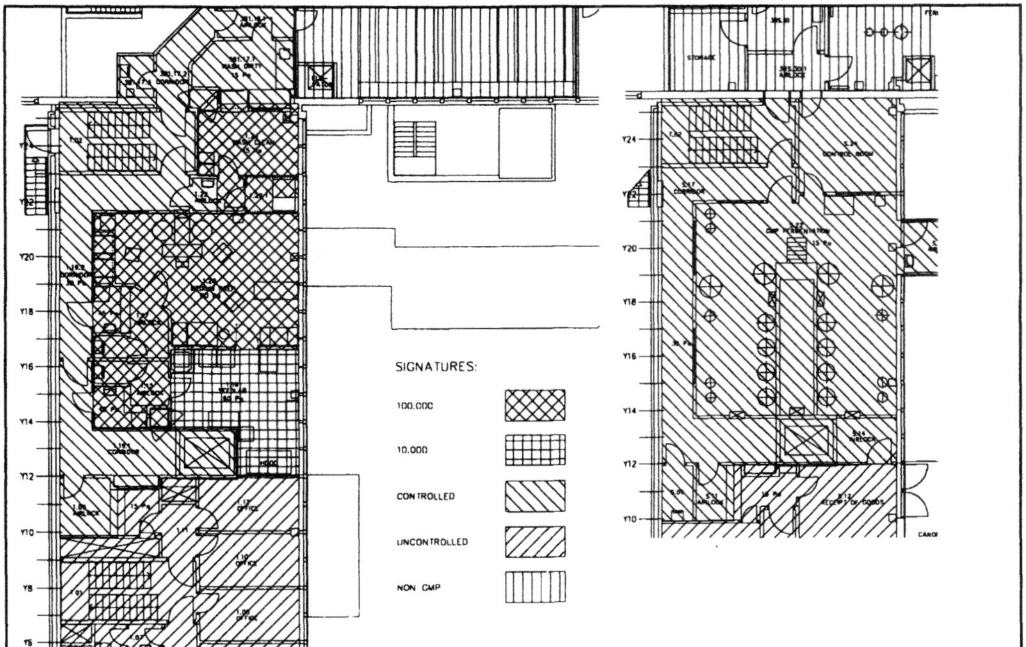

Figure 1A, 1st floor: Tanks in medium preparation are 500 Litre tanks

Figure 1B, ground floor: Fermenters are 50 and 500 Litre. Medium and harvest tanks are 100 - 500 Litre tanks.

Room	Floor	Walls	Ceiling
Controlled corridors	Epoxy	Acrylic paint	Suspended, painted
Inventories		Acrylic painted	Suspended acrylic paint
Dirty wash	Vinyl	Acrylic paint	Suspended acrylic paint
Clean wash	Vinyl	Vinyl	Vinyl
Medium preparation	Epoxy	Vinyl	Vinyl
Seed laboratory	Vinyl	Vinyl	Vinyl
Fermentation	Epoxy	Acrylic painted	Concrete acrylic painted

Table 2: Room finish.

An intensified sampling plan will be followed for three weeks in the classified rooms and four weeks for the critical utilities. After that a less intensive programme will be followed for 12 months. For the water system capacity and sanitation efficiency will be tested during PQ. The CIP and SIP procedures will be tested in triplicate for all tank types. TOC, pH, LAL and conductivity will be measured on rinse and swab samples and temperature distribution recorded during SIP. Design reviews were conducted at the end of each of the major project phases described and an inspection by the local medicines agency on behalf of all the EU countries is expected before manufacturing of the first clinical batch in the facility.

Chapter X. ANIMAL OR PATIENT DERIVED CELLS AND TISSUES FOR THERAPEUTIC APPLICATIONS

HUMAN KERATINOCYTE CULTURE ON MACROPOROUS CARRIERS AT THE AIR-LIQUID INTERFACE

E.K. LINNAU*, A.M. BURT+, H. KATINGER*, A. HANDA-CORRIGAN#
*Institute of Applied Microbiology, University of Agricultural Sciences, Vienna, Austria, +Ludwig Institute for Cancer Research, London, UK, #University of Surrey, Guildford, UK
Muthgasse 18, A – 1190 Vienna, Austria

1. Introduction

High calcium medium containing foetal calf serum or low calcium, serum-free medium have both been used successfully for culture of keratinocytes on flat surfaces submerged in tissue culture medium or alternatively, for culture of cells exposed to the air-liquid interface [1, 2, 3]. In this paper we present morphological studies of high density cell cultures of human keratinocytes on macroporous discs called Porocell™ (Porvair Sciences, UK) that float at the air-liquid interface of stirred, surface-aerated cultures [4, 5].

2. Materials and Methods

2.1. INOCULATION AND CULTURE OF KERATINOCYTES ON POROCELL™

Primary cultures of keratinocytes were prepared by enzymatic digestion of human breast skin. The cells were first cultured in a serum-supplemented medium, weaned in to serum-free keratinocyte medium (Sigma, UK) after 48 hours and subcultured after 5-10 days. Each Porocell™ disc was seeded with $5x10^4$ cells in 50 µl of either serum-supplemented or serum-free medium. To enable cell attachment to Porocell™ batches of six discs were incubated in 6-well-plates at 37°C overnight in a humidified, 5% CO_2/air-atmosphere prior to transfer to stirred flasks containing 50 ml of medium and incubation for several weeks.

2.2. SCANNING ELECTRON MICROSCOPY OF CULTURED KERATINOCYTES

Porocell™ discs were fixed in 2% buffered glutaraldehyde solution, serially dehydrated in ethanol and finally in acetone prior to critical point drying in CO_2. The discs were then cut manually with a stainless steel blade, mounted on metal stubs and coated with gold. The samples were viewed with a Hitachi S-3200N electron microscope.

2.3. GLUCOSE AND LACTATE-ANALYSES

Cell growth was monitored by measurement of glucose uptake and lactate production using an Analox Micro-Stat analyser.

A. Bernard et al. (eds.), Animal Cell Technology: Products from Cells, Cells as Products, 535–537.

3. Results and Discussion

3.1. MORPHOLOGY OF KERATINOCYTE CULTURES ON POROCELL™

Keratinocytes that are cultured submerged in medium proliferate as monolayers of widely spaced cells [2, 6]. In contrast, human keratinocytes cultured in a low-calcium, serum-free medium on Porocell™ showed a striking degree of multilayered growth and stratification throughout the discs. In the first week of culture, large clumps of mainly rounded cells were seen uniformly attached to the porous matrix. After 19 days in culture flattened, hexagonally shaped cells as well as clumps of rounded cells were observed densely populating the carriers (data not shown). At 43 days, extended multi-layered sheets of cells had grown inside and over the porous matrix. The sheets were covered with small clumps of rounded cells and cellular debris that appeared to be sloughing off the underlying layers (Fig. 1.). In the high calcium, serum-supplemented medium the Porocell™ surface was covered with large flattened cells and clumps of rounded cells only one week post inoculation. After three weeks, multiple layers of cells stretched across the pores such that some 75% of the porous matrix were covered by a skin-like tissue. At day 43, the carriers appeared to be covered with sheets of dead, fibrous tissue that was also observed inside the pores of the discs (data not shown). The scanning electron micrographs obtained in this study show that an epidermal, tissue-like morphology was achieved by culturing human keratinocytes at the air-liquid interface of serum-free medium on the macroporous carrier, Porocell™. The differentiation process leading to the formation of sheets of tissue was faster in the serum-supplemented medium but in long-term cultures, dermal fibroblasts appear to have overgrown the epidermal keratinocytes.

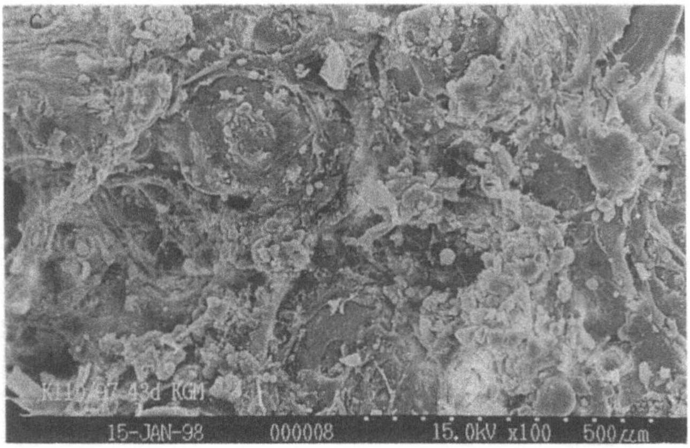

Figure 1. Human keratinocytes cultured in serum-free medium on Porocell™ 43 days post inoculation.

3.2. GLUCOSE UPTAKE AND LACTATE PRODUCTION RATES OF KERATINOCYTES CULTURED IN SERUM-FREE AND SERUM-SUPPLEMENTED MEDIA

We compared the glucose utilization and lactate production rates of human keratinocytes cultured on Porocell™. Our data show that these parameters are not affected by medium composition and properties. The glucose uptake rate of Porocell™-cultures inoculated with $5x10^4$ cells per disc was 0.125 ± 0.009 mmol/L/day in serum-supplemented medium compared with 0.125 ± 0.017 mmol/L/day in serum-free medium. The lactate production rate was 0.182 ± 0.011 mmol/L/day in serum-supplemented and 0.210 ± 0.034 mmol/L/day in serum-free medium (Fig. 2.).

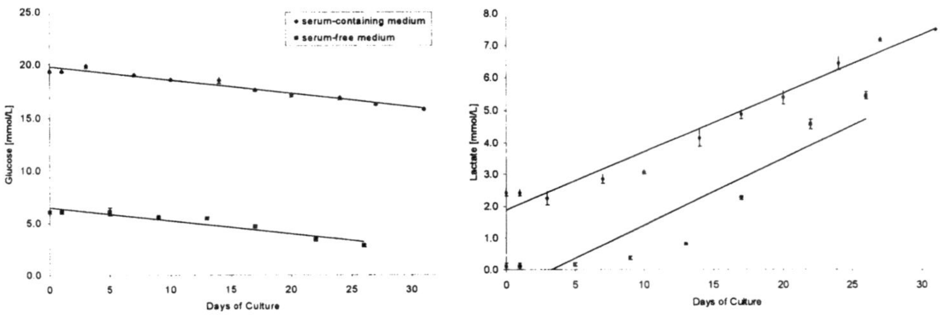

Figure 2. Glucose uptake and lactate production rates of human keratinocytes cultured on Porocell™.

4. Conclusions

Human keratinocytes are able to form tissue-like layers of cells on the macroporous carrier Porocell™. The Porocell™-keratinocyte discs can be cultured at the air-liquid interface of stirred cultures in either serum-supplemented or serum-free media. Compared with existing *in vitro* skin models, where keratinocytes are cultured on flat surfaces [7, 8], Porocell™ discs offer a high surface area per unit volume and allow high density culture of keratinocytes. A combination of microscopy techniques, biochemical analyses and culture optimisation strategies may be used to study the long-term growth and differentiation of keratinocytes on Porocell™. In the future, we aim to further assess the differentiated state of Porocell™-keratinocyte cultures and to develop and standardise assay methods that may be used specifically for Porocell™-keratinocyte applications in toxicological and pharmacological testing.

5. References

1. Daniels, J.T., Kearney, J.N. and Ingham, E. (1996) Human keratinocyte isolation and cell culture: a survey of current practices in the UK. *Burns* **22**(1), 35-39.
2. Navsaria, H.A., Myers, S.R., Leigh, I.M. and McKay, I.A. (1995) Culturing skin in vitro for wound therapy. *Tibtech* **13**, 91-100.
3. Pruniéras, M., Régnier, M. and Woodley, D. (1983) Methods for cultivation of keratinocytes with an air-liquid interface. *J Invest Dermatol* **81**, 28-33.
4. Handa-Corrigan, A., Traynor, R.M., Adamopoulos, E. and Salway, J. (1998) High density culture of the human hepatoma cell line HepG2: long-term culture for in vitro toxicology, in O.W. Merten, P. Perrin and B. Griffiths (eds), *New Developments and New Applications in Animal Cell Technology*, Kluwer Academic Publishers, Dordrecht, pp. 721-724.
5. Handa-Corrigan, A., Hayavi, S., Ghebeh, H., Mussa, N.A. and Chadd, M. (1998) Novel porous matrix and bioreactors for high density cultures of insulinoma cell lines: insulin secretion and response to glucose. *J Chem Technol Biotechnol* **71**, 52-56.
6. Castro-Muñozledo, F., Hernandez-Quintero, M., Marsch-Moreno, M. and Kuri-Harcuch, W. (1997) Cultivation, serial transfer, and differentiation of epidermal keratinocytes in serum-free medium. *Biochem Biophys Res Commun* **236**(1), 167-72.
7. Bell, E., Parenteau, N., Gay, R., Nolte, C., Kemp, P., Bilbo, P., Ekstein, B. and Johnson, E. (1991) The living skin equivalent: its manufacture, its organotypic properties and its responses to irritants. *Toxicology in Vitro* **5**, 591-596.
8. Rosdy, M. and Clauss, L.-C. (1990) Terminal epidermal differentiation of human keratinocytes grown in chemically definded medium on inert filter substrates at the air-liquid interface. *J Invest Dermatol* **95**, 409-414.

COMPARISON OF VIABILITY AND FUNCTION OF PRIMARY CELLS WITH TRANSFORMED CELLS FOR USE IN A BIOARTIFICIAL LIVER SUPPORT SYSTEM

L. WANG, L. LI, J. SUN, D. MEARS, M. HORVAT, A.G.R. SHEIL[+], C. HARBOUR[*]
[+] *Australian National Liver Transplantation Unit*
Royal Prince Alfred Hospital, Department of Surgery
[*] *Department of Infectious Diseases*
University of Sydney, NSW 2006, Australia

1. Abstract

Bioartificial liver support systems (BALSS) have been developed which incorporate either primary cells, e.g. porcine hepatocytes (PPH) or transformed cell lines, e.g. the human hepatoma cell line (C3A). The use of each cell type for a BALSS has both advantages and disadvantages and the aim of this study was to compare PPH and C3A survival and function in a variety of media that could be encountered in the preparation and operation of the BALSS. It was found that for short-term culture, maintenance of PPH viability was essential for retention of P450IA1 activity and effective ammonia clearance. When cell viability was below 60% P450IA1 activity could not be recorded and nitrogen elimination activity significantly diminished. In contrast to PPH, ammonia levels were markedly increased for C3A cells in all culture media tested and ammonia increase correlated with C3A viability. PPH metabolic function was superior to that of the C3A cell line when evaluated by ammonia removal and amino acid metabolism.

2. Introduction

Unlike kidney function, the complex functions of the liver cannot, as yet, be replaced by purely artificial systems. Thus several BALSS have been developed to perform as temporary replacements for liver function during liver failure. These BALSS utilise either primary liver cells, usually porcine hepatocytes (PPH), or transformed cell lines such as the human hepatoma cell line, C3A. Each has advantages and disadvantages and the aim of this study was to compare the survival and function of C3A cells with PPH in standard culture medium and in human plasma (HP) and human serum (HS) which the cells would encounter during clinical use of the BALSS.

3. Materials and Methods

A. Bernard et al. (eds.), Animal Cell Technology: Products from Cells, Cells as Products, 539–541.

3.1 PREPARATION OF LIVER CELLS

Pig liver cells were isolated from white landrace pigs weighing 6-8 Kg with a 3-step collagenase digestion technique.

3.1.1 *Cell Culture and Experimental Groups*

PPH were cultured in 96 and 24 well plates in Dubecco's Modified Eagle's medium (DMEM) supplemented with 10% foetal bovine serum, 100 μg/ml penicillin, 100 μg/ml streptomycin, 650 μ units/ml insulin and 10^{-8} M dexamethasone for 15 hr, followed by 24 hr culture in DMEM, human plasma (HP) or human serum (HS). The C3A cell line was purchased from the ATCC and in experiments cultured as described for PPH.

3.1.2 *Assessment of Cell Viability, Apoptosis and Functions*

Cell viability was assessed by 1 hr quantitative assay. Cell apoptosis was evaluated by cellular DNA fragmentation ELISA (Boehringer Mannheim GmbH). Cell functions were assessed by P4501A1 microassay. Hepatocytes in DMEM containing 2μM dicoumarol (Sigma) and 8μM 7-ethoxyresorufin (Sigma) were incubated in 24-well plates at 37°C for 10 mins. Fluorescence of supernatant was measured at 530nm excitation and 590nm emission. Resorufin was used to prepare a standard curve.

Ammonia levels were measured using a computerised autoanalyser system.

4. Results and Discussion

Cell viability of PPH was significantly lower than that of C3A whether cultured in DMEM, HP or HS ($P<0.01$) as shown in Fig. 1a. There were more apoptotic C3A cells in DMEM than in other media ($P<0.01$). P4501A1 activity of PPH or C3A in HP was lower than in DMEM and HS ($P<0.01$) as shown in Figure 1b. Ammonia levels were markedly higher in C3A cultures particularly in DMEM and HS as shown in Figure 1c. This study has shown that neither C3A cells or PPH function well in HP but C3A cells did show superior viability and P4301A1 function in all media. However apoptosis was seen less frequently in PPH cultures and, importantly, only PPH in DMEM or HS achieved decreases in ammonia levels which may be a key requirement for effective functioning of BAL systems.

5. Acknowledgements

This work was supported by a grant from the NHMRC of Australia and C. Wang, K. Woodman, B. Johnston, N. Koutalistras, J. Watson, L. Wu, and D. Verran also participated in the work.

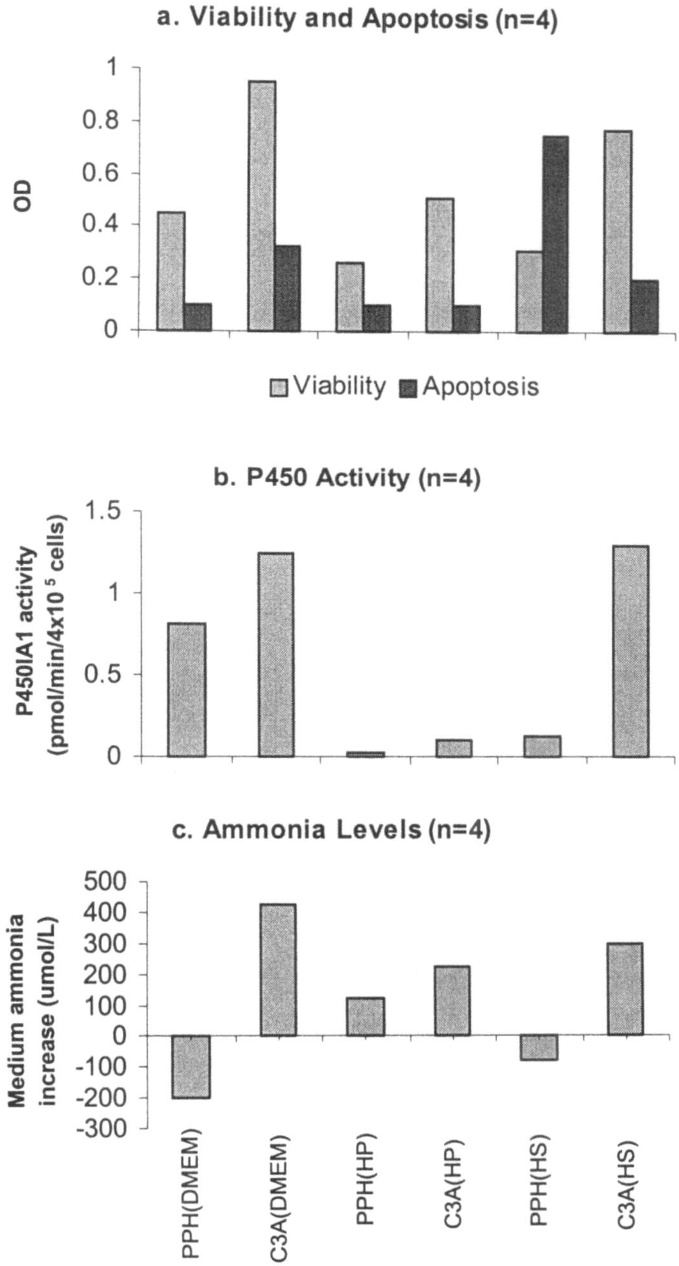

Figure 1. Cell viability, P450 activity and ammonia levels in different culture media

DETERMINING OPTIMAL TRANSPORT CONDITIONS OF IN VITRO CULTURED HUMAN CHONDROCYTES PREPARED FOR AUTOLOGOUS RE-IMPLANTATION USING FLOW CYTOMETRY

G. WOZNIAK*ξ, N. VELIKONJA*ξ, M KNEŽEVIĆ*ξ, M. URBAJS*, M. JERAS*, D. RADOSAVLJEVIĆ+ AND P.ROŽMAN*

*Blood Transfusion Centre, Šlajmerjeva 6, Ljubljana, Slovenia

ξ Educell d.o.o., Teslova 30, 1000 Ljubljana

+University Medical Centre, Department of Orthopaedic Surgery, Zaloška 9, Ljubljana, Slovenia

1. Introduction

Re-implantation of in vitro cultured chondrocytes is one of the most promising methods for curing deep aseptic lesions of articular cartilage. After cultivation, the cell suspension is submitted to a department of orthopaedic surgery. Transport conditions should guarantee over 90% viability of the shipped cell quantity after 72 hours.

Traditionally, trypan blue exclusion test was performed in determining viability and even apoptosis of cultured cells. However, only a limited number of cells can be counted comparing to the size of the sample that can be inspected by flow cytometer analysis.

Many flow cytometry studies have focused on the investigation of features of cells that have either died by apoptosis or by necrosis [2, 3]. Non-viable cells can be distinguished from live cells on flow cytometer due to changes in scattering properties and can be excluded from measurements only by proper gating [3].When cells are cultured in vitro and undergo a severe processing procedure such as trypsinization before measurements, these differences are not clear enough to permit accurate determination of viability.

Various fluorescent indicators of membrane integrity have been used for improved dead cell discrimination [1]. Propidium iodide is the most widely used non-vital DNA dye. Like trypan blue, it enters only permeabilised cells. It labels dead cells' DNA, intercalating into nucleic acid molecules. Fluorescein diacetate, on the other hand, as a near neutral molecule diffuses freely into the cell. Once inside, it is converted by intracellular esterases into a fluorescent product that is retained in cells with intact plasma membranes. Both substrates and their products leak rapidly from cells with damaged membranes.

In the assay, we attempted to optimise a procedure for determining viability of cultured chondrocytes intended for autologous transplantations. Double labelling with propidium iodide and fluorescein diacetate was used to differentiate between viable and non-viable cells. The viability was then the criterion for choosing optimal transport medium and temperature for the cell transport. In addition, for any combination of transport factors tested the level of apoptosis was determined using annexin V labelling.

A. Bernard et al. (eds.), Animal Cell Technology: Products from Cells, Cells as Products, 543–545.

2. Materials and methods

Primary cultures of human articular chondrocytes were established in F12/DMEM supplemented with 15% fetal bovine serum, gentamycine, fungizone and ascorbic acid. The cells were harvested in early passages (3-5) of the culture when grown to approximately 80% confluency. The cells were trypsinized and filtered through 40 μm nylon mesh cell strainers. Transport conditions were simulated in 1.7 ml plastic eppendorf tubes containing 500μl of transport medium with at least 1x 10^6cells. They were kept at 20-25 or 37 °C for several days. The transport media tested were F12/DMEM free of serum and F12/DMEM supplemented with 10 or 20/ FBS. Cell viability was determined after 2, 3 and 6 days of storage.
The reagents were prepared as follows: a 10 mg/ml stock solution of propidium iodide (PI) was prepared in 1x PBS and stored at 4 °C. It was further diluted immediately before use. One mg of fluorescein diacetate (FDA) powder was dissolved in 1000 μl of ice-cold acetone. A fresh stock was prepared for each experiment immediately before labelling.
For determination of apoptosis, we used a commercially available FITC conjugate of annexin V(Becton Dickinson). For negative control unlabelled reagent was used..
Fluorescent staining of least 1 x 10^5 PBS-washed cells was performed in a 12 x 75 mm-tube. Then 100 μl PBS with the addition of 1 μl of FDA stock solution was added and mixed well. Cells were then incubated and protected from light for 15 min at 37°C. For staining of DNA, cells were added a solution of RNase (17.6 Kunitz units = 200μg/ml) and PI (5 μg/ml) respectively. All samples were kept at 4°C in the dark until analysed by flow cytometer.Finally, 1x10^5 cells were washed in 1xPBS buffer and then resuspended in 100 μl annexin V binding buffer. Labelling with annexin V was performed exactly as instructed by the manufacturer (Becton Dickinson).
Samples were then analysed on a **FACScan flow cytometer** (Becton Dickinson Immunocytometry Systems). Green FDA and annexin V-FITC fluorescence was collected after a 530/30 nm band-pass filter. Orange emission from PI was filtered through a 585/42 nm band pass filter. Residual spectral overlap was then removed.. Photomultiplier tube voltage and spectral compensation were set initially using single FDA or PI stained cells. Settings were then further optimised on samples that were stained with both dyes to compensate for dye-dye interactions. FDA and PI fluorescence were shown on four decade log scales. A minimum of 3000 events were evaluated in each sample. All tests were done at least eight times and the significance of each transport factor studied was determined using Mann-Whitney U-test.

3. Results and discussion

Forward versus side scatter dot plots clearly showed that scatter discrimination between live and dead cells was clear-cut as the overlap of live cells with dead cells was minimal. The gating was more problematic i populations with low viability, where numerous dead and vitaly deteriorated cells caused shading in the lower values of side scatter.

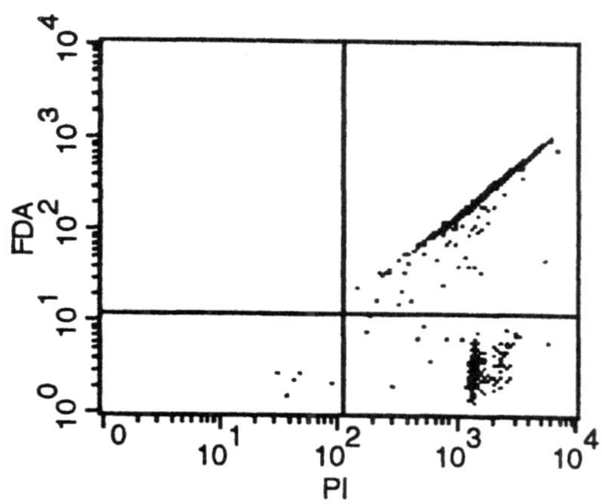

Figure 1. FDA/PI indicated viability of chondrocytes after 2 days of storage in 20% FBS at 4°C.

To optimize the detection of live, propidium iodide negative cells, we tested various concentrations of propidium iodide. In the range of 20 µg/ml to 2 µg/ml clear separation between live and dead cells was persisting. Also when lowering the concentration of FDA we observed a dimmer signal at concentration below 2 µg/ml, and hence this was chosen for the minimal concentration required for adequate staining.

The level of apoptosis in all assays was kept under 1.5%, which was not significantly different from the level of apoptosis determined in cellular suspensions obtained immediately after trypsinization of monolayer cultures .

In general, higher percentages of serum added to the medium, increased the viability of chondrocytes, stored both at 4 or 20-25 °C. When stored up to three days in serum free medium, the cells survived significantly longer at room temperature than at 4 °C. After 6 days the situation was reversed probably because a lower level of metabolism at lower temperature slowed down the depletion of transport medium. When the medium was supplemented with 10% FBS, the survival of cells kept at 4 °C or at room temperature was comparable up to three days. However after 6 days of storage, 4 °C again proved to be a more favorable temperature. If the medium with 20% of serum, the cells survived significantly better at 4 °C than at room temperature. In general, increased percentage of serum in medium significantly improved cell survival at and above room temperatures in all time intervals. cultured articular chondrocytes can be safely transported at ambient temperature for up to three days time, without a critical decrease in viability.

4. References

1. **Haughland, R. P.** (1994) Spectra of fluorescent dyes use in flow cytometry, *Methods Cell Biol* **42 PtB**, 641-663.
2. **Glander, H. J. and Schaller, J.** (1999) Binding of annexin V to plasma membranes of human spermatozoa: a rapid assay of membrane hanges after cryostorage. *Mol Hum Reprod* **5 (2)**, 109-15
3. **Schmid, I. et al**: (1999) Flow cytometric analysis of live cell proliferation and phenotype in populations with low viability. *Cytometry* **35**, 64-74

EFFECTS OF COLLAGENASE ON HUMAN HEMATOPOIETIC CELLS

N. Jelinek, S. Schmidt, S. Thoma, C. Wandrey, M. Biselli
Institute of Biotechnology 2, Forschungszentrum Jülich GmbH
D-52425 Jülich, Germany

ABSTRACT

Human hematopoietic cells could be successfully cultivated immobilized on collagen microcarriers in a fixed-bed reactor. In this stroma dependent system the cells are harvested by collagenase disruption of the carriers. Therefore the effect of the enzyme on hematopoietic cells was investigated. The surface molecules CD33, CD34, CD38, CD45, CDw123, HLA-DR, and Thy-1 (CD90) are resistant to collagenase. The enzyme has a treatment time dependent toxicity on the surface marker c-kit (CD117) and the CFC- (Colony forming cells) and CAFC- (Cobblestone area forming cells) populations.

1. INTRODUCTION

Ex vivo expansion of human hematopoietic cells is important in different medical fields like cancer treatment or gene therapy. Many aspects of maintenance and proliferation of primary hematopoietic cells in cell culture systems are still unknown. Therefore it is a promising approach to imitate the natural compartment of hematopoiesis, the bone marrow. The following aspects reflect the similarity of a cultivation in a fixed-bed reactor and the bone marrow: First of all the cocultivation with stromal cells obtains some important biological effects like production of cytokines and adhesion. Moreover the threedimensional structure of bone marrow is simulated by the immobilization of the cells on macroporous collagen microcarrier. And finally the continuous perfusion in the reactor ensures the optimal feeding of the cells.

The aim of ex vivo expansion is the production of cells. The cell harvest, done by enzymatic disruption of the microcarriers with collagenase, plays an important role in the development of clinical culture systems.

2. MATERIALS AND METHODS

Cells used in the collagenase experiments were primary $CD34^+$-cells after 7 days of cultivation and the $CD34^+$ cell line KG-1 [1]. These cells were analysed by FACS, the CFC-assay (Colony forming cells) and the CAFC-assay (Cobblestone area forming cells).

As carriers weighted macroporous collagen microcarriers (Figure 1) with 400-700 µm in diameter and a pore size of 40-50 µm (Cellex Biosciences, Inc.) were used.

A. Bernard et al. (eds.), Animal Cell Technology: Products from Cells, Cells as Products, 547–549.

They are degradable by collagenase type IA (Sigma, c = 2 mg/mL, solved in PBS). Immobilized cells on carriers or cells in solution ($1*10^6$ cells) were washed with PBS. After the enzymatic treatment with 1 mL Collagenase at 37°C for 30 to 120 min the enzyme was inactivated with PBS/EDTA. Following the cell-analysis the supernatant containing the cells was centrifuged (200*g, 10 min).

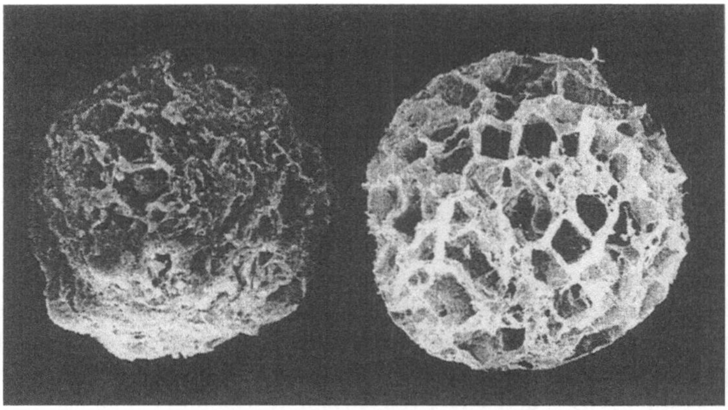

Figure 1. Carrier without and with immobilized stromal and hematopoietic cells

3. RESULTS

Collagenase is known to disrupt some lymphocyte surface molecules [2]. The effect of collagenase on surface markers typically considered for FACS analysis of hematopoietic cells was examined with the cell line KG-1 and primary $CD34^+$-cells. The analysed markers on the two different cell types are shown in Table 1. A significant influence of the enzyme treatment on the expression of c-kit (CD117) could be observed. c-kit is the receptor for SCF (stem cell factor) a cytokine acting on very early hematopoietic cells. Collagenase has no influence on this surface molecule after 30 min of enzyme treatment while a prolonged incubation time of 120 min shows a decreased level of c-kit^+-cells (87% of the control). For all other analysed surface molecules no effect of collagenase was found.

The CFC-assay was used to detect hematopoietic progenitor cells. We observed a decrease in the cloning efficiency after 30 min of treatment with collagenase (77% of the control) and even more after 2 h of incubation (34% of the control). In the control (30 min of incubation with PBS) the cloning efficiency was 17.3% ± 0.9.

The CAFC-assay was applied to trace even earlier progenitors. After 30 min of enzymatic treatment the percentage of CAFC is preserved (Figure 2). The storage of the cells in PBS at 37°C for 2 h has a reducing effect on the CAFC (80% relative to the control) and the addition of collagenase enforces this influence (77% relative to the control). Within the experiments a great variability of the donor material could be observed (value of control: 2,4% CAFC ± 0,5).

TABLE 1. Analysed surface markers on KG-1 and primary CD34$^+$-cells

Surface marker	Analysed Cells
CD33	KG-1
CD34	Primary CD34$^+$-cells and KG-1
CD38	KG-1
CD45	KG-1
Thy-1 (CD90)	Primary CD34$^+$-cells
c-kit (CD117)	Primary CD34$^+$-cells
HLA-DR	KG-1

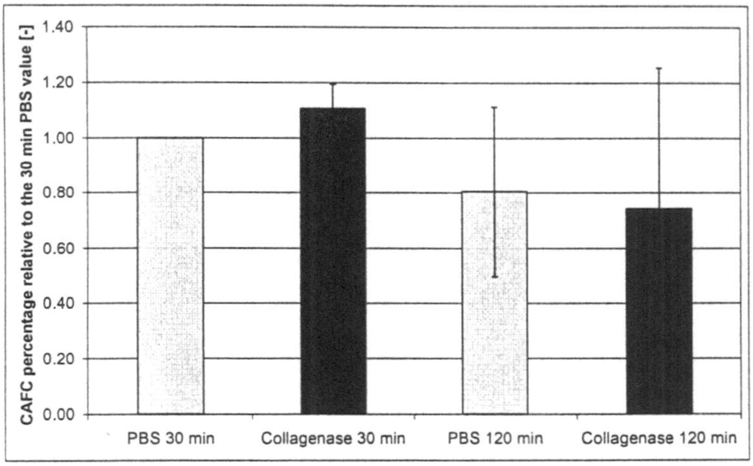

Figure 2. Effect of collagenase on the CAFC population of primary hematopoietic cells (n = 4)

4. CONCLUSIONS

The observed time dependent negative effects of collagenase on CFC and CAFC could be the result of degradation of special cytokine receptors. For the SCF receptor c-kit (CD117) this effect could be proved and possibly other receptors react like c-kit.

5. ACKNOWLEDGEMENTS

We like to thank Cornelia Herfurth and Hermann Bohnenkamp for their help and MainGen Biotechnologie GmbH, Frankfurt/M., Germany for supporting our work.

6. REFERENCES

[1] Koeffler, H. P. and D. W. Golde (1978) Acute Myelogenous Leukemia: A Human Cell Line Responsive to Colony-Stimulating Activity, *Science* **200**, 1153-1154.

[2] Abuzakouk, M., Feighery, C., O'Farrelly (1995) Collagenase and Dispase enzymes disrupt lymphocyte surface molecules, *J Immunol Meth* **194**, 211-216.

VALIDATION OF THE CACO-2 TEST FOR INTESTINAL PERMEABILITY ASSESSMENT OF DRUGS: THE USEFULNESS OF PAPP, THE APPARENT PERMEABILITY COEFFICIENT

F. FASSIO, M.L. NOLLI*, L. GOLZIO, M. BERTOLINO, S. CANALI and D. BARONE
LCG-RBM S.p.A.–Via Ribes, 1 –10010 Colleretto Giacosa (Torino), Italy
** Biotech consultant*

INTRODUCTION

The bioavailability of orally administered drugs depends to a great extent on their capability of being transported across the intestinal barrier. Caco-2 cells, from a human colon adenocarcinoma [1] achieve a high degree of enterocytic differentiation and can be used as an *in vitro* model for the investigation of drug transport through the intestinal epithelium.The Caco-2 test is based on a static model using a cell monolayer lying on the filter base of a transwell between the apical and the basolateral compartments, both filled with culture medium. These cells, grown onto collagen-coated polycarbonate membrane (transwell), form a monolayer of polarized epithelial cells, which represent a relevant model for the small intestinal epithelium. The process of differentiation starting at cell confluence leads to the formation of a brush border with well-developed microvilli, tight apical junctions, and a polarized distribution of membrane components, including enzymes, receptors, transport systems, ion channels and lipid molecules. The achievement of complete cell differentiation can be monitored by physical (Transepithelial Electrical Resistance, TEER) and chemical methods.

MATERIAL AND METHODS

Cell culture. The Caco-2 cells were cultured in Dulbecco Modified Eagles's Medium (DMEM) supplemented with 10% FCS, 2% L-glutamine 200mM and 1% non-essential amino acids solution.

Cell culture preparation

Cell bank. The cells were stored frozen in cryotubes under liquid nitrogen in FBS and 10% dimethyl sulfoxide (DMSO). Absence of contamination by mycoplasmas was assessed using the Boehringer Mannheim Mycoplasma Detection kit.
Cell culture thawing. Frozen Caco-2 cells were thawed rapidly at 37°C, added, drop by drop, to 10 ml of complete medium and centrifuged at 200g x 7 minutes. The supernatant was removed, the cell pellet reconstituted in the medium and seeded into 75 cm^2 flasks containing medium. The flasks were incubated at 37±0.5°C in an atmosphere of 5% CO_2.
Cell passage. The cells were serially subcultured when near-confluent monolayers were obtained. The medium of each flask was removed and the monolayer washed with 10-15 ml of phosphate buffered solution (PBS).Trypsin-EDTA solution was added to the cell monolayer, incubated at 37°C and tapped gently at intervals to dislodge the cells. Complete detachment and disgregation of the cell monolayer were confirmed by micro-

A. Bernard et al. (eds.), Animal Cell Technology: Products from Cells, Cells as Products, 551–553.

scopic examination. The cells were then resuspended in 10 ml of complete medium and centrifuged at 200 g for 7 minutes. The supernatant was discarded, the cells resuspended in culture medium and plated at $2.5 \times 10^5 \pm 5\%$ cells/ml in 75 or 175 cm^2 flasks.

Transwell cell culture insert preparation. The near-confluent monolayers were detached and disgregated by treatment with trypsin as described above. The cells were then resuspended in culture medium and counted. The cell suspension was diluted with medium to give about 3.5×10^6 cells/ml and 300 µl of cell suspension were seeded on the apical compartment of each transwell (6.5 mm ∅, 0.4 µm pore size). 600 µl of culture medium were put into the basolateral compartment. The plates were incubated at 37±0.5°C in a humidified atmosphere of 5% CO_2 in air for 15-21 days, changing the medium every 48-72 hours. The following compounds were used to validate this model: caffeine (CA, anhydrous), diclofenac (DI, sodium), metopropol (ME, tartrate), physostigmine (PH, hemisulfate) and ranitidine (RA, hydrochloride), which are known to have good intestinal permeability, and acebutolol (AC, hydrochloride), atenolol (AT), nadolol (NA), neostigmine (NE, sulfate) and thiocolchicoside (TH) which, on the contrary, are known to have poor, or very poor, intestinal absorbability. AT was dissolved in 1% DMSO in Hanks' balanced salt solution (HBSS) while DI and AC were in 1% ethanol in HBSS. All the other drugs were dissolved in HBSS. Final drug concentrations in the apical compartment were: 125, 250, 500 and 1000 µg/ml (standard range used, if no cytotoxic effects are seen in a preliminary cytotoxicity assay [Cell Titer 96™ Assay], based on the conversion, by Caco-2 cells, of a tetrazolium salt into a formazan product and detected using an ELISA plate reader at 570 nm).

Test procedure. The test was performed in duplicate wells of Caco-2 cells (three independent trials). The medium was removed from each of the apical chambers and replaced with 200 µl of different concentrations of test compounds in HBSS. Plates were then incubated at 37±0.5°C in a humidified atmosphere of 5% CO_2 in air for 2 hours. At the end of the incubation the % of drug passing from the apical to the basolateral side was evaluated using appropriate analytical procedures (LC-MS-MS). On the basis of this data the apical to basolateral apparent permeability coefficient (Papp) was evaluated using the following equation: $Papp = dA/(dt \cdot S \cdot C_o)$, where dA/dt is the permeability rate, S is the surface area of the cell monolayer and C_o is the initial drug concentration in the apical compartment. Papp value is expressed as cm/sec.

Cell monolayer integrity check. Of the three techniques (TEER, passage through the monolayer of ^{3}H-mannitol and fluorescein tests) most commonly used to check that complete cell differentiation has been achieved and, consequently, that the cell monolayer holds together, for our routine we chose TEER on the cell monolayers of all wells, both prior to and following incubation with the test compounds. The check at the end of incubation allows the integrity of the monolayer to be assessed for any *cytotoxicity damage* or for any technical trauma that may have occurred during handling.

Transepithelial Electrical Resistance (TEER) test. Transepithelial Electrical Resistance (TEER), expressed as ohms x cm^2, was measured in the transwells using a Millicell-ERS (Millipore). This device measures the TEER, which indicates cell confluence and the perfect holding together of the hermetic intercellular junctions. The monolayer was considered well differentiated when TEER value was higher than 800 ohms x cm^2 (generally after two weeks of culturing).

RESULTS

The results of drug permeability through the Caco-2 monolayers, expressed as % passage at the different concentrations tested and Papp values, are reported in the following table:

Compounds	% passage (mean ± SD, n=6)/compound concentrations				Papp Values
	125 µg/ml	250 µg/ml	500 µg/ml	1000 µg/ml	(cm/sec)
Caffeine (CA)	45.71 ± 10.21	39.99 ± 12.74	37.79 ± 6.60	35.16 ± 7.30	**28.2 ± 5.1**
Diclofenac (DI)	29.08 ± 8.36	44.13 ± 4.42	25.03 ± 8.86	33.60 ± 10.86	**27.2 ± 5.7**
Metoprolol (ME)	17.01 ± 3.75	18.42 ± 2.95	17.87 ± 4.38	12.80 ± 2.82	**9.9 ± 3.0**
Physostigmine (PH)	27.51 ± 3.12	24.28 ± 4.89	21.74 ± 4.27	20.61 ± 3.01	**16.3 ± 2.0**
Ranitidine (RA)	0.06 ± 0.06	0.05 ± 0.04	0.02 ± 0.02	0.01 ± 0.01	**0.169 ± 0.015**
Acebutolol (AC)	0.14 ± 0.04	0.18 ± 0.11	0.26 ± 0.10	0.21 ± 0.10	**0.184 ± 0.09**
Atenolol (AT)	0.24 ± 0.16	0.25 ± 0.10	0.40 ± 0.09	0.55 ± 0.10	**0.491 ± 0.09**
Nadolol (NA)	0.07 ± 0.04	0.09 ± 0.09	0.12 ± 0.05	0.22 ± 0.01	**0.208 ± 0.01**
Neostigmine (NE)	0.19 ± 0.20	0.09 ± 0.06	0.14 ± 0.12	0.10 ± 0.08	**0.082 ± 0.065**
Thiocolchicoside (TH)	0.30 ± 0.09	0.16 ± 0.06	0.24 ± 0.04	0.20 ± 0.04	**0.013 ± 0.005**

DISCUSSION AND CONCLUSIONS

The static version of Caco-2 monolayer proved to be able to select drugs with high potential absorbability such as CA, DI, ME and PH whose high % passage and Papp values at all concentrations tested reflect the good rate of oral absorption in humans, known from literature (2). At the same time, the high % passage and Papp values observed are in agreement with the data found in literature [2] on the Caco-2 model for some of these substances. AC, AT, NA, NE and TH, that are known to have poor, or very poor, absorption in humans, showed low % passage and Papp values at all concentrations tested. Among the selected compounds, RA, known to have a good oral absorption and a bioavailability of about 50%, was expected to show a good permeability in Caco-2 model. On the contrary, % passage and Papp value indicate that it is poorly absorbed through the monolayer. This finding is in agreement with the observations on Caco-2 cells challenged with RA by other investigators [3], RA being absorbed through alternative paracellular pathway(s) across the tight junctions. On the basis of the overall results of trials performed in our laboratory, we can assume that the process is reliable for assessing the apical to basolateral permeability of various compounds applied to the Caco-2 cell monolayer. However, it should be borne in mind that some compounds, such as RA, are excluded from the transcellular passage, being absorbed through alternative paracellular pathway(s), i.e. across the tight junctions, resulting in false negative permeability through Caco-2 cells. While this work provides results which are only partly original, on the other end it fully achieves its aim which was to give an example of methodology validation which must be done before a new method is introduced into routine testing. The method validated here proved to be reliable and repeatable with low inter and intra-assay variability, fluctuating around 8 and 12% (data not shown), respectively, when calculated from the data accumulated in the experience acquired after its implementation in our Institute.

REFERENCES

1. Fogh, J., Fogh, J.M. and Orfeo, T. (1977). One hundred and twenty seven cultured human tumor cell lines producing tumors in nude mice. J. Natl. Cancer Institute, 59: pages 221-226
2. Chong S., Dando S.A., and Morrison R.A. (1997). Evaluation of Biocoat intestinal epithelium differentiation environment as an absorption screening model with improved productivity. Pharm. Res., 14: pages 1835-1837
3. Gan, LS., Hsyu PH., Pritchard JF., Thakker D. (1993) Mechanism of intestinal absorption of ranitidine and ondasentron: transport across Caco-2 cell monolayers. Pharm. Res., 10: 12, pages 1722-1725

Author Index

Subject Index

MIX
Papier aus verantwortungsvollen Quellen
Paper from responsible sources
FSC® C105338

If you have any concerns about our products,
you can contact us on
ProductSafety@springernature.com

In case Publisher is established outside the EU,
the EU authorized representative is:
Springer Nature Customer Service Center GmbH
Europaplatz 3, 69115 Heidelberg, Germany

Printed by Libri Plureos GmbH
in Hamburg, Germany